MW01293567

The BIOLOGY *of* SEA TURTLES

Volume II

CRC

MARINE BIOLOGY

SERIES

Peter L. Lutz, Editor

Published Titles

Biology of Marine Birds
E. A. Schreiber and Joanna Burger

Biology of the Spotted Seatrout
Stephen A. Bortone

The BIOLOGY *of* SEA TURTLES

Volume II

Edited by
Peter L. Lutz
John A. Musick
Jeanette Wyneken

CRC PRESS

Boca Raton London New York Washington, D.C.

Library of Congress Cataloging-in-Publication Data

The biology of sea turtles / edited by Peter L. Lutz and John A. Musick.
p. cm.--(CRC marine science series)
Includes bibliographical references (p.) and index.
ISBN 0-8493-1123-3
1. Sea turtles. I. Lutz, Peter L. II. Musick, John A. III. Series: Marine science series.
QL666.C536B56 1996
597.92—dc20 96-36432
CIP

Direct all inquiries to CRC Press LLC, 2000 N.W. Corporate Blvd., Boca Raton, Florida 33431.

International Standard Book Number 0-8493-1123-3
Library of Congress Card Number 96-36432
Printed in the United States of America 2 3 4 5 6 7 8 9 0
Printed on acid-free paper

Preface

The success of the first volume of *The Biology of Sea Turtles* revealed a need for broad but comprehensive reviews of recent major advances in sea turtle biology. At that time, book size constraints as well as the fast-paced changes in some fields dictated that this need could be only partially addressed in a single volume. Many important topics were not covered and were left for future volumes. Volume II emphasizes practical aspects of biology that relate to sea turtle management and changes in marine and coastal ecosystems. These topics include the interactions of humans and sea turtles, an introduction to sea turtle anatomy, sensory and reproductive biology, sea turtle habitat use and ecology, stress and health, and the maintenance of captive animals. This volume provides both historical and up-to-press-time information. The field is growing dramatically as established scientists expand their views and fine new scientists bring their novel ideas, techniques, and perspectives to the understanding and application of the biology of marine turtles.

Preface

Acknowledgments

The encouragement, support, and suggestions of colleagues, especially those at the annual meetings of the International Sea Turtle Symposium, the Harbor Branch Oceanographic Institute course in sea turtle biology, and the Duke University classes in sea turtle biology and conservation inspired this volume. We are extremely grateful to Melanie Harbin for editorial assistance and to those who contributed chapters as well as those who gave their precious time as reviewers. We thank John Sulzycki, CRC Press senior editor, for his support and patience. Erika Dery, production manager and Amy Rodriguez, project editor, CRC Press, helped shepherd the book through. Steven Lutz again provided a wonderful photo to serve as the cover to this volume.

This book is dedicated to sea turtle biologists worldwide who are playing vital roles in preventing these fascinating marine reptiles from vanishing into oblivion.

Peter L. Lutz
Florida Atlantic University

John A. (Jack) Musick
Virginia Institute of Marine Science, College of William and Mary

Jeanette Wyneken
Florida Atlantic University

Editors

Peter L. Lutz, Ph.D., holds the McGinty Eminent Scholar Chair in Marine Biology at Florida Atlantic University. Dr. Lutz received both his B.Sc. (Honors) and Ph.D. from Glasgow University, Scotland. After earning his Ph.D. in 1970 he became a research associate with Dr. Knut Schmidt-Nielsen at Duke University, with whom he worked on avian physiology. He has held university faculty positions in Nigeria (University of Ife), England (Bath University), and the United States. In 1975 he joined the Department of Marine Biology and Fisheries at the Rosenstiel School of Marine and Atmospheric Science, University of Miami, and became chair of that department in 1983, a post he held until he took his present position in 1991. As a comparative physiologist Dr. Lutz has worked on the physiology of a wide variety of organisms, from liver flukes to duck-billed platypuses. His current interests focus on turtles, particularly stress, diving, and hypoxia. Dr. Lutz is an editor for the *Journal for Experimental Biology* and is series editor for the Marine Biology Series published by CRC Press. He was a governing council member of the Bahamas National Trust and a fellow of the Explorers Club. He has authored more than 150 research papers and 4 books.

John A. (Jack) Musick, Ph.D., holds the Marshall Acuff Chair in Marine Science at the Virginia Institute of Marine Science (VIMS), College of William and Mary, where he has served on the faculty since 1967. He earned his B.A. in Biology from Rutgers University in 1962 and his M.A. and Ph.D. in Biology from Harvard University in 1964 and 1969, respectively. While at VIMS he has successfully mentored 32 masters and 39 Ph.D. students. Dr. Musick has been awarded the Thomas Ashley Graves Award for Sustained Excellence in Teaching from the College of William and Mary and the Outstanding Faculty Award from the State Council on Higher Education in Virginia. He has published more than 100 scientific papers and 7 books focused on the ecology of sea turtles, sharks, and other marine fishes. In 1985 he was elected a Fellow by the American Association for the Advancement of Science. He has received Distinguished Service Awards from both the American Fisheries Society and the American Elasmobranch Society, for which he has served as president. Dr. Musick also has served as president of the Annual Sea Turtle Symposium (now the International Sea Turtle Society), and as member of the World Conservation Union (IUCN) Marine Turtle Specialist Group. Dr. Musick currently serves as co-chair of the IUCN Shark Specialist Group, and on two national, five regional, and five state scientific advisory committees concerned with marine resource management and conservation.

Jeanette Wyneken, Ph.D., is an Assistant Professor at Florida Atlantic University. Dr. Wyneken earned a B.A. from Illinois Wesleyan University and her Ph.D. from the University of Illinois in 1988. She was a research associate at the University of Illinois from 1988 through 1989, then took a position at Florida Atlantic University in 1990 where she became a research assistant professor until assuming her current position in 2000. Dr. Wyneken has successfully mentored seven masters students while at Florida Atlantic University. She developed and has taught the Biology of Sea Turtles course at Harbor Branch Oceanographic Institution since 1996. Along with Selina Heppell and Larry Crowder, she helped develop and teach Duke University's Biology and Conservation of Sea Turtles course. She maintains professional affiliations with Duke University Marine Laboratory and Mote Marine Laboratory. Dr. Wyneken is a former president of the Annual Symposium on Sea Turtle Biology and Conservation (now the International Sea Turtle Society). As a functional and evolutionary morphologist, Dr. Wyneken has studied a variety of lower vertebrates; much of her attention is focused on the integration of anatomy, physiology, and behavior in understanding marine turtle biology. Her work emphasizes the integral roles these play in the conservation and management of marine turtles. She has authored more than 20 research papers and one book. She is a member of a number of professional organizations, including the IUCN Marine Turtle Specialists Group, the Society for Integrative and Comparative Biology, Sigma Xi, and the Association for Reptilian and Amphibian Veterinarians.

Contributors

Karen A. Bjorndal
Archie Carr Center for Sea Turtle Research
Department of Zoology
University of Florida
Gainesville, Florida

Alan B. Bolten
Archie Carr Center for Sea Turtle Research
Department of Zoology
University of Florida
Gainesville, Florida

Lisa M. Campbell
Department of Geography
University of Western Ontario
London, Ontario, Canada

Larry B. Crowder
Nicholas School of the Environment
Duke University Marine Laboratory
Beaufort, North Carolina

Sheryan P. Epperly
NOAA
National Marine Fisheries Service
Miami, Florida

Nat B. Frazer
Wildlife Ecology and Conservation
University of Florida
Gainesville, Florida

Jack Frazier
Conservation and Research Center
Smithsonian Instititution
Front Royal, Virginia

Mark Hamann
Key Centre for Tropical Wildlife Management
Northern Territory University
Darwin, Northern Territory

Selina S. Heppell
Department of Fisheries and Wildlife
Oregon State University
Corvallis, Oregon

Lawrence H. Herbst
Institute for Animal Studies
Albert Einstein College of Medicine
Bronx, New York

Benjamin M. Higgins
NOAA
National Marine Fisheries Service
Southeast Fisheries Science Center
Galveston, Texas

Jeremy B.C. Jackson
Scripps Institution of Oceanography
University of California, San Diego
La Jolla, California

Elliott R. Jacobson
Department of Small Animal Clinical Sciences
College of Veterinary Medicine
University of Florida
Gainesville, Florida

Colin J. Limpus
Queensland Parks and Wildlife Service
Conservation Strategy Branch
Queensland Department of Environment and Heritage
Brisbane, Australia

Peter L. Lutz
Department of Biological Sciences
Florida Atlantic University
Boca Raton, Florida

Jeffrey D. Miller
Queensland Parks and Wildlife Service
Northern Region
Cairns, Queensland, Australia

Sarah L. Milton
Department of Biological Sciences
Florida Atlantic University
Boca Raton, Florida

Soraya Moein Bartol
Biology Department
Woods Hole Oceanographic Institute
Woods Hole, Massachusetts

Jack Musick
Fisheries Science Laboratory
Fisheries Science/VIMS
Gloucester Point, Virginia

David W. Owens
Graduate Program in Marine Biology
Grice Marine Lab, University of Charleston
Charleston, South Carolina

Pamela Plotkin
Office of Research and Sponsored Programs
East Tennessee University
Johnson City, Tennessee

Melissa L. Snover
Nicholas School of the Environment
Duke University Marine Laboratory
Beaufort, North Carolina

Thane Wibbels
Biology, Campbell Hall
University of Alabama
Birmingham, Alabama

Blair E. Witherington
Florida Fish and Wildlife Conservation Commission
Florida Marine Research Institute
Melbourne Beach, Florida

Jeanette Wyneken
Department of Biological Sciences
Florida Atlantic University
Boca Raton, Florida

Table of Contents

1 Prehistoric and Ancient Historic Interactions between Humans and Marine Turtles

Jack Frazier

CONTENTS

1.1 INTRODUCTION

A usual opening for descriptions about marine turtles is to announce proudly that they have been around for hundreds of millions of years, and have outlived even the dinosaurs. Our species, of course, has been on the planet for much less time, just a few hundred thousand years; commonly there has been a tacit assumption that interactions between these ancient reptiles and people have really only become significant in historic or contemporary times. The object of this chapter is to explore the antiquity and complexity of human–turtle relations by providing a summary

0-8493-1123-3/03/$0.00+$1.50

and synthesis of widely scattered information on prehistoric and ancient historic interactions between marine turtles and people in different societies, in different parts of the world. The information is grouped into three principal themes: zooarchaeology (or archaeozoology), cultural artifacts, and ancient textual accounts.

The following summaries do not pretend to be complete; they focus on two separate geographic regions for which relatively large amounts of information are readily available: one is the western Indian Ocean basin, particularly the Arabian Peninsula; the other is the southeast U.S., Caribbean, and Yucatán Peninsula areas. As a result, vast regions have been omitted or very superficially covered in this review, namely the shores of Africa, Asia, Mediterranean, Oceania, and much of South and Central America. For the time period, it was arbitrarily decided to use the European conquest of the Americas as an end point. This means that there is a gap of half a millennium between this chapter and Chapter 12, which focuses on contemporary issues. As a result, valuable ethnographies that detail and elucidate human–turtle interactions in numerous societies around the world are omitted. Accounts of societies in which marine turtles play, or have played, central roles — the Carancahua of coastal Texas (Hammond, 1891), the Miskito of Caribbean Honduras and Nicaragua (Nietschmann, 1973; 1979), the Seri (or Comcaac) of the Sea of Cortez (Felger and Moser, 1991; Nabhan et al., 1999), the Vezo of western Madagascar (Astuti, 1995), Renaissance Christian-pagan comportments in the Mediterranean (Maffei, 1995), and nineteenth-century ladies of high society in Buenos Aires (Gudiño, 1986), to name but a few — are simply not included here. However, Thorbjarnarson et al. (2000) provide a recent review of human use of marine turtles from around the world, focusing mainly on the last few centuries.

As Campbell (Chapter 12) explains, it is much easier to appreciate "culture" when it is distinct from one's own. Another common conception of culture is that it is something to be displayed, as in a museum. The accounts in this chapter are very much about "other" cultures: those that were part of societies that no longer exist, and those that produced curious artifacts from bygone days that are routinely put on display. If we can appreciate better the diversity of ways in which people and marine turtles have interacted over the ages, by focusing on the other cultures, perhaps it will enlighten our view of contemporary and future relations between *Homo sapiens* and these marine reptiles.

1.2 ZOOARCHAEOLOGY: ARCHAEOLOGICAL SPECIMENS OF MARINE TURTLES

As zooarchaeologists have explained, "animal remains from archaeological sites represent not only a piece of nature but more significantly, the relationship between culture and the natural environment" (Hamblin, 1984:17). The data allow for analyses of types and intensities of utilization, in relation to past cultures and their environments, providing insights into prehistoric subsistence strategies (Wing and Reitz, 1982: 14; see also Daly, 1969: 146). Not only are marine turtle remains distinctive, but the bones are large and relatively robust, making them resistant to degradation after death, and hence available to zooarchaeological investigations.

However, there has been no systematic compilation of marine turtle remains in archaeological sites. Even so, a rapid review of the literature clearly illustrates diverse prehistoric human interactions with these animals from around the world. For the present study, we will focus on only a few select geographic areas.

Typically, fragments of the bony carapace dominate the zooarchaeological specimens that are recovered, which may or may not have relevance to cultural aspects at the time when the bones were deposited. One way or the other, these specimens are rarely identifiable beyond the family level. To date, nearly all — if not all — zooarchaeological specimens have been identified as belonging to the taxonomic family of hard-shelled marine turtles, Cheloniidae; there appear to be few, if any, records from the family Dermochelyidae.

1.2.1 Zooarchaeological Remains in the Mediterranean, Arabian Peninsula, and Indian Ocean Basin

Some of the oldest evidence of interactions between humans and marine turtles comes from the Arabian Peninsula and other nearby localities (Table 1.1). Dalma Island, United Arab Emirates (UAE), has a habitational site dating to about 5000 B.C.E.,[1] and fragments of cheloniid turtles were reported from several layers (Beech, 2000). Another Ubaid[2] site of about the same age, at As-Sabiyah, Kuwait, has also yielded marine turtle bones (Beech, 2002). Mediterranean remains, although not quite as old, also date back millennia; a fragment of the plastron of a cheloniid turtle (possibly *Chelonia mydas*, the green turtle) was recovered from the Early Bronze age site (about 3600–3300 B.C.E.) of Afridar on the Israeli coast near Ashkelon (Whitcher-Kansa, in press).

Several sites in eastern Arabia have yielded bones of marine turtles, often in relatively large numbers. Characteristically on the seashore, many of these sites have been identified as Bronze Age, and some have been dated at more than 3000 years B.C.E. Among the oldest are Abu Khamis, at Ras az-Zor (or Ras as-Zoror), Saudi Arabia, one of a number of Ubaid sites on the Arabian Gulf (Masry, 1974), and Ra's al-Hamra, on the Batnah coast of Oman, both dated to about 3500 B.C.E. At the latter site, remains of marine turtles are found on the ancient surface of the site, and shallow graves often include animal remains, usually fish or turtle: "The shell of the green turtle (*Chelonia mydas*) was more common in the graves than the remains of any other animal, and thus seems to have had particular significance for the ancient inhabitants of Ra's al-Hamra" (Potts, 1990: 71; see also Durante and Tosi, 1977).

Elsewhere in Oman there are relatively large numbers of marine turtle remains at Ras al-Junayz that date from 4000 to 2000 B.C.E. (Bökönyi, 1992) and at Ra's al-Hadd, a Bronze age site dating to about 2000 years B.C.E. (Mosseri-Marlio, 1998).

---

[1] The time scale is divided into B.C.E., before the common era, or before 2000 years ago, and C.E., common era.

[2] The term "Ubaid," which derives from the small ancient site of Al' Ubaid (or "Tell al Ubaid") near the Sumerian city of Ur along the banks of the Euphrates River in present-day Iraq, is used to refer to a particular pottery style, as well as an associated cultural period in Mesopotamia and the Arabian Gulf, and usually dated from about 5500–4000 C.D.E. (Oates, *et. al.*, 1977; Moorey, 1994; xix; Potts, 1997:xi-xii; Postgate, *in litt.* 4 June 2002).

TABLE 1.1
Summary of Marine Turtle Remains in the Arabian Gulf, Mediterranean, and Western Indian Ocean

Site	Period and Estimated Date	Species*	Remarks	Source
Arabian Gulf	**Ubaid**			
Dalma Is. UAE	5000 B.C.E.	chl	Habitational site	Beech, 2000
As-Sabiya, Kuwait	~5000 B.C.E.	chl		Beech, 2002
Mediterranean	**Bronze Age**			
Afridar, Israel	3600–3300 B.C.E.	*Cm*?		Whitcher-Kansa, in press
Arabian Gulf	**Bronze Age**			
Abu Khamas, Saudi Arabia	3500 B.C.E.		Reported as "Ubaid"	Masry, 1974
Ra's al-Hamra, Oman	3500 B.C.E.	*Cm*	Most common animal remains in graves	Durante and Tosi, 1980; Potts, 1990
Ra's al-Junayz, Oman	4000–2000 B.C.E.	*Cm*?	>3400 turtle bones	Bökönyi, 1992
Ra's al-Hadd, Oman	2000 B.C.E.	*Cm*?	>3000 fragments; signs of butchery and burning	Mosseri-Marlio, 1998
Umm an-Nar, Abu Dhabi	2700 B.C.E.	*Cm*?	> 4000 turtle bones	Hoch, 1979; 1995
Arabian Gulf	**Dilmun**			
Tell Abraq, Abu Dhabi	2200–300 B.C.E.	*Cm*	>3100 fragments	Stephan, 1995; Potts, 2000; Uerpmann, 2001
Ghanda Island, UAE	~2300 B.C.E.	chl		Yasin Al-Tikkriti, 1985
Qala'at al Bahrain, Bahrain	2150–1900 B.C.E.	*Cm*, *Ei*	> 400 fragments	Uerpmann and Uerpmann, 1994; 1997
Saar, Bahrain	2100 B.C.E.		Marine turtle bones rare	Killick et al., 1991; Uerpmann and Uerpmann, 1999
West Indian Ocean				
Dembeni, Mayotte, Comoro Islands	800–900 C.E.	*Cm*	Signs of burning	Redding and Goodman, 1984

* *Notes:* chl = cheloniid; *Cm* = Chelonia mydas; *Ei* = Eretmochelys imbricata.

At both sites it was assumed that the remains were of *C. mydas*. At Ra's al-Hadd cut marks on "both shell and bone" were evidence of butchery; much of the material was reportedly burned and it was hypothesized that the burning resulted from cooking and also rubbish disposal, and that both turtles and dolphins were taken, not only for the meat, but also as sources of fat, possibly used for protecting boats made from reed bundles (Mosseri-Marlio, 1998; 2000).

At Umm an-Nar, a site on a small island off the coast of Abu Dhabi that may date back as late as 2700 B.C.E., more than 4000 bones of marine turtles showed variation in body size, indicating that some turtles were not adult and were thus captured in the water, and not on nesting beaches. On the basis of present-day distributions it was *assumed* that the turtles were *C. mydas* (Hoch, 1979; 1995).

Farther north along the coast of Abu Dhabi, the settlement at Tell Abraq, dated at 2200–300 B.C.E., has also yielded a significant amount of marine turtle bone (some 3107 identified fragments). As usual, most of the fragments are from carapace or plastron bones, so they are difficult to identify to species, but in a few cases the imprints of edges of keratinous scutes or suture patterns provided evidence of *C. mydas* (Stephan, 1995: 58; Potts, 2000: 60–61; Uerpmann, 2001). Also in the UAE, turtle bones were reported to be one of the most common animal remains from Ghanda Island, a site dating from the second half of the third millennium (Yasin Al-Tikkriti, 1985).

On the island of Bahrain, the site of Qala'at al-Bahrain dates to the period 2150–1900 B.C.E. This Dilmun[3] capital had diverse and abundant faunal remains, of which bones of marine turtles were especially numerous. *C. mydas* and *Eretmochelys imbricata*, hawksbill turtle, were identified, and a humerus of the latter had cut marks indicating butchery. Bones from individuals smaller than adults indicated that turtles were hunted in the water. Turtle bones were encountered in strata spanning nearly two millennia, and there was an increase in the ratio of turtle to fish remains at the transition from Period I to Period II (between 2150 and 1900 B.C.E.), which was thought to be from changes in bone deposition, rather than from changes in turtle exploitation (Uerpmann and Uerpmann, 1994; 1997). However, in a related study of fish bones from this site, it was suggested that the relative differences in the abundance of remains of certain fish species at different periods at Qala'at al-Bahrain might relate to changes in fishing techniques to be more effective in deeper waters (Van Neer and Uerpmann, 1994: 450 f.f.). If this change did occur, it could also explain differences in relative abundance of turtle bones over time.

Seven kilometers southwest and inland of Qala'at al Bahrain is another Dilmun settlement at Saar, which is thought to have been contemporaneous. At both sites fish bones were abundant, showing a heavy and consistent dependence on marine resources, but very few marine turtle remains were found at Saar (Killick et al., 1991: 134). It was suggested that the lack of turtle remains was because slaughtering was done at the coast, and only meat was transported inland (Uerpmann and Uerpmann, 1999: 639). The fact

[3] The term "Dilmun" (or "Tilmun") comes from Sumerian texts, and was originally a geographic reference to present-day Bahrain and adacent parts of the Arabian Peninsula; it is also used to refer to the period of about 2000-1000 B.C.E. in Bahraini archaeology (Potts, 2000: 119; Postgate, *in litt.* 4 June 2002).

that seals from Saar include marine turtle motifs indicates that the reptiles not only were known, but also were important to the people there (see Section 1.3.1).

Evidence from outside the Arabian and Middle Eastern area is dispersed and not well known. For example, remains of marine turtles were found at Dembeni, Mayotte Island, Comoro Archipelago, and dated at 800–900 C.E. One bone was identified as *C. mydas*, and some of the bones were affected by fire (Redding and Goodman, 1984).

There have been several inferences about overkill of marine turtles during prehistoric times, resulting in the extirpation of local populations from eastern Arabia (e.g., Hoch, 1979; Mosseri-Marlio, 2000: 31). The data are sketchy (indeed, species identifications are unconfirmed), and arguments that there is no longer ready abundance of marine turtles near to the archaeological localities need to be substantiated.

1.2.2 Zooarchaeological Remains in the Western Hemisphere

In the Western Hemisphere concentrations of records exist from the southeastern U.S. (Table 1.2), Caribbean (Table 1.3), and the Yucatán Peninsula area (inhabited by the Mayan civilization) (Table 1.4). The oldest record seems to be a passing comment in a paper on nesting activity: archaeological evidence suggests that *Caretta caretta*, the loggerhead turtle, was utilized, supposedly as a food source, on Kiawah Island, South Carolina, since at least 2000 B.C.E. (Combs in Talbert et al., 1980). Wing (1977: 84) emphasized the importance of these reptiles to prehistoric societies by referring to the "sea turtle harvesting constellation."

Larson (1980: 128, 132–133) compiled information from nine published and five unpublished sources reporting on marine turtle bones in the southeastern coastal plain of the U.S. (Table 1.2). This included four sites in Georgia and more than ten sites in Florida that ranged from the late Archaic to Mississippi period, including three cultural epochs of the latter: Deptford, Weeden Island, and Glades II. Hence, marine turtle remains in just the Southeast may date from 3000 B.C.E. to 1500 C.E. Although in most cases species identifications were not made, three species were identified: *C. caretta* was reported from two sites, *C. mydas* was reported from four sites, and *Lepidochelys kempii* was reported from one site. Bones of immatures were reported from Wash Island, Florida, and at the Jungerman site in Florida bones from at least 12 individuals were found.

Wing and Reitz (1982) summarized information from the Caribbean, and included 16 sites from Florida where marine turtle remains have been reported (Table 1.3); three of these sites had been reported in Larson's earlier compilation for the southeastern coastal plane (Table 1.2). Schaffer (2001) further updated the information, reporting turtle remains from 24 sites in Florida, six in Georgia, and one in North Carolina, involving the same three species reported previously. However, he reported that there was rarely evidence of more than one individual turtle per site. A more detailed study (Schaffer and Thunen, in press) concluded that the majority of sites in Florida were in the south and southwest coasts of the state, which were coastal areas formerly inhabited by the Tequesta and Calusa peoples, respectively. In contrast, Timucua middens, from northeastern Florida, showed relatively few remains of marine turtles, although these people evidently had access to turtles (Reitz and Scarry, 1985, cited in Schaffer and Thunen, in press). It was argued that

TABLE 1.2
Summary of Marine Turtle Remains in the Southeastern U.S. (See also Table 1.3)

Site*	Period and Estimated Date*	Species*	Remarks	Source
Kiawah Island, SC	2000 B.C.E.?	*Cc*?		Combs, in Talbert et al., 1980
Mound C, GA	Etowah, 1500 C.E.?	*Ei*	Hairpin, from Calusa?	Larson, 1993
Kenan, Sapelo Is., GA	MS: 800–1500 C.E.	*Cm*		Crook, 1978, in Larson, 1980
Shell ring, Sapelo Is., GA	LA: 3000–500 B.C.E.	*Cc*?	Humerus and part of carapace	Waring and Larson, 1968; Larson, 1980
St. Simons Is., GA		chl		Martínez, 1975, in Larson, 1980
Cumberland Is., GA	DP: 500–700	chl		Milanich, 1971, in Larson, 1980
Garden Patch, FL		*Lk*	Two bones from one individual	Kohler, 1975, in Larson, 1980
Garden Patch, FL	WE: 200–1000 C.E.	*Cc*	One individual	Kohler, 1975, in Larson, 1980
Madden, Dade County, FL	G II: 750–1200 C.E.	?		Laxson, 1957a; Larson, 1980
Dade County, FL	G II: 750–1200 C.E.	?		Laxson, 1957b; Larson, 1980
Dade County, FL	G II: 750–1200 C.E.	?		Laxson, 1957c, in Larson, 1980
Dade County, FL	G II: 750–1200 C.E.	?		Laxson, 1959; Larson, 1980
Wash Island, FL	MS: 800–1500 C.E.	*Cm*, chl	≥12 immature individuals	Wing, 1963a; Larson, 1980
Jungerman site, FL		chl	≥12 individuals	Wing, 1963b; Larson, 1980
Goodman site, FL		chl	One individual	Wing, 1963b; Larson, 1980

(continued)

TABLE 1.2 (continued)
Summary of Marine Turtle Remains in the Southeastern U.S. (See also Table 1.3)

Sanibel Is., FL	*Lk*, chl		Fradkin, 1976
Cushing site, Marco Is., FL	*Cm*?	≥4 individuals	Wing, 1965; Larson, 1980
Van Beck, Marco Is., FL	*Cm*?	≥26 individuals	Wing, 1965; Larson, 1980
Belle Glade site, FL	*Cm*		Willey, 1949: 61
Opa Locka, FL	Sea turtle		Willey, 1949: 87
Miami Circle, FL			Schaffer and Ashley, 2001
Florida, various	*Ei*	Used in animal effigies	Schaffer and Ashley, 2001
Florida, various			See Table 1.3

**Notes:* FL = Florida; GA = Georgia; SC = South Carolina; DP = Deptfort; G II = Glades II; LA = late Archaic; MS = Mississippi period; WE = Weeden Island; chl = cheloniid; *Cc* = *Caretta caretta*; *Cm* = *Chelonia mydas*; *Ei* = *Eretmochelys imbricata*; *Lk* = *Lepidochelys kempii*.

TABLE 1.3
Summary of Marine Turtle Remains in the Caribbean (See also Tables 1.2 and 1.4)

Site*	Period and Estimated Date*	Species*	Source*
Boynton, FL (east)	Glades III	*Cm, Ei,* chl	Coggin, in W & R, 1982
Jungerman, FL (east)	St. Johns I	chl	Jordan et al., 1963, in W & R, 1982
Jupiter Inlet, FL (east)	1000–1600 C.E.	chl	Sears, in W & R, 1982
McLarty, FL (east)	1400–1715 C.E.	chl	Wing, 1978a, in W & R, 1982
Summer Haven, FL (east)	1380 ± B.C.E.	chl	Bullen and Bullen, 1961a, in W & R, 1982
Bear Lake, FL (south)	Glades I-III	*Cm*	Coggin, 1950b, in W & R, 1982
Snapper Creek, FL (south)	Glades I	chl	Coggin, 1950a, in W & R, 1982
Granada, FL (Biscayne Bay)	Glades I-III	*Cm*, chl	Wing and Loucks, in W & R, 1982
Burtine Is., FL (west)	500 B.C.E.–500 C.E.	chl	Bullen, 1966, in W & R, 1982
Key Marco, FL (west)	Glades II and III	*Cm*	Van Bech, 1965; Wing, 1965a, in W & R, 1982
Palmer, FL (west)	2000–1000 B.C.E.	chl	Bullen, 1961, in W & R, 1982
Shired Is, FL (west)	Deptford	chl	Coldburt, 1966, in W & R, 1982
Wash Island, FL (west)	WE SH	chl	Bullen and Bullen, 1961b; and Wing, 1963, in W & R, 1982
Weeden Is., FL (west)	300–1200 C.E.	chl	Sears, 1971, in W & R, 1982
Ta 1, FL (northwest)	200–1000 C.E.	chl	Phelps, in W & R, 1982
Tucker, FL (northwest)	500 B.C.E.–1500 C.E.	chl	Sears, 1963, in W & R, 1982
MC 6, Middle Caicos, Bahamas	750–1500 C.E.	chl	Sullivan, 1981, in W & R, 1982; Wing and Scudder, 1983
Palmetto Grove, Bahamas	850–1100 C.E.	chl	Wing, 1969; Hoffman, 1970, in W & R, 1982
Grand Turk, Turks and Caicos		*Cm*, chl	Scudder, *in litt.*, 5 March 2002
Virgin Is., various sites	700–1500 C.E.	chl	Bullen, 1962; Sleight, 1962, in W & R, 1982
Dominican Republic			Scudder, *in litt.*, 5 March 2002
Bellevue, Jamaica	900–1000 C.E.		Medhurst, 1976; 1977; Wing, 1977, in W & R, 1982

(continued)

TABLE 1.3 (continued)
Summary of Marine Turtle Remains in the Caribbean (See also Tables 1.2 and 1.4)

Cinnamon Hill, Jamaica	900–1500 C.E.	chl	Johnson, 1976; Osbourne and Lee, 1976, in W & R, 1982
Rio Bueno, Jamaica		chl	Osbourne; and Vanderwall, in W & R, 1982
Rio Nuevo, Jamaica		chl	Wing, 1977; Osbourne, in W & R, 1982
Cancún Is., México	Formative	chl	Wing, 1974; Andrews, 1965, in W & R, 1982
Nueva Cadiz, Cubagua Is., Venezul	Historic	chl	Cruxent and Rouse, 1958; Wing, 1961, in W & R, 1982
Cedros, Trinidad	1–300 C.E.	chl	Bullbrook, 1953, in W & R, 1982
Erin, Trinidad	300–1500 C.E.	chl	Bullbrook, 1953, in W & R, 1982
Palo Seco, Trinidad	300–700 C.E.	chl	Bullbrook, 1953, in W & R, 1982
Quinam, Trinidad	300–1500 C.E.	chl	Bullbrook, 1953, in W & R, 1982
Mill Reef, Antigua	600–1500 C.E.	chl	Wing et al., 1968, in W & R, 1982
Barbados, various sites	380–1500 C.E.	chl	Bullen and Bullen, 1967, in W & R, 1982
Granada, various sites	0–1500 C.E.	chl	Bullen, 1964, in W & R, 1982
St. Eustatius, Netherlands Antilles	500 C.E.	Ei	Versteeg and Effert, 1987
Sugar Canyon, St. Kitts	1–300 C.E.	chl	Wing and Scudder, 1980, in W & R, 1982
Sugar Factory, St. Kitts	700–1000 C.E.	chl	Rouse, 1964; Wing, 1973; Goodwin, in W & R, 1982
Grand Anse, St. Lucia	0–1500 C.E.	*Cm, Ei*, chl	Haag and Bullen, in W & R, 1982
Folle Anse, Maria Galante, Gdlp	220 B.C.E.	chl	Barbotin, in W & R, 1982
Taliseronde, Marie Galante, Gdlp	300–600 C.E.	chl	Emond, in W & R, 1982

* *Notes:* FL = Florida; Venezul = Venezuela; Gdlp = Guadeloupe; WE SH = Weeden Island–Safety Harbor; chl = cheloniid; *Cm* = *Chelonia mydas*; *Ei* = *Eretmochelys imbricata*; W & R, 1982 = Wing and Reitz, 1982.

TABLE 1.4
Summary of Marine Turtle Remains in the Yucatán Peninsula Area (See also Table 1.3)

Site*	Period and Estimated Date	Species*	Remarks	Source
Mexico				
Cancún Is., QR	≤630 B.C.E.	chl	2250 carapace fragments	Andrews et al., 1974; Wing, 1974
Cozumel, QR	Late Postclassic–historic	chl	19 marine turtle fragments of 22,649 total	Hamblin, 1984
El Meco, QR	Classic: 300–600 C.E.	chl		Andrews, 1986; Andrews and Robles, 1986
Tancah–Tulum, QR	100 B.C.E.–1500 C.E.			Miller, 1982
Chitzén Itzá, Yucatán		chl	In Sacred Cenote	Carr, *in litt.*, 20 April 2002
Dzibilchaltun, Yucatán		*Cm*		Wing and Steadman, 1980
Isla Cerritos, Yucatán	100 B.C.E.–1500 C.E.	chl	>1000 turtle fragments; cut marks and burning	Carr, 1989a; 1989b
Belize				
Cerros	50 B.C.E.–150 C.E.	*Cc*, chl	5 marine turtle fragments	Carr, 1985, 1986a, 1986b
Kakalche	100–300 C.E.		2 marine turtle fragments	Graham, 1994
Saktunja	600–1500 C.E.		1 marine turtle fragment	Barr, 2000
Santa Rita Corozal	Postclassic			Morton, 1988
Watson Is.	100–300 C.E.		79 marine turtle fragments	Graham, 1994

**Notes:* QR = Quintana Roo; chl = cheloniid; *Cc* = *Caretta caretta*; *Cm* = *Chelonia mydas*.

marine turtles were commonly used for food by some aboriginal peoples of southern Florida, although it is not clear whether the hawksbill (*E. imbricata*) was commonly eaten; bone fragments of hawksbills from three Calusa sites in southwest Florida showed no evidence of burning or cut marks (Schaffer and Ashley, in press).

In addition to providing a food source, parts of marine turtles were evidently used for a variety of purposes in the southeast. Marine turtle shells have been found in funerary situations in various places in Florida, notably in Calusa and Tequesta sites. At Miami Circle, a Tequesta site, an entire marine turtle carapace was found aligned east to west, as was the custom for human burials (Schaffer and Thunen, in press, and citations therein). Net gauges were made from peripheral and other bones of marine turtles, and game and divining pieces were also made from marine turtle bones. There is also a report of "hollow shaving-blades" made from the mandible of a marine turtle (Cushing, 1897: 378–379).

Cultural artifacts made of tortoiseshell[4] have been found at Mound C in northern Georgia, a site of the Etowah culture. Of particular note is what appears to be an effigy of a crested bird, evidently used as a hairpin or other form of head ornament (Figure 1.1), found in a funerary context. The style closely resembles

FIGURE 1.1 Tortoiseshell pin from burial no. 109, Mound C, Bartow County, Georgia, USA; overall length is 24 cm; shaft is 18 cm long; greatest width is 1.1 cm (see Larson, 1993: 179–181; photo courtesy of L. Larson).

[4] In contemporary times the term "tortoiseshell" refers to the keratinous, or epidermal, scutes of the hawksbill sea turtle, *Eretmochelys imbricata;* tortoiseshell is used as a raw material for various crafts. There is, however, evidence in ancient Classical Greek texts that the keratinous scutes of land tortoises (family Testundinidae) were also used in certain crafts (see Casson, 1989; 102, 168; and discussion below).

that found in crested bird pins attributed to the Calusa of southern Florida, estimated to be from the period after contact with Europeans (i.e., after 1492, and probably after 1521; Larson, 1993).

Tortoiseshell was used in various ways for ceremonial and ornamental objects. These include hairpins, inserts to portray the pupil of the eye in various wooden carvings of animals, and wings for bird effigies (Schaffer and Ashley, in press). The shell and scutes of *E. imbricata* are thought to have been objects of religious importance for the Calusa (Schaffer and Thunen, in press, and citations therein).

The fact that tortoiseshell, as well as other mortuary objects such as sharks' teeth, Dover flint, and certain ceramics, is not found locally at inland Etowah sites indicates that an exchange or trade system in the southeastern U.S. brought exotic raw materials and craft products to southern Georgia. It has even been suggested that tortoiseshell was a trade item used by the Calusa throughout the southeastern region (Larson, 1993). In fact, it appears that like marine shells (Foster, 1874: 234), hawksbill materials and crafts entered into extensive trade during pre-Columbian times, because — along with marine shells — they were found in Hopewell sites (about 200 B.C.E.–400 C.E.) as distant from the sea as Illinois, Indiana, and Ohio (Shetrone and Greenman, 1931: Figures 57–59; Fagan, 1995: 411 f.f., 420).

The period between the Kiawah Island and Etowah sites represents a span of nearly three and a half millennia. However, there seems to be little specific information on the dates of marine turtle materials from the southeastern U.S.

Another region where marine turtle remains are common is the Caribbean basin (Table 1.3). Wing and Reitz (1982) reported relicts of these animals in association with human sites, continental and insular, that date from 1380 B.C.E. to 1715 C.E. Marine turtle remains were recorded in 37 of 47 collections (16 of which were on peninsular Florida), and they were the principal vertebrate remains identified at some sites. Remains from turtles of various sizes, including small-sized individuals (such as at Middle Caicos, see Wing and Scudder, 1983: 209), and abundant remains at sites that are not now close to known nesting beaches, provide evidence that Caribbean peoples took turtles from feeding grounds, and not just from nesting beaches (Wing and Reitz, 1982: 20–21). Evidently marine turtles were an important part of the diet and culture of many of these past societies. On the other hand, several hypotheses were offered to explain an apparent under-representation of marine turtle remains at some sites: seasonal habitation of coastal areas at that time of year when the turtles were not present; a taboo, or active avoidance, on consuming turtles; and the effect of transporting meat from a large animal, and leaving large amounts of the bony remains outside of archaeological sites (Wing and Reitz, 1982: 21).

The remains of "many bones of large sea-turtles" at Golden Rock, St. Eustatius, are reportedly from the largest pre-Columbian excavation in the Lesser Antilles, a site dated at about 500 C.E., or 1500 years ago. An intact marine turtle, except for the cranium, was assumed to have been decapitated before burial; and, because sponge spicules were found in the body cavity, it was assumed to be a hawksbill, the species known to specialize on sponges (Versteeg and Effert, 1987: 11, 18). No explanation was offered for the burial of a decapitated turtle, and any number of hypotheses are possible, including some form of ceremonial burial.

Some of the most detailed interpretations of animal remains have been done for sites in the Caribbean by Wing (2001a; 2001b). Using a series of estimates of body size, minimum number of individuals, biomass contribution, and trophic level, she provided a number of lines of evidence that nonindustrial, indigenous peoples made serious impacts on the animal populations that they exploited, essentially "fishing down the food web." This phenomenon, described in detail by Pauly et al. (1998) for modern fisheries, involves a trend in taking relatively fewer select species (e.g., top predators) and relatively more lower trophic level species (e.g., herbivores) over time; it is a clear indication of overfishing and nonsustainable resource use. Wing's analysis (2001a: Tables 5–12) indicates that the relative contribution of marine turtle biomass declined markedly over time in two of the four sites that were evaluated.

In addition to direct consumption, there is evidence from the Caribbean that tortoiseshell was fashioned into fishhooks during pre-Columbian times (Price, 1966: 1364; Wing and Reitz, 1982: 24). The use of tortoiseshell by peoples of the southeastern plain has been mentioned above.

Numerous archaeological sites along the shores of the Yucatán Peninsula have been identified as Maya (e.g., Andrews et al., 1974; Miller, 1982; Hamblin, 1984; Andrews and Robles, 1986; Carr, 1989a; 1989b). Given the antiquity, permanence, and level of sophistication of the Mayan civilization (Drew, 1999), together with the fact that the Yucatán Peninsula is a major nesting area for three species of marine turtle (Frazier, 1993), it is expected that archaeological remains of these animals should be common and widespread. However, there seem to be remarkably few reports — much less systematic studies — of marine turtle remains in Maya sites (Table 1.4).

A large coastal midden on Cancún Island was dated at no older than 630 B.C.E. (Andrews et al., 1974: 157, 166). The vertebrate remains were exclusively marine, and marine turtles were the most abundant taxa, with more than 2250 carapace fragments recovered. It was estimated that "at least nine large individuals" were represented, but species identification was not possible (Wing, 1974: 187). Remains of *C. mydas* were recorded at Dzibilchaltun, a site just north of Mérida that was occupied for millennia; it was argued that marine life was not important in the diet, but was important for religious and ritual activities of the Maya (Wing and Steadman, 1980). However, there were also suggestions that marine resources were underrepresented because of transport practices (Andrews in Wing and Steadman, 1980: 331, footnote). Miller (1982: 6), reporting from Tancah-Tulum, on the Caribbean coast of the Peninsula, states that marine subsistence patterns did not change between the Maya Preclassic and Colonial periods (a span of about 1.5 millennia). He indicates that turtle carapaces were consistently found in different levels; however, no specific information was provided.

Hamblin (1984) performed one of the most detailed studies of animal remains in the Maya area, working with a remarkable volume of animal bone (some 20,649 specimens) and diversity of species (at least 77) from the island of Cozumel. Included were many species not usually identified, particularly animals of small body size and with relatively delicate bones. Although turtle bones were abundant, there were just 19 bones from marine turtles (only about 1% of all turtle bones found, the vast majority being from smaller, freshwater species). At least two thirds of the total faunal sample was from the late Postclassic to historic periods, and the few marine turtle bones identified fit this pattern. Remarkably, it was concluded that "marine animals represented the basic food resource of the Cozumel Maya" (p. 174). In trying to explain the relative rarity of marine turtle remains, but the abundance of nesting marine turtles on eastern Cozumel and their widely recognized food value, at least during contemporary times, Hamblin suggested that there may have been a taboo among the Maya on taking these animals, but no further evidence was presented.

A relatively small ceremonial–habitational (and commercial) site is at El Meco, on the coast of Quintana Roo, about 2.5 km north of Puerto Juárez/Cancún City, and to the immediate west of Isla Mujeres. The vertebrate remains are varied in species, and more than half of them were identified as marine turtle, some of which were burned and had cut marks. Many of these turtle remains dated from the early Classic period (300–600 C.E.; Andrews, 1986: 70). Nevertheless, it was during the Postclassic period that the site seems to have had an especially active level of marine exploitation, and it was then (1100–1500 C.E.) that it was evidently used as a commercial port, linking the interior of the Peninsula with Isla Mujeres (Andrews and Robles, 1986: 132). It was proposed that the turtle carapaces — and even the skulls — were used as containers (Andrews, 1986: 69).

The use of marine turtle shells as vessels is not unusual in the Maya area or elsewhere, but there is no known evidence for using the skulls as containers, or anything else other than hunting trophies. Although El Meco is on the coast, at present the closest shore is rocky limestone, unsuitable for nesting by marine turtles. However, immediately north and south of the solitary rock outcrop on which it is situated there are vast sandy beaches, which in the 1950s and 1960s were used for nesting by large numbers of *C. mydas.* Moreover, the protected waters enclosed between the mainland coast at El Meco and Isla Mujeres were renowned half a century ago as a major place for turtle fishing (Andrews, *in litt.*, 10 March 2002). If this coast were in the same condition for the last 1.5 millennia, the turtles could have been caught and transported to El Meco from nearby beaches and offshore foraging areas. However, the marine turtle remains from El Meco have not been analyzed for size composition, so it is not known whether animals below adult size were in the middens, and hence whether there is evidence that turtles were captured on feeding grounds, rather than on nesting beaches.

Carr (1989a) reported on a collection of 4000 animal fragments from Isla Cerritos, off the north coast of Yucatán, and about 5 km west of San Felipe, at the mouth of Ria Lagartos. The ceramic remains on the island show that it was occupied from late Preclassic (about 100 B.C.E.) to Early Colonial times (sixteenth century), with the peak in human population coinciding with the height of development at Chichén Itzá, approximately 900–1200 C.E.; earthworks and other evidence showed that the island, about 200 m in diameter and about 500 m offshore, "is largely artificial" and that it served as an important port for widespread trade in the Mesoamerican region (Andrews et al., 1988). Nearly a third of the identified animal fragments were identified as turtles, and these were thought to be mainly marine turtles. The ubiquity of marine turtle remains in excavated pits on the island supported a hypothesis of year-round exploitation, involving the taking of animals both from nesting beaches and from the water (Carr, 1989b: 6). Remarkably, the proportion of identified marine turtle remains to total animal remains *increased* after the earliest sample and stayed relatively constant thereafter, over a period of centuries.

On the basis of the relative proportions of cranial and postcranial remains of marine fishes, it was suggested that an important industry on Isla Cerritos was the processing and exporting of marine fishes. These findings were compared with a collection of approximately 6000 animal remains from Chichén Itzá (located some 90 km inland from the north coast, and an important ceremonial center that experienced a peak in population size around the same time as Isla Cerritos). It was found that headless (i.e., processed) remains of marine fishes were abundant, indicating that Chichén Itzá was a site for importation. However, although there was ample evidence for an active fishery for marine turtles at Isla Cerritos, there were no identified remains of these animals at Chichén Itzá. Carr reasoned that no evidence was not negative evidence, and she suggested that turtle meat without the heavy bones was exported from Isla Cerritos. As she explained, this form of trade would be "invisible" to archaeological investigations that were based on animal remains (Carr, 1989a; 1989b: 19).

At Isla Cerritos four fragments of burned turtle shell were found with scrapes, surface gloss, or very even breaks. These were interpreted as evidence that the animals were roasted on an open fire and that cut marks were received during butchery (Carr, 1989b: 8, 13). Yet only 18% of reptile bones (which were marine turtles in the main) showed signs of burning (Carr, 1989b: 15). Scrapes on the outer surface of one piece were hypothesized to have resulted from "removal of the horny scutes for ornamental purposes" (Carr, 1989b: 13), but there is no evidence for Maya having used tortoiseshell, which at any rate would not likely have been removed by scraping. Carr (*in litt.*, 22 February 2002) identified two marine turtle fragments among the extensive remains from the Sacred Cenote (a sinkhole used for ceremonial offerings) at Chichén Itzá, but no special significance was attributed to these because there was no way of knowing whether the turtle bones had been ritually deposited (Carr, *in litt.*, 20 April 2002).

Observing that there were only "trace amounts of definite and possible cranial material," Carr (1989b: 10) speculated that marine turtles were killed by decapitation. However, there is no other evidence for this practice, which would be inconsistent with contemporary customs, worldwide, for butchering marine turtles.

In contrast, a number of other studies of faunal relics in the Yucatán have reported no evidence at all of marine turtles, despite reporting a variety of other animals, often with an abundance of marine species (e.g., Pollock and Ray, 1957; Flannery, 1982; Shaw, 1995). There has been no attempt to explain this apparent anomaly, but it is likely that the transport of boneless meat is at least partially responsible for the lack of marine turtle bones at many sites.

Faunal remains have been reported at several coastal sites in Belize, also part of the Maya area (Table 1.4). However, there is apparently only one thorough study of faunal material, which is as yet unpublished (Carr, 1986a). There are four sites from where marine turtle bones have been recorded. Cerros, on the southeast shore of Corozal Bay, was evidently a trading center. A large number of remains of marine fishes and mud turtles (Kinosternidae) were identified in a collection of some 16,000 bones. Only five marine turtle fragments have been identified to date (including one *C. caretta*), and most of these came from the Tulix Phase, dated from 50 B.C.E. to 150 C.E. It is possible that this paucity is due, at least in part, to transport practices, although it is not known why marine turtle remains are so rare in a site where they should have been available as a ready source of food (Carr, 1985; 1986a; 1986b, *in litt.*, 22 February 2002).

Saktunja, also on the north coast, was evidently a salt production site. Dated from at least 600 to 1500 C.E., the faunal remains are dominated by fish, with turtle shell less abundant but significant. Only one marine turtle bone (a jaw) has been identified, although the turtle shell has not been fully studied; it has been argued that although Saktunja is on the sea, marine turtles do not come into the mangrove areas where the site is located and hence are not readily available (Barr, 2000; *in litt.*, 12 and 14 March 2002).

The most abundant animal remains from the late Postclassic at Santa Rita Corozal, Belize, were fragments of turtle shells, but there was no further identification to species. There was evidence of charring and of one piece having been worked into a disk (Morton, 1988: 119). Kakalche and Watson Island are coastal sites in Stann Creek District, Belize, and both date to about 100–300 C.E. Carapace fragments of marine turtles were reported from both sites (2 and 79 fragments, respectively), but no further details on the turtles were given. Marine fauna were well represented at both sites (Graham, 1994: 37, 55, 250, 252, 256).

There seem to be few recorded marine turtle remains from archaeological sites in South America. Strauss (1992: 87) indicated that marine turtles were part of the diet during the "meso-indian" period, from 5000 to 1000 B.C.E. in what is now the coast of Sucre and Anzoátegui states, Venezuela; however, no details were provided. A multiauthor volume on the prehistory of South America, with several chapters on coastal sites (see Meggers, 1992), makes no mention at all of marine turtle remains. However, the archaeological collections at the University of Florida include records of marine turtles from four sites in Ecuador (Scudder, *in litt.*, 5 March 2002).

The carapace of a large marine turtle, evidently *C. caretta*, is at Misiones, Argentina (located some 800 km from the coast) and was apparently used by the Guaraní people, evidently as a shield (Richard, *in litt.*, 3 December 1999). However, no further information is available.

The situation in Peru is remarkable. Despite millennia of human habitation along the coast, including complex and sophisticated societies that left large and diverse middens in an environment ideal for preserving archaeological materials, there is an anomalous paucity of marine turtle relicts. There are a few fragments at five coastal sites that are of different ages. From north to south they are: Quebrada de Siches, 5980–3605 B.C.E.; Pariñas, 300 B.C.E.–400 C.E.; Huaca Prieta, 2257–1550 B.C.E.; Los Gavilanes, 2869–1908 B.C.E.; and Sto. Domingo, Paracas, 4000–2000 B.C.E. In addition, there are remains from three coastal sites in Chile: Playa Miller; Los Verdes, 930–1070 C.E.; and Playa Vicente Mena, 1000–600 B.C.E. At the last-named beach, two tombs were each covered with a carapace. Several explanations were offered about the remarkable scarcity of marine turtle remains from what otherwise appears to be an ideal area in which to find them: marine turtles did not occur in significant numbers in pre-Columbian Peru; the technology for capturing them was not available; there was a taboo on taking turtles; preparation and transport methods resulted in remains that are not detectable; and/or archaeologists have not reported the remains systematically (Frazier and Bonavia, 2000; in preparation).

1.2.3 Summary of Zooarchaeological Remains

In fact, any one of the above explanations could apply for many of the archaeological sites discussed above. In many cases, the lack, or low number, of marine turtle bones that have been reported may be from the "schlep effect" — that is, "the larger the animal and the farther from the point of consumption it is killed, the fewer of its bones will get 'schlepped' back to camp, village, or other area" (Daly, 1969: 149; see also Perkins and Daly, 1968: 104). Olijdam (2001: 200) hypothesized the same thing in regard to large turtles' being relatively underrepresented from Arabian sites.

Another problem is the way that archaeologists have treated animal bones. In some cases they have knowingly discarded some of the osseous material, keeping just the more intact or readily recognized bones, not understanding that the zooarchaeologist needs to have access to as much material as possible; on the other hand, animal bones that have been culturally modified are occasionally kept from the zooarchaeologist (Wing, *in litt.*, 28 January 2002). As Daly (1969: 146) explained, there has been a tendency to treat animal remains from archaeological sites as "nonartifactual" and of rather second-class status. This is not to mention the biases that can be caused by different sampling (e.g., sieving) techniques: in some cases, such as at Saar, Bahrain, small mesh sieves have been used in meticulous efforts to retrieve even minute faunal material, but in other — particularly older — studies, if sieving were carried out it was with large-mesh screens. Another common problem is that the faunal remains are often the last to be worked up from archaeological studies, and in many cases the information available is only preliminary.

The methods used for capturing specimens documented in zooarchaeological studies are rarely decipherable. On beaches, nesting turtles (or possibly individuals that were sunning) likely have been captured by turning. The implements that are most likely to have been used for catching marine turtles at sea include arrows, harpoons, spears, tridents, nets, and traps (Wing and Reitz, 1982; Price, 1966). One of the more curious methods of catching marine turtles was with wooden decoys, which could be combined with set nets in which the turtles became entangled (Shaw, 1933: 68, 70). Previously used in various parts of the Caribbean, decoys seem to have been developed well before the Spanish conquest. There are even reports in the early French literature about Carib Indians catching copulating turtles by slipping a noose around a flipper or the neck, or even grabbing them (see Price, 1966: 1365). In this light, there is an account "of an Indian fishing slave, apparently Island Carib, who was dragged for two days by a frisky sea turtle" (see Price, 1966: 1368).

Not only is there ample archaeological evidence of widespread exploitation of marine turtles, evidently for consumption, but there is also evidence of using parts of marine turtles for purposes other than food. Two carapaces in coastal Chile were covering funerary urns, and it appears that carapaces have been used as shields both in Yucatán and on the Atlantic coast of South America. Tortoiseshell (the epidermal scutes of *E. imbricata*) seems to have been especially useful for cultural artifacts. It was used for fishhooks in the Caribbean, and also for hairpins and other body ornaments, as well as for religious objects, in the southeastern U.S.

1.3 ARCHAEOLOGY: CULTURAL ARTIFACTS RELATED TO MARINE TURTLES

As with the zooarchaeological material, some of the oldest records of cultural artifacts seem to come from Arabia and nearby. However, in many cases there is confusion about whether marine turtles were actually being depicted. Chelonians "are not often represented in [ancient Mesopotamian] art, and even when they are pictured there is as a rule no clear distinction between the salt-water turtle and the land tortoise" (van Buren, 1939: 103). Some interpretations of chelonians — for example, on cylinder seals — have been subsequently reinterpreted as representing totally different animals, for example, hedgehogs (van Buren, 1939: 103). It is not valid to assume that if there are zooarchaeological records of marine turtles there should also be cultural artifacts. For example, it was concluded that "Although dugong and turtle dominated among the faunal remains, they appear not to have been the dominant creatures in the Umm an-Narians' world of art and religion" (Hoch, 1979: 606).

There are numerous representations of chelonians in Asia, particularly from India and China, as well as in ancient Greece and Oceania. Some of these clearly represent marine turtles (e.g., Wingert, 1953: 73), but a thorough review of these voluminous materials is beyond the scope of the present chapter. (See also Molina, 1981.)

1.3.1 Cultural Artifacts of Marine Turtles in Arabia, the Middle East, and the Mediterranean

Engraved cylinders, or seals, are known from a variety of sites in Mesopotamia and neighboring areas. One such seal from the nineteenth-century collections of Layard, attributed to the Mesopotamian site of Nimrod (or "Nimroud"), depicts a large turtle with a large bird, before a human figure (Layard, 1853: 604). The chelonian is thought to be a marine turtle (Albenda, 1983: 27), and the relatively large size of the flippers is consistent with this interpretation.

Seals and seal impressions that include motifs of turtles have been recorded at Saar, Bahrain, and dated to 2000–1900 B.C.E., the early Dilmun period. A stamp seal found in a grave at Hamad Town (Figure 1.2) depicts a fishing scene in which two people are standing in a boat, one of whom is apparently lifting (or spearing) a turtle with large flippers, using a pole (or possibly a gaff or harpoon) of some sort (Vine, 1993: 53). At least three other seals from Saar include chelonians, and in two of these the animals clearly appear to have flippers; either one or both of these may represent a marine turtle. In one case the turtle accompanies an erotic scene (Figure 1.3), but erotic depictions on Dilmun seals are not unusual, so the significance of the turtle is unclear (Killick, *in litt.*, 13 February 2002).

FIGURE 1.2 A Dilmun Period stamp seal, found in a grave at Hamad Town, Bahrain. (From Vine, P. (ed.) 1993. *Bahrain National Museum.* Immel Publishing; London. p. 53. With permission.)

In categorizing the crafts of the Early Isin Period (from about 2000 to 1600 B.C.E.), van de Mieroop (1987: 37) stated that "Tortoise shell (ba-sig$_4$) is delivered for a throne in one text (*BIN* 9: no 182), probably for inlay, and as such was possibly used by the carpenters." However, the term "ba-sig$_4$" is not regarded as referring unequivocally to tortoiseshell (see Section 1.4 on ancient historic accounts). Leemans (1960: 25, footnote 4) did a comprehensive evaluation of foreign trade during the old Babylonian period, and on encountering this term he questioned whether it refers to tortoiseshell, explaining that there was no evidence of tortoiseshell from Ur.

A round stamp from ancient Mesopotamia, molded in clay, was apparently used to impress an image on cakes of lactic products; it has a flat face in incuse showing a turtle surrounded by wavy lines (symbolizing water) with a ring of three birds alternating with three fish around the periphery (van Buren, 1939: Figure 95). The form of this turtle is remarkably similar to the turtles shown on the reliefs in King Sargon's palace at Khorsabad (see below). A second, larger stamp is described as having a scene with a plain-shelled turtle in the middle, surrounded by birds and fish on the outer edge and "three small tortoises with carefully patterned carapaces" in the midfield (van Buren, 1939: 104, Figure 95). In both cases, the chelonians are aquatic, and they may be meant to represent marine species.

The reliefs on the walls of Sennacherib's palace at Nineveh show scenes of the King's victorious campaigns, including the sea with various creatures, such as turtles, crabs, and fish. Another slab is thought to depict a swiftly flowing river with a turtle and fish (van Buren, 1939: 104; see note 3 for primary citations), which would

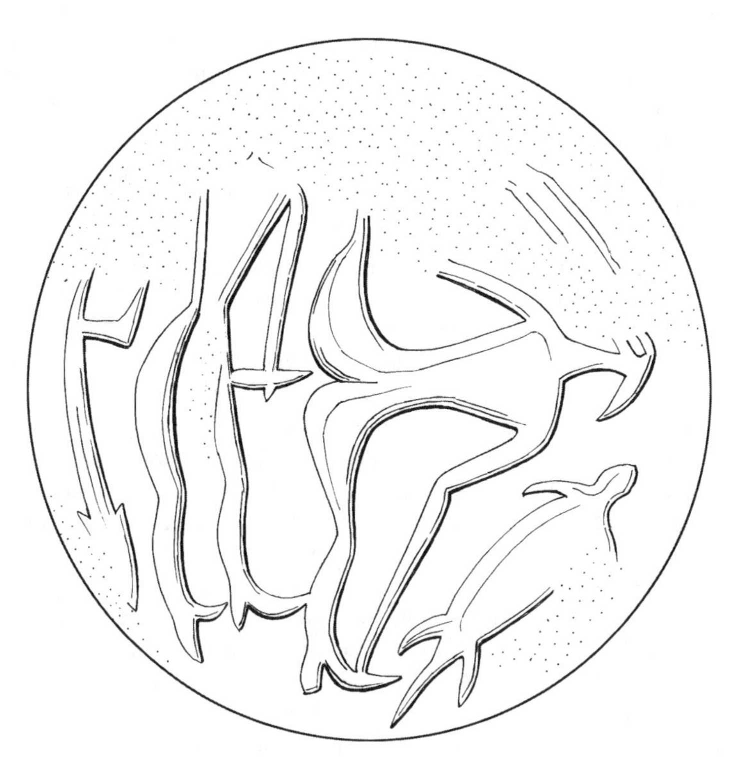

FIGURE 1.3 An Early Dilmun Period (2000–1900 B.C.E.) stamp seal from Saar, Bahrain. (Courtesy of London-Bahrain Archaeological Expedition and Robert Killick.)

appear to distinguish the freshwater creatures from those in the marine scene. Wall reliefs in King Sargon's palace at Khorsabad have been categorized as "monumental sculptural arts." Dating from 721–705 B.C.E., the relief on the southeast wall of a prominent courtroom in the palace shows an aquatic scene emphasizing the transport of timber; the sculpture may have originally measured 3 m high by 14 m wide. Interpreted as a seascape of the Mediterranean coast, it seems to have been an important geopolitical statement about Assyria's domination over trade routes and states along the eastern shore of the Mediterranean (Albenda, 1983). Several turtles are depicted in the aquatic setting. Their form, particularly the limbs that are *not* flipper-shaped, but rather like webbed feet, each with five claws, is not consistent with that of a marine turtle. Moreover, one of the turtles seems to have been added at the Museum at the Louvre during a restoration of the wall (Albenda, 1983: 6). However, because Assyrian depictions of turtles in other situations also show characteristics that deviate from real features, it has been argued that the turtles on the wall at Khorsabad are marine turtles (Albenda, 1983: 27).

For decades it has been claimed that tortoiseshell was used in ancient Egypt (Lucas, 1948: 50; see also Parsons, 1972), so it has been argued that tortoiseshell was also used in ancient Mesopotamia (Leemans, 1960, in particular footnote 4, pp. 25–26). Yet, there seems to have been confusion among contemporary scholars of both Egypt and Mesopotamia in distinguishing between the shell of a land tortoise and the epidermal scutes of the hawksbill sea turtle, known as "tortoiseshell." For example, the primary literature for a number of ancient Egyptian artifacts that were reportedly made of "tortoise-shell" clearly expresses doubt about whether the material was horn or tortoiseshell, or the original description indicated that the material was evidently *not* made of the scutes of the hawksbill sea turtle (cf. British Museum, 1904: 73; 1922: 31; Carnarvon and Carter, 1912: 76; Brunton, 1937: 5, 24, 30, 53, 57, 88, 110, 141, 142, 146, 147). At Abu Salabikh, Iraq (an urban settlement about 2500–2350 B.C.E.), a "most unusual object … [was] a complete tortoise shell lying upside down next to a conical bowl on the floor" (Postgate and Moorey, 1976: 167; see also Postgate, 1977: 275). This account from Abu Salabikh has been used by some authors as evidence of an active Mesopotamian trade in tortoiseshell (e.g., Olijdam, 2001). In developing his own argument for trade, however, Moorey (1994: 128–129) was careful to explain that no specimen of tortoiseshell (i.e., hawksbill turtle) has ever been found in Mesopotamia,[5] and he suggested that the shell of more than one kind of chelonian may have been used in trade. There is some evidence in ancient historic records for the use of, and trade in, shell other than that of the hawksbill marine turtle, evidently involving land tortoises (e.g., Lucas, 1948: 50; Casson, 1989: 102, 168; see Section 1.4).

In this light, there are detailed lists of materials used in ancient sites that specifically do *not* include tortoiseshell, "when it might have been expected" (Moorey, 1994: 129). This is the case for Ras Sharma, a second-millennium B.C.E. site in Syria (Caubet and Poplin, 1987: 289). There are strong arguments for seafaring merchants from Mesopotamia and for ancient activities of exploitation and exchange with eastern Arabia (e.g., Leemans, 1960; Oates et al., 1977), yet despite tantalizing bits of evidence, there is no unequivocal proof that tortoiseshell was included in the ancient trade from this region.

In the Mediterranean region, Aeginetans are thought to have been the first Greeks to have struck coins, and the earliest coin from Aegina (about 700–650 B.C.E.) is regarded to be one depicting a marine turtle. The second coin (after 650 B.C.E.) shows a turtle with front flippers and relatively large head, all consistent with a marine species, with a series of five raised circles running down the median axis of the carapace (Lorch, 1999), perhaps representing the five vertebral scutes of the usual turtle carapace. The use of the marine turtle, or later the land tortoise, on early Greek coins is thought to have been a type of "peloglyphic" legend, in which a symbol (turtle in this case) represented a particular sign in a prealphabetical syllabic script (Lorch, 1999).

The "Metope," a relief of a man crouched on a chelonian nearly his size, is regarded to depict an ancient Greek legend involving a wandering seafarer, evidently being aided by a marine turtle (Venizelos, *in litt.*, 30 May 2002).

1.3.2 Cultural Artifacts of Marine Turtles in the Western Hemisphere

Turtle effigies, particularly those depicting hawksbills, were evidently used to guard charnel houses, or mortuary areas, of the Calusa of south Florida. Figureheads,

[5] For example, Potts (1997) does not even mention "tortoiseshell" once in his book *Mesopotamian Civilization: The Material Foundations.*

thought to represent marine turtles (evidently hawksbills), are also known from the Calusa area. Historic accounts indicate that the Calusa of Key Marco, southwest Florida, carved and painted masks representing "Hawksbill Man" (or at least "Turtle-man") and "Hawksbill Spirit" (or at least "turtle spirit") (Schaffer and Ashley, in press; Schaffer and Thunen, in press, and citations therein).

The turtle has been of great importance in Maya and other Mesoamerican cultures, and occurs frequently in diverse media including stone, ceramic, stucco, cliff carvings, and parchment (Taube, 1988). However, it is not always clear if the animal depicted was meant to be a marine turtle. For example, the most frequent animal image at Mayapan was the turtle, but not one of the 18 carved limestone artifacts illustrated (Proskouriakoff, 1962: 331–333, Figures 1 and 2) can be identified as portraying a marine species: although Mayapan is not on the coast, a variety of marine artifacts were found (e.g., Proskouriakoff, 1962), indicating that exchange/trade in marine products — and knowledge of them — did not limit the ability to depict marine turtles.

There are several depictions in ceramics, figurines, stone altars, and other media where the carapace of a turtle appears to represent the earth. This includes the much-reproduced vessel number 117 (Robicsek and Hales, 1981: 91) showing a youthful Maize God being reborn out of a cracked turtle carapace, not to mention a number of stone altars interpreted as depicting a turtle that seems to symbolize the earth (Taube, 1988: 189, 193–198).

Analyses of painted and carved capstones found in corbelled vaults of Chenes and Puuc buildings of Yucatán, as well as ancient and contemporary Maya houses, added further evidence of the importance of turtles in the cosmology of the Maya (Carrasco and Hull, 2002; Hull and Carrasco, in preparation). It was concluded that the roof of some Maya structures was symbolically regarded as the shell of the cosmic turtle. Structurally redundant crossbeams in Maya houses are named *cap-aac* and *chan cap-aac* ("turtle arm" and "little turtle arm," respectively), and this was concluded to be "part of the zoomorphic symbolism of the Maya house."

God N[6] (phonetically "Pauahtun") was one of the major gods of the ancient Maya, the sky-bearer or world-bearer. A contemporary name is *Mam*, and the Nahuatl ("Aztec") counterpart was *Tzitzimitl*. God N is often depicted as wearing a carapace; he also seems to be the god of thunder, mountains, and the interior of the Earth (Taube, 1992: 92 ff., table; Miller and Taube, 1993: 132, 175). By the same token, the god Yahul of the Mixtecs and Zapotecs is often portrayed wearing a turtle shell, and this supernatural being may be related to thunder and lightning (Miller and Taube, 1993: 175, 191).

The turtle was used to depict the "earth platform," for example, in showing the passage of a deceased figure into the underworld (Stone, 1995: 22, Figure 3.8). In Maya cosmology, the "main event of Creation was the appearance of this turtle shell" (Freidel et al., 1993: 65). It has been concluded that there is a symbolism of a circular earth, "a form apparently conceptualised as a great turtle surrounded by

[6] The use of letters from the Roman alphabet to name Maya gods was developed by Schellhas, who previously referred to God N as the "God of the End of the Year" because of the hieroglyphic symbols denoting the end of the year that are associated with this representation (Schellhas, 1904: 37).

the sustaining sea" (Taube, 1988: 199; see also Miller and Taube, 1993: 69, 175). With this in mind, it could be argued that many of the chelonians depicted by Maya, from Preclassic to Colonial times, refer in one way or another to marine turtles. Further evidence for the singular importance of marine turtles is found in the Maya term *mak*, which has considerable symbolic importance and has been defined as meaning "sea turtle" or "turtle shell" (Hull and Carrasco, in preparation).

A detailed analysis of the Paris Codex emphasized the importance of chelonians to Maya culture, notably in the Maya constellation *ak ek'*, or turtle star (Orion in western astrology) (Love, 1994: 90, 95–101). Often the turtle shell carries the "K'an cross," which has been interpreted as a sort of "X marks the spot" where rebirth and creation took place (Freidel et al., 1993: 94, 281–283; see also Carrasco and Hull, 2002). It was concluded that *ak ek'* represents a giant marine turtle (Love, 1994: 98), but this is open to interpretation.

In their compilation of animal figures in the Maya codices, Tozzer and Allen (1910: 321–323, plate 14, no. 4) reported a dozen representations of turtles. No less than three of these depictions were thought to represent marine turtles. One of these was thought to be the shell of a marine turtle being used as a shield (p. 322, plate 14, 4), which seems to be consistent with historic accounts from the time of the Spanish conquest (see below). Turtles, together with about a dozen other animals, are key astronomical symbols; and turtles are thought to represent the summer solstice (Tozzer and Allen, 1910: 287, 323; see also Freidel et al., 1993; Love, 1994).

In lowland Maya symbolism, turtles, both marine and terrestrial forms, were often identified with water (Tozzer and Allen, 1910: 291; Miller and Taube, 1993: 174; Stone, 1995: 28, 238), a critical commodity for survival and agricultural success in that region. Turtles are central to the theme of rebirth, which was symbolized by bloodletting. For example, there are two scenes in the Madrid codex that associate turtles with ceremonial bloodletting: in one scene no less than five gods standing around a central turtle altar are strung together by a rope that perforates their penises (Lee, 1985: 94). Carrasco and Hull (2002) and Hull and Carrasco (in preparation) determine that the central turtle in this scene also represents the roof of a celestial building with a corbelled vault. A cache of 25 ceramic figures from Santa Rita Corozal, Belize, includes four late Postclassic figurines, each depicting an aged man engaged in penis perforation, standing on a turtle (Chase and Chase, 1986: p. 12, bottom figure, p. 17, right-hand figures; Taube, 1988: 193). The shapes and relative sizes of shell, head, and limbs of the turtles are consistent with marine species.

In his detailed compilation of turtles in cultural artifacts of pre-Hispanic Maya, Taube (1988) explained how these animals symbolize not just the round earth, but also one of the major units of time for the Maya, the *k'atun* — the cycle of 7200 days, or nearly 20 years. There are several possible symbolic links: the *k'atun* wheel, which often takes the shape of a turtle, is composed of a cycle of 13 *k'atuns*, the same number of large scutes on the carapace of a turtle. In addition, 20 years approximates the maturation period for many turtles.

The complexities of cultural interpretations are illustrated by a study at Tancah-Tulum, on the southern coast of Quintana Roo. Remains of marine turtles from secondary deposits in all levels from the Preclassic to the Postclassic, from about 300 B.C.E. to 1500 C.E., led to the conclusion that these animals were consistently an important source of food over a period of approximately two millennia. But a turtle effigy censer found at the site prompted the suggestion that the turtles were not simply killed and eaten. It was hypothesized that the seasonal appearance of nesting marine turtles on the easternmost coast of the Yucatán peninsula that emerged from the east (a sacred direction for the Maya) to complete their reproductive cycle may have been incorporated into a cult. Miller (1982: 7, 61, 62) suggested that in addition to being a food source that had been exploited for millennia, marine turtles were also a symbol used by the Maya to celebrate the cycle of life, rebirth, and renewal, and also to revere the sacred east.

In this light, it is interesting to note that a literary interpretation of the *Popol Vuh*, the sacred book of the Quiche Maya of the Guatemalan highlands, suggests that the wife of one of the four founding fathers of the human race was named "Red Sea Turtle," and that these fundamental people in Maya cosmology arrived from the east (Tedlock, 1993: 1, 234). However, other than this story, there seems to be little supporting evidence for Red Sea Turtle Mother's being part of the Maya creation myth.

Numerous ceramic pieces from various periods of Maya culture clearly depict turtles (e.g., Robicsek and Hales, 1981), but there are no clear representations of marine turtles. For example, incensario supports from Palenque showing realistic human figures standing on realistic turtles have been given considerable attention (Rands et al., 1979: Figures 3, 4, and 5; Taube, 1988: 193), yet from the form of their appendages and shell, the turtles depicted are clearly freshwater species.

There are at least three Moche (200 B.C.E.–700 C.E.) ceramics, from coastal Peru, that unquestionably depict marine turtles (Frazier and Bonavia, 2000; in preparation).

Petroglyphs that appear to represent marine turtles have been recorded from El Yunque, Puerto Rico, and Jamaica (Schaffer and Ashley, in press). At Quebrada El Médano, south of Antofagasta, Chile, there are clear petroglyph scenes of turtles being harpooned from small boats (Frazier and Bonavia, in preparation).

1.3.3 Summary of Cultural Artifacts of Marine Turtles

Marine turtles have been represented in a wide variety of media by numerous cultures around the world. In some cases there appears to be a religious, mythical, or spiritual context, but in many cases the "cultural motivations" that prompted the peoples of bygone times to depict marine turtles are not at all clear. One may need look no further than the fulfillment of human curiosity and the omnipresent drive to create and manipulate, while at the same time adorning one's immediate environment. Postmodern societies do not have the exclusive claim to appreciate cultural artifacts as works of art. There is no reasonable justification in negating the possibility that the crafters of Dilmun seals, Calusa hairpins, Maya ceramics, Moche vases, or Chilean petroglyphs, who incorporated marine turtles in their creations, were striving to fulfill basic human emotions. How marine turtles relate to that is, however, not at all clear.

1.4 ANCIENT HISTORIC ACCOUNTS OF HUMAN–TURTLE INTERACTIONS

Some of the oldest written accounts interpreted as referring to turtles are inscribed on cuneiform tablets from Sumerian cities of the Ur III Period, about 2100–2000 B.C.E. (Owen, 1981; Englund, 1990), and even as far back as the late Uruk Period (3500–3000 B.C.E.) there appear to be archaic texts that record an active trade in marine products, including turtles (Englund, 1998). An analysis of Ur texts at the time of the early Larsa kings (about 2000–1600 B.C.E.) showed the importance of trade and transport, including seafaring merchants from Mesopotamia (Leemans, 1960); and an analysis of pottery from eastern Arabia supported the conclusion that there were active merchants and exploiters from Ur operating in the Arabian–Persian Gulf (Oates et al., 1977).

Perhaps the most influential archaic text that has been used to suggest the importance of marine turtles to ancient Mesopotamian societies is the record of religious tithe. Among the various gifts given to the goddess Ningal of Ur were "30 finger-shaped(?) pieces of 'ba-sig$_4$'." Leemans (1960: 25, footnote 4, 34) explained that "ba is a tortoise" (chelonian) and "sig$_4$ may denote things like spine or case." From this interpretation, and from the fact that the use of tortoiseshell was reported in Egypt "from an early date" (cf. Lucas, 1948), Leemans deduced that it is "unlikely that tortoise shell should have been entirely unknown in southern Mesopotamia."

In fact, the sign "ba" is interpreted to be a short form of the term *bal-gi* (or *balgi*), which refers to "turtle." The sign "sig$_4$" (meaning "brick," pronounced "sig," and reading "murgu") is taken to mean "back" or "shoulder area" (Englund, *in litt.*, 2, 3, 10, 12 June 2002). With these interpretations of the cuneiform signs it seems plausible that tortoiseshell was described, but it is far from confirmed.

Farber (1974) did a detailed linguistic study of "ba," and he clarified that there is no substantiated species identification for this term. His analysis showed that in pre-Sargonic texts the sign "ba" was listed together with marine fishes; however, in Sargonic times and later it was listed as its own item. In some cases, the numbers of "ba" reported were in the thousands, and this gave rise to the question whether eggs were being referred to.

Textual references from Mesopotamia indicate that turtles were used for food, medicine, ritual food, and even food for royalty; before the Sargonic Period (i.e., earlier than 2350 B.C.E.) records of turtles were "not uncommon." On the basis of both textual and archaeological information from as early as "Old Babylonian levels" (i.e., about 2000–1600 B.C.E.), it was concluded that "various types of turtles played a significant and familiar role in the daily life of the inhabitants of Mesopotamia and elsewhere in the Near East" (Owen, 1981: 41, 42, and references therein).

In most cases, however, it is not clear what types of turtles are being referred to, or even if they are marine species. Nonetheless, Englund (1998) reasoned that in the context of the late Uruk references to unspecified turtles, it was most likely that marine turtles were reported. His earlier study (1990) indicated that in many of the Ur III texts turtles were mentioned in the context of marine fisheries, and the animals were delivered to state larders in a processed or conserved form. Olijdam

(2001) developed a series of arguments not only for the presence of an established fishery for marine turtles in the early Dilmun period (2100–1900 B.C.E.), but also for the occurrence of an organized trade in tortoiseshell, involving import centers in Mesopotamia.

Although the textual evidence is speculative and not supported unequivocally by the archeological data, the numbers of both eggs and turtles reported in the ancient texts — sometimes in the thousands — need to be considered in the light of biological understanding. The only turtles in the Arabian area that are known to occur, and nest, in densities adequate to make the collection and transportation of thousands of animals or their eggs an economically viable venture are marine turtles. Hence, several lines of evidence indicate that the ancient Mesopotamian texts refer to an organized trade in marine turtles; the evidence for tortoiseshell, however, is less apparent.

A very different writing system was developed in ancient China. The Chinese character for turtle, *gui*, clearly shows a pictogram of a chelonian, and indicates that writing about turtles may be as old as Chinese writing itself. Texts with the turtle character may date to the Quin (221–207 B.C.E.) and Han (206 B.C.E.–220 C.E.) dynasties (Dwe, 1981: 678).

Certain accounts by Greek authors are essential for understanding human–turtle relations more than three millennia ago. Agatharchides of Cnidus, writing in the third century B.C.E., described the habits of "turtle eaters" (*Chelonophagi*), a primitive group of people who lived on islands, apparently in the southern extreme of the Red Sea. Huge turtles were said to be common in the waters around the islands, where the people caught them at sea, pulled them onto shore, cooked their innards by the heat of the sun, and from single shells were said to make shelters, vessels for holding water, and boats (Burstein, 1989: 85–87). This account was repeated by Pliny (1940:187), but the claim about turtles being made into boats may derive from misunderstanding about skin boats (Burstein, *in litt.,* 25 February 2002), or an exaggeration: remarkably, there are contemporary stories of children in southern Arabia using turtle shells as "boats" (Pilcher, *in litt.*, 8 June 2002).

An unknown author, evidently an Egyptian Greek, writing in the middle of the first century C.E., produced what is essentially a traders' handbook for the Indian Ocean, particularly for luxury goods, known today as the *Periplus Maris Erythraei* (Mathew, 1975; Casson, 1989: 6, 7, 15). This describes in considerable detail well-established, highly organized commercial enterprises, involving a great diversity of activities and commodities in trade. One of the most important items, for shipping back to the Egypt and the Mediterranean, was tortoiseshell: it was traded in all the major ports that were described, including those in the Red Sea, the horn and east coast of Africa, the southern coast of Arabia, as well as India, Sri Lanka, and Malay or Sumatra (Casson, 1989: 17, 101). Indeed, "Tortoise shell[7] receives more mention in the *Periplus* than any other object of trade," and "the finest quality was brought to Muziris/Nelkynda all the way from Malay to be made available to Western

[7] The term used in the *Periplus* was "chelone," which commonly refers to the animal, but in the context it is interpreted as referring to the product of the turtle (Casson, personal comunication; Margaritoulis, *in litt.*, 13 June 2002).

merchants" (Casson, 1989: 17, 101). When the tortoiseshell trade began is not clear, but there had been wide-ranging trade in the Indian Ocean for at least two millennia before the *Periplus* was written (Casson, 1989: 11). The *Periplus* also provides other interesting details, such as at Menuthias Island, where dugout canoes and sewn boats were used for catching turtles (Casson, 1989: 59).

Although tortoiseshell is the most frequently mentioned article of trade in the *Periplus*, the document also records other types of chelonians ("a little land tortoise, and a light-colored variety with rather small shields," and "mountain tortoise"), thus indicating that these "other chelonians" were of interest and evidently traded (Casson, 1989: 51, 59, 69). This, in addition to other ancient records, led Casson (1989: 102, 168) to develop an argument that terrestrial and aquatic turtles were used since ancient times, not only for food but also as sources of raw materials for various crafts, and that the scutes of land tortoises were used for adorning especially large objects (perhaps this was the original derivation of the term *tortoiseshell*, now used to refer to the scutes of the hawksbill sea turtle). On the basis of the same textual evidence in the *Periplus*, Lucas (1948: 50), writing about ancient Egyptian materials and industries, had also proposed that "in ancient times probably the plates of more than one kind of turtle and also of the land tortoise were used."

A millennium after the *Periplus*, other historic documents, including tenth-century Arab and later Chinese accounts, reported the trade in tortoiseshell that ranged across much of the Indian Ocean (Al-Mas'udi in Freeman-Grenville, 1962: 15; Trimingham, 1975: 133; Wheatley, 1975: 107). Indeed, it was proposed that marine turtles, particularly the shell, were an important exchange item in obtaining imported ceramics for 800–900 C.E. societies in the Comoro Islands (Wright, 1984: 14, 57).

The extent, spatial and temporal, of this Indian Ocean trade has been used by some authors (e.g., Uerpmann and Uerpmann, 1994; Mosseri-Marlio, 2000; Olijdam, 2001) to argue that there had to be an established tortoiseshell trade between Mesopotamia and the Gulf. However, although there were clearly extensive trade routes in this region from ancient times, the evidence for tortoiseshell trade is unclear.

A number of scholars have provided invaluable resumes of diverse historic documents (e.g., Freemen-Grenville, 1962; Wheatley, 1975). However, there are likely to be many more ancient literary records of human–turtle interactions from other centers of civilization such as Arabia, China, India, and Greece, but the linguistic challenges may have impeded the advancement of investigations.

Ancient "written" information is also available from the Western Hemisphere. The analysis of Maya glyphs provides evidence from diverse sources of the importance of turtles to a culture that dates back more than three millennia, and as discussed above, marine turtles were evidently referred to in many cases. For example, inscriptions on a stela read as "was seen, the first turtle image, great god lord" are interpreted to refer to the rebirth of the First Father, the God of Maize, through the cracked shell of a turtle; the "main event of Creation was the appearance of this turtle shell" (Freidel et al., 1993: 65).

In the well-established trade between Aztecs, from the central valley of Mexico, and lowland Maya populations, the Chontal merchants of present-day Tabasco state were essential intermediaries. According to some authors, among the more prized trade objects originating from Yucatán was tortoiseshell, a product that the Aztecs took back home (Scholes and Roys, 1968: 29).

At the time preceding the Spanish conquest of the Yucatán peninsula, in the early sixteenth century, there were reports of Maya warriors carrying shields made of marine turtle shells (Diaz del Castillo, 1908). Just after the conquest, Friar Diego de Landa recorded the great size and good taste of marine turtles; eggs were also mentioned, but it was not specified whether those were also eaten (Gates, 1937: 99; Tozzer, 1941: 192; Pagden, 1975: 145; Landa, 2001: 136). Shortly after the conquest it was also reported by the Spanish chroniclers that turtles were among the various marine animals taken from the Gulf of Mexico by the peoples of present-day Tabasco state (Scholes and Roys, 1968: 30). After the Spanish became established in present-day Mexico, it was stated that "turtle fishing was a lucrative industry" and that a variety of highly prized items were manufactured from tortoiseshell by the Tixchel people for the Spaniards during the latter half of the sixteenth century and into the seventeenth century (Scholes and Roys, 1968: 244, 302, 329, 336). In addition, at the beginning of the Colonial era, Oviedo described the use of dragnets to capture turtles in the new Spanish colony (Wing, 1974: 187).

1.5 CONCLUSIONS AND DISCUSSION

Marine turtles have captivated the human imagination for millennia, for many and diverse reasons. Textual references to turtles, including marine turtles, seem to occur in many civilizations, and date back millennia. Providing nutritional, economic, and spiritual sustenance to peoples around the world, these marine reptiles are part of the cultural fabric of many coastal communities. There is no question that human societies and human cultures have been impacted by these animals. Whether there have been critical events in human prehistory and ancient history directly influenced by marine turtles is moot. Scholarly discourses on the relationship between societies, culture, and environment, and how they relate to such basic activities as acquiring food, are nothing new; classical Greek authors, from before the time of Christ, deliberated this issue (Burstein, 1989: 29). If marine turtles provided critical sources of nutrients, for example at remote settlements on desert stretches of the Arabian coast, or on isolated islands of the Caribbean, it is possible that access to these reptiles made the difference between starvation and survival for certain communities.

Given the importance that some marine turtle products have had in ancient and prehistoric trade, for example tortoiseshell throughout the Indian Ocean, or across North America, it also seems that marine turtles may have provided the raw materials on which certain ancient or prehistoric institutions were founded or sustained. There is no doubt that marine turtles, at different places and in different times, have enriched the human spirit in countless ways.

Chapter 12 examines the contemporary aspects of human–turtle interactions, through the lens of cultural and social analysis, while Chapter 14 delves into the philosophical aspects of value systems related to the ways that contemporary societies view marine turtles and the habitats on which they depend. As is made clear in both of these chapters, there is no single or simple logic for rationalizing human–turtle interactions, and often there is tension between developed and undeveloped societies. This philosophical debate has existed since antiquity. Concerns for the corrupting influence of material goods from "civilized" societies on "uncorrupted noncivilized peoples," with a certain romanticism for the "primitive customs," have been expressed since the time of classical Greek writers (Burstein, 1989: 76, footnote 2). In the end, the roots of the concepts and rationales presented by Campbell (Chapter 12) and Witherington and Frazer (Chapter 14) can be found deep in the prehistoric and ancient relationships between humans and marine turtles.

There is another side to this discourse: how have marine turtles been affected by humans? The obvious discussions about intense exploitation, overfishing, and subsequent decline and collapse of marine turtle populations need no belaboring. It would be simplistic, however, to portray the impact of humans on marine turtles as nothing more than a question of overfishing. Bjorndal and Jackson (Chapter 10) explore the roles that these reptiles may once have played in marine and coastal environments. If the ecological functions of marine turtles have been drastically altered because of human actions (Jackson et al., 2001), then human impacts directed at the turtles have clearly gone far beyond the reptiles themselves, and have had much more profound implications on vast and complex marine ecosystems.

This raises the issue of the "pristine myth," which has been debated by scholars from a variety of disciplines. Contrary to popular beliefs that immense areas of the planet were in pristine conditions until European technology arrived, it is clear that since prehistory and antiquity humans have played major roles in shaping the form and function of diverse and vast terrestrial ecosystems (e.g., Lewis, 1980; Hughes, 1985: 302 f.f.; Diamond, 1986; Chapman et al., 1989; Bowden, 1992; Denevan, 1992; 1996; Turner and Butzer, 1992; Wilson, 1992; McDonnell and Pickett, 1993; Kay, 1995; Hames, 1996; Miller et al., 1999). However, there has been an enduring and widespread misconception that marine ecosystems — with relatively few known extinctions of marine species — are not subject to significant human impact (Roberts and Hawkins, 1999). Clearly, they are subject to such impact, and as Bjorndal and Jackson (Chapter 10) explain, the structure and function of marine and coastal ecosystems, notably regarding the ecological roles of marine turtles, have been greatly altered as a result of human activities.

Hence, although they are poorly appreciated and little studied, human–turtle interactions must include much more than changes in number, densities, and geographic distributions of the reptiles. Behaviors such as timing, periodicity, and location of nesting; timing and location of feeding activities; and mating activities may all have been substantially affected by generations of human predation and perturbation. Whether anatomical and physiological aspects of marine turtles have been influenced by millennia of interactions with humans is unknown. Nevertheless, there are unanswered questions: to what extent have human activities over past millennia influenced the biology of marine turtles? To what extent is contemporary marine turtle biology a product of interaction with human beings?

ACKNOWLEDGMENTS

Valuable assistance with obtaining and interpreting information used in this chapter was provided by B. Blair, S.M. Burstein, L. Casson, Betty Faust, N'omi B. Greber, Nancy Hamblin, A. Hutchinson, B. Love, Natalie Monro, P.R.S. Moorey, E. Olijdam, D.I. Owen, G. Peters, Sylvia Scudder, Margarethe Uerpmann, Lily E. Venizelos, Sarah Whitcher-Kansa, Elizabeth W. Wing, H. Wright, and Melania Yánez; numerous other colleagues not only provided valuable information but made constructive comments on earlier versions of this paper: Anthony P. Andrews, Guy Bar-Oz, Mark Beech, H. Sorayya Carr, Robert K. Englund, Arlene Fradkin, Brendan Godley, Robert G. Killick, Lewis Larson, Christine Mosseri-Marlio, Nick Pilcher, Ina Plug, Nicholas Postgate, Dan T. Potts, Chuck Schaffer, Karl A. Taube, and Wendy Teeter. Support for the study was provided by Friends of the National Zoo (FONZ).

REFERENCES

Albenda, P. 1983. A Mediterranean seascape from Khorsabad. *Assur* (*Monogr. J. Near East*) 3(3): 1–34 + 13 plates.

Andrews, A.P. 1986. La fauna arqueológico de El Meco. Pp. 67–75. In: A.P. Andrews and F. Robles C. (coordinators). *Excavaciones Arqueológicas en El Meco, Quintana Roo, 1977.* Instituto Nacional de Antropología e Historia. Serie Arqueolgía.

Andrews, A.P. and F. Robles C. 1986. Proyecto El Meco. 1977; Conclusiones. Pp. 131–134. In: A.P. Andrews and F. Robles C. (coordinators). *Excavaciones Arqueológicas en El Meco, Quintana Roo, 1977.* Instituto Nacional de Antropología e Historia. Serie Arqueolgía.

Andrews, A.P. et al. 1988. Isla Cerritos: an Itzá trading port on the north coast of Yucatán, Mexico. *Natl. Geogr. Res.* 4(2): 196–207.

Andrews, E.W. IV et al. 1974. Excavation of an early shell midden on Isla Cancun, Quintana Roo, Mexico. *Archaeological Investigations on the Yucatan Peninsula.* National Geographic Society — Tulane University Program of Research on the Yucatan Peninsula. Middle American Research Institute, Tulane University; New Orleans. Publication 31. (1975) pp. 147–197.

Astuti, R. 1995. *People of the Sea: Identity and Descent Among the Vezo of Madagascar.* Cambridge University Press; New York. x + 188 pp.

Barr, B. 2000. Preliminary analysis of Saktunja fauna. Pp. 56–62. In: S.B. Mock (ed.). A view from the lagoons: Late Maya Classic-Postclassic settlement on the northern Belize coast. The Northern Belize Coastal Project (NBCP): Interim Report, 1999. Saktunja. Report to the Department of Archaeology, Belmopan, Belize and Foundation for the Advancement of Mesoamerican Studies (FAMSI). Institute of Texas Cultures, University of Texas, San Antonio, TX.

Beech, M. 2000. Preliminary report on the faunal remains from an Ubaid settlement on Dalma Island, United Arab Emirates. Pp. 68–78. In: M. Mashkour et al. (eds.). *Archaeozoology of the Near East.* IV b. ARC - Publicatie 32, Groningen; The Netherlands.

Beech, M. 2002. Faunal remains from site H3, As-Sabiyah: an Arabian Neolithic site in Kuwait. Paper presented at the 6th ICAZ Archaeozoologists of Southwest Asia (ASWA) Working Group Meeting, held at the Institute of Archaeology, University College, London (August 30–September 1, 2002).

Bökönyi, S. 1992. Preliminary information on the faunal remains from excavations at Ras al-Junayz (Oman). Pp. 45–48. In: J.P. Gerry and R.H. Meadow (eds.). *South Asian Archaeology 1989. Monographs in World Prehistory*, 14; Madison.

Bowden, M.J. 1992. The invention of American tradition. *J. Hist. Geogr.* 18(1): 3–26.

British Museum. 1904. A guide to the third and fourth Egyptian rooms. Predynastic antiquities, mummied birds and animals, portrait statues, figures of gods, tools, implements and weapons, scarabs, amulets, jewellery, and other objects connected with the funeral rites of the ancient Egyptians. British Museum, Department of Egyptian and Assyrian Antiquities; London.

British Museum. 1922. A guide to the fourth, fifth and sixth Egyptian rooms, and the Coptic room. British Museum; London.

Brunton, G. 1937. Mostagedda and the Tasian culture. In: *British Museum Expedition to Middle Egypt, First and Second Years, 1928, 1929.* Bernard Quaritch Ltd.; London. viii + 163 pp. + 84 plates.

Burstein, S.M. 1989. *Agatharchides of Cnidus: on the Erythraean Sea.* The Hakluyt Society; London. xii + 202 pp.

Carnarvon, G.E.S.M.H., 5th Earl of, and H. Carter. 1912. *Five Years' Explorations at Thebes; a Record of Work Done 1907–1911.* H. Frowde; New York.

Carr, H.S. 1985. Subsistence and ceremony: faunal utilization in a late Preclassic community at Cerros, Belize. Pp. 115–132. In: M. Pohl (ed.). *Prehistoric Lowland Maya Environment and Subsistence Economy.* Papers of the Peabody Museum, 77. Harvard University; Cambridge, MA.

Carr, H.S. 1986a. Faunal utilization in a late Preclassic Maya community at Cerros, Belize. Unpublished Ph.D. dissertation, Tulane University; New Orleans.

Carr, H.S. 1986b. Preliminary results of analysis of fauna. Pp. 127–146. In: R.A. Robertson and D.A. Freidel (eds.). *Archaeology at Cerros, Belize, Central America. Volume I: an Interim Report.* Southern Methodist University Press; Dallas, TX.

Carr, H.S. 1989a. Patterns of exploitation and exchange of subsistence goods in late Classic–early Postclassic Yucatan: a zooarchaeological perspective. Paper presented at the 54th Annual Meeting of the Society for American Archaeology, 5–9 April 1989, 12 pp.

Carr, H.S. 1989b. Non-molluscan faunal remains (of Isla Cerritos, Yucatan). Manuscript submitted to Isla Cerritos Project, 1989.

Carrasco, M.D. and K. Hull. 2002. The cosmogonic symbolism of the corbelled vault in Maya architecture. *Mexicon.* 24(2): 26–32.

Casson, L. 1989. *The Periplus Maris Erythraei.* Princeton University Press; Princeton, NJ. xvii + 320 pp.

Caubet, A. and F. Poplin. 1987. Les objets de matière dure animale: étude du matériau. Pp. 273–306. In: M. Yon (ed.). *Ras-Shamra-Ougarit III.* Le centre de la ville. 38e–44e campagnes (1978–1984). ADPF; Paris.

Chapman, J., H.R. Delcourt, and P.A. Delcourt. 1989. Strawberry fields, almost forever. *Nat. Hist.* 9: 50–59.

Chase, D.Z. and A.F. Chase. 1986. *Offerings to the Gods: Maya Archaeology at Santa Rita Corozal.* University of Central Florida; Orlando, FL, 22 pp.

Cushing, F.H. 1897. Exploration of the ancient key dwellers remains on the Gulf Coast of Florida. *Proc. Am. Philos. Soc.* 35(153): 329–448.

Daly, P. 1969. Approaches to faunal analysis in archaeology. *Am. Antiq.* 34(2): 146–153.

Denevan, W.M. 1992. The pristine myth: The landscape of the Americas in 1492. *Ann. Assoc. Am. Geogr.* 82(3): 369–385.

Denevan, W.M. 1996. Pristine myth. Pp. 1034–1036. In: D. Levinson and M. Ember (eds.). *Encyclopedia of Cultural Anthropology.* Henry Holt and Co.; New York.

Diamond, J. M. 1986. The environmentalist myth. *Nature.* 324: 19–20.

Diaz del Castillo, B. 1908. *The Conquest of New Spain.*

Drew, D. 1999. *The Lost Chronicles of the Maya Kings.* Weidenfeld & Nicolson; London. xvi + 450 pp.

Durante, S. and M. Tosi. 1977. The aceramic shell middens of Ra's Al-Hamra: a preliminary note. *J. Oman Stud.* 3(2): 137–162.

Dwe Yu Ze. 1981. *Shou Wen Jie Zi.* Shanghai Chinese Classics Publishing House; Shanghai. 867 + various pages.

Englund, R.K. 1990. Organisation un Verwaltung der Ur III-Fischerei. (Berliner Beiträge zum Vordern Orient, 10) Berlin.

Englund, R.K. 1998. Texts from the late Uruk period. In: J. Bauer, R.K. Englund, and M. Krebernik (eds.), *Mesopotamien: Späturuk- und Frühdynastische Zeit.* Orbis Bibliocus et Orientalis; Freiburg, Switzerland. 160(1).

Fagan, B.M. 1995. *Ancient North America: The Archaeology of a Continent.* Thames and Hudson; New York. 528 pp.

Farber, W. 1974. Von ba und anderen Wassertieren: testudines sargonicae? *J. Cuniform Stud.* 26: 195–207.

Felger, R.S. and M.B. Moser. 1991. *People of the Desert and Sea: Ethnobotany of the Seri Indians.* The University of Arizona Press; Tucson, AZ. xv + 438 pp.

Flannery, K.V. 1982. *Maya Subsistence: Studies in Memory of Dennis E. Puleston.* Academic Press; New York. xxiii + 368 pp.

Foster, J.W. 1874. *Pre-Historic Races of the United States of America.* S.C. Griggs & Co.; Chicago. xv + 415 pp.

Fradkin, A. 1976. The Wightman Site: a study of prehistoric culture and environment on Sanibel Island, Lee County, Florida. Unpublished master's thesis, Department of Anthropology, University of Florida, FL.

Frazier, J. (chief ed.). 1993. *Memorias del IV Taller Regional Sobre Programas de Conservación de Tortugas Marinas en la Península de Yucatán.* Universidad Autónoma de Yucatán, Mérida. iii + 212 pp.

Frazier, J. and D. Bonavia. 2000. Prehispanic marine turtles in Peru: where were they? Pp. 243–245. In: *Proceedings of the Eighteenth International Symposium on Sea Turtle Biology and Conservation.* U.S. Department of Commerce, National Oceanic and Atmospheric Administration, National Marine Fisheries Service, Southeast Fisheries Science Center; Miami, FL. NOAA Technical Memorandum NMFS-SEFSC-436.

Freeman-Grenville, G.S.P. 1962. *The East Africa Coast: Select Documents from the First to the Earlier Nineteenth Century.* Clarendon; Oxford. xi + 314 pp.

Freidel, D., L. Schele, and J. Parker. 1993. *Maya Cosmos: Three Thousand Years on the Shaman's Path.* William Morrow and Co., Inc.; New York. 543 pp.

Gates, W. (transl. and ed.) 1937. *Yucatan Before and After the Conquest* (based on De Landa, Friar Diego. 1566. *Relacion de las Cosas de Yucatán*). The Maya Society; Baltimore. No. 20. (Republished 1978 by Dover Publications, Inc.; New York. xv + 162 pp.).

Graham, E. 1994. The highlands of the lowlands: environment and archaeology in Stann Creek District, Belize, Central America. In: *Monographs in World Archaeology*, No. 19. Prehistory Press; Madison, WI. xx + 372 pp.

Gudiño Kieffer, E. 1986. *El Penetón.* Ediciones de Arte Gaglianone; Buenos Aires, Argentina. 108 pp.

Hamblin, N.L. 1984. *Animal Use by the Cozumel Maya.* The University of Arizona Press; Tucson, AZ. vii + 206 pp.

Hames, R. 1996. Game conservation or efficient hunting? Pp. 92–107. In: B.J. McCay and J.M. Acheson. *The Question of the Commons: The Culture and Ecology of Communal Resources.* University of Arizona, Tucson, AZ.

Hammond, C.A. 1891. The Carancahua Tribe of Indians. Peabody Museum of American Archaeology and Ethnology, Papers. Vol. 1: 73–77.

Hoch, E. 1979. Reflections on prehistoric life at Umm an-Nar (Trucial Oman) based on faunal remains from the third millennium BC. Pp. 589–638. In: M. Taddei (ed.). *South Asian Archaeology 1977.* Seminario di Studi Asiatici, Series Minor, 6, Naples.

Hoch, E. 1995. Animal bones from the Umm an-Nar settlement. Pp. 249–256. In: K. Frifelt, *The Third Millennium Settlement.* P. Mortensen (ed.) *The Island of Umm an-Nar Vol. 2.* Jutland Archaeological Society Publications; Aarhus, Denmark. 26/2.

Hughes, J.D. 1985. Theophrastus as ecologist. *Environ. Rev.* 9(4): 296–306.

Hull, K. and M.D. Carrasco. *Mak*-'portal' rituals uncovered: an approach to interpreting symbolic architecture and creation of sacred space among the Maya. In preparation.

Jackson, J.B.C. et al. 2001. Historical overfishing and the recent collapse of coastal ecosystems. *Science.* 293: 629–638.

Kay, C.E. 1995. Aboriginal overkill and native burning: implications for modern ecosystem management. *West. J. Appl. For.* 10(4): 121–126.

Killick, R.G. et al. 1991. London-Bahrain archaeological expedition: 1990 excavations at Saar. *Arab. Archaeol. Epigr.* 2: 107–137.

Landa, D. de. 2001. *Relacion de las cosas de Yucatan.* Editorial Dante, S.A. de C.V.; Mérida, Mexico. 153 + (iv) pp.

Larson, L.H. 1980. Aboriginal subsistence technology on the southeastern coastal plain during the late Prehistoric period. *Ripley P. Bullen Monographs in Anthropology and History,* University of Florida; Gainesville, FL. No. 2. xii + 260 pp.

Larson, L. 1993. An examination of the significance of a tortoise-shell pin from Etowah site. In: Archaeology of Eastern North America: Papers in honor of Stephen Williams. *Archaeological Report* No. 25: 169–185.

Laxson, D.D. 1957a. The Madden site. *Fla. Anthropol.* 10(1/2): 1–16.

Laxson, D.D. 1957b. Three small Dade County sites. *Fla. Anthropol.* 10(1/2): 17–22.

Laxson, D.D. 1957c. The Arch Creek site. *Fla. Anthropol.* 10(3/4): 1–10.

Laxson, D.D. 1959. Excavations in Dade County during 1957. *Fl. Anthropol.* 12(1): 1–7.

Layard, A.H. 1853. *Discoveries among the Ruins of Nineveh and Babylon; with Travels in Armenia, Kurdistan, and the Desert: Being the Result of a Second Expedition Undertaken for the Trusties of the British Museum.* G.P. Putnam and Co.; New York. xxiii + 686 pp.

Lee, T.A. 1985. *Los Códices Mayas.* Universidad Autónoma de Chiapas; San Cristóbal de las Casas, Chiapas, México. 216 pp.

Leemans, W.F. 1960. *Foreign Trade in the Old Babylonian Period as Revealed by Texts from Southern Mesopotamia.* E.J. Brill; Leiden.

Lewis, H.T. 1980. Indian fires of spring. *Nat. Hist.* Jan: 76–78, 82–83.

Lorch, F.B. 1999. Sea turtles and the ancient Greeks (a reassessment). *Archeol. Arts.* 73: 97–98.

Love, B. 1994. *The Paris Codex: Handbook for a Maya Priest.* University of Texas Press; Austin, TX. xviii + 124 pp.

Lucas, A. 1948. *Ancient Egyptian Materials & Industries.* 3rd ed. Edward Arnold Ltd.; London. xi + 570 pp.

Maffei, M.M. 1995. Achille e la tartaruga miti, usanze e rituali nella pesca della tartaruga nel Mediterraneo. WWF Atti 4° Seminario Ecosistema Marino Sperlonga-Gaeta-Capri, Giugno. 127–152.

Masry, A.H. 1974. *Prehistory in Northern Arabia: The Problem of Inter-Regional Interaction.* Field Research Projects; Miami, Coconut Grove. (Republished 1997, Kegan Paul International; London, England and New York; xi + 252 pp.)

Mathew, G. 1975. The dating and significance of the *Periplus of the Erythrean Sea.* Pp. 147–163. In: H.N. Chittick and R.I. Rotberg (eds.) *East Africa and the Orient: Cultural Syntheses in Pre-Colonial Times.* Africana; New York.

McDonnell, M.J. and S.T.A. Pickett (eds.). 1993. *Humans of Components of Ecosystems: The Ecology of Subtle Human Effects and Populated Areas.* Springer; New York. xxi + 364 pp.

Meggers, B.J. (ed.). 1992. *Prehistoria Sudamericana: Nuevas Perspectivas.* Taraxacum; Chile/Washington. 381 pp.

Miller, A.G. 1982. *On the Edge of the Sea. Mural Painting at Tancah-Tulum, Quintana Roo, Mexico.* Dumbarton Oaks; Washington, DC. xiv + 133 pp. + 47 plates.

Miller, G.H. et al. 1999. Pleistocene extinction of *Genyornis newtoni*: human impact on Australian megafauna. *Science.* 283: 205–208.

Miller, M. and K. Taube. 1993. *The Gods and Symbols of Ancient Mexico and the Maya: An Illustrated Dictionary of Mesoamerican Religion.* Thames and Hudson; New York. 216 pp.

Molina, S. 1981. *Leyendo en la Tortuga (Recopilación).* Martin Casillas Editores; Mexico. 173 pp.

Moorey, P.R.S. 1994. *Ancient Mesopotamian Materials and Industries: The Archaeological Evidence.* Clarendon Press; Oxford. xxiii + 414 pp.

Morton, J.D. 1988. A preliminary report on the faunal remains from Santa Rita Corozal, Belize. Appendix IV. Pp. 118–122. In: D.Z. Chase and A.F. Chase, (eds.). *A Post-classic Perspective: Excavations at the Maya Site of Santa Rita Corozal, Belize.* Monograph 4. Pre-Columbian Art Research Institute; San Francisco, CA.

Mosseri-Marlio, C. 1998. Marine turtle exploitation in Bronze Age Oman. *Mar. Turtle Newsl.* 81: 7–9.

Mosseri-Marlio, C. 2000. The Ancient distribution of sea turtle nesting beaches: an archaeological perspective from Arabia. *Testudo.* 5(2): 31–36.

Nabhan, G. et al. 1999. Sea turtle workshop for indigenous Seri tribe. *Mar. Turtle Newsl.* 86: 14.

Nietschmann, B. 1973. *Between Land and Water: The Subsistence Ecology of the Miskito Indians, Eastern Nicaragua.* Seminar Press; New York. xiv + 279 pp.

Nietschmann, B. 1979. *Caribbean Edge: The Coming of Modern Times to Isolated People and Wildlife.* The Bobbs-Merrill Company; New York. xvi + 280 pp.

Oates, J. et al. 1977. Seafaring Merchants of UR? *Antiquity.* 51: 221–234.

Olijdam, E. 2001. Exploitation of sea turtles in the early Dilmun period (c. 2100–1900 BC). *Proceedings of the Seminar for Arabian Studies*, Vol. 31: 195–202.

Owen, D.I. 1981. Of birds, eggs and turtles. *Z. Assyriol.* 71: 29–47.

Pagden, A.R. (ed. and transl.). 1975. *Diego de Landa's Account of the Affairs of Yucatán. The Maya.* J. Philip O'Hara, Inc.; Chicago, IL. 191 pp.

Parsons, J.J. 1972. Etudes de géographie tropicale offertes à Pierre Gourou. In: *Ecole Pratique desa Hautes Etudes.* Sorbonne; Paris.

Pauly, D. et al. 1998. Fishing down marine food webs. *Science.* 279: 860–863.

Perkins, D., Jr. and P. Daly. 1968. A hunters' village in Neolithic Turkey: were the inhabitants of Suberde herdsmen or hunters? The analysis of animal bones at the site shows not only that they were hunters but also that they probably "schlepped" meat home in animal skins. *Sci. Am.* 219(5): 97–104, 106.

Pollock, H.E.D. and C.E. Ray. 1957. Notes on vertebrate animal remains from Mayapan. Pp. 633–656. In: Carnegie Institution Report No. 41. Department of Archaeology, Carnegie Institution, Washington, DC. .

Postgate, J.N. 1977. Excavations at Abu Salabikh, 1976. *Iraq*. 39: 269–300.

Postgate, J.N. and P.R.S. Moorey. 1976. Excavations at Abu Salabikh, 1975. *Iraq*. 38(2): 133–169.

Potts, D.T. 1990. The Arabian Gulf in antiquity. Vol. 1. *From Prehistory to the Fall of the Achaemenid Empire*. Clarendon Press; Oxford. xxvii + 419 pp

Potts, D.T. 1997. *Mesopotamian Civilization: The Material Foundations*. Cornell University Press; Ithaca, NY. xxi + 366 pp.

Potts, D.T. 2000. *Ancient Magan*: *The Secrets of Tell Abraq*. Trident Press Ltd.; London. 144 pp.

Price, R. 1966. Caribbean fishing and fishermen: a historical sketch. *Am. Anthropol.* 68(6): 1363–1383.

Proskouriakoff, T. 1962. The artifacts of Mayapan. Pp. 321–442 + 52 figs. In: H.E.D. Pollock et al. (ed.). *Mayapan Yucatan Mexico*. Publication 619, Carnegie Institution of Washington; Washington, DC.

Rands, R.L., R.L. Bishop, and G. Harbottle. 1979. Thematic and compositional variation in Palenque-region incensarios. Pp. 19–30. In: M.G. Robertson (ed.). *Tercera Mesa Redonda de Palenque*, Vol. 4. Herald Printers; Monterey, CA.

Redding, R.W. and S.M. Goodman. 1984. Reptile, bird and mammal remains. In: H.T. Wright. Early Seafarers of the Comoro Islands: the Dembeni Phase of the IXth-Xth Centuries AD. *AZANIA* 19: 51–54.

Roberts, C.M. and J.P. Hawkins. 1999. Extinction risk in the sea. *Trends Ecol. Evol.* 14(6): 241–246.

Robicsek, F. and D.M. Hales. 1981. *The Maya Book of the Dead: The Ceramic Codex. The Corpus of Codex Style Ceramics of the Late Classic Period.* Yale University; New Haven, CT. xxi + 257 pp.

Schaffer, C. 2001. Chelonian zooarchaeology of Florida and the adjacent southeast coastal plain. In: *Program of the Joint Meeting of the Herpetologists' League and the Society for the Study of Amphibians and Reptiles*, 27–31, July 2001, Indianapolis, IN (abstract).

Schaffer, C. and K. Ashley, in press. The cultural context of the hawksbill sea turtle (*Eretmochelys imbricata*) in Calusa Society. Proceedings of the 22nd Annual Symposium on Sea Turtle Biology and Conservation.

Schaffer, C. and R. Thunen. In press. Florida marine chelonian zooarcheology and ethnozoology. In: *Proceedings of the 21st Annual Symposium on Sea Turtle Biology and Conservation.*

Schellhas, P. 1904. Representations of deities of the Maya Manuscripts. Papers of the Peabody Museum of American Archaeology and Ethnology, Harvard University; Boston. 4(1): 1–50.

Scholes, F.V. and R.L. Roys. 1968. *The Maya Chontal Indians of Acalan-Tixchel. A Contribution to the History of the Ethnography of the Yucatan Peninsula.* University of Oklahoma Press; Norman, OK. xiii + 565 pp.

Shaw, E.B. 1933. The fishing industry of the Virgin Islands of the United States. *Geogr. Soc. Philadelph. Bull.* 31: 61–72.

Shaw, L.C. 1995. Analysis of faunal materials from Ek Luum. Pp. 175–181. In: T.H. Gunderjan and J.F. Garber (eds.). *Maya Maritime Trade, Settlement, and Populations on Ambergris Caye, Belize.* Maya Research Program and Labyrinthos; San Antonio, TX, and Lancaster, CA.

Shetrone, H.C. and E.F. Greenman. 1931. Explorations of the Seip group of prehistoric earthworks. *Ohio Archaeol. Hist. Q.* 40(3): 349–509.

Stephan, E. 1995. Preliminary report on the faunal remains of the first two seasons of Tell Abraq/Umm al Quwain/United Arab Emirates. Pp. 52–63. In: H. Buitenhuis and H.–P. Uerpmann (eds.). *Archaeozoology of the Near East II.* Backhuys; Leiden.

Stone, A.J. 1995. *Images from the Underworld. Naj Tunich and the Traditions of Maya Cave Painting.* University of Texas Press; Austin, TX. xi + 284 pp.

Strauss, K. and Rafael A. 1992. *El Tiempo Prehispánico de Venezuela.* Fundación Eugenio Mendoza; Caracas, Venezuela. 279 pp.

Talbert, O.R., Jr. et al. 1980. Nesting activity of the loggerhead turtle. *Caretta caretta* in South Carolina I: a rookery in transition. *Copeia.* 1980(4): 709–718.

Taube, K.A. 1988. A prehispanic Maya Katun wheel. *J. Anthropol. Res.* 44(2): 183–203.

Taube, K.A. 1992. *Major Gods of Ancient Yucatan.* Dunbarton Oaks Research Library and Collection, Studies in Pre-Columbian Art & Archaeology. No. 32: 160 pp.

Tedlock, D. 1993. *Breath on the Mirror: Mythic Voices & Visions of the Living Maya.* Harper; San Francisco. xii + 256 pp.

Thorbjarnarson, J. et al. 2000. Human use of turtles: a worldwide perspective. Pp. 33–84. In: M.W. Klemens (ed.). *Turtle Conservation.* Smithsonian Institution; Washington, DC.

Tozzer, A.M. 1941. Landa's Relación de las cosas de Yucatán. Papers of the Peabody Museum of American Archaeology and Ethnology, Harvard University; Boston. Vol. 18.

Tozzer, A.M. and G.M. Allen. 1910. Animal figures in the Maya codices. Papers of the Peabody Museum of American Archaeology and Ethnology. IV(3): 276–374 pp. + 39 plates.

Trimingham, J.S. 1975. The Arab geographers and the East African coast. Pp. 115–146. In: H.N. Chittick and R.I. Rotberg (eds.). *East Africa and the Orient: Cultural Syntheses in Pre-Colonial Times.* Africana; New York.

Turner, B.L. and K.W. Butzer. 1992. The Colombian encounter and land-use change. *Environment.* 34(8): 16–20, 37–44.

Uerpmann, M. 2001. Remarks on the animal economy of Tell Abraq (Emirates of Sharjah and Umm al-Qaywayn, UAE). *Proceedings of the Seminar for Arabian Studies* 31: 227–233.

Uerpmann, M. and H.-P. Uerpmann. 1994. Animal bone finds from Excavation 520 at Qala'at al Baharin. In: F. Højlund and H.H. Anderson (eds.). *Qaláat al-Baharin,* Vol. 1: The Northern City Wall and the Islamic Fortress. Jutland Archaeological Society Publications; Aarhus, Denmark. 30(1): 417–444.

Uerpmann, M. and H.-P. Uerpmann. 1997. Animal bone finds from Excavation 519 at Qala'at al Baharin. In: F. Højlund and H.H. Anderson (eds.). *Qaláat al-Baharin,* Vol. 2: The Central Monumental Buildings. Jutland Archaeological Society Publications; Aarhus, Denmark. 30(2): 235–264.

Uerpmann, M. and H.-P. Uerpmann. 1999. The animal economy of ancient Dilumn in the light of faunal remains from excavations at Saar and Qala'at al-Bahrain. ISIMU (Revista sobre Oriente Próximo y Egipto en la antigüedad; Madrid). 2: 635–646.

van Buren, E.D. 1939. The fauna of ancient Mesopotamia as represented in art. Pontificium Institutum Biblicum (Analecta orientalia: commentationes scientificae de rebus orientus antiqui); Rome. No. 18.

van de Mieroop, M. 1987. Crafts in the early Isin period: a study of the Isin Craft archive from the Reigns of Is̮bi-Erra and S̮ū-Ilis̮u. Orientalia Lovaniensia Analecta 24: xii + 157 pp.

van Neer, W. and M. Uerpmann. 1994. Fish remains from excavation 520 at Qala'at al-Bahrain. Pp. 445–454. In: F. Højlund and H.H. Anderson (eds.). *Qaláat al-Baharin,* Vol. 1: The Northern City Wall and the Islamic Fortress. Jutland Archaeological Society Publications; Aarhus, Denmark.

Versteeg, A.H. and F.R. Effert. 1987. Golden Rock: the first Indian village on St. Eustatius. *St. Eust. Hist. Found.* No. 1: 21 pp.

Vine, P. (ed.). 1993. *Bahrain National Museum.* Immel Publishing, Ltd.; London. x + 177 pp.

Waring, A.J., Jr. and L.H. Larson, Jr. 1968. The Shell Ring on Sapelo Island. In: S. Williams (ed.). *The Waring Papers: The Collected Works of Antonio J. Waring, Jr.* Papers of the Peabody Museum of Archaeology and Ethnology, Harvard University. 58: 261–278.

Wheatley, P. 1975. Analecta Sino-Africana Recensa. Pp. 76–114. In: H.N. Chittick and R.I. Rotberg (eds.). *East Africa and the Orient: Cultural Syntheses in Pre-colonial Times.* Africana; New York.

Whitcher, K.S. In press. Animal exploitation at early Bronze I Afridar: what the bones tell us (Initial analysis of the animal bones from areas E, F and G). *Atiqot.*

Willey, G.R. 1949. Excavations in southeast Florida. *Yale Univ. Publ. Anthropol.* 42: 137 pp. + 16 plates.

Wilson, S.M. 1992. "That unmanned wild country." *Nat. Hist.* 5: 16–17.

Wing, E.S. 1963a. Vertebrate remains from the Wash Island site. *Fla. Anthropol.* 16(3): 93–96.

Wing, E.S. 1963b. Vertebrates from the Jungerman and Goodman sites near the east coast of Florida. In: D. Fordan, E.S. Wing, and A.K. Bullen (eds.). Contributions of the Florida State Museum Social Sciences 10: 51–60.

Wing, E.S. 1965. Animal bones associated with two Indian sites on Marco Island, Florida. *Fla. Anthropol.* 18(1): 21–28.

Wing, E.S. 1974. Vertebrate faunal remains. Pp. 186–188. In: Andrews, E.W. IV et al. (ed.). Excavation of an early shell midden on Isla Cancun, Quintana Roo, Mexico. *Archaeological Investigations on the Yucatan Peninsula.* National Geographic Society — Tulane University Program of Research on the Yucatan Peninsula. Middle American Research Institute, Tulane University; New Orleans. Publication 31.

Wing, E.S. 1977. Subsistence systems in the Southeast. *Fla. Anthropol.* 30(1): 81–87.

Wing, E.S. 2001a. The sustainability of resources used by native Americans on four Caribbean islands. *Int. J. Osteoarchaeol.* 11: 112–126.

Wing, E.S. 2001b. Native American use of animals in the Caribbean. Pp. 481–518. In: Charles A. Woods and Florence E. Sergile (eds.). *Biogeography of the West Indies: Patterns and Perspectives*, 2nd ed. CRC Press; Washington, DC.

Wing, E.S. and E.J. Reitz. 1982. Prehistoric fishing economies of the Caribbean. *New World Archaeol.* 5(2): 13–22.

Wing, E.S. and S.J. Scudder. 1983. Animal exploitation by prehistoric people living on a tropical marine edge. Pp. 197–210. In: C. Grigson and J. Clutton-Brock (eds.). *Animals and Archaeology.* BAR International Series 183; Oxford.

Wing, E.S. and D. Steadman. 1980. Vertebrate faunal remains from Dzibilchaltun. In: E.W. Andrews IV and E.W. Andrews V (eds.). Excavations at Dzibilchaltun, Yucatan, Mexico. Tulane University; New Orleans. Middle American Research Institute, Pub. 48: 326–331.

Wingert, P.S. 1953. The Art of the South Pacific Islands: A Loan Exhibition. H.H. De Young Memorial Museum; San Francisco, California. 64 pp. + 102 plates.

Wright, H.T. 1984. Early seafarers of the Comoro Islands: the Dembeni Phase of the IXth-Xth Centuries AD. *AZANIA* 19: 12–59.

Yasin Al-Tikkriti, W. 1985. The archaeological investigations on Ghanda Island 1982–1984: further evidence for the coastal Umm an-Nar culture. *Archaeol. United Arab Emir.* 4. Al Ain: 9–19.

2 The External Morphology, Musculoskeletal System, and Neuro–Anatomy of Sea Turtles

Jeanette Wyneken

CONTENTS

0-8493-1123-3/03/$0.00+$1.50

2.1 INTRODUCTION

The anatomy of an animal defines how it can live within and interact with its environment. The form of an animal gives fundamental clues about its behavior, ecology, and physiology: how it functions. Its shape and overall design dictate how the animal can move and feed. The shapes of muscles, bones, and the nervous system, which controls responses and movements, allow function to match form given the animal's ecology.

For the biologist, anatomy provides the raw materials for evolutionary, taxonomic, and population studies (providing both the morphological data and the tissues that are sources of biochemical samples). Comparisons between normal anatomy, in contrast with anatomy that reflects disease or injury, provide information about acute and chronic changes in health. In addition, it is the remnants of the anatomy, as fossils or artifacts at archaeological sites (such as shells and skulls described in other chapters), that provide historical perspective about when and where species occurred, how they changed through time, and how they were used and valued by ancient peoples.

Extant marine turtles are morphologically distinct from other turtles. They are characterized by their large adult body size, hypertrophied forelimbs (flippers), and streamlined shells. In the following sections, external anatomy and gross morphology of the skeletal, muscular, and nervous systems are discussed because these components together perhaps are linked conceptually more so than any other combination of systems. The external morphology is not simply about scutes and skin, but provides fundamental information about how the animal interacts with its environment. Body design, color, and ontogenetic changes belie ecological and behavioral differences as well as provide taxonomic material. A marine turtle gains much of its overall form from the underlying musculoskeletal system. The homologies of structures, as well as their functions (including both motor patterns and movements), are dependent on innervations of the muscles. The form of the skeleton not only shapes what movements are possible, but also defines what and how organs receive protection. These functional and historical interrelationships of systems provide the links that bind them together for their description and discussion here.

2.2 EXTERNAL ANATOMY

The external anatomy includes the scales of the head (and body), the scutes (or lack of scutes) of the shell, and the form of the limbs and body. All of these characteristics are frequently used in species identification. Coloration and body design or form also provide key information about form and function in marine turtles.

Dermochelys coriacea V., a long-distance migrant and deep-diving species, has a streamlined body that tapers from the shoulders to the caudal tip, and extremely long clawless fore limbs. The body is black with varying degrees of white speckling. Five dorsal ridges run the length of the carapace, two ridges form the margins, and few ridges occur ventrally (Figure 2.1A). The head is oval with large eyes. A notch occurs in each side of the upper jaw (Pritchard, 1979).

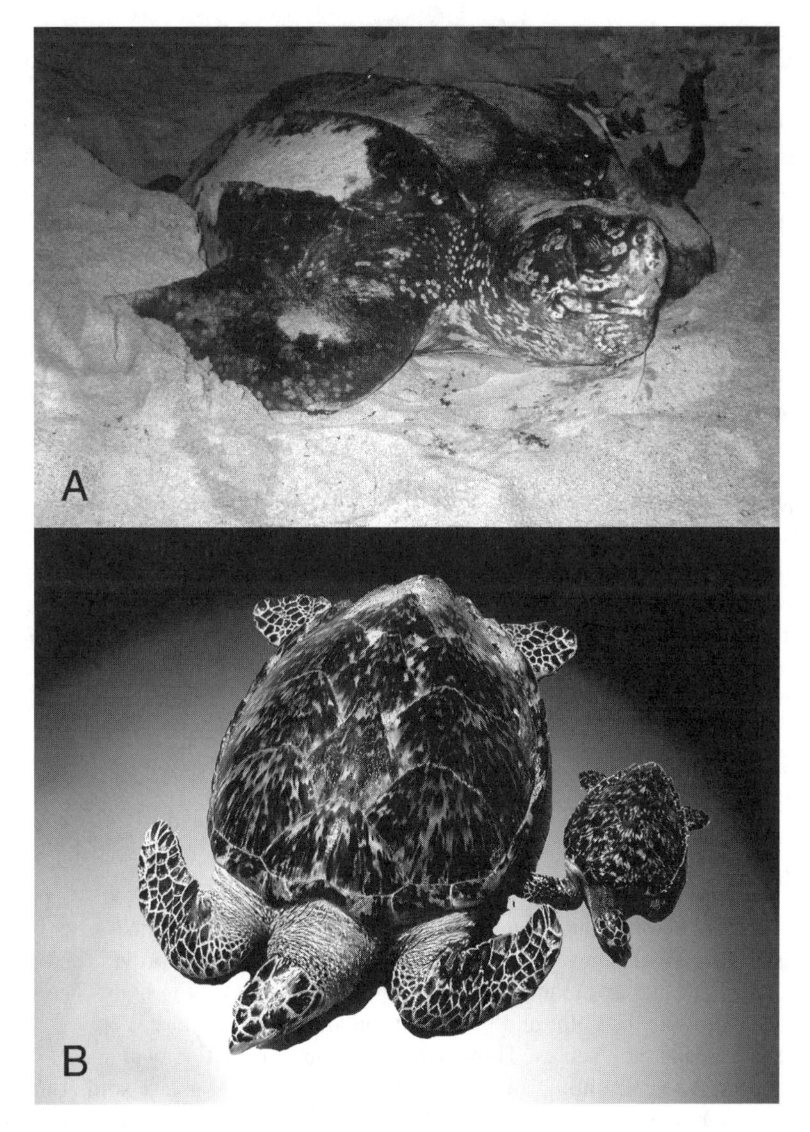

FIGURE 2.1 *Dermochelys coriacea*, the leathery shelled species (A), and *Eretmochelys imbricata*, a cheloniid (B), contrast the two major types of body plans typical of extant sea turtles.

Leatherback hatchlings have a black carapace with white-tipped scales along the ridges; the plastron has longitudinal black and white stripes.

Cheloniids (Figure 2.1B) have shells composed of bone overlaid by keratinous plates or scutes. The margins of the scutes and scales do not usually align with bony sutures. The bony shell is reduced compared to those of most other turtles.

All marine turtles have streamlined body forms that facilitate long-distance, relatively low-cost migrations (fusiform body, minimal head and limb pockets that minimize drag, and flippers that produce thrust on both their anterodorsal and posteroventral movements) (Wyneken, 1997).

Species are distinguished from one another based upon a combination of characteristics that include color, the form of the jaws, the scales on top of the snout (e.g., prefrontals), and the scutes on the carapace. The green turtle, *Chelonia mydas* L., and flatback, *Natator depressus* G., each have one pair of prefrontal scales; the other species have two pairs of prefrontals (Pritchard and Mortimer, 1999). Colors change with age. Most develop a cream-colored plastron as adults (except for an eastern Pacific race of *C. mydas* known as the black turtle). As hatchlings, *C. mydas* are black dorsally and white ventrally. With growth, the carapace becomes streaked with brown, tan, and black. Flatback hatchlings are gray and the margin of each carapacial scute starts out dark gray, surrounding a lighter gray center. The dark color becomes predominant until they are nearly mature, when the colors shift to lighter gray. The plastron shifts from white or pale gray to creamy as the animals age. Hatchlings of *Caretta caretta* L., *Eretmochelys imbricata* L., *Lepidochelys kempii* G., and *Lepidochelys olivacea* E. are brown to dark gray in color dorsally; the plastra vary from pale to dark brown. Juveniles and adults have carapaces that vary from browns (*C. caretta*); to grays (*L. kempii*, *L. olivacea*, and *N. depressus*); to the brown, tan, and yellow (*E. imbricata*); and gray-green to mottled shades of brown and black (*C. mydas*) (Pritchard, 1979; Pritchard and Mortimer, 1999; Wyneken, 2001). The plastra shift to yellow or light tan (*C. caretta*), to white (*Lepidochelys*) or, in *E. imbricata*, cream-colored scutes, sometimes each with a brown spot. Color provides not only taxonomic information, but also clues about adaptations for survival in different environments, probably through different forms of crypsis (Cott, 1966; Owen, 1982). Color may also be used in recognition of conspecifics; however, there are no studies of marine turtles addressing this hypothesis.

2.2.1 Scales and Scutes

Scutes are keratinous plates found on the shell. Scales are thickened areas of epidermis and keratin that cover the skin and head. Scutes and scales provide taxonomic information and act as landmarks for describing location on the body. The scutes and scales provide the color and texture of the animals as well.

Leatherbacks lack scutes. This species has small round or oval scales on the shell and throat as hatchlings. These are shed as the animal grows so that there is little or no hint of scales in the adult. The cheloniids have keratinous scutes covering the skeletal shell. The number and arrangement of these scutes are species-specific. The scutes are designated by position and number (Figures 2.2 and 2.3). The carapacial scutes along the midline are the *vertebrals*. The *nuchal* is located most anteriorly along the midline. Lateral to the vertebral scutes are *laterals* (or *costals*), which abut the peripherally arranged *marginals* (Figure 2.3). The scutes connecting the carapace and plastron are the *inframarginals*. The *plastral* scutes include the single *intergular*, closest to the throat, then from anterior to posterior, the paired *gular*, *humeral*, *pectoral*, *abdominal*, *femoral*, and *anal* scutes (Figure 2.3). Some individuals have a single unpaired *interanal* scute between or just posterior to the anal scutes (Pritchard, 1979; Pritchard and Mortimer, 1999; Wyneken, 2001).

FIGURE 2.2 Carapacial scutes of the same type (in this case, laterals) are numbered from anterior to posterior. (Adapted from Wyneken, J. *The Anatomy of Sea Turtles*, U.S. Department of Commerce, NOAA Technical Memorandum NMFS-SEFSC-470, 2001. With permission.)

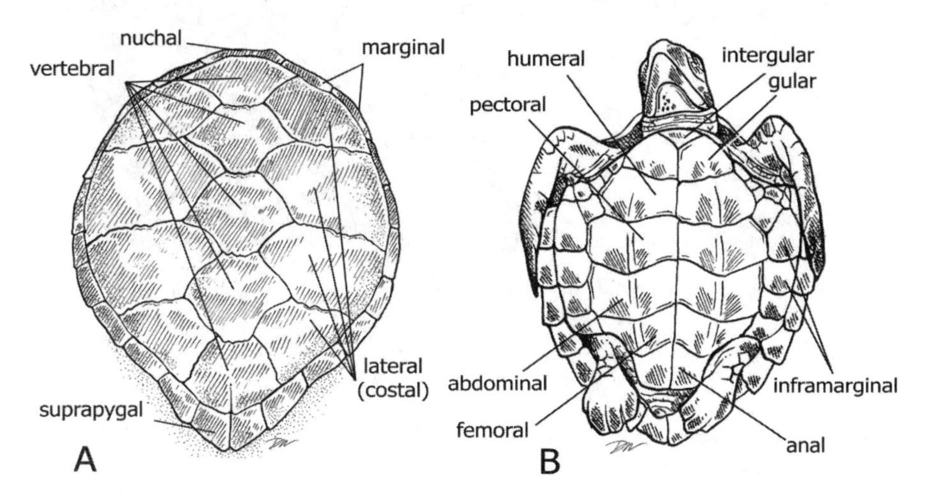

FIGURE 2.3 Scutes of the carapace (A) and the plastron (B) are identified by type and position. (Adapted from Wyneken, J. *The Anatomy of Sea Turtles*, U.S. Department of Commerce NOAA Technical Memorandum NMFS-SEFSC-470, 2001. With permission.)

2.2.2 Axial Body: Shell, Head, Neck, and Tail

Dermochelyids are characterized by a leathery shell composed of skin overlying a mosaic of thin bony plates (Figure 2.4A). Deep to the bony plates is a blubber layer of dense fibrous connective tissue and fat (Deraniyagala, 1939). There are well-developed longitudinal keels running the length of the carapace, and in adults the ridges and margins of the carapace are adorned with rows of knobs. The

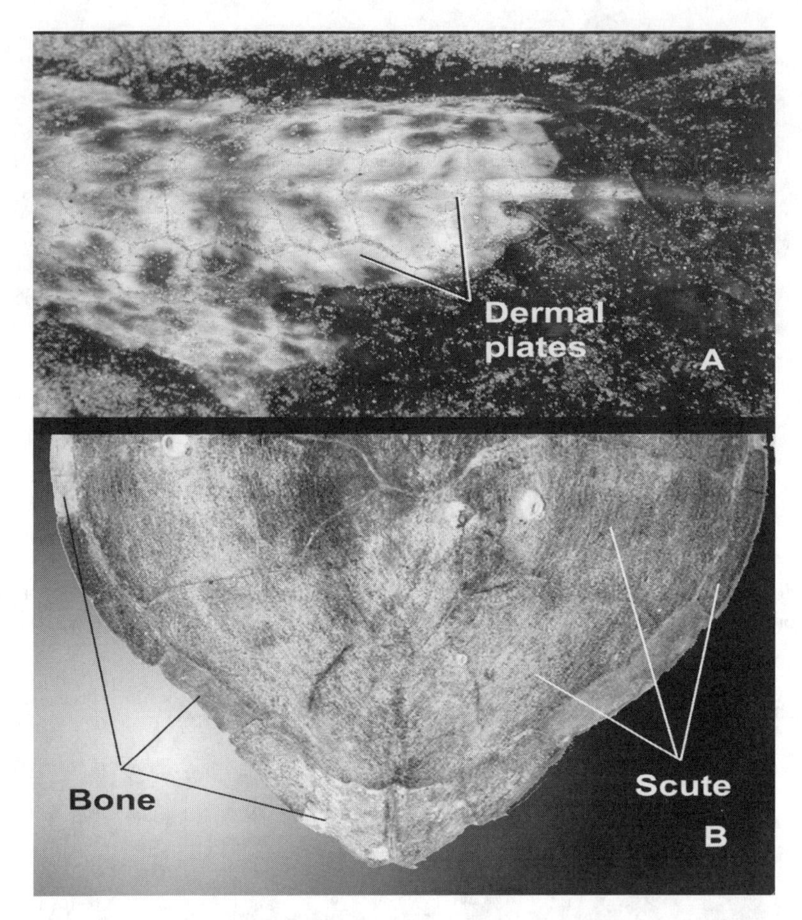

FIGURE 2.4 A mosaic of thin bony plates (A) several centimeters in diameter form a layer of thin bony armor deep to the skin in leatherbacks. The individual bony plates can be seen in the middorsal ridge of the leatherback. In cheloniids, keratinous scutes cover the bony skeleton (B). The posterior of this hawksbill shows both the remnants of scutes and the overlying bone.

plastron has weakly developed ridges. The functional significance of the unique leatherback shell design has not been studied.

Cheloniids have a bony shell covered with distinctive scutes (Figure 2.4B). The overall form of cheloniid shells is grossly similar. Unlike the leatherback shell, there is no blubber layer present. Cheloniid species differ in their arrangements of scutes. Although scute counts can vary among individuals, the majority of animals within a species will have a species-specific number and arrangement of scutes. The green turtle, hawksbill, and flatback typically have four pairs of lateral scutes, and the nuchal scute abuts the vertebral but not the lateral scutes. The scutes of *E. imbricata* tend to overlap one another along their posterior margins; those of *C. mydas* and *N. depressus* do not (Pritchard, 1979; Pritchard, 1997). The scutes of the flatback are thin, smooth, and waxy (Zangerl et al., 1988). The loggerhead and Kemp's ridley typically have five pairs of lateral scutes, and the nuchal abuts the first lateral as well as the first vertebral. The olive ridley has more than five normal vertebral scutes (not including irregular or supernumerary scutes) and usually six or more pairs of lateral scutes (Pritchard, 1979; Pritchard, 1997; Wyneken, 2001).

The head of the leatherback appears to be covered with smooth skin, except in hatchlings, which have small round scales on the lateral face and throat. Cheloniids have large scales covering the dorsal and lateral head (Figure 2.5), but the neck and throat are covered in moderately keratinized skin (Wyneken, 2001).

The rhamphotheci are the keratinous beaks of the upper and lower jaws in cheloniids. Their form differs with diet and they can be used to identify species (Ruckdeschel et al., 2000; Schumacher, 1973). Leatherbacks lack a distinct rhamphotheca on either jaw, although the skin of the jaws is more heavily keratinized than other parts of the animal. In *E. imbricata,* a reef omnivore (Bjorndal, 1997; Witzell, 1983) that is more of a specialist on sponges and tunicates at many sites (Meylan, 1988), the rhamphotheci are long and narrow with sharp cutting (triturating) edges (Figure 2.6). The palatal surface of the upper rhamphotheca is mostly smooth. The lower rhamphotheca is smooth with a triangular process extending anteriorly from the posterior (buccal) margin (Ruckdeschel et al., 2000). In contrast, the rhamphotheci of the omnivorous *C. caretta*, a species that feeds upon heavily armored prey (Youngkin, 2001), are robustly constructed (Figure 2.6). The upper and lower jaws are pointed in young loggerheads, but these points are typically worn away in large juveniles and adults. In all size classes, the palatal portion is wide and forms a crushing surface inside the mouth. The lower rhamphotheca is troughlike with a thick crushing surface and U-shaped cutting surface along its posterior margin. Similarly, the rhamphotheci of *L. kempii* and *L. olivacea* are

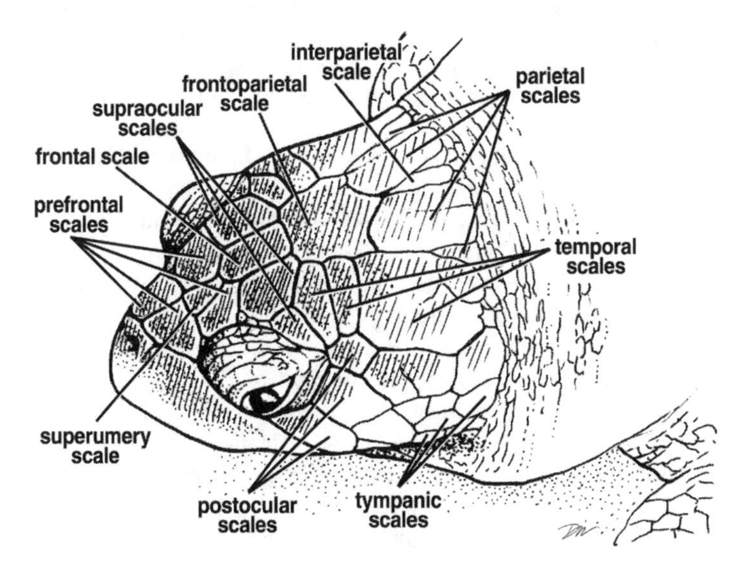

FIGURE 2.5 The scales of the head follow species-specific patterns in number and position. The counts of specific scales may differ, but the pattern is species-specific. (Adapted from Wyneken, J. *The Anatomy of Sea Turtles*, U.S. Department of Commerce, NOAA Technical Memorandum NMFS-SEFSC-470, 2001. With permission.)

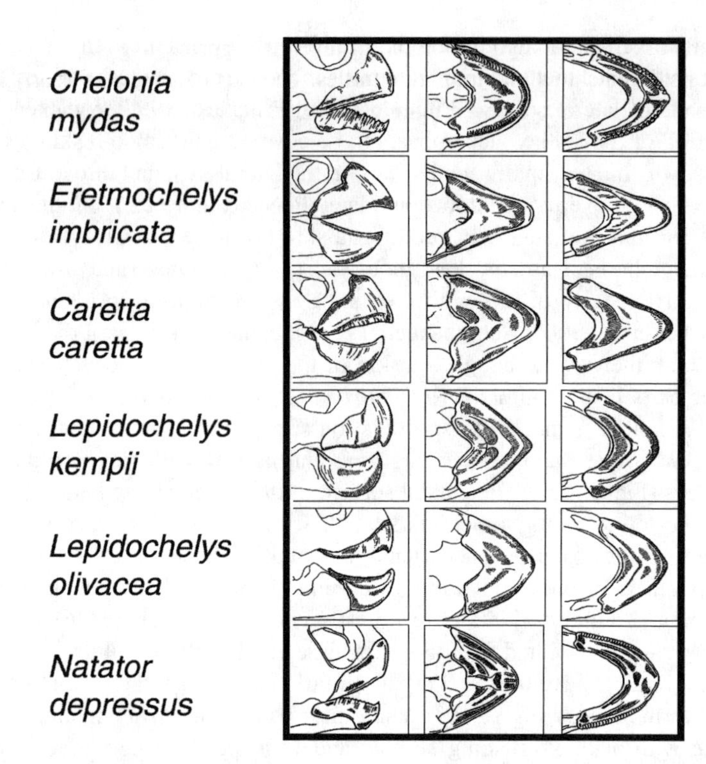

FIGURE 2.6 Rhamphotheci (or beaks) form the keratinous cutting and crushing surfaces of the mouth. Their form is species-specific. The rhamphotheci are shown by species in three columns (left to right): lateral view, the upper rhamphotheca, and the lower rhamphotheca, respectively. (Adapted from Wyneken, J. *The Anatomy of Sea Turtles*, U.S. Department of Commerce, NOAA Technical Memorandum NMFS-SEFSC-470, 2001. With permission.)

robust with thick, sharp-edged alveolar surfaces and are pointed anteriorly at the midline (Figure 2.6). In *L. kempii*, a mollusk and crustacean feeder, the palatal portion has a ridge bilaterally extending just anterior to the internal choanae. The lower rhamphotheca has a sharp, wide, V-shaped ridge running posteriorly along the buccal margin. In *L. olivacea*, an omnivore that favors crustaceans, the upper rhamphotheca has large cusps bilaterally on its palatal portion. The lower rhamphotheca is trough-like with two depressions that receive the palatal cusps, and a sharp U-shaped ridge is found along the posterior border.

In *C. mydas,* an herbivore for most of its life, the edge of the lower rhamphotheca is outlined by serrations and spike-like cusps or denticles (Figure 2.6). The lower jaw has bilateral inner ridges adorned with small cusps; the ridges are aligned parallel to the alveolar edge and connected midventrally (Wyneken, 2001).

The flatback, *N. depressus* (an omnivore that includes hard-bodied prey in its diet), has jaws that are covered by robust rhamphothecal plates with a sharp cutting edge. The upper rhamphotheca hosts a ridge on the palatal surface that parallels the margins of the upper jaw (Figure 2.6). The lower rhamphotheca has a robust flattened plate just internal to the cutting edge of each side of the mandibles (Zangerl et al., 1988).

The tails of sea turtles are dimorphic in juveniles approaching adulthood and in adults, but not in hatchling and immature turtles. The tails of mature or nearly mature males extend beyond the carapace margin, and the cloacal openings are found near the tail's tip. The tails of immature or female cheloniids normally do not extend beyond the carapace's posteriormost margin, and the cloacal opening is closer to the plastron than to the tail's tip (Wyneken, 2001). In *D. coriacea*, the tails of females sometimes extend beyond the caudal peduncle of the carapace (especially if the terminal end of the carapace has been damaged). However, the cloacal openings of adult female and immature leatherbacks are closer to the plastron than those of immature males. The position of the cloaca is a reliable indicator of sex in mature leatherbacks.

2.2.3 Appendages: Flippers and Hind Limbs

The forelimbs of all species are elongated as winglike flippers. The hind limbs are less elongated and more paddle-like in form. Leatherbacks have proportionately longer flippers than cheloniids of all size classes (Pritchard, 1979; Wyneken, 2001), lacking claws and scales. Cheloniids have claws on fore and hind limbs, and flippers are covered with scales. All cheloniids have three or more large scales along the trailing edge of each flipper near the axilla. These distinctive scales are often used as a site for flipper tag application. Hind limbs have few or no large scales.

The number of claws on the forelimbs is the same as on the hind limbs. *Chelonia* and adult *Natator* have one claw (on the first digit) on each foot. The remaining cheloniids (*Caretta*, *Eretmochelys*, *Lepidochelys*, and some immature *Natator*) have two claws (Pritchard, 1979; Zangerl et al., 1988). Claw I is usually larger than claw II (on digit II), and in adult males, the first claw curves strongly toward the ventral surface.

2.3 SKELETAL ANATOMY

The skeleton is composed of bones and cartilages, and is partitioned into three parts: the skull, axial skeleton, and appendicular skeleton. In sea turtles, each of these bony groups is a composite of several structures. The skull includes the braincase, jaws, and hyoid apparatus. The axial skeleton is composed of the carapace (*nuchal*, *dorsal*, *neural*, *pleural*, *suprapygal*, *pygal*, and *peripheral* bones, Figure 2.7) and plastron (*epiplastron*, *entoplastron*, *hyoplastron, hypoplastron*, and *xiphiplastron* bones, Figure 2.8), which includes ventral derivatives of the vertebrate axial skeleton as well as unique bones (Zangerl, 1939; Gilbert et al., 2001). The appendicular skeleton includes the flippers, hind limbs, and their supporting structures (the pectoral and pelvic girdles). The skeleton at the hatchling stage, like that of many vertebrates, is not completely ossified.

In chelonid hatchlings and all size classes of *D. coriacea*, the carapace is composed of ribs, a nuchal bone, and vertebrae (illustrated by Deraniyagala, 1939). In cheloniids, the shell becomes increasingly ossified with increase in size. Intermembranous bone hypertrophies between the ribs and grows laterally, further ossifying the carapace. The ribs grow laterally to meet the peripheral bones in cheloniids (Pritchard, 1979; Gilbert et al., 2001; Wyneken, 2001). Hatchlings have distinct flattened ribs, with little intermembranous bone between adjacent ribs. Individual

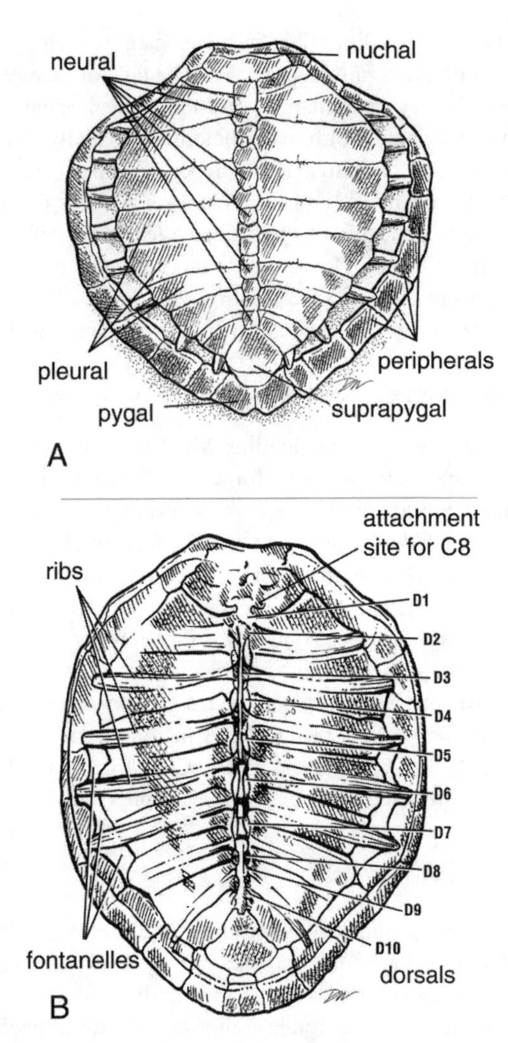

FIGURE 2.7 The bony carapace of a cheloniid is formed from a composite of bones. These and key landmarks are shown in dorsal view (A) and ventral view (B). (Adapted from Wyneken, J. *The Anatomy of Sea Turtles*, U.S. Department of Commerce, NOAA Technical Memorandum NMFS-SEFSC-470, 2001. With permission.)

bones of the plastron and limb girdles initially articulate with one another via cartilage or connective tissue. As cheloniids grow, bone fills in between the ribs, producing pleural bones that are ankylosed along their abutting borders; the hyoplastron and hypoplastron bones usually fuse by adulthood, as do the bones of the pelvic and pectoral girdles. Cartilages are found primarily on joint surfaces.

The skeleton of adult leatherbacks remains neotenic in form, with the bones retaining substantial cartilaginous processes, flexible articulations, and large vascular cartilages and cartilage-replacement bone on the limb bone ends (Rhodin, 1985).

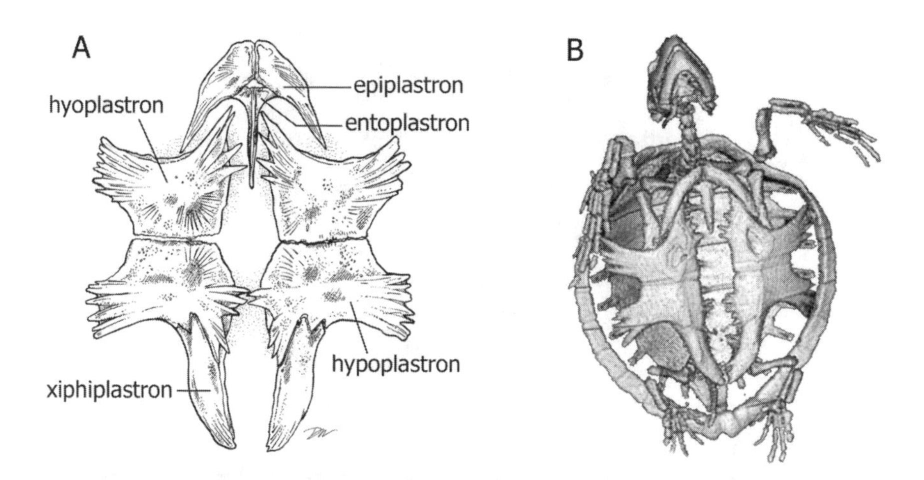

FIGURE 2.8 The cheloniid plastron is formed from a composite of bones with a medial fontanelle (A). One or more lateral processes of the hyoplastron and hypoplastron articulate with the peripheral bones of the carapace. The relationship of the plastron to the carapace can be seen in this computed tomography image (B) of an immature *L. kempii*. (Adapted from Wyneken, J. *The Anatomy of Sea Turtles*, U.S. Department of Commerce, NOAA Technical Memorandum NMFS-SEFSC-470, 2001. With permission.)

2.3.1 The Skull

The skull is formed by an inner braincase (the *neurocranium*) housing the brain, and an outer set of bones (the *splanchnocranium*). The braincase is found along the midline, internal to the skull roof, snout, and jaws of the splanchnocranium. The splanchnocranium houses the sense organs and provides the attachment sites for jaw, throat, and neck muscles (Hyman and Wake, 1979; Kardong, 2002). The eyes each contain a ring of bones (scleral ossicles) that, in life, are found deep to the iris. Skull shape plus the form and the patterns of bones on the roof of the mouth are diagnostic for species identification (Pritchard, 1979).

The jaws (mandibles) of sea turtles are composed of several articulating bones and Meckel's cartilage, within the lower mandible (Figure 2.9). Rhamphotheci cover the maxillary, premaxillary, and vomer bones of the upper jaw, and the dentary bones of the lower jaw in cheloniids.

Leatherback skulls are wide posteriorly, are rounded anteriorly, and possess large orbits; there are no parietal notches. The skull is composed of loosely articulated bones. There is no prominent supraoccipital process (Figure 2.10A). The jaws are characterized by the presence of a notch in each maxillary bone (Figure 2.10B). The lower jaw has a large cartilaginous portion, medial to the dentary. Anteriorly, the two mandibles articulate at the midline to form a sharp point.

Cheloniid skulls are formed by tightly articulating bones. There is a partial secondary palate present (Figure 2.11). Overall skull form and the details of the palate differ among species. In *C. mydas*, the skull is rounded with a short snout and shallow parietal notches. The palate, between the margins of the upper jaw and the internal nares, has a pair of ridges that run parallel to the outer edge of the jaw.

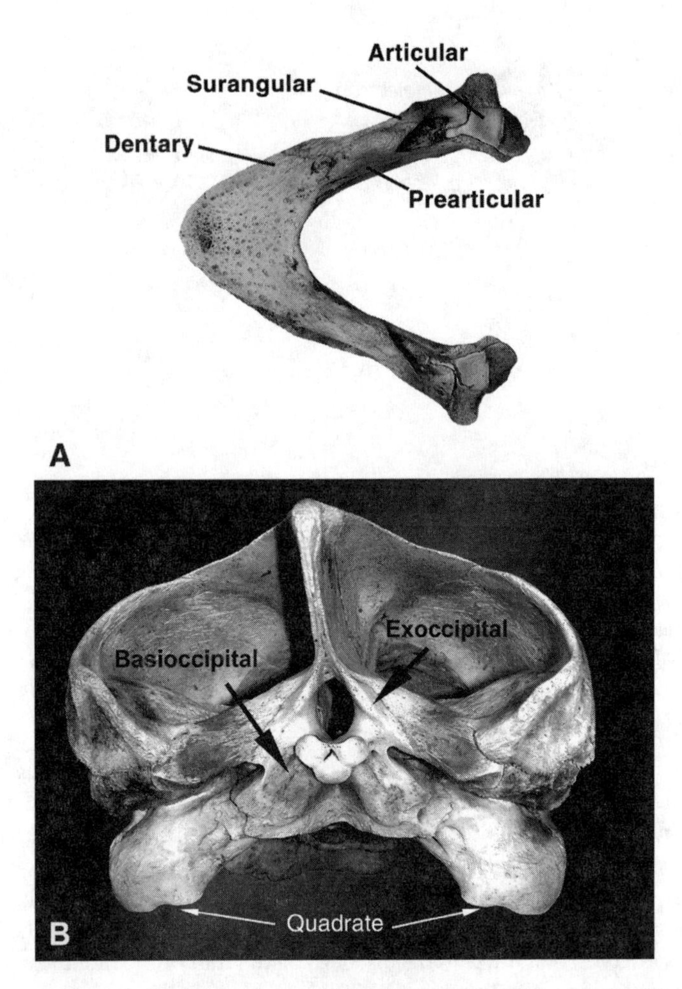

FIGURE 2.9 The lower jaw (A) is a composite of bones and a core of cartilage (Meckel's cartilage). The rhamphotheci have been removed from this jaw. The lower jaw articulates via the articular with the quadrate bone of the skull. The posterior view of the skull (B) shows the quadrate. The trilobed occipital condyle ventral to the vertical supraoccipital process and the foramen magnum (between the exoccipital and basiocipital) articulates with the first cervical vertebra (the atlas). (Part A adapted from Wyneken, J. *The Anatomy of Sea Turtles*, U.S. Department of Commerce, NOAA Technical Memorandum NMFS-SEFSC-470, 2001. With permission.)

The dentary may have many small cusps in young turtles; these are usually reduced or absent in older animals. The skulls of *C. caretta*, *L. olivacea*, *L. kempii*, and *N. depressus* are relatively large, and wide posteriorly, with a snout that tapers anterior to the orbits. Wide emarginations (parietal notches) are found along the posterior borders of the squamosal, parietal, and supraoccipital bones. The jaws are robust, and their buccal outline traces a wide "V." Loggerheads have a relatively long secondary palate that lacks alveolar ridges. The two maxillary bones articulate along

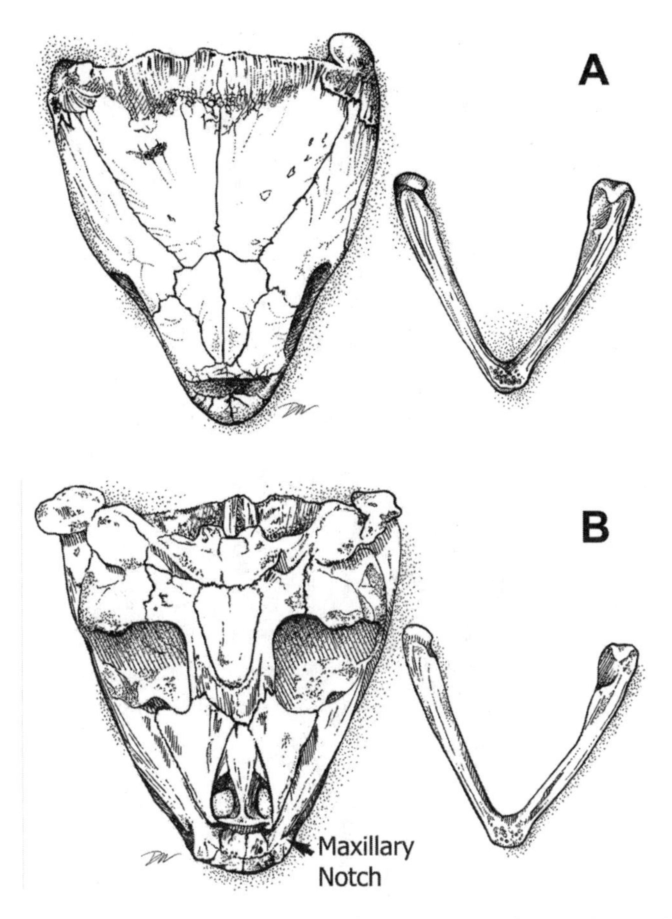

FIGURE 2.10 The skull and lower jaw of the leatherback in dorsal (A) and ventral (B) views. (Adapted from Wyneken, J. *The Anatomy of Sea Turtles*, U.S. Department of Commerce NOAA Technical Memorandum NMFS-SEFSC-470, 2001. With permission.)

the palate's midline, anterior to the vomer. In both *Lepidochelys* species, the two maxillary bones are separated by the vomer, which extends anteriorly to articulate with the premaxillary bones (Figure 2.11). The palate of *L. kempii* has longitudinal alveolar ridges, whereas that of *L. olivacea* lacks alveolar ridges. The pterygoid bones of the olive ridley are proportionately wider and the pterygoid processes are more pronounced than in the Kemp's ridley.

The hawksbill head is long and narrow (Figure 2.1B) in all but very young turtles; its length is approximately equal to twice the width (measured at the skull's widest part). Hawksbill skulls have deep parietal notches, and the snout tapers to a point. The secondary palate is well developed. As a result the internal nares are situated slightly more posteriorly than in other cheloniids.

The skull of sea turtles articulates with the cervical vertebrae via a single, trilobed occipital condyle formed by three occipital bones (Figure 2.9B).

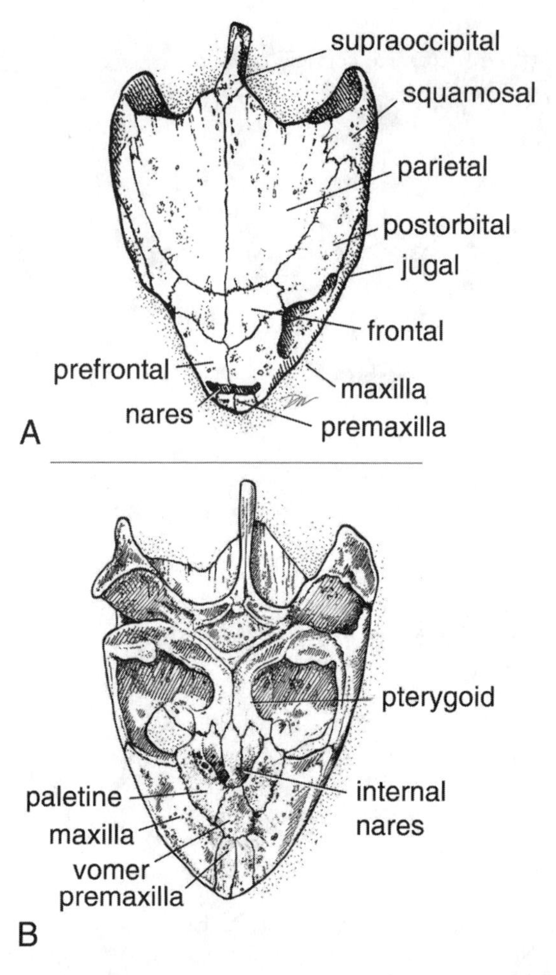

FIGURE 2.11 A cheloniid skull in dorsal (A) and ventral (B) views. The external bones of the splanchnocranium and neurocranium are identified. (Adapted from Wyneken, J. *The Anatomy of Sea Turtles*, U.S. Department of Commerce, NOAA Technical Memorandum NMFS-SEFSC-470, 2001. With permission.)

2.3.2 Axial Skeleton

Like all turtles, sea turtles have eight cervical vertebrae, with the atlas (C1) articulating with the skull and the C8 cervical vertebra attaching to the carapace. There are 10 dorsal (or trunk) vertebrae, 2–3 sacral vertebrae, and 12 or more caudal vertebrae (Figure 2.7). The first cervical vertebra (the atlas) is a three-part bone, which articulates with the occipital condyle at the posterior of the skull. Posteriorly, the atlas articulates with the axis. Next are three similarly shaped, opisthocoelous vertebrae with convex centra or vertebral bodies anteriorly and concave centra posteriorly. Cervical vertebrae 6 and 7 articulate with one another via flat-faced centra (Pritchard, personal communication) and are occasionally fused (Zangerl et al., 1988). The eighth cervical vertebra attaches to the carapace (Figure 2.7) and is limited in its range of movement. The vertebral arches of the successive cervical vertebrae have articulating processes with sliding joints that allow limited dorsal–ventral bending of the neck, but little twisting.

Each dorsal (or trunk) vertebra is composed of a separate dorsal arch and ventral vertebral body, which articulates bilaterally with a pair of ribs. Rib heads are aligned with the junctions of adjacent vertebral bodies. In *Dermochelys*, there is no hypertrophy of intermembranous bone between the ribs of the carapace or formation of peripheral bones. The bony carapace skeleton remains composed solely of an expanded nuchal bone, ribs, and vertebrae. Cheloniid vertebrae and ribs are connected via intermembranous bone to produce the carapacial bones. Neural bones are incorporated into the vertebral arch elements; the ribs and their intermembranous expansions form pleurals, and peripheral bones form the margins of the carapace. In *L. kempii*, the peripheral bones also widen with age and increasing body size. The spaces between the ribs and the carapace, the fontanelles, are closed by a thick membranes of cartilage and fibrous connective tissue underlying the scutes. The anterior-most bone is the nuchal, and the posterior-most of peripheral bones are the pygals. Between the last neural bone and the pygal are the suprapygals (one to three; usually two), which lack any vertebral fusion (Figure 2.7). The lateral processes of the sacral vertebrae, deep to the suprapygals, are not fused to the carapace. These lateral processes are formed by rib-like processes that articulate with the ilium. The caudal vertebrae of females are short and decrease in size distally; those of mature males are large with robust lateral and dorsal processes (Figure 2.12).

The cheloniid plastron is composed of four pairs of bones: epiplastron, hyoplastron, hypoplastron, and xiphiplastron, and one unpaired bone, the entoplastron (Figure 2.8). The leatherback plastron is formed from a peripherally located ring of reduced, paired plastron bones. No entoplastron is present (Deraniyagala, 1939; Pritchard, 1997).

2.3.3 Appendicular Skeleton

The appendicular skeleton of turtles differs from that of other reptiles in that the shoulder and pelvic bones are located inside the ribs and ventral to the vertebrae. This morphological difference results in rotation and repositioning of many bony structures in the limbs and of the associated soft tissues (muscles, nerves, blood vessels). The anterior appendicular skeleton is composed of flippers (Figure 2.13) and triradiate pectoral girdles (Figure 2.14). The latter are formed of two bones, the scapula, with its ventral acromion process, and the procoracoid (also called a coracoid). The scapula is aligned dorsoventrally and attaches to the carapace adjacent to the first trunk vertebra. Ventrolaterally, the two bones form the glenoid fossa, which articulates with the head of the humerus. An acromion process extends medially from each scapula to articulate with the entoplastron via a plastroacromial ligament. The fan-shaped procoracoid bones extend posteromedially. Each terminates in a crescent-shaped coracoid cartilage. A flat acromialcoracoid ligament connects the acromion and the posterior part of the coracoid on each side.

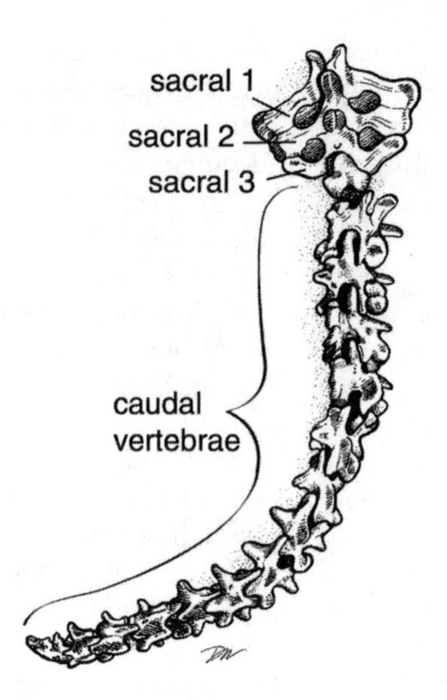

FIGURE 2.12 The sacral and caudal vertebrae of an adult male green turtle. The caudal vertebrae of a female would be much shorter with smaller lateral processes. (Adapted from Wyneken, J. *The Anatomy of Sea Turtles*, U.S. Department of Commerce, NOAA Technical Memorandum NMFS-SEFSC-470, 2001. With permission.)

The forelimb is composed of the humerus, radius, ulna, carpals, metacarpals, and five phalanges (Figure 2.13). The radius and ulna are short in sea turtles and, in adults, functionally fused by fibrous connective tissue. The flipper blade is formed by widening and flattening of the wrist bones (radiale, ulnare, centrale, pisiform, distal carpals) and elongation of the digits (metacarpals and phalanges) (Figure 2.13). The cheloniid humerus (Figure 2.15) articulates with the shoulder at the glenoid fossa (Figure 2.14); it is flattened with its head offset from the bone's shaft (Walker, 1973). A large medial process extends proximally beyond the humeral head and is a major attachment site for flipper abductor and extensor muscles. Distal to the head and almost diagonally opposite is the lateral process (deltoid crest), to which attach flipper protractor muscles. In *Dermochelys*, the humerus is extremely flattened, with large portions of the humeral head, distal condyles, and medial process formed by highly vascular articular cartilages (Figure 2.16).

The Dermochelyidae and Cheloniidae diverge in the types of bone and cartilage that form their appendicular skeleton. Extensive vascular channels in the cartilage ends of leatherback long bones (Figure 2.16) are indicative of chondro-osseus bone formation (Rhodin, 1985; Rhodin et al., 1981). This is unlike cheloniid bone, which is formed by deposition of relatively thick layers (lamellae) of cortical bone around a cellular bony core (cancellous bone). These two bone formation types are illustrated comparatively in gross form by Wyneken (2001) and histologically by Rhodin (1985).

The pelvis is composed of three paired bones: the pubis, ischium, and ilium. The two ilia are oriented dorsoventrally and articulate with the sacral vertebrae. The ischia, pubic bones, and cartilages form the largely flat ventral surface of the pelvic girdle. All

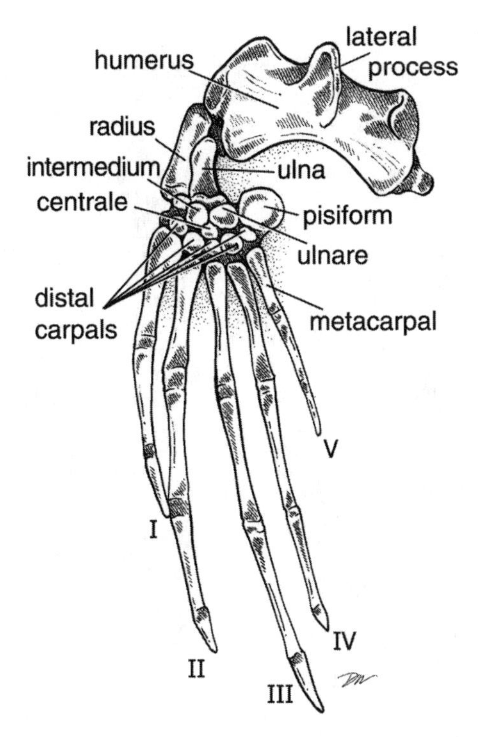

FIGURE 2.13 Dorsal view of a leatherback flipper showing the elongated digits, flattened wrist, and shortened antebrachium (radius and ulna). In *Dermochelys*, the lateral process is displaced distally on the humerus compared with its position in cheloniids. (Adapted from Wyneken, J. *The Anatomy of Sea Turtles*, U.S. Department of Commerce, NOAA Technical Memorandum NMFS-SEFSC-470, 2001. With permission.)

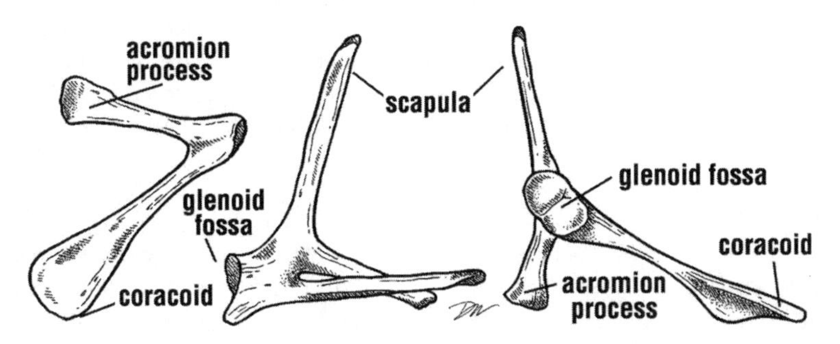

FIGURE 2.14 The left shoulder girdles of sea turtles (ventral, medial, and lateral views). The scapula is oriented roughly dorsoventrally, the acromion process is aligned mediolaterally, and the coracoid is positioned anterolaterally to posteromedially. (Adapted from Wyneken, J. *The Anatomy of Sea Turtles*, U.S. Department of Commerce, NOAA Technical Memorandum NMFS-SEFSC-470, 2001. With permission.)

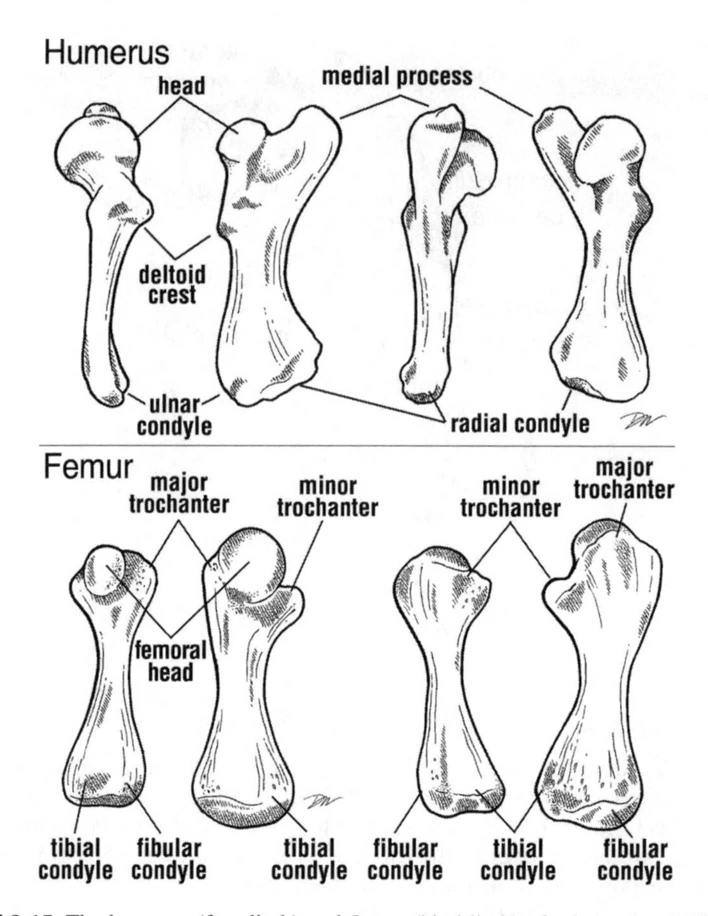

FIGURE 2.15 The humerus (fore limb) and femur (hind limb) of a loggerhead. The lateral and medial (deltoid crest) processes are the sites of attachment of many locomotor muscles. The femur has two trochanters that are more symmetrical to either side of the femoral head. The differences in the muscle attachment site forms and positions reflect intrinsic differences in the positioning of the bones, but also the functional differences between the fore and hind limbs in sea turtles. (Adapted from Wyneken, J. *The Anatomy of Sea Turtles*, U.S. Department of Commerce, NOAA Technical Memorandum NMFS-SEFSC-470, 2001. With permission.)

three bones contribute to form the hip socket (acetabulum). In young cheloniids and all *D. coriacea*, the three bones are separate and joined by cartilage. In maturing and mature cheloniids, they ossify and ankylose to form a single structure (Figure 2.17).

The hind limb is somewhat flattened as a rudderlike structure. The femur has a relatively straight shaft with a strongly offset head that articulates with the pelvis at the acetabulum. Two processes to either side of the head: major and minor trochanters (Figure 2.15) are for muscle attachment. The femur articulates with the tibia and fibula. The short, flat ankle consists of the calcaneum, astragalus, and tarsals. There are five digits. The first and fifth metatarsals are wide and flat, and the phalanges are extended, adding breadth to the distal hind limb area (Figure 2.18).

FIGURE 2.16 The cut end of a leatherback bone shows the vascular channels embedded within the white cartilage. (Adapted from Wyneken, J. *The Anatomy of Sea Turtles*, U.S. Department of Commerce, NOAA Technical Memorandum NMFS-SEFSC-470, 2001. With permission.)

2.4 MUSCULAR ANATOMY

The muscles are responsible for movement, modifying the action of other muscles, and stabilizing the joints. They originate and insert via tendons. A muscle's origin is its fixed point, and its insertion is typically the point that moves. Muscles can attach via tendons to bones, other muscles, other tendons, skin, or eyes. Muscle actions, as they apply to marine turtles, include: *flexion*, bending one part relative to another at a joint; *extension*, straightening of those parts; *protraction*, moving one part (usually a limb) outward and forward; *retraction*, moving that part inward and back; *abduction*, moving a part away from the plastron; and *adduction*, bringing the part toward the body's ventral surface. *Rotation* turns a structure about its axis. *Depression* (a special form of abduction) opens the jaws, whereas e*levation* closes the jaws (a kind of adduction).

Muscles are frequently grouped by their embryonic origin and position (e.g., hypaxial, epaxial, preaxial, or postaxial), extent (intrinsic or extrinsic), major innervations (e.g., trigeminal, cervical plexus, or sacral plexus), actions (e.g., extensors, abductors), and/or position in the body (e.g., axial, appendicular, pectoral, or pelvic). Here, muscles are discussed in groups by position because those in close proximity often act together and are often dissected and viewed by location. Pectoral and pelvic muscles are associated with the plastron as well as the limb girdles. Forelimb muscles are those of the flippers, pectoral girdles, and

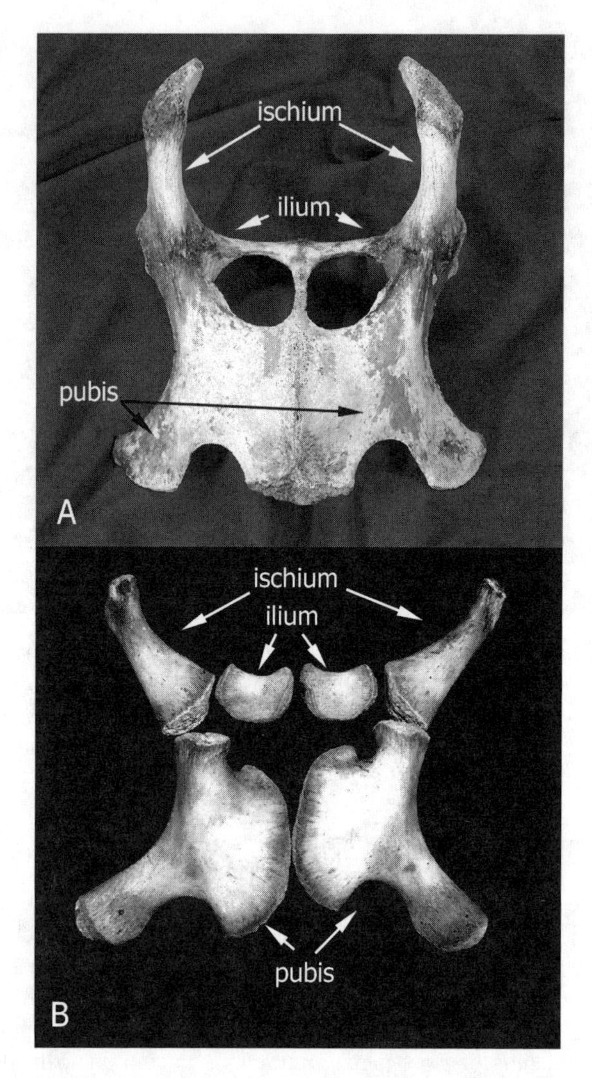

FIGURE 2.17 The pelvis of an adult cheloniid (A) and *Dermochelys* (B). The pubis, ilium, and ischium fuse in adult cheloniids, such as *C. caretta*, but they remain articulated by cartilage in *Dermochelys*. During skeletal preparation, the bones disarticulate from one another. (Adapted from Wyneken, J. *The Anatomy of Sea Turtles*, U.S. Department of Commerce, NOAA Technical Memorandum NMFS-SEFSC-470, 2001. With permission.)

anterior carapace, involved in flipper movements and, in a few cases, breathing. Posterior muscles are the large postaxial muscles of the hip, thigh, and lower leg.

2.4.1 Muscles of the Head and Neck

The major head muscles function in jaw opening and closing. They can be divided into jaw opening muscles innervated by the trigeminal nerve, and the jaw closing

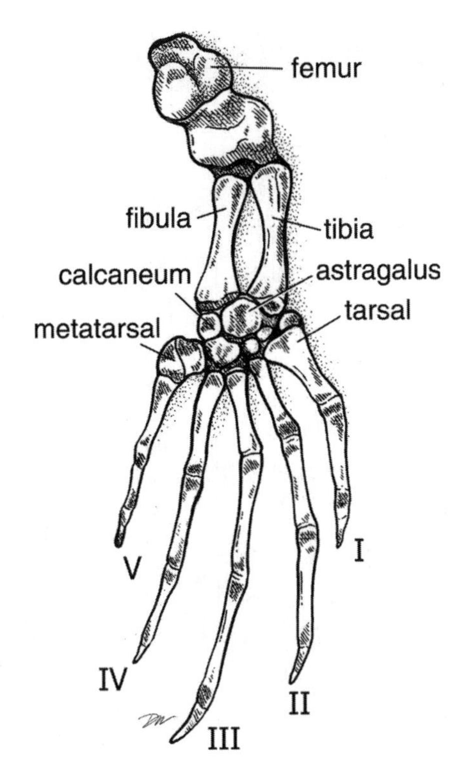

FIGURE 2.18 Ventral view of the leatherback hind limb. The femur is the bony element of the thigh; the tibia and fibula are the bony elements of the shank. The ends of these bones are cartilaginous. The ankle is somewhat flattened and laterally expanded, resulting in wide placement of the digits. This architecture contributes to the rudderlike shape of the hind limb. (Adapted from Wyneken, J. *The Anatomy of Sea Turtles,* U.S. Department of Commerce, NOAA Technical Memorandum NMFS-SEFC-470, 2001. With permission.)

muscles innervated by the facial nerve (Schumacher, 1973). The jaw muscles of turtles are mostly located inside the skull. Unlike mammals, turtles lack a mandibularis muscle; instead, they have an adductor mandibulae with several heads. The muscles of the head and throat and their specific locations (origins and insertions) and innervations are summarized in Table 2.1 and Figures 2.19–2.21. Muscles of the tongue, innervated by the hypoglossal and the glossopharyngeal nerves, are not described here but have been described by Schumacher (1973).

Neck muscles (Table 2.1) are superficial (Figure 2.19) or deep and are innervated by cervical nerves. The deep neck muscles include the neck extensors, the *longus colli* (short, segmentally arranged muscles that travel obliquely between successive cervical vertebrae), and the neck flexor and retractor (*retrahens colli*) muscles (extending from the cervical vertebrae to the dorsal vertebral elements of the carapace).

TABLE 2.1
Major Muscles of Sea Turtles: Locations and Innervations

Muscle	Origin	Insertion	Innervation
Pectoral Muscles			
Pectoralis major	Plastron	Lateral process and shaft of the humerus	Supracoracoid nerve (ventral brachial plexus)
Deltoideus (ventral part)	Ventral scapula, acromion, and anterior plastron bones	Lateral process of the humerus	Supracoracoid nerve (ventral brachial plexus)
Supracoracoideus (anterior part)	Acromion	Lateral process of the humerus	Supracoracoid nerve (ventral brachial plexus)
Biceps brachii (superficial head)[a]	Posterior coracoid	Via a long tendon to radius and ulna; sometimes pisiform of wrist	Inferior brachial nerve (median nerve)
Biceps brachii (profundus head)	Posterior coracoid ventral to the biceps superficialis	Via a tendon with the brachialis on radius and ulna	Inferior brachial nerve (median nerve)
Brachialis inferior	Distal humerus	With the biceps brachii profundus on radius and ulna	Inferior brachial nerve
Flexor carpi ulnaris	Ulnar epicondyle, humerus, ulnar border	Pisiform, skin of ulnar side of flipper blade	Median nerve
Flexor digitorum longus and palmaris longus	Humerus, ulna, and ulnar epicondyle	Ventral surface of the flipper phalanges	Median nerve
Flexor carpi radialis	Ulnar epicondyle	Radius and metacarpal I	na
Coracobrachialis magnus	Dorsal side of coracoid process	Medial process of the humerus	Supracoracoid nerve
Supracoracoideus (posterior part)	Procoracoid and coracoid cartilage	Lateral process of the humerus	Supracoracoid nerve
Latissimus dorsi/teres major complex	Scapula and carapace from scapular ligament, along first pleural bone to anterior peripheral bone	Via a common tendon just distal to the head of the humerus	Deltoid nerve (brachial plexus)
Deltoideus (scapular head)	Anterior scapula	Lateral process and shaft of the humerus	Deltoid nerve (brachial plexus)

Subscapularis	Medial and posterior scapula	Medial process and shaft of the humerus	Deltoid nerve (brachial plexus)
Triceps brachii (triceps superficialis) humeral head	Humerus	Via common tendon with scapular head on proximal ulna	Superficial radial nerve (brachial plexus)
Triceps brachii (triceps superficialis) scapular head[a]	Scapula	Via common tendon with humeral head on proximal ulna	Superficial radial nerve (brachial plexus)
Extensor digitorum extensor radialis[b] tractor radii	Radius and ulna	Dorsal carpals, metacarpals, and proximal phalanges	Ulnar nerve
Pelvic Muscles			
Rectus abdominis	Lateral pubis	Plastron	na
Puboischiofemoralis externus	Ventral pubis, ischium, and membrane covering the thyroid fenestrae	Minor trochanter	Obturator and tibial nerves (sacral plexus)
Puboischiofemoralis internus[a]	Dorsolateral pubis, ilium, and the sacral vertebrae	Major trochanter	Obturator and tibial nerves (sacral plexus)
Flexor tibialis complex			
Pubotibialis[a]	Pubic symphysis and lateral pubis above and sacral vertebrae	Tibia	Obturator and tibial nerves (sacral plexus)
Flexor tibialis internus Flexor tibialis externus	Postsacral vertebrae dorsal head arises from the ilium and ventral head from posterior ischium	Tibia, gastrocnemius muscle, skin, and connective tissue of shank	Obdurator and tibial nerves (sacral plexus)
Adductor femoris	Lateral ischium	Posterior femoral shaft	Obturator nerve (sacral plexus)
Ischiotrochantericus	Anterior pubis and pubic symphysis	Major trochanter	Obturator nerve (sacral plexus)
Ambiens	Pubioischiadic ligament	Patellar tendon and anterior tibia	Peroneal, tibial, and branch of femoral nerves (sacral plexus)
Femorotibialis	Dorsal and anteroventral femur	Patellar tendon	Peroneal and femoral nerves (sacral plexus)
Iliotibialis	Dorsal ilium	Patellar tendon	Peroneal and femoral nerves (sacral plexus)

(continued)

TABLE 2.1 (continued)
Major Muscles of Sea Turtles: Locations and Innervations

Hind foot extensors	Dorsal and lateral femur	Dorsal and anterior fibula and digits	Tibial nerve
Respiratory Muscles			
Testocoracoideus	Carapace near anterior inframarginals	Dorsal coracoid	Cervical spinal nerves
Testoscapularis	Carapace posterior to the latissimus dorsi	Dorsal scapula and the scapular attachment to the carapace	Cervical spinal nerves
Neck Muscles			
Transverse cervical	Parietal and supraoccipital	Pleural 1	Cervical spinal nerves
Biventer cervical	Supraoccipital	Cervical vertebra 8 and adjacent carapace	Cervical spinal nerves
Longus colli	Cervical vertebrae	Next successive cervical vertebra	Dorsal cervical nerve
Retrahens colli	Cervical vertebrae	Dorsal vertebral elements of the carapace	Dorsal cervical nerve
Head Muscles			
Intermandibularis	Medial dentary bones	Midline raphe tendon of pseudotemporalis	Trigeminal nerve
Constrictor colli	Dorsolateral cervical tendon	Midline tendon (raphe)	Facial nerve
Geniohyoideus	Medial dentary deep to intermandibularis	Hyoid	Facial nerve
Coracohyoideus	Coracoid	Hyoid	Facial nerve
Adductor mandibulae (parietal part)	Parietal	Dentary, with small insertions on the squamosal bone	Trigeminal nerve
Adductor mandibulae (supraoccipital part)	Supraoccipital	Dentary, with small insertions on the squamosal bone	Trigeminal nerve
Adductor mandibulae (quadrate part)	Quadrate	Dentary, with small insertions on the squamosal bone	Trigeminal nerve

Adductor mandibulae (prootic part)	Prootic	Dentary, with small insertions on the squamosal bone	Trigeminal nerve
Adductor mandibulae (opisthotic part)	Opisthotic	Dentary, with small insertions on the squamosal bone	Trigeminal nerve
Pseudotemporalis	Parietal	Medial internal tendon to lower jaw	Trigeminal nerve
Tongue muscles	na	na	Hypoglossal nerve and glossopharyngeal
Depressor mandibulae complex	Quadrate, quadratojugal, and squamosal bones	Articular of the lower jaw (in *Dermochelys* a portion inserts on the auditory tube)	Facial nerve

Note: na = where innervations could not be found or could not be confirmed.

[a]May be absent in *D. coriacea.*
[b]Very reduced.

Source: Based on Fürbringer, M., Zur vergleichenden anatomie der schultermuskeln (The comparative anatomy of shoulder muscles), *Jena Zschr.* 8, 175–280, 1874; Gadow, H., A contribution to the myology of the hind extremity in reptiles, *Morphologisches Jahrbuch*, 7, 329–466, 1882; Walker, W.F., Jr., The locomotor apparatus of testudines, in *Biology of the Reptilia*, Gans, C. and Parsons, T.S., Eds., Academic Press, New York, 1973, pp. 1–100; and Wyneken, J. *The Anatomy of Sea Turtles*, U.S. Department of Commerce NOAA Technical Memorandum NMFS-SEFSC-470, 2001, 172 pp.

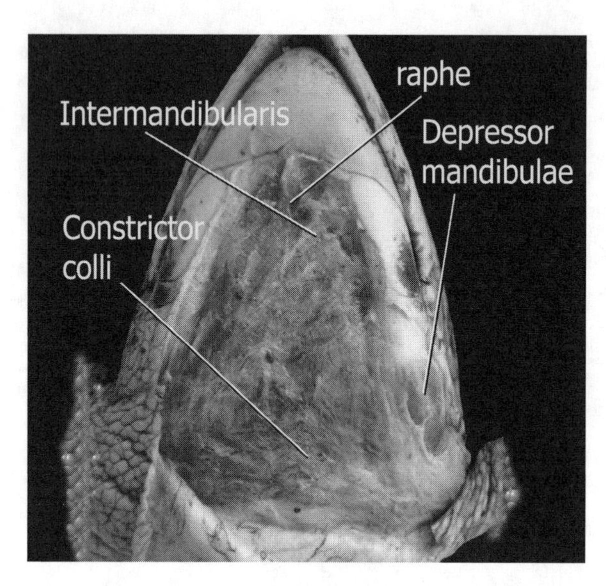

FIGURE 2.19 Superficial ventral throat and neck muscles. Connective tissue loosely attaches the muscle to the skin. The midline raphe (tendon) is visible along the anterior half of the muscle. (Adapted from Wyneken, J. *The Anatomy of Sea Turtles*, U.S. Department of Commerce NOAA Technical Memorandum NMFS-SEFSC-470, 2001. With permission.)

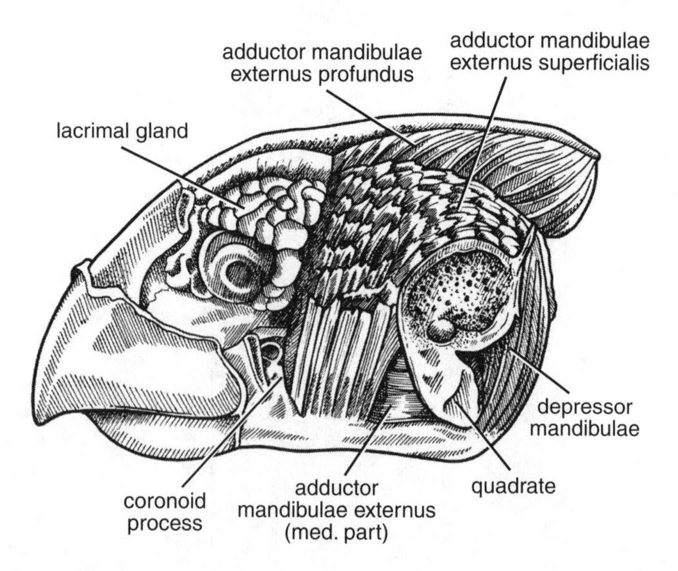

FIGURE 2.20 Lateral and dorsal jaw muscles are located inside the bony skull. The dissection of a hawksbill, shows the jaw adductors and depressors, which have multiple parts. The large external tendon (and internal tendon, not shown) serves as muscle attachment sites. The internal tendon slides over the trochlear process of the neurocranium. (Adapted from Schumacher, H. In: *Biology of the Reptilia,* C. Gans, Ed., Vol. 4, Academic Press, 1973. With permission.)

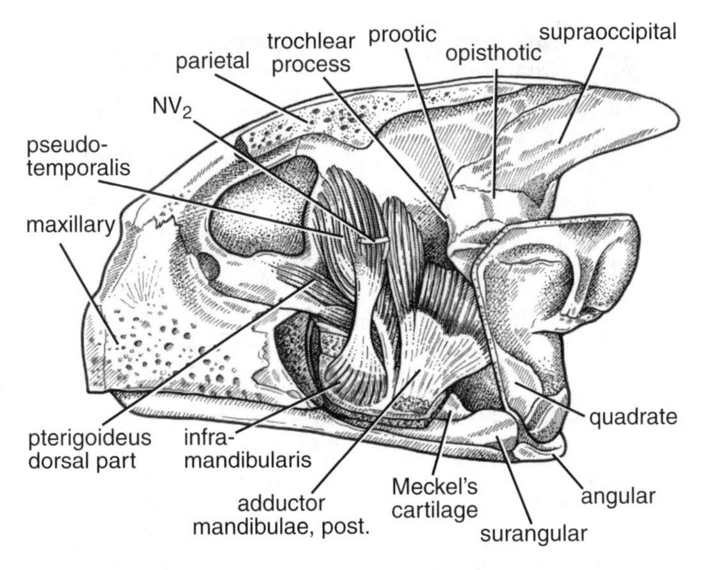

FIGURE 2.21 Deep jaw adductors and depressors of a loggerhead. The trochlear process of the neurocranium, (formed by the prootic an opisthotic bones) over which the larger jaw adductors and depressors move, is shown. NV_2 = Branches of the trigeminal nerve. (Adapted from Schumacher, H. In: *Biology of the Reptilia,* C. Gans, Ed., Vol. 4, Academic Press, 1973. With permission.)

2.4.2 Appendicular Muscles

The muscles of the fore limbs (including the shoulders) and hind limbs (including the hips) are the appendicular muscles.

2.4.2.1 Pectoral Muscles

The massive ventral musculature is dominated by pectoral muscles that adduct, retract, rotate, and flex the flippers (Wyneken, 1988), whereas the dorsal pectoral muscles are primarily flipper abductors, extensors, and protractors. The pectoral muscles, innervated by the nerves of the brachial plexus, are described in Tables 2.1 and 2.2 and illustrated in Figure 2.22. The major shoulder muscles are the *latissimus dorsi–teres major* complex, the scapular head of the *deltoideus*, and the *subscapularis* (Figure 2.22 and Table 2.1). These are responsible for flipper protraction, retraction, abduction, and adduction.

The remaining dorsal shoulder muscles are the *triceps brachii* (*triceps superficialis*), biceps complex (*biceps superficialis* and *biceps profundus*), and *brachialis* (a flipper flexor). The biceps muscles extend across the shoulder and the elbow and act as retractors of the humerus or flexors of the flipper at the elbow. Walker (1973) reports that in *Dermochelys* and *Lepidochelys*, there is often just a single head, the *biceps superficialis*, inserting on the radius and ulna.

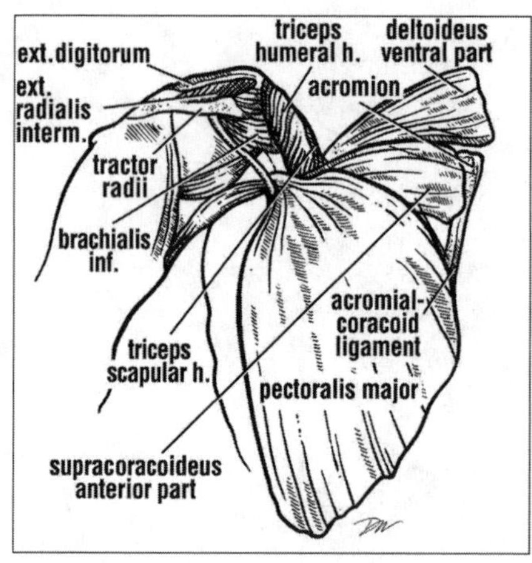

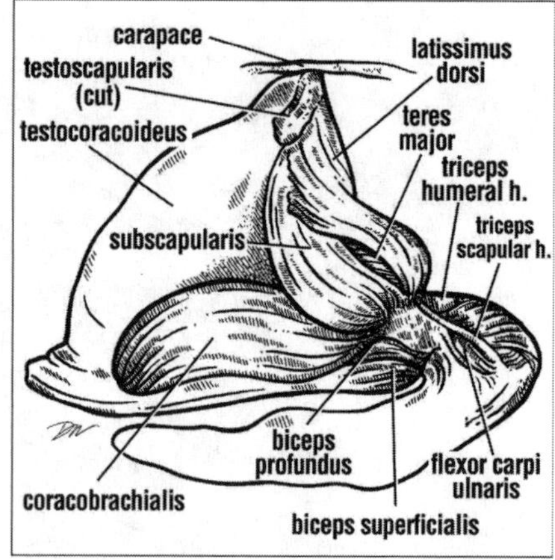

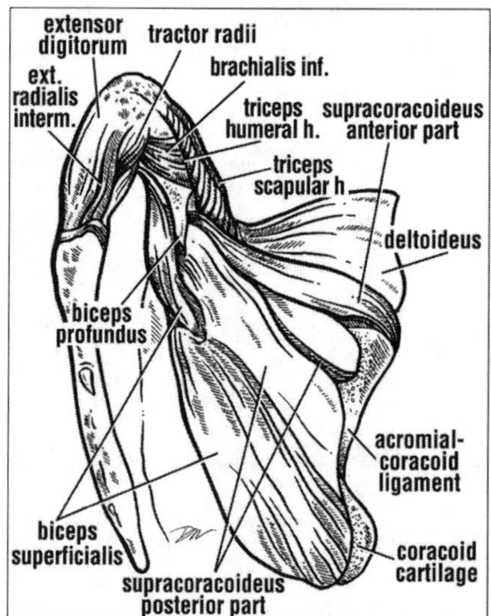

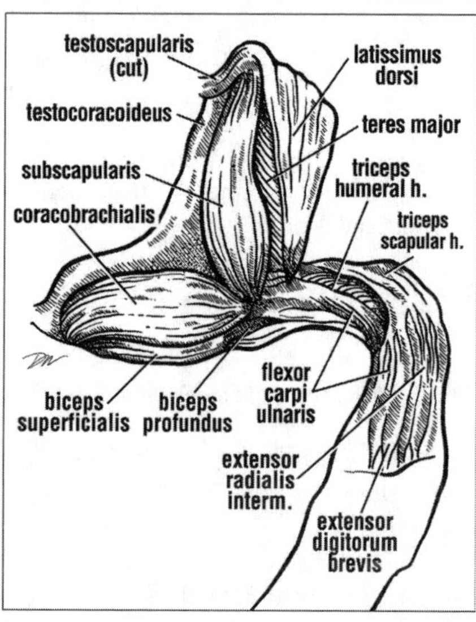

FIGURE 2.22 Diagrams of cheloniid right shoulder muscles, including locomotor and respiratory muscles. Superficial ventral muscles (top left), deep ventral muscles (bottom left), lateral muscles (top right), and posterior muscles (bottom right). The *extensor digitorum*, *extensor radiali intermedius*, *tractor radii*, and *flexor carpi* control the extension and flexion of the flipper blade. Ext. = extensor; h. = head; inf. = inferior; interm. = intermedialis. The large *subscapularis* covers most of the scapula; it protracts the flipper. The *latissimus dorsi*, a large sheetlike muscle, plus the *teres major* and *deltoideus* scapular head abduct and sometimes protract the flipper. The large *coracobrachialis* is seen ventrally, covering much of the procoracoid. The *biceps superficialis* extends from the shoulder (mostly the procoracoid) to the pisiform bone of the wrist, and probably helps control the twist or rotation of the flipper blade. The *biceps profundus* acts as a flipper retractor and a flexor of the flipper blade at the elbow. The two heads of the *triceps brachii* (triceps scapular head and triceps humeral head) are forelimb adductors. (Adapted from Wyneken, J. *The Anatomy of Sea Turtles*, U.S. Department of Commerce, NOAA Technical Memorandum NMFS-SEFSC-470, 2001. With permission.)

TABLE 2.2
Peripheral Nerve Innervations

Brachial Plexus Nerves	Muscle or Muscle Group
Superior brachial nerve	Tractor radii
Superficial radial nerve	Latissimus dorsi
Deep radial nerve	Latissimus dorsi, supracoracoideus, testoscapularis
Supracoracoideus nerve	Supracoracoideus, pectoralis major, biceps brachii (profundus and superficialis)
Subscapular nerve	Subscapularis
Axillary (deltoideus) nerve	Deltoideus (ventral parts), brachialis
Radial nerve	Latissimus dorsi, teres major, tractor radii, triceps brachii (humeral head), respiratory muscles
Ulnar nerve	Deltoideus (dorsal head), latissimus dorsi, subscapularis, extensor radialis, medial flipper muscles, extensors of digits
Median nerve	Coracobrachialis, flexor carpi ulnaris, flexors of digits
Sacral Plexus Nerves	
Crural nerve	Inguinal muscles, thigh protractors (triceps femoris complex)
Femoral nerve	*Puboischiofemoralis*, dorsal hip muscles
Obturator nerve	Ventral hip muscles, *caudi-ilioformoralis, ischiotrochantericus, adductor femoris, flexor tibialis (internus* and *externus), pubotibialis complex*
Ischiadicus nerve	Posterodorsal hip muscles
Sciatic nerve	*Gastrocnemius, iliofemoralis,* ventrolateral foot extensors
Peroneal nerve	*Triceps femoris (ambiens, femorotibialis, iliotibialis), gastrocnemius,* hind foot flexors
Tibial nerve	*Flexor tibialis (internus and externus), ambiens, pubotibialis,* inguinal muscles, foot extensors

2.4.2.2 Respiratory Muscles

Two sheetlike respiratory muscles (*testocoracoideus* and *testoscapularis*), located dorsally in the shoulder region, function with the posterior part of the *rectus abdominus* in changing body volume during ventilation. They are innervated by spinal nerves.

2.4.2.3 Pelvic Muscles

The major posterior muscles include the *puboischiofemoralis externus* (a thigh adductor, with parts that act as a protractor or retractor of the leg), the *puboischiofemoralis internus*, the *pubotibialis*, the *flexor tibialis* complex (thigh protractors), and the *ambiens*, which is a protractor and abductor (Table 2.1). The *puboishiofemoralis internus* and *pubotibialis* may be absent in *Dermochelys* (Walker, 1973). The *iliofemoralis* replaces in function and position the former (Figure 2.23). The *adductor femoris* (Figure 2.23) is a thigh adductor. The *ischiotrochantericus* is a thigh and leg retractor (Table 2.1). The dorsal hip and thigh muscles include the hip abductors: *iliotibialis*, *femorotibialis*, and sometimes a portion of the *ambiens*. The hind foot extensors flex the lower leg or extend the digits (Walker, 1973).

2.5 NERVOUS SYSTEM

The nervous system is composed of two structural/functional subdivisions: the central nervous system (the brain or CNS), and the peripheral nervous system (the spinal cord and its nerves). The brain of sea turtles is a longitudinally arranged tubular structure situated along the midline of the skull. It is housed in a tubular braincase composed anteriorly of the *ethmoid*, *epiotic*, *prootic*, *opisthotic*, *basisphenoid*, *laterosphenoid*, and *otic* bones and posteriorly by the *basioccipital*, *exoccipital*, and *supraoccipital* (Figure 2.9B). The parietal and frontal bones form the roof of the braincase.

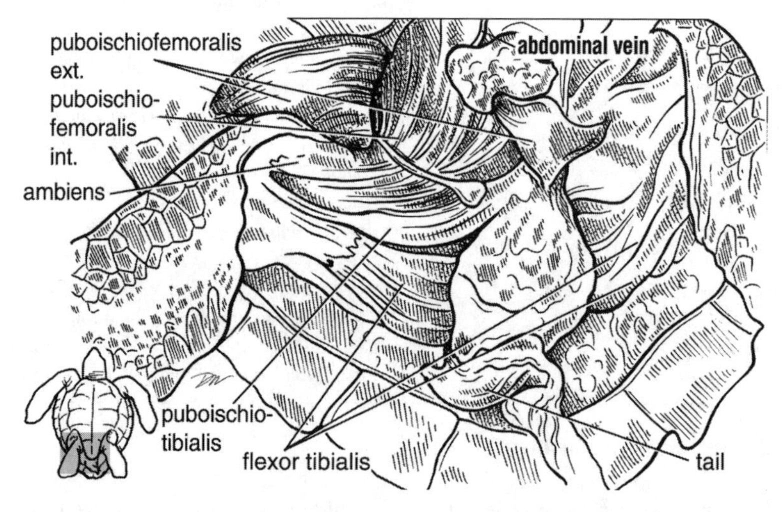

FIGURE 2.23 Superficial ventral hip and hind limb muscles. The *puboischiofemoralis externus* and *internus* (the anterior ventral portions shown) abduct the leg. The *flexor tibialis* complex, including the *pubotibialis,* retracts the leg and controls the shape of the trailing edge of the foot, perhaps during steering. More anteriorly, the *ambiens* is a weak adductor and protractor of the hind leg and can extend the shank. The deeper ventral hip muscles (*adductor femoris* and *puboischiofemoralis internus*, not shown) are antagonistic muscles that adduct and abduct the thigh, respectively. (Adapted from Wyneken, J. *The Anatomy of Sea Turtles*, U.S. Department of Commerce, NOAA Technical Memorandum NMFS-SEFSC-470, 2001. With permission.)

The spinal nerves leave the spinal cord as paired dorsal and ventral nerve roots and exit the vertebrae via intervertebral foramina. The dorsal nerve roots are composed of somatic and visceral sensory nerve fibers and may contain motor fibers as well; the ventral roots are generally composed of both somatic and visceral motor nerve fibers. These nerves function as the autonomic nervous system. The autonomic nervous system of turtles has both sympathetic and parasympathetic components. However, these are not anatomically segregated as thoracolumbar sympathetic and craniosacral parasympathetic regions, as in mammals (Kardong, 2002). Hence, nerves arising along the length of the spinal cord may have both sympathetic and parasympathetic components (Kardong, 2002). Two networks of interconnected spinal nerves, the brachial and sacral (lumbosacral) plexes, are associated with control of the limbs. They are poorly described in the literature on sea turtles, but are covered generally in freshwater species (Ashley, 1962) and other reptiles (Grasse, 1948).

Two tissue layers, the meninges ("menix" is singular), cover the brain and spinal cord. The outer menix is the tough *dura mater.* The *pia mater* or *leptomenix* is more delicate and lies directly on the brain's surface (Kardong, 2002). There are spaces within the braincase that are subdural (beneath the dura mater) and epidural (above the dura mater). Epimeningeal veins occupy some of the epidural space. The brain is bathed in clear cerebrospinal fluid within the subdural space. It is produced by the *tela choroidea* (a vascular region of the brain) and the pia mater dorsal to the medulla and fourth ventricle.

2.5.1 Peripheral Nervous System

In cheloniids, the plexes are formed by ventral nerve roots and their branches. The brachial plexus (Figure 2.24) arises at the level of cervical vertebrae VI–VIII in sea turtles (Grasse, 1948; Wyneken, 2001). These cervical nerves form a complex network innervating the shoulder, brachial (humerus), and flipper muscles (Table 2.2); they also send separate branches to the respiratory muscles. Most muscles receive innervation from more than one branch of the plexus, so multiple innervations are common. The cervical plexus (designated as simply the brachial plexus by some authors) develops as dorsal and ventral parts. The dorsal cervical plexus branches form the *deltoid*, *superior brachial*, *superficial radial*, and *deep radial* nerves to the forelimb adductors and extensors (Walker, 1973). The ventral sets of branches form the *supracoracoid* and the *inferior brachial* nerves (the latter subdivides into the *ulnar* and *median* nerves). A ventral branch of cervical spinal nerve VI makes a large contribution to the *median* nerve. Nerves VII and VIII give rise to the *superior brachial* nerve, which immediately divides to form the *superficial radial* nerve and the deep radial nerve to the anterior shoulder and dorsal flipper (Figure 2.24). The *supracoracoid, subscapular, inferior brachial*, and *ulnar* nerves travel to the supracoracoideus and subscapular pectoral muscles, the brachialis muscle, and the ventral side of the flipper. The *deltoideus* nerve, to the anterior shoulder muscles, arises primarily from cervical spinal nerves VI and VII. Not all cervical spinal nerves contribute to the plexes.

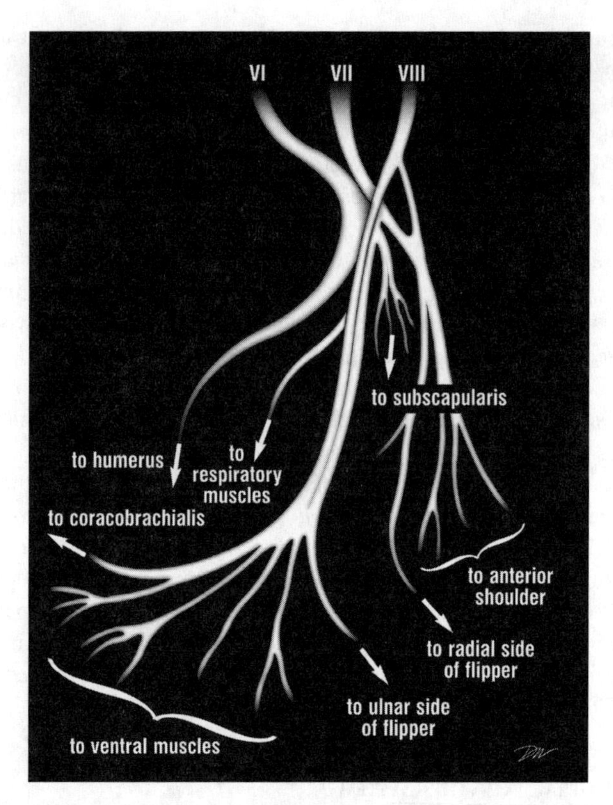

FIGURE 2.24 Brachial (cervical) plexus diagram. The three-dimensional brachial plexus is traced diagrammatically to its major innervation sites. (Adapted from Wyneken, J. *The Anatomy of Sea Turtles*, U.S. Department of Commerce, NOAA Technical Memorandum NMFS-SEFSC-470, 2001. With permission.)

The sacral plexus (Figure 2.25) arises as four (sometimes five to six) branches from spinal nerves XVII–XXI associated with the last trunk vertebra and the sacral vertebrae. The sacral plexus is poorly studied in turtles and described only grossly in sea turtles (Walker, 1973; Wyneken, 2001). These nerves interconnect and subdivide several times to send nerves to the inguinal, pelvic, and hind leg muscles (Table 2.2). The more posterior nerve roots give rise to the *obturator* nerve, which goes to the ventral pelvic muscles, and the *ichiadicus* nerve, which runs medial to the ilium and then divides to form the *peroneal* and *sciatic* nerves (Figure 2.25). The anterior two nerves interconnect to provide major innervations (via *crural*, *femoral*, and *tibial* nerves) to the inguinal muscles, thigh adductors, and leg extensors (Figure 2.25).

Like those nerves forming the cervical plexus, these sacral nerves arise in dorsal and ventral pairs, with the dorsal nerves generally innervating dorsal hind limb muscles and ventral nerves generally innervating ventral hind limb muscles. However, many pelvic and hind limb muscles receive multiple innervations. Hence, it is common for muscles to have innervations from multiple spinal nerves and have branches from both dorsal and ventral nerve tracks going to the same muscle.

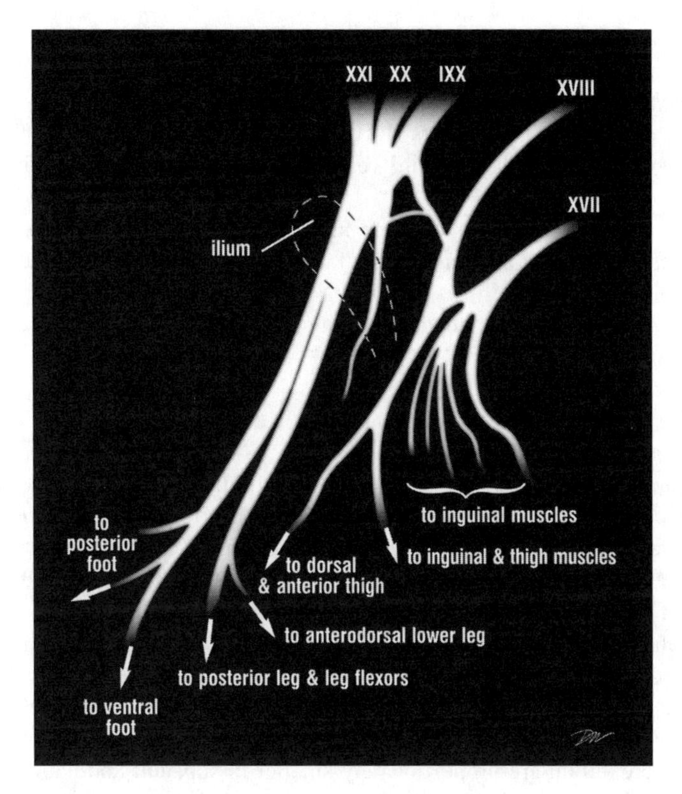

FIGURE 2.25 Sacral plexus diagram. The sacral plexus is shown in a simplified form. Its branches and major innervations in cheloniid sea turtles are traced. (Adapted from Wyneken, J. *The Anatomy of Sea Turtles*, U.S. Department of Commerce, NOAA Technical Memorandum NMFS-SEFSC-470, 2001. With permission.)

2.5.2 Central Nervous System

Traditionally, the brain (Figure 2.26) is described by three regions that arise as distinct vesicles during development: the forebrain, midbrain, and hindbrain. The following combinations of external and internal landmarks roughly identify these divisions. The forebrain extends from the nose to the posterior cerebrum. The midbrain extends from the eye to the posterior aspect of the optic lobes. The hindbrain extends from the ear to the posterior cerebellum (Romer and Parsons, 1986). These regions, in turn, are subdivided topographically and/or histochemically into principal divisions: the *telencephalon* and *diencephalon* of the forebrain, the *mesencephalon* of the midbrain, and the *metencephalon* and *myelencephalon* of the hindbrain (Hyman and Wake, 1979).

The divisions of the brain (Figures 2.26 and 2.27) and their major components are as follows:

Telencephalon: cranial nerve I (olfactory nerve), olfactory bulbs, cerebral hemispheres, lateral ventricles

Diencephalon: hypothalamus, thalamus, infundibulum and pituitary, pineal, optic chiasma, cranial nerves II–III (optic and oculomotor nerves)

Mesencephalon: optic lobes, third ventricle, cerebral aqueduct, cranial nerve IV (trochlear nerve)

Metencephalon: cerebellum, anterior part of medulla, fourth ventricle, cranial nerves V–X (trigeminal, abducens, facial, statoacoustic, glossopharyngeal, and vagus, respectively)

Myelencephalon: most of medulla, cranial nerves XI–XII (spinal accessory and hypoglossal)

Most of the cranial nerves arise ventrally and laterally. Cranial nerve innervation sites are summarized in Table 2.3.

During sea turtle development, the brain initially forms as a tube, and then undergoes considerable regionalization, torsion, and differential expansion to form the adult brain. Remnants of the nerve tube cavity persist as the lateral ventricles of the cerebral hemispheres, the third ventricle and cerebral aqueduct, and the fourth ventricle of the cerebellum and medulla (Figure 2.27).

The relative size of parts of the brain varies through ontogeny. The brain (as well as other parts of the body such as the forelimbs and head) is proportionately larger in hatchlings and juveniles than in subadults and adults (Wyneken, 2001). The olfactory nerves are proportionately longer and the cerebral hemispheres, optic lobes, and cerebellum are proportionately smaller in subadult and adult turtles.

Specific landmarks identifying the locations of the parts of the brain differ slightly among cheloniids, and even more so when compared with *Dermochelys*

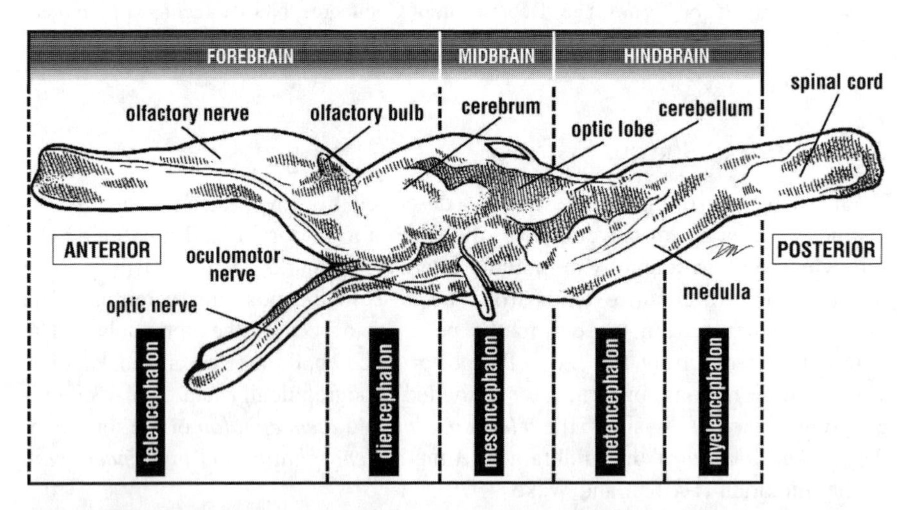

FIGURE 2.26 The sea turtle brain and its divisions. Anterior is to the left. (Adapted from Wyneken, J. *The Anatomy of Sea Turtles*, U.S. Department of Commerce, NOAA Technical Memorandum NMFS-SEFSC-470, 2001. With permission.)

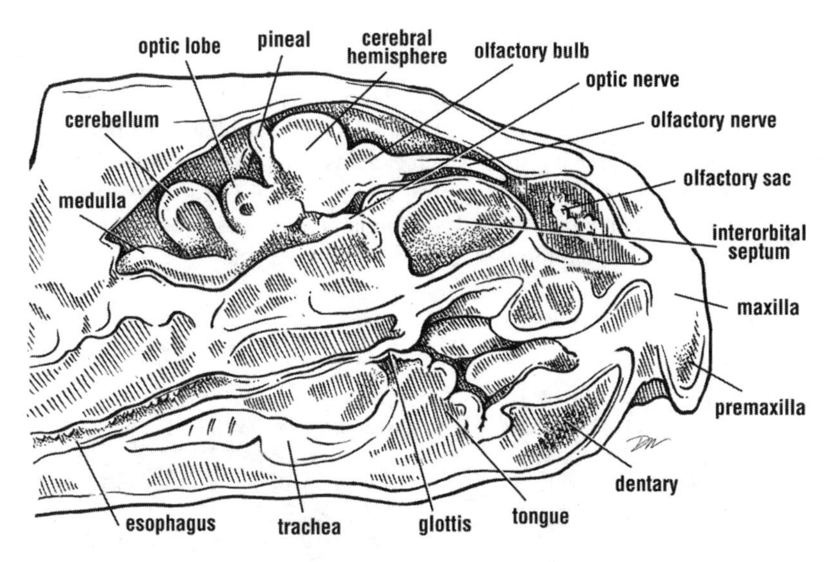

FIGURE 2.27 Cross section through a sea turtle head showing brain. This sagittal section is cut immediately lateral to the midline. The brain is bisected. (Adapted from Wyneken, J. *The Anatomy of Sea Turtles*, U.S. Department of Commerce, NOAA Technical Memorandum NMFS-SEFSC-470, 2001. With permission.)

TABLE 2.3
Cranial Nerves, Functional Components, and Innervations

Nerve Number	Nerve Name	Innervations
0	Accessory Olfactory (SS, VS)	Vasculature of olfactory sac
I	Olfactory (SS)	Olfactory cells of olfactory sac
II	Optic (SS)	Retina
III	Oculomotor (SM, VM)	Superior rectus, medial rectus, and inferior oblique eye muscles, iris, and ciliary body
IV	Trochlear (SM)	Superior oblique eye muscles
V	Trigeminal (SS, VM)	Skin of head and mouth, jaw muscles
VI	Abducens (SM)	Lateral rectus eye muscles
VII	Facial (SS, VM, VS)	Throat muscles, tongue, and taste buds
VIII	Auditory = Statoacoustic (SS)	Inner ear (balance and hearing)
IX	Glossopharyngeal (SS, VM, VS)	Taste buds, pharyngeal linings, throat muscles
X	Vagus (SS, VS, VM)	Esophagus, stomach, anterior intestines, heart
XI	Spinal Accessory (VM)	Pharynx, glottis
XII	Hypoglossal (SM)	Throat and tongue muscles

Notes: SM = somatic motor; SS = somatic sensory; VM = visceral motor; and VS = visceral sensory. Marine turtle cranial nerves have not received sufficient analysis to warrant further subdivision into general or specific functions.

(Figure 2.28). Among the cheloniids, the brain is closest to the skull roof in *L. kempii*. It is farthest from the skull roof in adult *C. caretta* and *E. imbricata* (Wyneken, 2001). Scalation patterns on the head and the position of the ear provide species-specific landmarks for some structures (Figure 2.28).The leatherback brain is housed deeply, except for the pineal organ, which extends dorsally in a cartilaginous cone-like chamber adjacent to the pink spot on the middorsal surface of the head. The pineal effects pigmentation, responds to photoperiod, and modulates biological rhythms. It may play a role in regulating reproductive cycles.

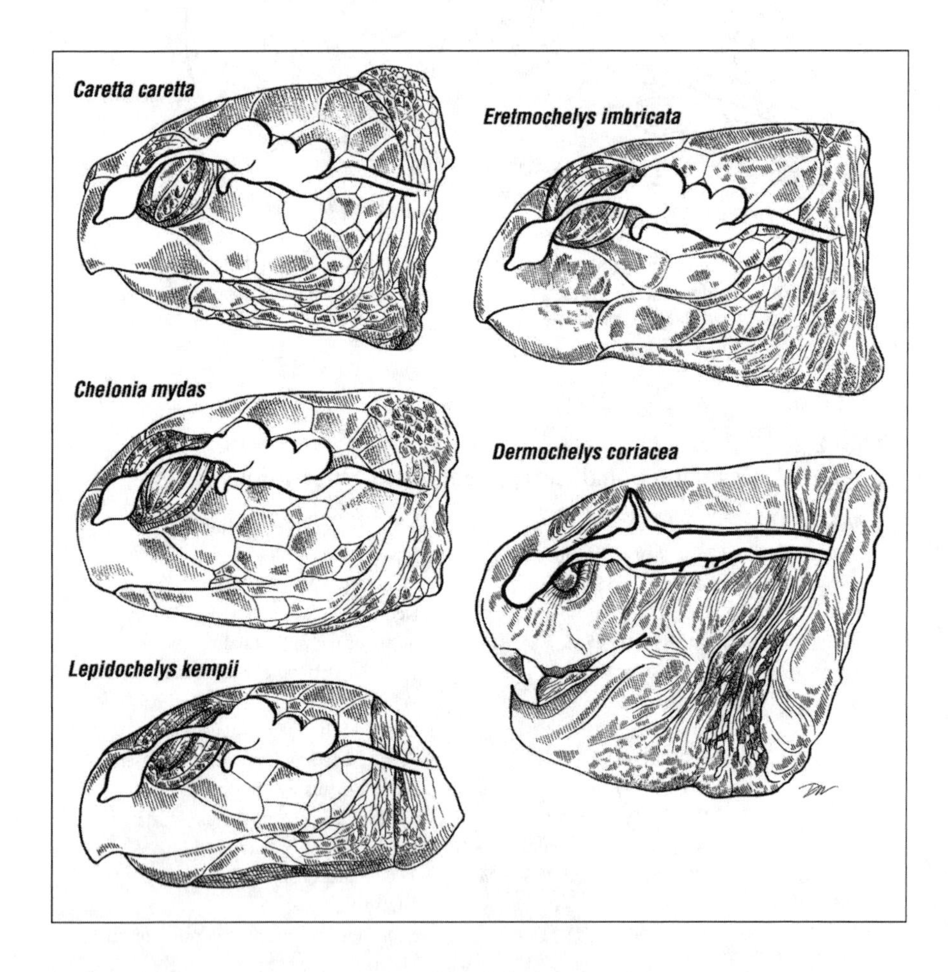

FIGURE 2.28 The position of the brain and braincase relative to head landmarks is shown for five species for which reliable data were available. The species vary in the extent of dorsoventral torsion and the depth of the brain relative to the roof of the skull. (From Wyneken, J. *The Anatomy of Sea Turtles*, U.S. Department of Commerce, NOAA Technical Memorandum NMFS-SEFSC-470, 2001. With permission.)

2.6 CONCLUSIONS

This chapter provides the foundations and references to make the gross structure and the bone–muscle–nerve complexes of marine turtles accessible and understandable. Although descriptive anatomy will always have its place in science, it is what the anatomy tells us that is particularly interesting. Suites of structures interact with one another and the turtle's environment. The structures define what is physically possible for an animal and what is not. With the marine turtles, this is illustrated most clearly by their body form. The head and limbs cannot be tucked within the protective carapace; however, *that* functional loss enhanced streamlining, an essential feature for highly migratory marine animals. Similarly, the flipper blades, formed from almost ridiculously long digits, support little weight on land. Yet they are most effective acting as wings and paddles in the water, accelerating the water around them to move the turtles along. The patterns in which the flippers move are defined by their bones, muscles, and nerves. There is still much to learn about innervations.

Other aspects of anatomy such as body form and color betray the different behavioral and ecological strategies of different species. The countershading of young green turtles, the cryptic browns of young loggerheads nestled in browns and tans of flotsam, and the mottled black and white of leatherbacks in deep water make their outlines seem to disappear. The wealth of questions brought forward by the anatomy will remain a rich resource to help us better understand the integrated biology of marine turtles.

ACKNOWLEDGMENTS

Access to specimens was provided by R. Boettcher, M. Bresette, L. Csuzdi, N. DiMarco, S. Epperly, K. Fick, A. Foley, J. Foote, C. Harms, K. Hart, H. Horta, C. Johnson, C. Manire, J. Braun McNeill, N. Mette, J. Parsons, R. Prescott, P.C.H. Pritchard, T. Redlow, K. Rusenko, C. Ryder, K. Singel, M. Snover, T. Sposato, W. Teas, M. Walsh, P. Wells, L. Wood, the Cayman Turtle Farm, Chelonian Research Institute, Florida Fish and Wildlife Conservation Commission, Gumbo Limbo Environmental Center, Harbor Branch Oceanographic Institution, The Marinelife Center of Juno Beach, Mote Marine Laboratory, National Marine Fisheries Service — Beaufort Laboratory, National Marine Fisheries Service — Miami Laboratory, North Carolina Wildlife Commission, Philadelphia Academy of Sciences, and the U.S. Fish and Wildlife Service. Dr. F. Steinberg and University MRI, Inc. provided the machine time and technical expertise (D. Wilke) for the MRI and CT images. T.H. Frazzetta, P.C.H. Pritchard, M. Salmon, and K. Stewart provided thoughtful discussions. M. Salmon improved the quality of the manuscript and showed his endurance by reading through several drafts. Illustrations were created by D. Witherington. The research for this chapter was funded, in part, by grants or contracts to the author by the NMFS, the National Science Foundation (OISE 0124271), an FAU Research Initiation Award, and the Nelligan Sea Turtle Fund, which enabled components of this chapter to be written and/or refined. Personal funds were also used to complete this work.

REFERENCES

Ashley, L.M., *Laboratory Manual of the Turtle*, WCB/McGraw-Hill, Dubuque, IA, 1962.

Bjorndal, K.A., Foraging ecology and nutrition of sea turtles, in *The Biology of Sea Turtles*, Lutz, P.L. and Musick, J.A., Eds., CRC Press, Boca Raton, FL, 1997, pp. 199–231.

Cott, H.B., *Adaptive Coloration in Animals*, Methuen, London, 1966.

Deraniyagala, P.E.P., The tetrapod reptiles of Ceylon, *Ceylon J. Sci.*, 1, 1–242, 1939.

Fürbringer, M., Zur vergleichenden anatomie der schultermuskeln (The comparative anatomy of shoulder muscles), *Jena Zschr.* 8, 175–280, 1874.

Gadow, H., A contribution to the myology of the hind extremity in reptiles, *Morphologisches Jahrbuch*, 7, 329–466, 1882.

Gilbert, S.F., Loredo, G.A., Brukman, A., and Burke, A.C., Morphogenesis of the turtle shell: the development of a novel structure in tetrapod evolution, *Evol. Devel.*, 3, 47–58, 2001.

Grasse, P.-P., *Reptiles*, Vol. 14, *Traité de Zoologie; Anatomie Systématique, Biologie*, Masson, Paris, 1948, 836 pp.

Hyman, L.H. and Wake, M.H., *Hyman's Comparative Vertebrate Anatomy*, University of Chicago Press, Chicago, 1979.

Kardong, K.V., *Vertebrates: Comparative Anatomy, Function, Evolution*, McGraw-Hill, Boston, 2002.

Meylan, A., Spongivory in hawksbill turtles: a diet of glass. *Science*, 239, 393–395, 1988.

Owen, D.F., *Camouflage and Mimicry*, University of Chicago Press, Chicago, 1982.

Pritchard, P.C.H., *Encyclopedia of Turtles*, T.F.H., Neptune, NJ, 1979.

Pritchard, P.C.H., Evolution, phylogeny, and current status, in *The Biology of Sea Turtles*, Lutz, P.L. and Musick, J.A., Eds., CRC Press, Boca Raton, FL, 1997, pp. 1–28.

Pritchard, P.C.H. and Mortimer, J., Taxonomy, external morphology, and species identification, in *Research and Management Techniques for the Conservation of Sea Turtles*, Eckert, K.L. Bjorndal, K.A., Abreu-Grobois, F.A., and Donnelly, M., Eds., IUCN/SSC Marine Turtle Specialist Group Publication No. 4, 1999, pp. 21–38, Washington, D.C.

Rhodin, A.G.J., Comparative chondro-osseous development and growth of marine turtles, *Copeia*, 3, 752–771, 1985.

Rhodin, A.G.J., Ogden, J.A., and Conlogue, G.J., Chondro-osseous morphology of *Dermochelys coriacea*, a marine reptile with mammalian skeletal features, *Nature*, 290, 244–246, 1981.

Romer, A.S. and Parsons, T.S., *The Vertebrate Body*, Saunders College, Philadelphia, 1986.

Ruckdeschel, C., Shoop, C.R., and Zug, G.R., *Sea Turtles of the Georgia Coast*, Cumberland Island Museum, St. Marys, GA, 2000.

Schumacher, G.-H., Head muscles and hyolaryngeal skeleton of turtles and crocodilians, in *Biology of the Reptilia*, Gans, C. and Parsons, T.S., Eds., Academic Press, New York, 1973, pp. 101–199.

Walker, W.F., Jr., The locomotor apparatus of testudines, in *Biology of the Reptilia*, Gans, C. and Parsons, T.S., Eds., Academic Press, New York, 1973, pp. 1–100.

Witzell, W.N., Synopsis of Biological Data on the Hawksbill Turtle, *Eretmochelys imbricata* (Linneaeus, 1766), FAO Fisheries Synopsis, No. 137, 1983.

Wyneken, J., Comparative and functional considerations of locomotion in turtles, Ph.D. dissertation, Urbana-Champaign, University of Illinois, 1988.

Wyneken, J., *The Anatomy of Sea Turtles*, U.S. Department of Commerce, NOAA Technical Memorandum, NMFS-SEFSC-470, 2001, 172 pp. Miami, FL.

Wyneken, J., Sea turtle locomotion: mechanics, behavior, and energetics, in *The Biology of Sea Turtles*, Lutz, P.L. and Musick, J.A., Eds., CRC Press, Boca Raton, FL, 1997, pp. 165–198.

Youngkin, D., A long-term dietary analysis of loggerhead sea turtles (*Caretta caretta*) based on strandings from Cumberland Island, Georgia, Master's thesis, Florida Atlantic University, Boca Raton, FL, 2001.

Zangerl, R., Hendrickson, L.P., and Hendrickson, J.R., A redescription of the Australian flatback sea turtles, *Natator depressus*, *Bishop Museum Bull. Zool.*, 1, 1–69, 1988.

Zangerl, R., The homology of the shell elements in turtles, *J. Morphol.*, 65(3), 383–409, 1939.

3 Sensory Biology of Sea Turtles

Soraya Moein Bartol and John A. Musick

CONTENTS

0-8493-1123-3/03/$0.00+$1.50

3.1 INTRODUCTION

The study of sensory biology in sea turtles is still in its infancy. Even the basic morphology of the eye, ear, and nose of sea turtles has been described in detail in only one or two species. The same may be said for electrophysiological and behavioral studies of sea turtles' sensory systems. The ontogenetic and interspecific difference in the sensory biology of sea turtles has been little studied and the sensory biology of the leatherback (*Dermochelys coriacea*), a species whose ecology is greatly different from the cheloniids, is virtually unknown. The present chapter will focus on the current state of knowledge of the sensory biology of vision, hearing, and olfaction in sea turtles.

3.2 VISION

3.2.1 Morphology and Anatomy of the Eye

3.2.1.1 Main Structures of the Eye

The anatomy of the sea turtle eye appears to be typical of that found in all vertebrates (Granda, 1979; Walls, 1942). The eyeball is filled with two ocular fluids, aqueous and vitreous humors, and is organized into three layers: (1) the outermost layer, consisting of the sclera and cornea; (2) the middle layer, which includes the choroid, ciliary body, and iris; and (3) the inner layer, or the retina. The sclera is inelastic and is responsible for the eyeball's static shape, whereas the aqueous humor keeps this fibrous layer distended. The anterior portion of the sclera, the cornea, is transparent and responsible for much of the refraction of light in air, yet is virtually transparent in water. The choroid of the middle layer is highly pigmented and vascularized; the pigmentation deflects stray light from entering the eye and prevents internal reflections. The inner layer of the eyeball, the retina, contains the visual cells (rod and cone photoreceptor cells) and ganglion cells, and is continuous with the optic nerve (Walls, 1942; Copenhaver, 1964; Granda, 1979; Ali and Klyne, 1985; Bartol, 1999).

The lens of the green sea turtle (*Chelonia mydas*) is nearly spherical and rigid (Ehrenfeld and Koch, 1967; Granda, 1979; Walls, 1942), and appears to be quite different from that of freshwater turtles, which have developed an advanced means of accommodation through the manipulation of an extremely pliable lens. For sea turtles, however, ciliary processes do not reach the lens and the *ringwulst* is weakly developed, and thus active accommodation does not appear to be possible (Ehrenfeld and Koch, 1967). However, this type of spherical lens is ideal for underwater vision. In the absence of corneal refraction while underwater, the refractive index of the cornea is nearly identical to that of seawater, and the lens is the only structure responsible for the refraction of incoming light. The spherical lens has a high refractive index, which compensates for the lack of corneal refraction (Sivak, 1985; Fernald, 1990).

3.2.1.2 Cells of the Retina

The vertical organization of the retina has been examined in the juvenile loggerhead sea turtle (*Caretta caretta;* Bartol and Musick, 2001) (Figure 3.1). The layers of the

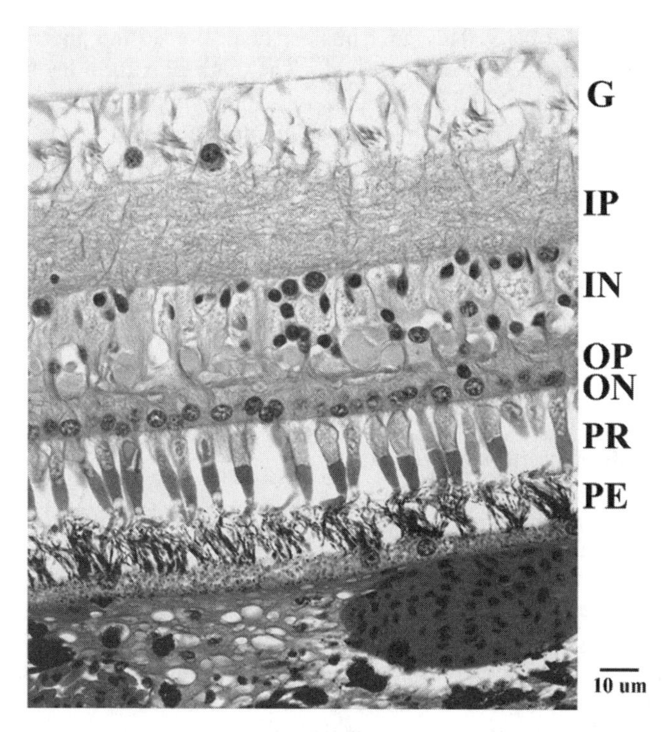

FIGURE 3.1 Light micrograph of the retina of a juvenile loggerhead sea turtle (*C. caretta*). Abbreviations: G = ganglion layer; IP = inner plexiform layer; IN = inner nuclear layer; OP = outer plexiform layer; ON = outer nuclear layer; PR = photoreceptor layer; PE = pigment epithelium. Scale bar equals 10 μm. (From Bartol, S.M. and Musick, J.A., Morphology and topographical organization of the retina of juvenile loggerhead sea turtles (*Caretta caretta*), *Copeia*, 3, 718, 2001. With permission.)

retina are consistent with the generalized vertebrate plan and consist of seven layers (from the center of the eye out to the edge): ganglion layer, inner plexiform layer, inner nuclear layer, outer plexiform layer, outer nuclear layer, photoreceptor layer, and the pigment epithelium. Bartol and Musick (2001) focused mainly on the photoreceptor layer, which contains the stimulus receptors, and found that it is duplex in nature, consisting of both rod and cone photoreceptors. These two types of photoreceptor cells are similar in diameter and height, yet the rod does not have an oil droplet above the ellipsoid element, and the outer segment of the rod photoreceptor is longer and more cylindrical than that of the cone photoreceptor. Homogeneity of photoreceptor cell types is unusual; typically rods are much longer and narrower than cones in vertebrate retinas. However, this same homogeneity of cells can be found in the retina of the common snapping turtle (*Chelydra serpentina*; Walls, 1942).

In the loggerhead, Bartol and Musick (2001) found that the pigment epithelium, the outermost layer of the retina, is firmly connected to the choroid, and contains heavy pigment-laden processes that intertwine with the outer segments of the photoreceptor cells. The outer nuclear layer houses the photoreceptor cell nuclei and is generally only one cell wide. The outer plexiform layer is homogenous, but in Bartol and Musick's preparations, the synaptic connections between the nuclear layers could not be identified. The inner nuclear layer is composed of the nuclei of bipolar, amacrine, and horizontal cells, although these cells were not differentiated in this study. The inner plexiform layer is similar to the outer plexiform layer and is composed of synaptic connections between the inner nuclear layer and ganglion layer. Finally, the innermost layer, the ganglion cell layer, is relatively thick (23% of the overall width of the retina) and is composed solely of the ganglion cells and their axons (Bartol and Musick, 2001).

3.2.2 Sensitivity to Color

3.2.2.1 Photopigments and Oil Droplets

The spectral sensitivity of sea turtles has been investigated using morphological, electrophysiological, and behavioral methods. Liebman and Granda (1971) examined the visual pigments associated with photoreceptor cells of the red-eared freshwater turtle (*Pseudemys scripta elegans*) and green turtle (*C. mydas*). Microspectrophotometric measurements were performed on preparations of these cells to determine the absorption spectra of these light-absorbing visual pigments. Both species have a duplex retina containing both rod and cone photoreceptor cells. For the green turtle, the rod photosensitive pigments absorbed light maximally at 500–505 nm. This retinal pigment was indistinguishable from the rhodopsin identified in frog preparations. Three photopigments were found associated with cone photoreceptors for *C. mydas*. The most common pigment, identified as iodopsin, absorbed light maximally at 562 nm. The two other cone visual pigments identified absorbed light maximally at 440 and 502 nm (Figure 3.2). Note that one cone photoreceptor visual pigment was identical to that of the rod visual pigment. The authors hypothesized that the cone that absorbs at 502 nm is actually the accessory cone of a double cone pair. The double cones of *C. mydas* have been found to have a principal receptor (full-sized cone with oil droplet) and a secondary receptor (the non-oil droplet member) (Walls, 1942; Liebman and Granda, 1971). Liebman and Granda (1971) suggest that the accessory cone actually contains the rhodopsin pigment of the rod photoreceptor. The freshwater turtle (*P. scripta elegans*) examined in this study contained visual pigments that absorb longer wavelengths than those found in *C. mydas*; rods absorbed maximally at 518 nm and cones contained photopigments that absorbed 450, 518, and 620 nm maximally (Figure 3.2). The authors concluded that the light-absorbing visual pigments in both the freshwater and marine turtle were suitable for the environments in which the animals reside (seawater transmits shorter wavelengths than freshwater) (Liebman and Granda, 1971; Granda, 1979).

3.2.2.2 Electrophysiology

The spectral sensitivity of *C. mydas* has also been investigated through the collection of electroretinograms (ERGs) from dark-adapted eyes (Granda and O'Shea, 1972).

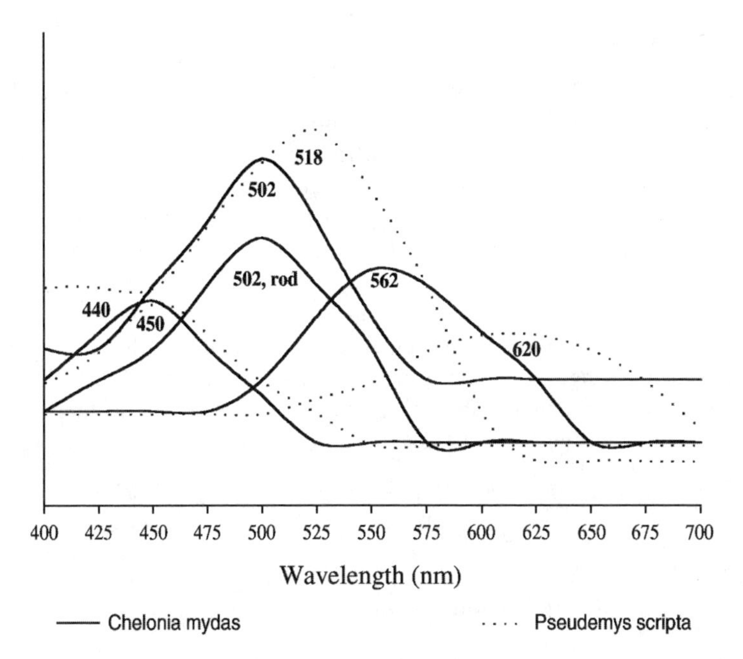

FIGURE 3.2 Visual pigment measurements, using microspectrophotometric techniques, of rod and cone photoreceptors for both *C. mydas* (solid lines) and *P. scripta* (dotted lines). (Data redrawn from Liebman, P.A. and Granda, A.M., Microspectrophotometric measurements of visual pigments in two species of turtle, *Vision Res.*, 11, 105, 1971.)

An ERG is a recording of rapid action potentials between the cornea and retina when the eye is stimulated, and is a robust measurement of early retinal stages in the visual pathway (preganglion cell responses) (Davson, 1972; Riggs and Wooten, 1972; Ali and Klyne, 1985). Granda and O'Shea (1972) found the spectral sensitivity for *C. mydas* to peak at 520 nm, with secondary peaks at 450–460 and 600 nm. The spectral sensitivities recorded using these methods were longer (except for the shortest wavelength) than those found through light microspectrophotometric measurements (440, 502, and 562 nm; Leibman and Granda, 1971), and the discrepancy of wavelength measurements is attributed to the interaction of the visual pigments and the cone oil droplets (Granda and O'Shea, 1972). For cone photoreceptors, light must first pass through oil droplets before it reaches and excites the photopigments. In *C. mydas*, the cone oil droplets are saturated oil globules that can be clear, yellow, or orange. The orange and yellow droplets are optically dense and can act as filters, shifting the wavelength that excites the photopigments (Granda and O'Shea, 1972; Granda and Dvorak, 1977; Peterson, 1992). Specific colored oil droplets appear to be paired with a specific photopigment: the clear oil droplet appears to be associated with the 440 nm photopigment (no shift in absorbed spectral sensitivity), the yellow oil droplet with the 502 nm photopigment (shifting the absorbed spectral sensitivity to 520 nm), and the orange oil droplet with the 562 nm photopigment (shifting the absorbed spectral sensitivity to 600 nm) (Granda and O'Shea, 1972; Peterson, 1992).

3.2.2.3 Behavior

Behavior studies on sea turtles performed in the aqueous setting are limited because of the difficulties associated with training turtles to respond to specific stimuli. Fehring (1972), however, used the sea turtle's ability to detect colors to develop a hue discrimination behavioral study. Broadband hues were used (deep blue, magenta, and red-orange) to determine whether loggerhead sea turtles (*C. caretta*) could be trained to use hue in search for food. The research study was not designed to test for an inherent hue preference, but rather was designed to test whether the turtles could be trained to pick one hue over another. Each animal was given a choice of two hues and, through training, was taught that only one of these hues would provide a food reward. Fehring found that these animals were easily trained, with relatively few errors, and thus concluded that sea turtles are able to use their ability to distinguish colors to find food (1972).

3.2.3 Visual Acuity

3.2.3.1 Topographical Organization of the Retina

Retinal morphology and topography research can describe the potential resolving power of an eye under differing illumination conditions. Within the retina itself, two factors can affect the ability of an animal to resolve items under varying light conditions: convergence of photoreceptor cells onto ganglion cells, and the topographical organization of photoreceptor cells along the surface of the retina (Walls, 1942; Davson, 1972; Ali and Klyne, 1985). Within the photoreceptor layer, the sea turtle has two types of cells: rods and cones. For most vertebrates, and sea turtles are no exception, the general function of the rod photoreceptor is to maximize sensitivity of the eye to dim stimuli, whereas the general function of the cone photoreceptor is to resolve details of a visual object (Copenhaver, 1964; Davson, 1972; Stell, 1972). Convergence of photoreceptor cells upon ganglion cells, otherwise termed summation, can prove to be both beneficial and disadvantageous. When the stimulus is weak (under dim light conditions), more than one rod photoreceptor cell converging onto a single ganglion cell will subsequently increase the strength of the neural signal, allowing the stimulus to be recognized. However, when summation occurs between cone photoreceptor cells and ganglion cells, the information relayed to the optic tectum is not characteristic of one cone, but rather a summation of many, resulting in reduced spatial resolution (Walls, 1942; Davson, 1972).

Topographical distribution of cone photoreceptor cells also can be an indication of the resolution ability of an animal. The retinas of many vertebrates have regions of higher cell densities, often called an area centralis or visual streak, which provides a region of increased visual acuity. The area centralis can vary in shape and location along the retina among species, and this variation is often indicative of behavior and life history attributes of the animal (Walls, 1942; Brown, 1969; Heuter, 1991).

Both summation and regional density of photoreceptor cells have been examined in both hatchling and juvenile sea turtles (Oliver et al., 2000; Bartol and Musick, 2001). Oliver et al. (2000) examined the ganglion cell densities of three species of sea turtle hatchlings: greens (*C. mydas*), loggerheads (*C. caretta*), and leatherbacks (*D. coriacea*). From plots of contour maps of ganglion cells, visual streaks were found for all three species; however, the streaks varied in shape. *Caretta mydas* was found to have a narrow and long streak, with a much higher cell concentration within the streak as opposed to areas outside the streak. Of the three turtles, *C. mydas* had the most characteristically horizontal streak. *Caretta caretta* had a wider streak dorsoventrally, with lower density counts than the green sea turtle. The retina of *D. coriacea* contained a distinct rounded area temporalis (a site of high cell counts) as well as a horizontal streak. Cell counts were the highest for the retina within this area temporalis. The authors attribute the differences among species to the environment that these hatchlings occupy. For example, as hatchlings, *C. mydas* may be found in clear water, feeding during the day as omnivores beneath the flat ocean surface, whereas *C. caretta* is typically found within sargassum mats, feeding in an environment with a less defined horizon. This behavior of feeding beneath a defined, flat surface helps explain why green sea turtles have a stronger horizontal streak than other sea turtles. *Dermochelys coriacea* hatchlings feed on gelatinous prey in the open ocean, an environment where an area temporalis would be more advantageous than a horizontal streak (Oliver et al., 2000).

Bartol and Musick (2001) examined the vertical organization of the main features of the retina as well as the spatial variation of the photoreceptor cells of large juvenile loggerhead sea turtles (*C. caretta*). On the basis of the properties of the neural layers, the vertical organization of the retina indicated a low degree of summation. In animals with a low summation level, the inner nuclear layer (composed of bipolar cells, horizontal cells, and amacrine cells) and the ganglion layer are thick relative to the rest of the retina, indicating a high number of neurons corresponding to each photoreceptor cell (Walls, 1942). In juvenile loggerheads, these two layers (out of the seven overall layers) comprised approximately 37% of the total retina (Bartol and Musick, 2001; see Figure 3.1). Bartol and Musick (2001) also examined the topography of the retina by plotting the counts of cone and rod photoreceptor cells and ganglion cells (Figure 3.3). Both cone photoreceptors and ganglion cells progressed from high to low density in a stair-step fashion from the back to the front of the eye. Rod photoreceptors, however, were more likely to maintain a constant density throughout the back half of the eye, rapidly decreasing in number near the cornea. Dorsal–ventral differences were also observed when the cell counts were plotted on a three-dimensional sphere. A horizontal streak of ganglion cells and cone photoreceptor cells in the dorsal hemisphere of the eye indicated a region of decreased summation and thus increased acuity. Rods, however, were found in lower numbers and ubiquitously throughout the two hemispheres, resulting in a constant sensitivity to low light situations. This regionalization of cells was hypothesized to aid the juvenile loggerhead in finding benthic slow-moving prey in their shallow water habitat (Bartol and Musick, 2001).

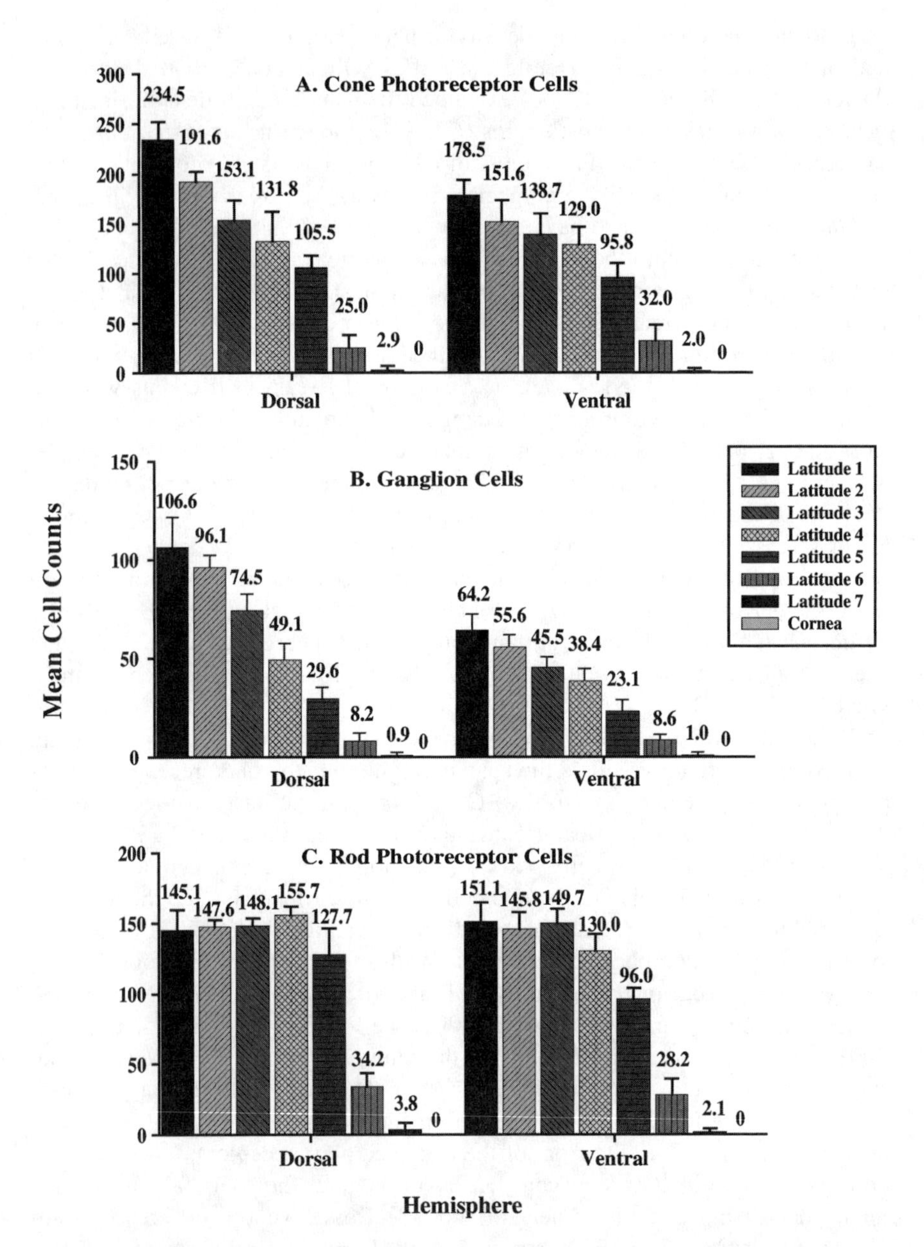

FIGURE 3.3 Mean cell counts, collected from the retinas of juvenile loggerhead sea turtles (*C. caretta*), for the eight latitudes of the eye in both the ventral and dorsal hemispheres. All error bars denote + 1 SD. (A) Cone photoreceptor cells. (B) Ganglion cells. (C) Rod photoreceptor cells. (From Bartol, S.M. and Musick, J.A., Morphology and topographical organization of the retina of juvenile loggerhead sea turtles (*Caretta caretta*), *Copeia*, 3, 718, 2001. With permission.)

3.2.3.2 Electrophysiology

Electrophysiological techniques have also been employed to investigate the visual acuity thresholds of sea turtles (Bartol et al., 2002). Electrical responses recorded from the visual system provide an objective measure of a variety of visual phenomena, including the dependence of a response on the character of the stimulus (Riggs and Wooten, 1972; Bullock et al., 1991). In the Bartol et al. (2002) study, the technique of visual evoked potentials (VEPs) was used. VEPs are compound field potentials of any neural tissue in the visual pathway and can be obtained from a subject animal by the use of surface electrodes placed on the head directly above the optic nerve and corresponding optic tectum. In this study, the researchers used a modified goggle filled with seawater over the stimulated eye. This apparatus allowed for the testing of underwater acuity. The stimuli were black and white striped patterns of decreasing size, yet always of equal brightness. One peak in the VEP recordings was found by the researchers to be present in all suprathreshold recordings, showing a dependence of peak amplitude on stimulus stripe size (Figure 3.4). From this peak, Bartol et al. (2002) were able to identify an acuity threshold level of 0.187 (visual angle = 5.34 min of arc) when data from all six turtles were pooled. This level of acuity would permit loggerheads to discern prey, such as horseshoe and blue crabs, as well as large predators, and is comparable to many species of marine fishes. Interestingly, these researchers were unable to collect any discernible VEP response when the turtles were tested with their eyes in air (i.e., without the water-filled goggle), suggesting that the sea turtle eye operates much differently in the two media (Bartol et al., 2002) (Figure 3.4).

3.2.3.3 Behavior

Psychophysical methods were used to investigate the visual acuity of juvenile loggerhead sea turtles (*C. caretta*) in the aquatic medium (Bartol, 1999). An operant conditioning method was developed to train juvenile loggerheads in a tank environment to identify a striped stimulus. The tank was set up with two response keys: one was located below a striped panel and the other below a gray panel. Turtles were trained by receiving a food reward only when the response key was chosen below the striped panel. Once training of these turtles was achieved, the stimulus was reduced in size until the turtle could no longer respond correctly. These turtles were found to be highly appropriate subject animals for an in-tank behavior study, and retained their training over time. From these trials, Bartol (1999) found the behavioral acuity threshold for juvenile loggerheads to be approximately 0.078 (visual angle of 12.89 min of arc), comparable to that found in the electrophysiology study (Bartol et al., 2001) and similar to the visual acuity of other benthic shallow-water marine species.

3.2.4 Visual Behavior on Land

The visual behavior of hatchling and nesting female sea turtles as they orient toward water while on land also has been studied. Vision has been identified in numerous articles as the primary sense used in sea-finding behavior of both hatchlings and

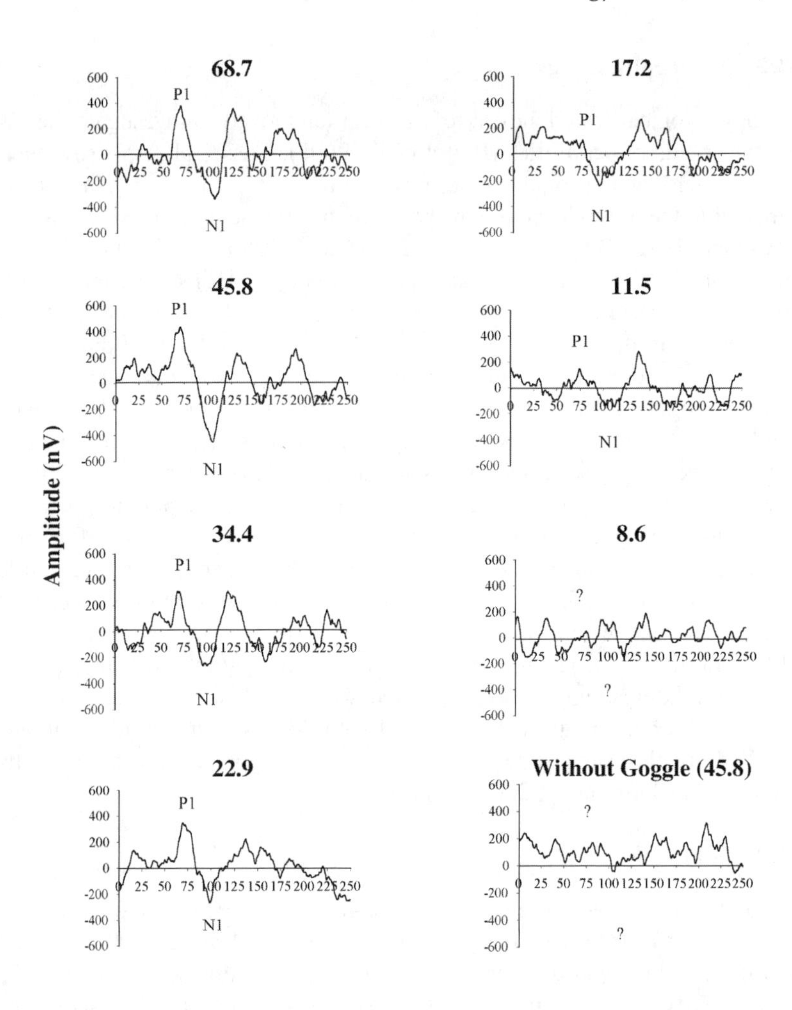

FIGURE 3.4 Visual evoked potential recordings for a session with one loggerhead sea turtle (*C. caretta*) using seven stimuli sizes ranging from 68.7 to 8.6 min of arc, visual angle and the recording for a trial without the goggle (in-air experiment) for 45.8 min of arc, visual angle. Notice that the amplitude difference between P1 and N1 decreases with a decrease in stripe size, until it can no longer be identified. Furthermore, for trials without the goggle, neither peak is identifiable, nor could the amplitude differences be measured. Each wave is an average of 500 responses; time zero is the start of stimulation. (Based on Bartol, S.M., Musick, J.A., and Ochs, A.L., Visual acuity thresholds of juvenile loggerhead sea turtles (*Caretta caretta*): an electrophysiological approach, *J. Comp. Physiol. A.*, 187, 953, 2002. With permission.)

adults. The type of visual stimuli used by sea turtles (whether shapes, colors, or brightness cues) has been the subject of many research articles (Ehrenfeld and Carr, 1967; Ehrenfeld, 1968; Mrosovsky and Shettleworth, 1968; Witherington and Bjorndal, 1991; Salmon and Wyneken, 1990; 1994). In some of the earliest studies, blindfolds were placed on the turtles to determine whether they could orient without visual input. Bilaterally blindfolded turtles were unable to find the sea at all (Daniel and Smith, 1947; Carr and Ogren, 1960; van Rhijn, 1979), and unilaterally blindfolded sea turtles circled toward the uncovered eye, suggesting that the sea turtle finds the sea using tropotactic behavior (comparing intensities in both eyes and moving accordingly) (Ehrenfeld, 1968; Mrosovsky and Shettleworth, 1968; Mrosovsky, 1972; Mrosovsky et al., 1979). These hatchling sea turtles are attracted to, and move toward, the brightest direction.

Shape identification, or the ability of a sea turtle to visualize objects on the beach, has also been investigated in the context of sea-finding behavior. The reaction by hatchlings to a horizon obstructed by objects found on or surrounding the beach has been documented in many studies (Parker, 1922; Limpus, 1971; Salmon et al., 1992). Salmon and Wyneken (1994) found that sea-finding for sea turtles depends on three rules when orienting toward the sea: (1) sea turtles move toward brighter regions, (2) sea turtles move away from high beach silhouettes (such as foliage or sand dunes), and (3) when these two cues are inconsistent, sea turtles move in relation to elevation (beach silhouettes), not brightness. Ehrenfeld and Carr (1967) tested the extent to which green sea turtles (*C. mydas*) visualize objects on the beach when making decisions about which direction to crawl. Adult turtles were fitted with an eye-covering apparatus that was designed to hold wax paper filters. The wax paper filter acted to soften sharp images by scattering light. The results showed that if the turtles were allowed to acclimate to the wax paper filter for 10 min, then their sea-finding ability was not hampered by a diffuse vision. The result of this research implies that *C. mydas* adults are not using sharp visual acuity to find water, but rather diffuse beach silhouettes.

Brightness level, a known stimulus to which sea turtles respond, is often a result of the wavelength characteristics of that stimulus. Therefore, wavelength preferences of turtles on the beach have also been investigated as a tool for finding the sea after hatching or a nesting event. Ehrenfeld and Carr (1967) found that adult female green sea turtles (*C. mydas*) wearing colored filters (red, green, and blue) were still able to find water better than those turtles that were blindfolded. However, some colors worked better than others. For example, sea turtles wearing a green filter performed as well as the control group (nonblindfolded turtles). However, turtles wearing the red filter showed a sharp decrease in performance, indicating a possible upper limit to spectral sensitivity.

Mrosovsky and Shettleworth (1968) found that green hatchling sea turtles had a preference for short wavelengths, even if the intensity of the longer wavelengths was stronger. Mrosovsky (1972) found that red wavelengths had very little effect on green sea turtles except when very bright, but turtles were attracted to blue light even at low energy levels. These studies indicate that green turtles have a preference for shorter wavelength light. Witherington and Bjorndal (1991) tested loggerhead (*C. caretta*) and green (*C. mydas*) sea turtle hatchlings for color preference in air using a V-maze, two-choice design. When placed in the maze, both species chose 360 (near-ultraviolet), 400 (violet), and 500 (blue-green) nm wavelengths over a constant light source, but did not choose 600 (yellow-orange) or 700 (red) nm wavelengths. Loggerheads actually moved away from 560 (green-yellow), 580 (yellow), and 600 (yellow-orange) nm wavelengths when the choice was color vs. a darkened window, but green sea turtles did not. These results indicate that loggerhead sea turtles are capable of seeing at least from 360 to 700 nm, whereas green sea turtles see wavelengths from 360 to 500 nm. Furthermore, loggerheads appear to be xanthophobic (averse to yellow-orange light) (Witherington and Bjorndal, 1991).

3.2.5 Concluding Remarks

Researchers are just beginning to develop a complete picture of the visual niche of sea turtles. The mechanisms by which sea turtles, as both hatchlings and adult females, return to the sea after hatching or nesting on land involve visual cues to find the ocean, though these cues seem to be restricted to diffuse images, and brightness levels and/or contrasts. This information has been invaluable in both defining the ecology of sea turtles on land and providing guidelines for the protection of these animals from anthropogenic light sources. The role of visual stimuli underwater for sea turtles also has been recently elucidated. From morphological studies, the roles of visual photoreceptor cells are being defined for both color vision and visual acuity. Retinal morphology studies may reveal the maximum capability of a visual system; certain cells and structures must be present for the retina of a typical vertebrate eye to process visual stimulation. Consequently, predictions have been made from identifying cell characteristics, describing pathways from one cell layer to the next, and mapping regions within the retina of high- and low-density cell counts. Electrophysiological studies on both color vision and visual acuity have supported the morphological work. Sea turtles have color vision, primarily in the shorter wavelengths (450–620 nm), and have the visual acuity to discern relatively small objects within the marine environment. Behavior studies further support these conclusions.

3.3 HEARING

3.3.1 Morphology and Anatomy of the Ear

3.3.1.1 Main Structures of the Middle and Inner Ear

Sea turtles do not have an external ear; in fact, the tympanum is simply a continuation of the facial tissue. The tympanum is posterior to the midline of the skull and is distinguishable only by palpation of the area. Beneath the tympanum is a thick layer of subtympanal fat, a feature that distinguishes sea turtles from both terrestrial and semiaquatic turtles. The middle ear cavity lies posterior to the tympanum; the eustachian tube connects the middle ear with the throat near the posteroventral edge of the middle ear cavity (Lenhardt et al., 1985; Wever, 1978) (Figure 3.5).

The ossicular mechanism of the sea turtle ear consists of two elements, the columella and the extracolumella. The extracolumella is a cartilaginous, mushroom-shaped disk under the tympanic membrane, which is attached by its posterior end firmly to the columella. The columella, a long rod with the majority of the mass concentrated at each end, travels through a bone channel, and expands within the

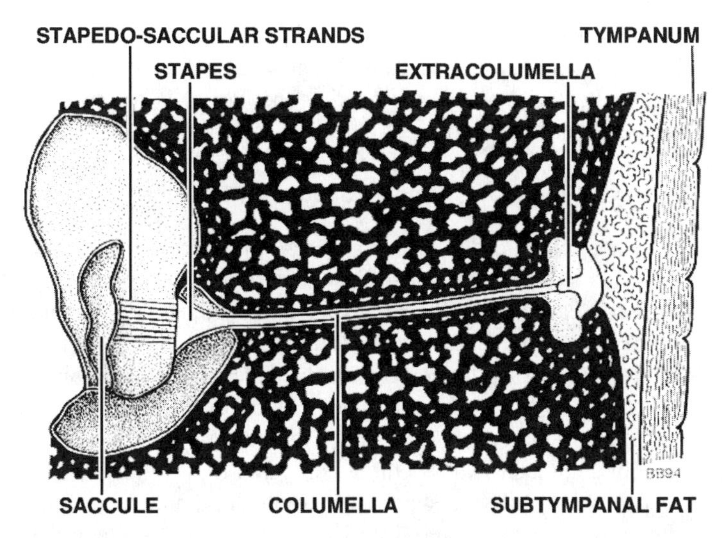

FIGURE 3.5 Schematic of middle ear anatomy of the juvenile loggerhead sea turtle. (From Moein, S.E., Auditory evoked potentials of the loggerhead sea turtle (*C. caretta*), master's thesis, College of William and Mary, Virginia Institute of Marine Science, Gloucester, VA, 1994. With permission.)

oval window to form a funnel-shaped stapes. The columella is free to move only longitudinally within this channel, so when the tympanum is depressed directly above the middle of the extracolumella, the columella moves readily in and out of the oval window, without any flexion of the columella. The stapes and oval window are connected to the saccular wall by fibrous strands, a unique feature of turtles. It is thought that these stapedo-saccular strands relay vibrational energy of the stapes to the saccule (Wever and Vernon, 1956; Lenhardt et al., 1985) (Figure 3.5).

We have not found any research on the inner ear of the sea turtle, but we can speculate from research performed on other species of turtles. The cochlea of turtles is thought to employ a reentrant fluid circuit for pressure relief (unlike most lizards, birds, and mammals, which release fluid pressure by means of protruding the round window membrane) (Turner, 1978; Wever, 1978). When the inward movements of the stapes displace the fluids of the inner ear, these fluids circle around the cochlear pathway, past the round window, back to the lateral side of the stapes (the direction of the fluid is reversed with an outward movement of the stapes). A limitation of this circular fluid motion is the added volume, from the displaced fluid, found at the site of the stapes that must be moved by alternating sound pressure. This fluid circuit may help describe the frequency range for turtles. Under these conditions of mass loading, the amount of sound pressure needed to move the columella increases with an increase in frequency, resulting in turtles' being insensitive to high frequencies. Loading does not present a problem at low frequencies, and sea turtles are thought to hear primarily in the low frequency range (Wever and Vernon, 1956; Turner, 1978; Wever, 1978).

The auditory ending, or sensory organ, within the inner ear of the reptilian cochlea is the basilar papilla (also known as the basilar membrane). The basilar membrane is a thin partition in the circular fluid pathway, which contains two basic cell types: hair cells and supporting cells. In most reptiles, and presumably in sea turtles as well, the tectorial membrane overlies the hair cells of the basilar papilla (Wever, 1978; Lewis et al., 1985).

3.3.1.2 Water Conduction vs. Bone Conduction Hearing

The functional morphology of the sea turtle ear is still under some debate. Lenhardt et al. (1985) postulated that the sea turtle ear is a poor aerial receptor. For the terrestrial vertebrate ear, the middle ear acts as an impedance transformer between the media by which the sound is propagated (air) and the media by which the receptor cells reside (fluid). Generally, this impedance mismatch can be overcome by having a high ratio of the area of the tympanic membrane to that of the oval window, and by employing a columella lever ratio. Lenhardt et al. (1985) found both of these ratios to be low in the loggerhead sea turtle compared to its terrestrial counterparts. They suggested, instead, that the shape of the columella and its interactions with the cochlea and saccule are not optimized for hearing in air, but rather are adapted for sound conduction through two media, bone and water. If the turtle uses bone conduction to process sound, sound flows through the bones and soft tissue to stimulate the inner ear. The tympanum would act as a release mechanism rather than a sound receptor. However, if the turtle uses water conduction to process sound, the tympanum and subtympanal fat could act as low-impedance channels for underwater sound, resulting in columellar displacement to stimulate the inner ear. Recent imaging data strongly suggest that the fats adjacent to the tympanal plates in at least three turtle species are highly specialized for underwater sound conduction (Ketten et al., 1999).

3.3.2 Electrophysiology

Electrophysiological studies on hearing have been conducted on juvenile green turtles (*C. mydas*) (Ridgeway et al., 1969) and on juvenile loggerheads (*C. caretta*) (Bartol, 1999). Ridgeway et al. (1969) used both aerial and vibrational stimuli to obtain auditory cochlear potentials. The active electrode was placed, using surgical techniques, in the perilymph spaces of the labyrinth. Sounds were presented either with a loudspeaker or with a mechanical vibrator. Absolute thresholds were not measured; instead, cochlear response curves of 0.1 μV potential were plotted for frequencies ranging from 50 to 2000 Hz. Green sea turtles detect a limited frequency range (200–700 Hz), with best sensitivity at the low tone region of about 400 Hz. Although this investigation examined two separate modes of sound reception (air conduction and bone conduction), sensitivity curves were relatively similar (Figure 3.6). These results suggest that the inner ear is the main structure for determining frequency sensitivity (Ridgeway et al., 1969).

Bartol et al. (1999) used a second technique for obtaining electrophysiological responses to sound stimuli from sea turtles, the collection of auditory brainstem responses (ABRs). ABRs are sequences of events originating in the brain stem and are generated by separate parts of the auditory pathway in the first 10 msec after

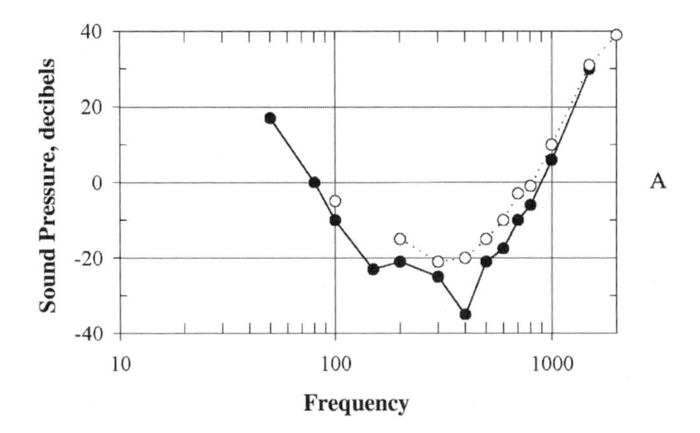

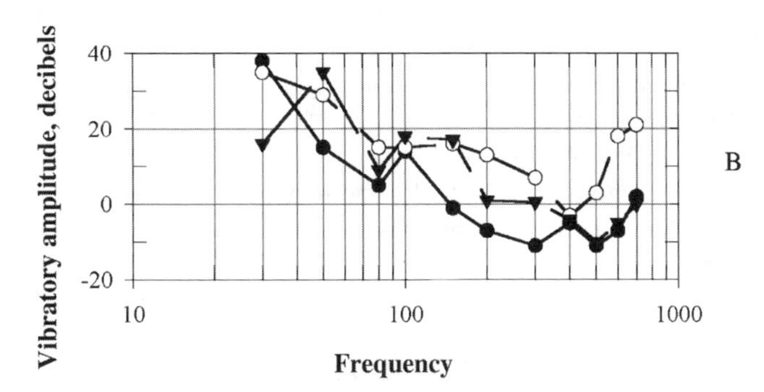

FIGURE 3.6 Hearing sensitivity curves obtained from green sea turtles (*C. mydas*). (A) Data collected from aerial sound stimuli. The sound pressure is shown in decibels relative to 1 dyne/cm^2 required to produce a cochlear potential of 0.1 μV. (B) Data collected from vibratory stimuli. The vibratory amplitude is shown in decibels relative to 1 mμ, required to produce a cochlear potential of 0.1 μV. (From Ridgeway, S.H. et al., Hearing in the giant sea turtle, *Chelonia mydas*, *Proc. Natl. Acad. Sci.*, 64, 884, 1969. With permission.)

stimulation. ABRs reflect the synchronous discharge of large populations of neurons within the auditory pathway, and therefore are useful monitors of the functioning of the throughput of the auditory system. Historically, ABRs have been used as a method for testing for audition and acoustic threshold in noncommunicative species. The technique is noninvasive, is rapid, and requires no overt training of the subjects. These recordings have been found to be consistent within species and similar across vertebrate classes in general form and origin, regardless of auditory apparatus (Corwin et al., 1982). Furthermore, the technique can be performed on awake subject animals (Bullock, 1991; Corwin et al., 1982). Bartol et al. (1999) recorded auditory evoked potentials from juvenile loggerheads using subdermal platinum electrodes implanted on awake animals. Vibratory stimuli, of known frequency, were delivered directly to the dermal plates over the sea turtle's tympanum. Signal averaging techniques were used to isolate the auditory evoked potentials from unrelated neural and muscular activity. Thresholds were recorded for both tonal and click stimuli. Best sensitivity was found in the low-frequency region of 250 –1000 Hz. The decline in sensitivity was rapid after 1000 Hz, and the most sensitive threshold tested was at 250 Hz (the lowest frequency tested), with a mean threshold of ~26.3 dB re 1 *g* root mean square (rms) + 2.3 dB standard deviation (SD).

3.3.3 Behavior

Two research studies have examined the response of juvenile loggerheads to sound in their natural environment (Moein et al., 1995; O'Hara and Wilcox, 1990). In both cases, these studies were initiated to assist in the development of an acoustic repelling device for sea turtles. O'Hara and Wilcox (1990) attempted to create a sound barrier for loggerhead turtles at the end of a canal of Florida Power & Light using seismic air guns. The test results indicated that at 140 kg/cm^2 the air guns were effective as a deterrent for a distance of about 30 m. The sound output of this system was characterized as approximately 220 dB re 1 μPa at 1 m in the 25–1000 Hz frequency range. However, this study did not account for the reflection of sound by the canal walls. Consequently, the stimulus frequency and intensity levels are ambiguous (O'Hara and Wilcox, 1990).

Moein et al. (1995) investigated the use of pneumatic energy sources (air guns) to repel juvenile loggerhead sea turtles from hopper dredges. A net enclosure (approximately 18 m × 61 m × 3.6 m) was erected in the York River, VA, to contain the turtles, and an air gun was stationed at each end of the net. A float attached to the posterior of the carapace was used to note the position of the turtle as the air guns fired. Sound frequencies of the air guns ranged from 100 to 1000 Hz (Zawila, 1995). Three decibel levels (175, 177, and 179 dB re 1 μPa at 1 m) were used every 5 sec for 5 min. Avoidance of the air guns was observed upon first exposure for the juvenile loggerheads. However, these animals also appeared to habituate to the sound stimuli. After three separate exposures to the air guns, the turtles no longer avoided the stimuli (Moein et al., 1995).

3.3.4 Concluding Remarks

These studies highlight the need for more research on the auditory capabilities of sea turtles. It is believed that physiological and behavioral adaptations may have evolved for sea turtles based on their selection of aquatic niches with each ontogenetic stage. For these three stages of life, the sensory environment also changes. Shallow-water habitats of the juvenile and adult stages are a much "noisier" world than the open ocean environment of the hatchling stage. Ambient noise in the inshore environment is heavily weighted to low-frequency sound (Hawkins and Myrberg, 1983). In highly developed areas (coastal waters) low-frequency noises associated with shipping lanes, recreational boat traffic, and biological organisms are prominent. Differences in functional morphology and behavioral hearing capabilities among species and life history stages have not been documented for sea turtles in the literature. In fact, only juvenile loggerhead and green sea turtles have undergone any auditory investigations. Both the middle and inner ear regions of sea turtles need to be reexamined using the latest laboratory techniques. Furthermore, behavioral responses by multiple life history stages of sea turtles to sound stimuli, in the form of behavioral audiograms, need to be pursued in future research studies.

3.4 CHEMORECEPTION

3.4.1 ANATOMY OF THE NASAL STRUCTURES

The structure of the sea turtle nose is relatively simple: it opens to the outside world through external nares and into the palate through the internal nares on the posterior end. The external nares are connected to the nasal cavity by a tubelike vestibulum, and the nasal cavity is connected to the palate by a long nasopharyngeal duct (Scott, 1979). The nasal cavity is divided into two regions: the intermediate region and the olfactory region (Figure 3.7). The intermediate region lies ventrally and is attached to both the vestibulum and the nasopharyngeal duct. The intermediate region is large, occupies $^3/_4$ of the nasal cavity, and has two pockets of sensory epithelium called the Jacobson's organs. The functional significance of the Jacobson's organ is unknown, and although it appears to be capable of chemoreception, it has been assumed that this region is nonolfactory in the anatomy literature. In the sea turtle, these Jacobson's organs receive information in the same manner as olfactory epithelium. However, the information from this sensory epithelium is sent to the accessory olfactory bulb and the trigeminal nerve system. Posterodorsally in the nasal cavity lies the olfactory region, which is small compared to the intermediate region. The olfactory region is lined with a second type of sensory epithelium, Bowman's glands, which send information directly to the main part of the olfactory bulb. The olfactory nerve arises from these two types of sensory epithelium of the nose and forms two groups of trunks that lead to distinct portions of the olfactory bulb and accessory bulb. In the sea turtle, both the olfactory and accessory bulbs are notably large for a vertebrate (Parsons, 1959; 1971; Scott, 1979).

Tucker (1971) discussed the nonolfactory response within the nasal cavity and argued that the intermediate region received chemical stimulation in a similar manner to the olfactory region. However, because the intermediate region is ventrally located within the nasal cavity, it is almost continually bathed with water. The olfactory region, on the other hand, could contain an air bubble because of its dorsal location and thus remain dry as the turtle draws water into the nasal cavity. Tucker (1971) also made the assumption that an air-breathing animal cannot smell underwater. Thus, only the region called the olfactory region, and not the intermediate region, could be responsible for olfactory, chemosensory reception. The intermediate region was assumed to be involved with nonolfactory chemoreception (Parsons, 1971; Tucker, 1971). These assumptions, based on anatomical descriptions, have been debunked by several behavioral studies, and in fact sea turtles have been shown to "smell" underwater (see Section 3.4.2.2). In addition, recent research on fishes (Walker et al., 1997) has found that the receptor organs for geomagnetic orientation are located in the olfactory epithelium and are innervated by the trigeminal system. Sea turtles have been shown to have an elegant geomagnetic sense (Lohmann and Lohmann, 1994; Lohmann et al., 1997). Could the Jacobson's organ be the location of geomagnetic receptors in sea turtles?

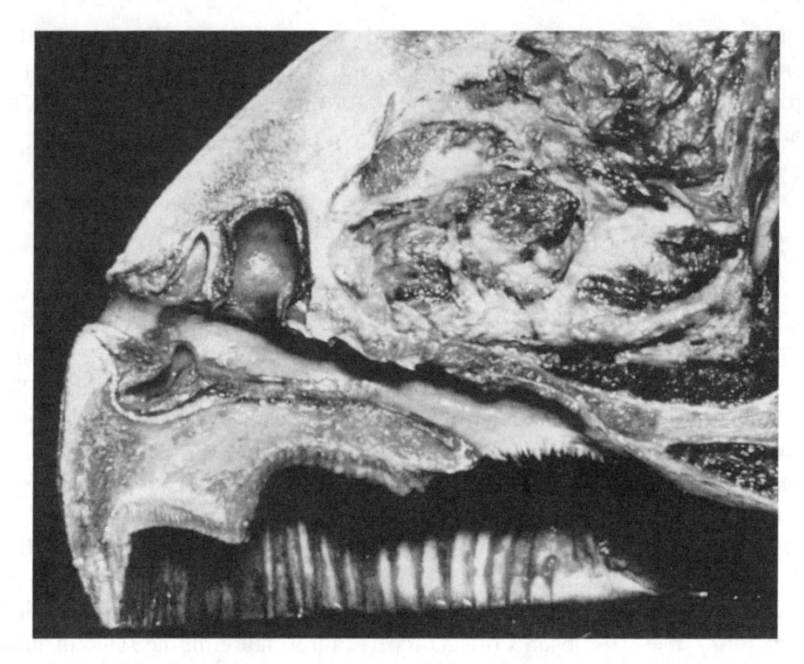

FIGURE 3.7 Right nasal cavity of green sea turtle (*C. mydas*). (From Parsons, T.S., Anatomy of nasal structures from a comparative viewpoint, in *Handbook of Sensory Physiology Vol. IV/I*, Beidler, L.M., Ed., Springer-Verlag, Berlin, 1971. With permission.)

3.4.2 Behavior

3.4.2.1 General Behavioral Observations

In a study that generally documented the sea turtle's natural behavior, Walker (1959) reported that sea turtles open their nostrils and move their mouths slowly open and closed while underwater. Walker postulated that this throat-pumping behavior moves water over the nostrils for olfaction, as had been suggested for many freshwater turtles (McCutcheon, 1943; Root, 1949). Throat-pumping has not been observed when sea turtles are resting or when they are breathing at the surface. This repetitive blowing of water out of the external nares while underwater occurs only while the animal is awake and active, and is postulated to be a mechanism for moving water over the chemoreceptor organs (Manton, 1979).

3.4.2.2 Odor Discrimination

Two operant conditioning studies examining underwater chemosensory behavior in green sea turtles (*C. mydas*) have been performed (Manton et al., 1972a; 1972b). Both studies used similar procedures. A tank was set up with two response keys suspended underwater; a light was mounted over each key (Figure 3.8). The turtles were able to swim freely within the tank environment. Turtles were first trained (using a food reward as reinforcement) to press either the right or the left key in

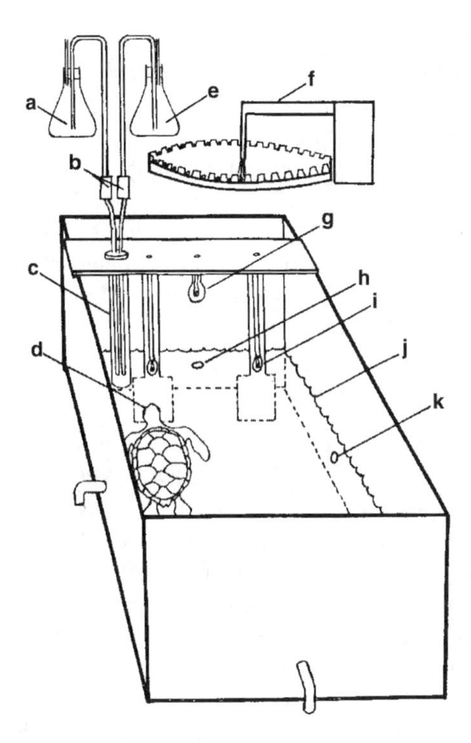

FIGURE 3.8 Diagram of experimental tank used to examine chemoreceptory ability of green sea turtles (*C. mydas*). (From Manton, M.L., Karr, A., and Ehrenfeld, D.W., Chemoreception in the migratory sea turtle, *Chelonia mydas*, *Biol. Bull.*, 143, 184, 1972. With permission.)

response to a light stimulus. Once the turtles were trained, the light signal was progressively reduced, and replaced with a chemical signal. For all remaining tests, the turtles first pressed the left key. If a chemical was released into the water, the turtles could then press the right key to receive a food reward. If no chemical was released into the water, and the turtles subsequently pressed the right key, this was marked as an incorrect response. All trials were completed with the turtles completely submerged underwater. This behavioral technique proved to be very successful, and once trained, the turtles completed the sequence rapidly. Habituation was never encountered (Manton et al., 1972a; 1972b).

The first of these two studies tested for underwater chemoreception (Manton et al., 1972a). The chemicals used for this study were organic compounds selected based on the chemosensory literature, and included such volatile compounds as phenethylalcohol and acetate, as well as two nonvolatile amino acids, serine and glycine. The control in this experiment was tank water. Except for the amino acids (which were not detected), the turtles responded to the chemicals with a mean correct detection of 89%, a much higher probability than for the control. When the chemical was released into the water, the turtles always directed their nostrils downward and performed the characteristic throat-pumping action (Manton et al., 1972a).

Although this study provides convincing evidence that sea turtles are capable of chemoreception, it does not distinguish between chemoreception by olfaction and taste. The same group of researchers also tested for olfaction by temporarily inducing anosmia (loss of the sense of smell) in their subject animals (Manton et al., 1972b). By exposing the internal nares to $ZnSO4$, while ensuring that the oral cavity did not come into contact with the chemical, they were able to temporarily render the olfactory sense inoperative. After treatment with $ZnSO4$, the turtles were unable to distinguish the chemical from the control, indicating that these animals were using olfaction and not taste for chemoreception. Chemosensory acuity was also estimated from the data. These turtles were found to be able to detect chemicals at a relatively low level; the threshold occurred at concentrations of approximately $5 \times 10–6$ to $5 \times 10–5$ M (Manton et al., 1972b).

3.4.3 Chemical Imprinting Hypothesis

Chemoreception has long been proffered as the basis for orientation and long-distance migration by sea turtles (Koch et al., 1969; Manton, 1979; Owens et al., 1982). Though there appears to be very little evidence that sea turtles use chemoreception to navigate long distances, some research has been performed on the role that chemical cues play in the identification of a natal beach by adult nesting female sea turtles. Grassman et al. (1984) explored the theory that these animals can retain olfactory information gathered from the nesting beach and surrounding waters as hatchlings (that is, they become imprinted) and store this information for many years until they return as nesting females. They used Kemp's ridley (*Lepidochelys kempii*) hatchlings collected from Rancho Nuevo, Mexico, during oviposition and moved the eggs to Padre Island National Seashore in Texas. The eggs were incubated in Padre Island sand until hatching; hatchlings were allowed to perform their natural crawl across the sand and enter the surf zone. These animals were recaptured, and raised in tanks. At 4 months old, these same turtles were tested in a multipartitioned arena. When placed in this arena, the turtles could choose among a section containing a solution of Padre Island sand and water; a section containing a solution of Galveston, TX, sand and water; and two sections containing untreated solutions. Turtles spent significantly more time in the Padre Island compartment than either the Galveston or untreated sections. Although the turtles entered the Galveston compartment frequently, they did not stay in the compartment any longer than when the turtles had entered the untreated sections. The authors interpreted this behavior as a preference for the Padre Island treatment (Grassman et al., 1984).

A second experiment investigated the behavioral responses of sea turtles exposed to two chemicals, morpholine and 2-phenylethanol (Grassman and Owen, 1987). These chemicals were chosen because they are not naturally occurring, yet from the previous operant conditioning studies (Manton et al., 1972b), the researchers knew that green turtles could detect low concentrations of similar organic chemicals. Eggs were collected; the artificial nest environment was moistened with either one of the two chemicals or with untreated water. When the sea turtles hatched, they were placed in holding tanks that were also treated with the same chemical as the nest for 3 months. The turtles were segregated into four treatments: (1) both the nest and the water were treated with a chemical, (2) only the nest was treated, (3) only the water was treated, and (4) both the nest and water were untreated. After 2 additional months of no exposure, the animals were placed in the same multipartitioned arena as in the previous study (Grassman et al., 1984). The only group of turtles that spent significantly more time in the chemically treated compartment, as opposed to the untreated compartment, was the group that was exposed to the chemicals both in the nest and in the water. These results suggested that not only the environment of the nest, but also the chemosensory environment of the water are important in the imprinting process (Grassman and Owens, 1987).

3.4.4 Concluding Remarks

Many of the inferences made from anatomical studies were based on the assumption that an air-breathing vertebrate could not detect chemicals underwater using olfaction. Behavioral studies have proved that this is not the case for sea turtles. The anatomy of the sea turtle olfactory system should be revisited with the behavioral data in mind.

REFERENCES

Ali, M.A. and Klyne, M.A., *Vision in Vertebrates*, Plenum Press, New York, 1985.

Bartol, S.M., Morphological, electrophysiological, and behavioral investigation of visual acuity of the juvenile loggerhead sea turtle (*Caretta caretta*), dissertation, College of William and Mary, Virginia Institute of Marine Science, Gloucester Point, VA, 1999.

Bartol, S.M., Musick, J.A., and Lenhardt, M., Auditory evoked potentials of the loggerhead sea turtle (*Caretta caretta*), *Copeia*, 3, 836, 1999.

Bartol, S.M. and Musick, J.A., Morphology and topographical organization of the retina of juvenile loggerhead sea turtles (*Caretta caretta*), *Copeia*, 3, 718, 2001.

Bartol, S.M., Musick, J.A., and Ochs, A.L., Visual acuity thresholds of juvenile loggerhead sea turtles (*Caretta caretta*): an electrophysiological approach, *J. Comp. Physiol. A.*, 187, 953, 2002.

Brown, K.T., A linear area centralis extending across the turtle retina and stabilized to the horizon by non-visual cues, *Vision Res.*, 9, 1053, 1969.

Bullock, T.H. et al., Dynamic properties of visual evoked potentials in the tectum of cartilaginous and bony fishes, with neuroethological implications, *J. Exp. Zool. Suppl.*, 5, 142, 1991.

Carr, A. and Ogren, L., The ecology and migration of sea turtles. 4. The green turtle in the Caribbean Sea, *Am. Mus. Nat. Hist. Bull.*, 121, 1, 1960.

Copenhaver, W.M., *Bailey's Textbook of Histology*, Williams & Wilkins, Baltimore, 1964.

Corwin, J.T., Bullock, T.H., and Schweitzer, J., The auditory brain stem response in five vertebrate classes, *Electroenceph. Clin. Neurophysiol.*, 54, 629, 1982.

Daniel, R.S. and Smith, K.U., The sea-approach behavior of the neonate loggerhead turtle (*Caretta caretta*), *J. Comp. Physiol. Psychol.*, 40, 413, 1947.

Davson, H., *The Physiology of the Eye*, Academic Press, New York, 1972.

Ehrenfeld, D.W., The role of vision in the sea-finding orientation of the green turtle (*Chelonia mydas*). 2. Orientation mechanism and range of spectral sensitivity, *Anim. Behav.*, 16, 281, 1968.

Ehrenfeld, D.W. and Carr A., The role of vision in the sea-finding orientation of the green turtle (*Chelonia mydas*), *Anim. Behav.*, 15, 25, 1967.

Ehrenfeld, D.W. and Koch, A.L., Visual accommodation in the green turtle, *Science*, 155, 827, 1967.

Fehring, W.K., Hue discrimination in hatchling loggerhead turtles (*Caretta caretta*), *Anim. Behav.*, 20, 632, 1972.

Fernald, R.D., The optical systems of fishes, in *The Visual System of Fish*, Douglas, R.H. and Djamgoz, M.B.A., Eds., Chapman & Hall, London, 1990.

Granda, A.M., Eyes and their sensitivity to light of differing wavelengths, in *Turtles: Perspectives and Research*, Harless, M. and Morlock, H., Eds., John Wiley & Sons, New York, 247, 1979.

Granda, A.M. and Dvorak, C.A., The visual system in vertebrates: vision in turtles, in *Handbook of Sensory Physiology*, Crescitelli, F., Ed., Springer-Verlag, Berlin, 1977, 451, 1977.

Granda, A.M. and O'Shea, P.J., Spectral sensitivity of the green turtle (*Chelonia mydas*) determined by electrical responses to heterochromatic light, *Brain Behav. Evol.*, 5, 143, 1972.

Grassman, M.A. et al., Olfactory-based orientation in artificially imprinting sea turtles, *Science*, 224, 83, 1984.

Grassman, M. and Owens, D., Chemosensory imprinting in juvenile green sea turtles, *Chelonia mydas*, *Anim. Behav.*, 35, 929, 1987.

Hawkins, A.D. and Myrberg, A.A., Jr., Hearing and sound communication under water, in *Bioacoustics: A Comparative Approach*, Lewis, B., Ed., Academic Press, London, 347, 1983.

Heuter, R.E., Adaptations for spatial vision in sharks, *J. Exp. Zool. Suppl.*, 5, 130, 1991.

Ketten, D.R. et al., Acoustic Fatheads: Parallel Evolution of Soft Tissue Conduction Mechanisms in Marine Mammals, Turtles, and Birds, paper presented at Acoustical Society of America/European Acoustics Association, Berlin, March 14–19, 1999.

Koch, A.L., Carr, A.F., and Ehrenfeld, D.W., The problem of open-sea navigation: the migration of the green turtle to Ascension Island, *J. Theor. Biol.*, 22, 163, 1969.

Liebman, P.A., Microspectrophotometry of phototreceptors, in *Photochemistry of Vision. Vol. VII/1. Handbook of Sensory Physiology*, Dartnall, H.J.A., Ed., Springer-Verlag, New York, 1972, p. 481.

Limpus, C., Sea turtle ocean finding behaviour, *Search*, 2, 385, 1971.

Lenhardt, M.L., Klinger, R.C., and Musick, J.A., Marine turtle middle-ear anatomy, *J. Aud. Res.*, 25, 66, 1985.

Lewis, E.R., Leverenz, E.L., and Bialek, W.S., *The Vertebrate Inner Ear*, CRC Press, Boca Raton, FL, 1985.

Lohmann, K.J. et al., Orientation, navigation, and natal beach homing in sea turtles, in *The Biology of Sea Turtles*, Lutz, P.L. and Musick, J.A., Eds., CRC Press, Boca Raton, FL., 107, 1997.

Lohmann, K.J. and Lohmann, C.M., Detection of magnetic inclination angle by sea turtles: a possible mechanism for determining latitude, *J. Exp. Biol.*, 194, 23, 1994.

Manton, M.L., Olfaction and behavior, in *Turtles: Perspectives and Research*, Harless, M. and Morlock, H., Eds., John Wiley & Sons, New York, 289, 1979.

Manton, M.L., Karr, A., and Ehrenfeld, D.W., An operant method for the study of chemoreception in the green turtle, *Chelonia mydas*, *Brain Behav. Evol.*, 5, 188, 1972a.

Manton, M.L., Karr, A., and Ehrenfeld, D.W., Chemoreception in the migratory sea turtle, *Chelonia mydas*, *Biol. Bull.*, 143, 184, 1972b.

McCutcheon, F.H., The respiratory mechanism in turtles, *Physiol. Zool.*, 16, 255, 1943.

Moein, S.E. et al., Evaluation of seismic sources for repelling sea turtles from hopper dredges, in *Sea Turtle Research Program: Summary Report*, Hales, L.Z., Ed., Prepared for U.S. Army Engineer Division, South Atlantic, Atlanta, GA, and U.S. Naval Submarine Base, Kings Bay, GA, Technical Report CERC-95-31, 90, 1995.

Mrosovsky, N., The water-finding ability of sea turtles, *Brain Behav. Evol.*, 5, 202, 1972.

Mrosovsky, N., Granda, A.M., and Hays, T., Seaward orientation of hatchling turtles: turning systems in the optic tectum, *Brain Behav. Evol.*, 16, 203, 1979.

Mrosovsky, N. and Shettleworth, S., Wavelength preferences and brightness cues in the water-finding behavior of sea turtles, *Behavior*, 32, 211, 1968.

O'Hara, J. and Wilcox, J.R., Avoidance responses of loggerhead turtles, *Caretta caretta*, to low frequency sound, *Copeia*, 2, 564, 1990.

Oliver, L.J. et al., Retinal anatomy of hatchling sea turtles: anatomical specializations and behavioral correlates, *Mar. Fresh. Behav. Physiol.*, 33, 233, 2000.

Owens, D.W., Grassman, M.A., and Hendrickson, J.R., The imprinting hypothesis and sea turtle reproduction, *Herpetologica*, 38, 124, 1982.

Parker, G.H., The crawling of young loggerhead turtles towards the sea, *J. Exp. Zool.*, 36, 323, 1922.

Parsons, T.S., Nasal anatomy and the phylogeny of reptiles, *Evolution*, 13, 175, 1959.

Parsons, T.S., Anatomy of nasal structures from a comparative viewpoint, in *Handbook of Sensory Physiology Vol. IV/I*, Beidler, L.M., Ed., Springer-Verlag, Berlin, 1, 1971.

Peterson, E.H., Retinal structure, in *Biology of the Reptilia: Sensorimotor Integration*, Gans, C. and Ulinski, P.S., Eds., The University of Chicago Press, Chicago, 1, 1992.

Ridgeway, S.H. et al., Hearing in the giant sea turtle, *Chelonia mydas*, *Proc. Nat. Acad. Sci.*, 64, 884, 1969.

Riggs, L.A. and Wooten, B.R., Electrical measures and psychophysical data on human vision, in *Handbook of Sensory Physiology, Visual Psychophysics, Vol. VII No. 4*, Jameson, D. and Hurvich, L.M., Eds., Springer-Verlag, Berlin, 690, 1972.

Root, R.W., Aquatic respiration in the musk turtle, *Physiol. Zool.*, 22, 172, 1949.

Salmon, M. and Wyneken, J., Do swimming loggerhead sea turtles (*Caretta caretta* L.) use light cues for offshore orientation, *Mar. Behav. Physiol.*, 17, 233, 1990.

Salmon, M. et al., Seafinding by hatchling sea turtles: role of brightness, silhouette and beach slope as orientation cues, *Behaviour*, 122, 56, 1992.

Salmon, M. and Wyneken, J., Orientation by hatchling sea turtles: mechanisms and implications, *Herp. Nat. Hist.*, 2, 13, 1994.

Scott, T.R., Jr., The chemical senses, in *Turtles: Perspectives and Research*, Harless, M. and Morlock, H., Eds., John Wiley & Sons, New York, 267, 1979.

Sivak, J.G., The Glenn A Fry award lecture: optics of the crystalline lens, *Am. J. Optom. Physiol. Optics*, 62, 299, 1985.

Stell, W.K., The morphological organization of the vertebrate retina, in *Handbook of Sensory Physiology, Physiology of Photoreceptor Organs, Vol. VII/2*, Fuortes, M.G.F., Ed., Springer-Verlag, Berlin, 111, 1972.

Tucker, D., Nonolfactory responses from the nasal cavity: Jacobson's organ and the trigeminal system, in *Handbook of Sensory Physiology Vol. IV/I*, Beidler, L.M., Ed., Springer-Verlag, Berlin, 151, 1971.

Turner, R.G., Physiology and bioacoustics in reptiles, in *Comparative Studies of Hearing in Vertebrates*, Popper, A.N., Ed., Springer-Verlag, New York, 205, 1978.

van Rhijn, F.A., Optic orientation in hatchlings of the sea turtle *Chelonia mydas*. I. Brightness: not the only optic cue in sea-finding orientation, *Mar. Behav. Physiol.*, 6, 105, 1979.

Walker, M.M. et al., Structure and function of the vertebrate magnetic sense, *Nature*, 390, 371, 1997.

Walker, W.F., Closure of the nostrils in the Atlantic loggerhead and other sea turtles, *Copeia*, 3, 257, 1959.

Walls, G.L., *The Vertebrate Eye and Its Adaptive Radiation*, The Cranbook Institute of Science, Bloomfield Hills, MI, 1942.

Wever, E.G., *The Reptile Ear: Its Structure and Function*, Princeton University Press, Princeton, NJ, 1978.

Wever, E.G. and Vernon, J.A., Sound transmission in the turtle's ear, *Proc. Nat. Acad. Sci.*, 42, 229, 1956.

Witherington, B.E. and Bjorndal, K.A., Influences of wavelength and intensity on hatchling sea turtle phototaxis: implications for sea-finding behavior, *Copeia*, 4, 1060, 1991.

Zawila, J.S., Characterization of a seismic air gun acoustic dispersal technique at the Virginia Institute of Marine Science sea turtle test site, in *Sea Turtle Research Program: Summary Report*. Hales, L.Z., Ed., Prepared for U.S. Army Engineer Division, South Atlantic, Atlanta, GA, and U.S. Naval Submarine Base, Kings Bay, GA. Technical Report CERC-95-31, 88, 1995.

4 Critical Approaches to Sex Determination in Sea Turtles

Thane Wibbels

CONTENTS

0-8493-1123-3/03/$0.00+$1.50

4.1 INTRODUCTION

Temperature-dependent sex determination (TSD) was first reported in 1966 in a lizard.[1] Since that time it has been shown to occur in a wide variety of reptiles, including all crocodilians, most turtles, some lizards, and the tuatara.[2–4] The occurrence of TSD in primitive groups of reptiles such as turtles, crocodilians, and tuatara has led some researchers to hypothesize that it may represent an ancestral form of sex determination from which avian and mammalian sex determination systems have evolved.[5] The adaptive advantage of TSD is not clear, but a number of proposed hypotheses might explain why many reptiles, including sea turtles, have retained TSD.[6–8]

In sea turtles, TSD was first documented in the loggerhead, *Caretta caretta*.[9] Since that time it has been shown to occur in all extant species: the green turtle, *Chelonia mydas*;[10–12] the olive ridley, *Lepidochelys olivacea*;[13–15] the leatherback, *Dermochelys coriacea*;[12,16,17] the hawksbill, *Eretmochelys imbricata*;[18–21] Kemp's ridley, *Lepidochelys kempi*;[22–24] the black turtle, *Chelonia agassizi*;[25–27] and the flatback, *Natator depressa*.[28] The occurrence of TSD in sea turtles generates a wide variety of questions regarding the physiological, ecological, and conservational implications of this form of sex determination. The purpose of this review is to summarize what is known about TSD in sea turtles and to use that information to address basic questions regarding the biology and conservation of marine turtles.

4.2 CHARACTERISTICS OF TSD IN SEA TURTLES

Although a wide variety of reptiles possess TSD, the effect of a particular temperature may vary depending on the species.[2,29] Several patterns of sex determination have been described regarding the effects of temperature on sex determination in reptiles.[2,29] All sea turtles examined to date appear to have a male–female (MF) pattern in which cooler incubation temperatures produce males and warmer incubation temperatures produce females (Figure 4.1). Several terms have been created to describe TSD in reptiles.[30] The transitional range of temperatures (TRT) is the range of temperatures in which sex ratios shift from 100% male to 100% female (Figure 4.1). In the case of the MF pattern in sea turtles, temperatures above the TRT will produce all females and temperatures below the TRT will produce all males. Within the TRT, there is temperature referred to as the pivotal temperature, which is the constant incubation temperature that will produce a 1:1 sex ratio (Figure 4.1). The pivotal temperature can vary between and even within a species.[31] It has also been reported that the TRT may vary among sea turtle populations.[32] Therefore, if one is interested in studying sex determination or estimating hatchling sex ratios in a given sea turtle population, it is optimal to determine these parameters for that particular population, rather than extending data from one population to another.

4.2.1 Accuracy of Temperature Estimates

Before delving into the specific temperature estimates from previous studies of TSD in sea turtles, it is necessary to address the potential problems associated with accuracy. The goal of many TSD studies is to estimate parameters that describe TSD

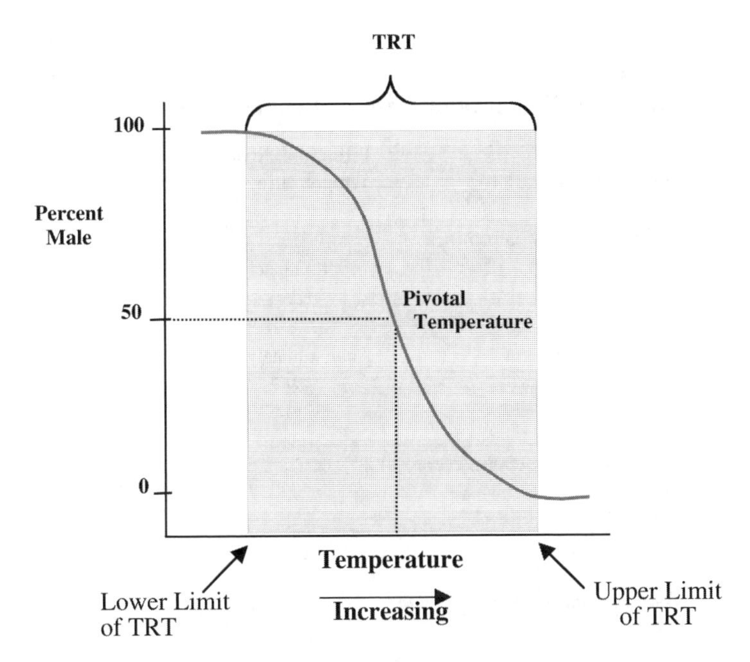

FIGURE 4.1 General pattern of TSD in sea turtles. Relatively cool temperatures produce males and relatively warm temperatures produce females. Pivotal temperatures produce a 1:1 sex ratio. The TRT is the range over which sex rations shift from all male to all female.

in a species or population (e.g., pivotal temperature, transitional range of temperatures, etc.). Subsequent studies then use those parameters for comparative purposes or even for predicting sex ratios. Therefore, the accuracy of the temperature recordings is of paramount importance. For example, a change of only a few tenths of a degree Celsius can have a significant effect on sex ratio if that change occurs near the pivotal temperature. The accuracy of temperature estimates reported in TSD studies is limited by factors such as the accuracy of the recording equipment, the stability of the incubators, and the experimental protocol. For example, the resolution of typical temperature data loggers is approximately 0.2–0.4°C, and a typical incubator may vary by several tenths of a degree around the selected temperature. Thus, when reviewing data on TSD, one should be cautious when comparing temperature estimates from studies that might use different experimental protocols and temperature recording equipment.

4.2.2 Pivotal Temperatures of Sea Turtles

A variety of studies have estimated pivotal temperatures in sea turtles (Table 4.1). In general, the reported pivotal temperatures occur over a relatively narrow temperature range from approximately 27.7 to 31°C, depending on the particular species and study, with the majority clustering in the 29.0 to 30.0°C range.[31] The reason for this narrow range is unknown, but it is plausible that it could relate to factors such as ecological or physiological constraints that may be selecting for this range of pivotal temperatures.

TABLE 4.1
Examples of Pivotal Temperatures Reported for Sea Turtles

Species and Location	Estimated Pivotal Temperature (°C)	Reference
Loggerhead (*Caretta caretta*)		
U.S.	Approximately 30	44
U.S.	29.0	34
Australia	27.7, 28.7	33
Australia	Approximately 29.0	36
S. Africa	29.7	38
Brazil	29.2	37
Leatherback (*Dermochelys coriacea*)		
Suriname and French Guiana	29.5	16
		39
Costa Rica	29.4	17
Hawksbill (*Eretmochelys imbricata*)		
Antigua	29.2	20
Brazil	29.6	101
Green (*Chelonia mydas*)		
Suriname	28.8	12
Costa Rica	Approximately 28.5–30.3	46
		47
Olive Ridley (*Lepidochelys olivacea*)		
Costa Rica	Approximately 30	13
Costa Rica	Approximately 31	15
Kemp's Ridley (*Lepidochelys kempii*)		
Mexico	30.2	23

Despite this relatively narrow range, the variation associated with reported pivotal temperatures often makes it difficult to assign a specific pivotal temperature to the tenth of a degree Celsius for a given population. Table 4.1 shows that the reported pivotal temperatures vary between species and populations. In addition, interclutch variation has been reported for sea turtles and other reptiles with TSD.[29,33–35] The variation in reported pivotal temperatures is exemplified by the loggerhead turtle, which has received more attention than any other species. Mrosovsky (1988) estimated pivotal temperatures for clutches of loggerhead eggs from three different locations along the coast of the U.S. That study found that all predicted pivotal temperatures were near 29.0°C, but estimates for individual clutches ranged from approximately 28.5 to 29.2°C.[34,36] Furthermore, a significant interclutch variation was detected between two clutches from the same nesting beach during that study.[34] A study of loggerhead turtles in Australia estimated pivotal temperatures of 27.7 and 28.7°C for two different nesting beaches, and a significant interclutch variation was detected from clutches from one of those beaches.[33] A later study suggested a pivotal temperature of approximately 29.0°C for one of those same nesting beaches. Other estimates for loggerhead pivotal temperatures include 29.2°C for a Brazilian nesting beach[37] and 29.7°C for a South African nesting beach.[38] Thus, pivotal temperatures within this species have been reported to vary by up to 1.0°C or more.

Although some of this variation may be associated with experimental error, differences in protocols, or divergences in statistical analyses of the data, these findings suggest that pivotal temperatures vary among and even within a population. A reported pivotal temperature simply represents an estimate of the mean pivotal temperature for a sample from a population. Considering the large number of sea turtle populations in the world, there have been only a limited number of pivotal temperature studies. Because of the endangered or threatened status of many of these populations, pivotal temperature studies have usually used a limited number of hatchlings and clutches. To better understand the variations in pivotal temperatures within and between populations, more comprehensive studies using large numbers of clutches would be optimal.

If one is interested in studying sex determination or estimating hatchling sex ratios in a given sea turtle population, it is advantageous to determine the pivotal temperature for that particular population, rather than estimating it based on data from other populations. Because interclutch variation can occur, the accuracy of the estimated pivotal temperature will be dependent on the number of clutches examined. Furthermore, because the estimated pivotal temperature will represent a mean, it should include statistics to describe the variance associated with the mean (see Section 4.2.4).

4.2.3 TRT of Sea Turtles

In addition to pivotal temperature, the TRT (Figure 4.1) is a useful parameter for characterizing sex determination in sea turtles. As defined earlier, the TRT represents the range of temperatures in which the sex ratio shifts from 100% male to 100% female.[30] For sea turtles (which have an MF pattern of sex determination), temperatures below the lower limit of the TRT will produce only males and temperatures above the upper limit of the TRT will produce only females. A TRT is more difficult to accurately estimate than a pivotal temperature because of the greater number of incubation temperatures necessary to clearly pinpoint the lower and upper bounds of the TRT curve. Most studies have not ruled out the possibility that the TRT could be larger or smaller than reported.

A general characterization of the TRT can be gleaned from past studies by examining maximum incubation temperatures that produce all males and minimum incubation temperatures that produce all females in a particular study (Table 4.2). Many past studies did not precisely determine the TRT because of the number of incubation temperatures and the specific temperatures utilized in the study. However, some studies included a reasonable number of appropriate incubation temperatures, and the results

TABLE 4.2
Examples of Incubation Temperatures Producing All Male or All Female Hatchlings in Previous Studies of Sea Turtle Sex Determination

Species and Location	Temperatures (°C) Producing		Reference
	100% Male	100% Female	
Loggerhead (*Caretta caretta*)			
U.S.	<27.5	30.4–30.5	34
Australia	26.0	30.0–32.0	33
Australia	<26.0	31.0	36
Brazil	28.0	30.6	37
Leatherback (*Dermochelys coriacea*)			
Suriname	28.75	29.75	39
Costa Rica	29.0	30.0	17
Hawksbill (*Eretmochelys imbricata*)			
Antigua	28.5–29.0	30.3	20
Brazil	28.4	30.4	101
Green (*Chelonia mydas*)			
Suriname	<27.75	>29.25–30.75	12
Costa Rica	<28.0	>30.5	46
Olive Ridley (*Lepidochelys olivacea*)			
Costa Rica	<28.0	32.0	13
Costa Rica	27.0	32.0	15
Kemp's Ridley (*Lepidochelys kempi*)			
Mexico	<29.0	31.0	23

Note: All studies are based on constant temperature studies with the exceptions of those for green turtles in Costa Rica, which were based on field studies of natural nests.

provide a general characterization of the TRT in some species. The findings from these studies indicate that the TRT can vary between species and populations.

As with pivotal temperatures, the values of the lower and upper limits of the TRT, as well as the width of the TRT, can vary between species and populations. For example, minimum temperatures producing all females ranged from approximately 29.75 to 32.0°C (Table 4.2), and maximum temperatures producing all males ranged from approximately 26.0 to 28.75°C (Table 4.2), depending on the species and population examined. Studies of loggerhead populations have suggested TRT widths as great as 2.0 to 3.0°C,[33,34] whereas a study of a leatherback population suggested a TRT width as narrow as approximately 1.0°C.[39] It was suggested that the TRT recorded in the latter study was significantly different from that reported for another leatherback population.[32]

Collectively, these data indicate that the upper and lower limits of the TRT can vary between sea turtle species and populations. Although data from previous studies are far from comprehensive, they suggest general ranges for the lower limits (approximately 26.0 to 28.75°C) and upper limits (approximately 29.75 to 32.0°C) of TRTs, and they provide a general range of widths for TRTs (approximately 1.0 to 3.0°C). This sort of information has major conservational and ecological implications. For example, this suggests that a temperature change of 3.0°C or less could potentially shift sex ratios from all male to all female (or vice versa). It also provides information on thermal ranges where small changes should not make a difference for sex ratios (e.g., extremely warm temperatures that are well above pivotal). In addition, these data indicate that physiologically, the sex determination cascade switches from male to female (or vice versa) over a relatively narrow temperature window.

4.2.4 Methods for Estimating Pivotal Temperature and TRT in a Sea Turtle Population

Before characterizing TSD in a sea turtle population, one should first address the reasons for doing so. Knowledge of pivotal temperatures and TRTs has a variety of uses ranging from highly applied to theoretical. First, it has direct applications to the conservation of endangered populations. If you know the TRT and pivotal temperature in a population, you can then predict hatchling sex ratios by simply monitoring incubation temperatures in nests. Furthermore, this knowledge provides the basis for potentially manipulating hatchling sex ratios in an effort to enhance reproductive output in a population.[40,41] Pivotal temperatures and TRTs are also of ecological, physiological, and evolutionary interest because they allow researchers to begin addressing questions regarding variation in TSD and how it might affect the reproductive ecology in a population.

Conservational considerations must also be taken into account before beginning a study of pivotal temperatures and TRTs. In most cases the study will involve an endangered or threatened sea turtle population. Currently, it is impossible to verify the sex of a sea turtle hatchling using external characteristics. The definitive method of determining sex is by histological analysis of the gonads, which requires that the hatchling be killed. Although this sounds counterintuitive to enhancing the recovery of an endangered population, it is quite possible that the knowledge gained will greatly enhance the recovery of the population.

When the effects of temperature on sex in sea turtles are studied, the resolution and accuracy of temperature recordings is of major importance. For example, data from previous studies suggest that pivotal temperatures can vary by tenths of degrees between populations and even between clutches (reviewed by Ref. 31). Therefore, accurate temperature measurements are a prerequisite to obtaining meaningful results. A variety of temperature recording devices are available commercially. For automated recording of temperatures, one of the most commonly used devices is the temperature data logger. These units are battery operated, have an internal and/or external temperature probe, and can be programmed to take temperature recordings at a wide variety of intervals. Many of these data loggers are small enough to be placed among the eggs in a nest or incubator.[42] The data loggers are connected to a computer to program the unit and to download data. When purchasing data loggers, one should consider factors such as resolution and accuracy of temperature recordings, amount of memory, size of data logger, battery life, and cost. In general, the resolution of many typical data loggers ranges from approximately 0.2 to 0.4°C. All data loggers and temperature probes should be calibrated before and after use. Calibration requires comparison to a thermometer that is traceable to National Institute of Standards and Technology (NIST) or its equivalent in other countries. Temperature recordings from each data logger and/or temperature probe should be taken simultaneously with the NIST-traceable thermometer in an incubator with a stable temperature in the typical range of sea turtle nest temperatures (e.g., approximately 26 to 35°C). If the values are not consistent, a correction factor should be determined for each data logger and temperature probe.

The majority of previous studies of pivotal temperatures and TRT in sea turtles have involved the incubation of eggs at constant temperatures in laboratory incubators. This represents an efficient means of examining the effects of constant temperatures on sex determination because a wide variety of specific temperatures can be examined. In such studies, the thermal characteristics of the incubator should be evaluated prior to the experiments because they can have a profound effect on the results. Factors such as the thermal stability of the incubator and thermal gradients within the incubator should be examined. For example, incubators often develop "thermal inertia" when the heating or cooling system is activated and will often overshoot the preset temperature. Incubators with the smallest amount of thermal inertia are preferred. An incubator that can maintain a preset temperature to plus or minus a few tenths of a degree should be adequate. The cost of the incubator does not necessarily reflect its effectiveness for this sort of study. It is possible to build incubators using polystyrene (Styrofoam) boxes and aquarium heaters for a relatively modest price that may outperform some of the commercial models.[43] Furthermore, the home-built incubators may be the only option if the experiments are performed in the field, where it is not practical to have large commercial incubators.[17]

Incubation temperatures for these studies should be chosen in an attempt to resolutely determine the lower and upper limits of the TRT and the pivotal temperature. Multiple temperatures (and thus incubators) will be needed to identify each parameter. Data from previous studies of other populations of the same species should provide insight on appropriate temperatures to examine. Ideally, eggs should be placed into incubators as soon as possible after they are laid. However, this is not a necessity, because previous studies indicate that the temperature-sensitive period of sex determination is approximately the middle third of the incubation period[44] (reviewed by Ref. 45). Given the increased sensitivity of sea turtle eggs to movement after the first day of incubation, however, it is advisable to place the eggs in the incubators as soon as possible after they are laid.

Because variation in pivotal temperatures has been detected between clutches of eggs, it is important to use multiple clutches of eggs for estimating pivotal temperatures. For example, it is far superior to use 20 eggs from each of five clutches than to use 100 eggs from a single clutch. Ideally, large samples from multiple clutches can be used to evaluate average pivotal temperature and interclutch variation. It is also better to use clutches that have all been laid on the same day, thus synchronizing the development of the eggs. Depending on the type of incubators, the eggs may need to be placed on a moist substrate (e.g., moistened vermiculite) and kept in covered containers to prevent desiccation.[34] Some of the homemade incubators maintain a very high humidity (because of a water-filled heat sink in the bottom of the incubator), so moist substrate is not necessary. Eggs from different clutches should be randomly assigned a position in the incubator to prevent the effect of location within the incubator being misinterpreted as clutch effects. The location of eggs should be rotated periodically (e.g., daily) within the incubator to minimize any position effects in the incubators. Temperature data loggers and/or temperature probes should be placed adjacent to eggs within the incubators, and temperatures should be recorded a minimum of several times per hour throughout the incubation. Upon hatching, the sex of each hatchling must be verified (see Section 4.3.7).

Although most studies have used constant temperatures to analyze pivotal temperature and TRT, studies using fluctuating (e.g., natural) temperature regimes are of distinct interest. Such studies are of particular relevance for sea turtle conservation programs that incubate eggs in natural or translocated nests on natural nesting beaches. Some studies have reported sex ratios relative to natural nest incubation temperatures.[11,46,47] This subject has been addressed experimentally by Georges et al.,[36] and their findings indicate that daily temperature fluctuations can have a profound effect on sex determination, but the effect is dependent upon the magnitude of the fluctuation. The depth of sea turtle nests limits the daily temperature fluctuations (e.g., maximum of 0.3–1.4°C in studies of loggerhead turtles).[48,49] On the basis of the model by Georges et al.,[36] fluctuations of this magnitude would alter the effect of temperature by a maximum of only a few tenths of a degree Celsius in comparison with constant temperatures. This suggests that data from constant-temperature studies are generally applicable to studies examining natural sea turtle nests with fluctuating temperatures. However, this concept is based on a single study that examined temperatures fluctuating around a single mean temperature.[36] Thus, it would be useful to have future studies addressing this subject in sea turtles.

Once sex ratio data are obtained from pivotal temperature and TRT studies, they must be analyzed. Analysis in previous studies has varied from simple estimates to elaborate statistical analyses. A rigorous and standardized statistical method of describing pivotal temperature and TRT has been reported.[50] In addition to describing TSD in a particular population, these statistics allow for comparison of TRT and pivotal temperatures among sea turtle populations.[32,50]

4.3 SEX RATIOS IN SEA TURTLE POPULATIONS

Evolutionary theory suggests that the primary sex ratio (i.e., the sex ratio of the hatchlings) should be 1:1 if parental investment in both sexes (among other factors) is equal.[51,52] However, TSD has the potential of producing a wide variety of sex ratios, and numerous questions arise regarding sex ratio produced from TSD. For example, do naturally occurring sex ratios conform to a 1:1 sex ratio, as suggested by evolutionary theory?[51,52] What range of sea turtle sex ratios occurs in nature? Do sex ratios vary over time and within a population (e.g., seasonal, yearly, by size class, etc.)? What effects does sex ratio have on reproductive success in a population?

Are there optimal sex ratios for the recovery of a population, and should sex ratios be manipulated in an effort to enhance the recovery of an endangered sea turtle population? These are just a few of the questions regarding TSD that confront sea turtle biologists. A prerequisite to answering these questions is the examination of natural sex ratios in sea turtle populations and in conservation programs.

4.3.1 Hatchling Sex Ratios from Nesting Beaches

It is beyond the intended scope of this chapter to provide a critical review of all previous reports of naturally occurring sea turtle sex ratios, however, a general overview will be presented. Examples of hatchling sex ratios that have been predicted for sea turtle nesting beaches are shown in Table 4.3. The reader should be cautioned that there is great variability in the methodology and scope of these studies. For example, some predictions are for a single nesting season, whereas others are for up to 14 different nesting seasons, and the amount of temperature data recorded and/or the amount of sexing data on hatchlings vary. Furthermore, some of these studies rely on pivotal temperature data from other sea turtle populations for sex ratio projections. Regardless, several general points can be derived from these data. First and foremost, the great majority of the predicted sex ratios do not conform to a 1:1 sex ratio suggested by evolutionary theory. Although a few of the sex ratios approach 50% female, there is an obvious predominance of beaches that produce female-biased sex ratios, and some of these biases are extreme (greater than 90% female). No reports exist of extreme male biases over an entire nesting season. On the basis of these data, we are confronted with the possibility that TSD in sea turtles may not conform to the predictions of evolutionary theory by Fisher;[51] rather, a predominance of female biases may exist. However, other explanations are possible. It is plausible that some criteria are not fulfilled regarding fisherian sex ratios, such as sex ratios not being at equilibrium.[31,53] Alternatively, because the sex ratios reported represent only a small sampling of all sea turtle populations and nesting beaches, the results could be affected by sampling bias. Thus, one should be cautious in extrapolating from the limited database that is currently available.

4.3.2 Nest Location and Hatchling Sex Ratios

Several studies have shown that nest location can have a profound influence on hatchling sex ratios. Nesting beaches can have several thermal zones (e.g., beach slope near the water, open beach flat, dune bordering beach, dune with vegetation, etc.) that can influence sex ratio.[46,47,54] Multiple nesting beaches on islands can also provide a variety of thermal environments. Depending on the specific beach chosen by green turtles nesting on a small island (Heron Island) on the Great Barrier Reef, hatchling sex ratios were shown to vary from 29.5 to 63.1% female.[33,55] A similar situation was predicted for green turtles on Ascension Island in which one nesting beach was 2.6°C warmer than another.[56] Indeed, sand color has been shown to be directly related to the thermal properties of nesting beaches.[57A] Thus, the specific location of nesting can significantly affect sex ratio.

TABLE 4.3
Examples of Hatchling Sex Ratios Predicted for Sea Turtle Nesting Beaches

Location	Method for Predicting Sex	Predicted Sex Ratio	Reference
Loggerhead *(Caretta caretta)*			
South Carolina and Georgia	GH	56.3% female	Mrosovsky et al., 1984
Florida	GH	>93% female	Mrosovsky and Provancha, 1988
Florida	GH, BT	87.0–99.9% female	Mrosovsky and Provancha, 1992
Brazil	ID	82.5% female	Marcovaldi et al., 1997
Florida	NT	>90% female	Hanson et al., 1998
Cyprus and Turkey	GH	Female-biased	Kaska et al., 1998
Cyprus	NT	Female-biased	Godley et al., 2001
Green *(Chelonia mydas)*			
Sarawak	ID	74% female	Standora and Spotila, 1985, based on data from Hendrickson, 1958
Suriname	GH, ID	53.9% female	Mrosovsky et al., 1984
Costa Rica	GH	67% female	Spotila et al., 1987
Suriname	GH, BT	68.4% female	Godfrey et al., 1996
Cyprus	GH	Female-biased	Kaska et al., 1998
Cyprus	NT	Highly female-biased	Broderick et al., 2000
Leatherback *(Dermochelys coriacea)*			
Suriname	GH, ID	44% female	Mrosovsky et al., 1984
French Guiana	GH	Nearly 1:1	Rimblot-Baly et al., 1987
Suriname	GH, BT	53.6% female	Godfrey et al., 1996
Costa Rica	BT, NT	93.5–100% female	Binckley et al., 1998
Hawksbill *(Eretmochelys imbricata)*			
Antigua	BT, ID	Not likely to be highly female-biased	Mrosovsky et al., 1992
U.S. Virgin Islands	GH	Female-biased	Wibbels et al., 1999
Brazil	ID	>90% female	Godfrey et al., 1999

Notes: The predicted sex ratio is based on the authors' prediction in percent female (if reported) or is the general description given by the authors (e.g., female bias, etc.). The methodology can vary greatly between these studies, so readers should refer to each individual study for information on number of seasons estimated, seasonal variation, interyear variation, etc. GH = gonadal histology; BT = beach temperature; ID = incubation duration; NT = nest temperature.

4.3.3 Seasonal Variation in Hatchling Sex Ratios

A number of previous studies also indicate that hatching sex ratios can vary over the nesting season. An initial study by Mrosovsky et al.[73] indicated that sex ratios of hatchling loggerheads in South Carolina and Georgia could vary from less than 10% female during the cooler portions of the nesting season to 80% female during the warmer months. Standora and Spotila[46] reviewed incubation duration data from Hendrickson[57B] and predicted seasonal changes in sex ratios of hatchling green turtles at Sarawak, with monsoon season producing predominantly males, whereas nests laid from April through November produced predominantly females. Other examples of seasonal changes in hatchling sex ratios from nesting beaches include loggerheads from Florida,[54] Brazil,[37] and Cyprus,[58] hawksbills from Brazil, leatherbacks from Suriname,[12,53] leatherbacks from French Guiana,[39] and green turtles from Ascension Island[59] and Suriname.[12] Of particular interest, several of these past studies have predicted that the temperature decreases associated with periods of rain can have a profound effect on sex ratios, resulting in the production of male biases.[12,53]

Although seasonal variation in sex ratios may occur on many nesting beaches, it may not occur on all beaches, depending on weather conditions and length of nesting season. For example, beach temperatures were relatively constant at Tortuguero, Costa Rica, during the 1980 nesting season for green turtles, suggesting no seasonal variation in sex ratios.[47] Furthermore, in some situations, incubation temperatures may be high enough that minor temperature fluctuations have little or no effect, and 100% females are produced, as suggested for the 1993–1994 nesting season of leatherbacks at Playa Grande, Costa Rica.[17]

4.3.4 Yearly Variation in Hatchling Sex Ratios

As one might predict, hatching sex ratios from a given nesting beach can also vary from year to year depending on factors such as weather and timing of nesting. Although yearly variations have been reported, these variations are often relatively small for a given nesting beach. For example, sex ratios of hatchling loggerheads from a Florida nesting beach were predicted to be 92.6–96.7% female during 1985, 94.7–99.9% female during 1986, and 87–89% female during 1987.[54] Godfrey et al.[53] predicted a 10% variation between two nesting seasons (1982 vs. 1993) for hatchling sex ratios of both leatherback and green turtles. A study of green turtle nesting beaches on Ascension Island predicted only small interannual differences in beach temperatures, on the basis of 14 years of data.[56]

Variations in annual rainfall have been suggested to affect yearly variations in sex ratios projected during 14 different nesting seasons of leatherback and green sea turtles in Suriname, with annual differences as great as 20–90% female for green turtles.[53] It also is possible that physical changes in a nesting beach can affect year-to-year variation. Because of displacement of beach sand by ocean currents, green and leatherback nesting beaches in Suriname and French Guiana have been moving to the west at approximately 2 km per year.[60] This change in beach location could obviously contribute to yearly variations reported for hatchling sex ratios.[53] Human alteration of beach characteristics could also affect sex ratios produced from a nesting beach. Beach nourishment projects have the potential of changing the thermal characteristics of the beach and thus incubation temperatures.[42] Furthermore, housing developments on nesting beaches have the potential of limiting the nesting area (and thus the range of thermal environments) and also shading nests.[42,61]

4.3.5 Sex Ratios in Immature and Adult Portions of a Population

Sex ratios in other size classes in sea turtle populations (e.g., immature and adult) are also of interest. These groups represent a condensation of many years of hatchling sex ratios, although it must be kept in mind that these sex ratios could also reflect potential sex-specific mortality during development or as adults. For example, energy expenditures associated with egg production and nesting, together with movements onto nesting beaches, could increase the mortality of adult females relative to males. It is also plausible that sex-specific hatchling and post-hatchling mortality could occur because of seasonal variation in hatchling sex ratios. For example, hatchlings produced early in the nesting season might be predominantly one sex and may experience different food availability and predation from hatchlings produced later in the nesting season (which could be predominantly the other sex). Knowledge of sex ratios in the various size–age classes within a population is necessary to begin understanding the potential dynamics of sex ratios within a population. Ultimately, such knowledge would provide insight on the long-term effects of sex ratios on the reproductive ecology of a population.

Our knowledge of adult sex ratios in sea turtle populations is limited because of a number of factors. First, it is logistically difficult to sample adult sea turtles in the ocean. Another problem relates to sex-specific migration patterns of adult males vs. females. For example, Henwood[75] found an approximate 1:1 sex ratio of adult loggerheads captured in waters near Cape Canaveral, FL; however, he recorded seasonal differences in the abundance of adult males vs. adult females. A study of green turtles of the southern Great Barrier Reef reported a 1:1 adult sex ratio, but suggested that the ratio may have been due to sex-biased migratory patterns in which adult males stayed closer to the breeding grounds (and sampling areas) than did the adult females.[62,63]

Human impacts may also affect adult sex ratios. In a study of adult green turtles in Oman, Ross[68] reported an overall sex ratio that was not significantly different from 1:1, but found an excess of females or males in several locations. In the case of excess of males, it was noted that fishermen in that area preferentially hunted for females.[68] Several studies have reported female-biased sex ratios of green turtles based on commercial catches,[64–67] but there is the potential of sampling bias because of fishing practices.[68A] Thus, a number of potential problems can affect the results of adult sex ratio studies. Accurate evaluation of adult sex ratios requires knowledge of migratory behavior (e.g., timing and specific migratory routes) as well as other factors that might impact the sex ratio of turtles captured in a particular sampling location. Regardless of the associated difficulties, studies of adult sex ratios are necessary to fully understand the potential variation and dynamics of sex ratios within a sea turtle population.

Studies of sex ratios in the immature portion of a population (i.e., juvenile to subadult turtles) may provide more accurate estimates of sex ratios, because these studies may not be hampered by sampling bias problems associated with adult breeding migrations (although it is possible that sex-specific behavior may be expressed by juvenile and subadult turtles). A number of studies have examined sex ratios in the immature portion of sea turtle populations, and the results provide insight on the potential variability of sex ratios within a sea turtle population. For loggerheads along the Atlantic coast of the U.S., one study used blood testosterone levels to sex 218 immature turtles in the Atlantic waters off Florida and found a 2:1 (female–male) sex ratio.[69] A similar study used blood testosterone levels to sex loggerheads from four different locations along the Atlantic coast of the U.S., and also found an approximate 2:1 (female–male) sex ratio.[70] Another study used necropsy to sex 139 immature loggerhead turtles that had stranded on the Georgia coast, and also found an approximate 2:1 (female–male) sex ratio.[71] These studies suggest that an approximate 2:1 (female–male) sex ratio may exist in the immature portion of the loggerhead population along the Atlantic coast of the U.S. An approximate 2:1 (female–male) sex ratio has also been reported for immature loggerheads in the Gulf of Mexico.[68B] Although distinctly female-biased, this 2:1 sex ratio is far less than the approximate 9:1 (female–male) hatchling sex ratios suggested for Florida nesting beaches.[54,72]

A number of suggested hypotheses could account for this discrepancy. First, the two Florida beaches examined for hatchling sex ratios may not be representative of all nesting beaches for the loggerheads in the U.S. It is possible that other loggerhead nesting beaches produce greater numbers of males. For example, a study of hatchling loggerhead sex ratios on South Carolina and Georgia beaches (which are part of the northern nesting subpopulation) suggested a sex ratio of approximately 1:1 (female–male).[73] However, the two beaches examined in Florida (Cape Canaveral and Hutchinson Island) represent major nesting beaches for the south Florida nesting subpopulation, which produces approximately 90% of all hatchling loggerheads from U.S. beaches.[74] If other beaches in the south Florida nesting subpopulation also produce hatchling sex ratios of approximately 9:1 (female–male), then even strong male biases on northern or Gulf of Mexico nesting beaches could not contribute enough hatchlings to account for an overall 2:1 (female–male) sex ratio that was suggested for the immature portion of the population. Therefore, other factors must account for the difference. One possibility is that some nesting beaches of the south Florida subpopulation may produce more males. It is also possible that some previous years had relatively cool periods that produced more males. Another hypothesis would be that males may be produced during the early portion of the nesting season[73] and have greater survival rates because of factors such as food availability or longer growth period prior to first winter. An additional possibility is that turtles captured off the U.S. Atlantic coast come from other nesting beaches (e.g., Mexico) that could produce more male-biased ratios. No attempts were made to assign beaches of origin to the samples of turtles in those studies. Adult sex ratios in this population have not been adequately addressed because of the logistical difficulty of sampling adult loggerheads in coastal waters, and also sampling biases associated with sex-specific migration patterns of adult sea turtles.[75] Regardless, the discrepancy between the sex ratios of hatchling vs. immature loggerheads in U.S. waters is intriguing and worthy of future studies.

A similar scenario may occur with the Kemp's ridley sea turtle. Data suggest a strong female-bias (possibly 80–90% female or greater) of hatchling sex ratios produced over 4 recent years,[76] and this could be indicative of previous years because almost all nests have been translocated to the same protected egg "corral" for more than 20 years. Blood testosterone levels were used to predict the sex of 39 immature Kemp's ridley turtles captured along the Gulf coast of Florida and predicted a 1.7:1.0 (female–male) sex ratio.[77] A similar study examined blood testosterone in 42 Kemp's ridley turtles captured along the Gulf coast of Florida and reported a sex ratio of approximately 3.7:1 (female–male).[78] Laparoscopy was used to sex 231 Kemp's ridley turtles captured off the coast of Texas and Louisiana, and found a sex ratio of 1.3:1 (female–male).[79] Necropsy was used to sex 89 stranded Kemp's ridley turtles from the Texas coast, and found a sex ratio of 3:1 (female–male).[80] A similar study examined 89 stranded immature Kemp's ridleys from the Texas coast and found an approximate 2.0:1.0 (female–male) sex ratio.[68B] Thus, the sex ratios reported for the juvenile portion of this population appear female-biased, but the bias is distinctly less than that predicted for recent hatchling sex ratios. As with the loggerhead sex ratios described above, the discrepancy between the sex ratios of hatchling vs. immature ridley turtles is worthy of future studies.

Although female-biased sex ratios have been reported in many populations, a near 1:1 sex ratio has been predicted for the immature portion of the Hawaiian green turtle population on the basis of two different studies. One study used blood testosterone levels to sex 63 juvenile turtles captured in waters off the coasts of Molokai and Hawaii, and predicted a sex ratio of approximately 1:1.[81] A second study used necropsy to examine 421 immature and adult turtles, and the sex ratio of juveniles did not significantly differ from a 1:1 sex ratio.[82] The Hawaiian green turtle represents an interesting population for sex ratio studies because it is an isolated group of green turtles,[83] and data suggest relatively cool incubation temperatures in comparison to those of other green turtle populations.[84] It is possible that this population represents an excellent example of how the thermal environment of a nesting beach may select for a specific pivotal temperature.

Blood testosterone levels were also used to predict the sex of 111 immature green turtles from Bermuda.[85] The results indicated a 1.4:1.0 (female–male) sex ratio, which did not significantly differ from a 1:1 ratio. Genetic data from that group of turtles suggest that the majority originated from Caribbean nesting beaches, particularly Tortuguero, Costa Rica.[86] A previous study at Tortuguero predicted an approximate 2:1 (female–male) hatchling sex ratio,[47] but a more recent study suggested a male-biased (2:3, female–male) hatchling sex ratio from Tortuguero.[87] Compared to the loggerhead and Kemp's ridley data reviewed above, these represent a relatively small difference between predicted hatchling and immature sex ratios; each study involved only a single nesting season and may not be indicative of all nesting seasons at Tortuguero. Nevertheless, this is another example of possible variation of sex ratios within a population.

Chaloupka and Limpus[63] conducted a comprehensive examination of sex ratios within the immature and adult portions of green and loggerhead turtles inhabiting the waters of the southern region of the Great Barrier Reef from 1985 through 1992. Sex ratios were determined for 954 green turtles and 271 loggerheads using laparoscopy. In greens, the sex ratio of immatures fluctuated over the 8-year study, but was consistently female biased. This was contrary to the unbiased sex ratio recorded for immature greens during an earlier study.[88] The yearly fluctuations in these immature sex ratios could be due to sampling bias, but could also reflect factors such as year-to-year variability in hatchling sex ratios. The consistent female bias of immature green turtles recorded over the 8-year study probably reflects an overall female bias of hatchlings produced from the southern Great Barrier Reef.[89] The adult sex ratio of greens from that 8-year study fluctuated significantly and was consistently male-biased, but it was speculated that the male bias could be due to sex-specific migratory patterns of adults, which could cause a sampling bias.[63] In contrast to the green turtle data, the sex ratios of both immature and adult loggerheads from that region remained relatively constant over the 8-year study, and both were significantly male biased.[63] Similarly, in a study on foraging grounds along the Queensland mainland, immature greens had a female-biased sex ratio and immature loggerheads had a male-biased sex ratio.[90,91A]

In summary, male-biased, female-biased, and unbiased sex ratios have been reported in the immature portions of sea turtle populations. Data from a long-term study of green sea turtles suggest that immature sex ratios within a population can fluctuate over time.[63,88] The sex ratio in the immature portion of the population most certainly reflects hatchling sex ratios from previous years, but the exact relationship is not clear because variations between immature and hatchling sex ratios are suggested in some populations. Potential factors such as sex-specific survival rates and sex-specific behavioral differences could enter into the equation. A thorough understanding of sex ratio dynamics within a population may require long-term evaluation of hatchling sex ratios coupled with long-term evaluation within various size classes of immature turtles within a population.

4.3.6 Problems Encountered When Estimating Sex Ratios in a Sea Turtle Population

A variety of hurdles must be surmounted to accurately predict a sex ratio in a sea turtle population. First, a basic need exists to accurately verify or predict the sex of individual turtles. Fortunately, this is not a problem with adult sea turtles because males develop a large muscular tail during puberty (see photo by Wibbels[91B]). However, one should be cautious when assigning sex to minimum-sized adults on the basis of tail length because it is possible to mistake large immature males for small mature females.[92] The primary problem associated with accurately predicting adult sex ratios is potential sampling bias due to sex-specific migratory behavior (specific examples were discussed previously). One might be able to decrease or avoid sampling bias by sampling on a foraging ground during a nonmigratory time of year (e.g., several months after nesting season). Regardless, when estimating adult sex ratios, one should attempt to interpret the results relative to what is known about the migratory behavior in a given population.

Evaluation of sex ratios in the immature portion of the population avoids the problem of sex-specific breeding migrations, but a difficulty arises in identifying the sex of live turtles. A variety of methods have been evaluated for sexing immature sea turtles (reviewed by Ref. 84). Laparoscopy represents a definitive method for verifying sex, but it is difficult in the field and requires surgical training.[88,93] The most widely used method uses testosterone levels in the blood (determined by radioimmunoassay) to predict the sex of individual turtles (reviewed by Ref. 84). The advantage of this method is that the assay is performed in the laboratory, so the field component is limited to obtaining and storing blood samples. Radioimmunoassays can vary among laboratories, so the accuracy of this method is dependent on the validation of a particular radioimmunoassay using samples from turtles of known sex (e.g., sex identified by laparoscopy). In that way, male and female ranges of testosterone levels can be determined and then used to predict the sex of other turtles. It is optimal to validate the assay with samples from turtles from the population being studied. Thus, for accurate predictions, it is imperative to validate the radioimmunoassay that will be used in a particular study.

4.3.7 Predicting Hatchling Sex Ratios

4.3.7.1 Direct Methods for Predicting Hatchling Sex Ratios

Several difficulties are associated with estimating hatchling sex ratios. Again, a major problem is identifying the sex of individual turtles because they do not possess external characteristics that can be used to distinguish sex. Basic techniques used to sex hatchlings have been reviewed in detail.[94] The traditional method has required the dissection of hatchlings and evaluation of the gonads. The most resolute way of doing this is through histological study, in which the gonads show sex-specific characteristics.[94,95] Female hatchlings have a well-developed cortex (i.e., a thickened cortical layer) and poorly organized medullary portion of the gonad, whereas males have little or no cortex and show organization in the medulla (Figure 4.2). Some researchers have suggested that gross morphology of the gonads can be used to determine the sex of hatchlings.[13,15,96] Others have suggested that histology is preferred over gross morphology to prevent errors.[41,97]

Although these methods can provide valuable data, they require either that hatchlings be killed or that hatchlings found dead in nests be used. The availability of an accurate and nonlethal sexing technique for hatchling sea turtles would greatly enhance our ability to investigate sex ratios in sea turtle populations. One method has been proposed for sexing hatchling loggerheads by examining both testosterone and estrogen levels in the blood or chorioallantoic fluid via radioimmunoassay.[98] That study found significantly higher estrogen–testosterone ratios in female hatchlings, with only a minor overlapping of male and female ranges. However, that method did not prove to be effective for sexing pre- and post-hatchling olive ridley turtles.[94] Thus, the accurate use of that technique requires further validation.

Ideally, a nonlethal hatchling sexing technique should be accurate and practical for sexing large numbers of hatchlings. Because sea turtles have TSD, there may not be underlying genetic differences in the DNA that can be used to identify phenotypic sex. Therefore, physiological differences may be the best avenue. Furthermore, if physiological differences can be detected in the blood (e.g., sex-specific

A

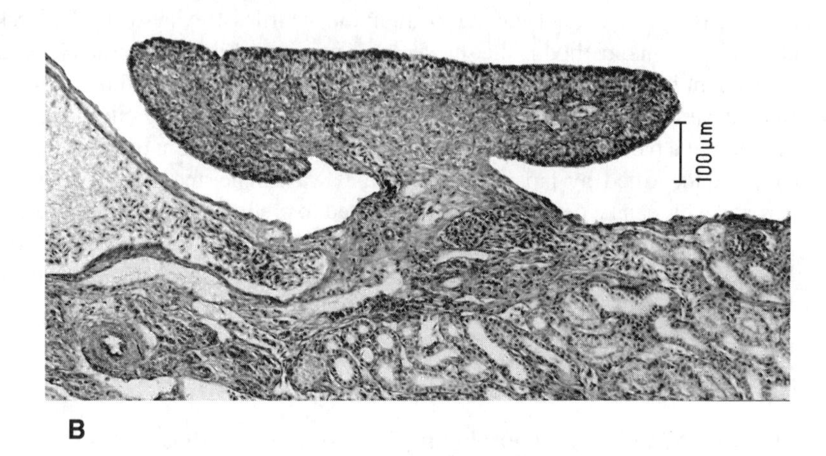

B

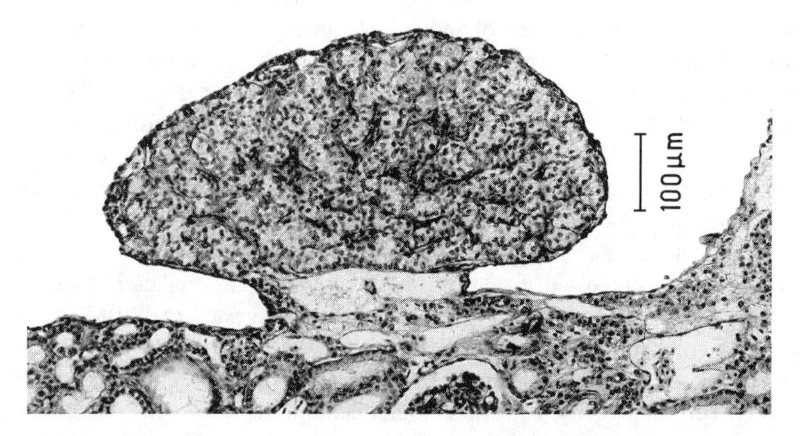

FIGURE 4.2 Hatchling ovary (A) vs. testis (B) in the olive ridley sea turtle. The ovary has a well-developed cortex that stains heavily. The medullary region of the ovary consists of degenerating sex cords. In contrast, in the testis the cortex has degenerated and the medulla has groups of cells (sex cords) that will continue to develop into seminiferous tubules.

hormones), it could prove logistically feasible to sex large numbers of hatchlings, because blood sampling from hatchlings is a relatively simple procedure.[99] However, for now, histological evaluation of the gonads appears to be the most accurate method for sexing individual hatchlings.

4.3.7.2 Indirect Methods for Predicting Hatchling Sex Ratios

The logistical difficulty of sexing individual hatchlings combined with the conservational dilemma associated with the killing of hatchlings for sex ratio studies has resulted in a variety of studies adopting indirect methods for predicting hatchling sex ratios. Beach temperatures, nest incubation temperatures, and incubation durations have been used in a variety of studies to predict hatchling sex ratios.[12,37,48,49,53,54,56,58,59,72,100–102] Previous studies indicate that temperature during the middle third of incubation determines sex in sea turtles.[44,103] Therefore, if the pivotal temperature and TRT are known for the population being studied, nest incubation temperature or even beach temperature (e.g., at nest depth) can be used to predict hatchling sex ratios. For example, an average incubation temperature during the middle third of incubation that is above the upper limit of the TRT would be predicted to produce all females, whereas an average incubation temperature below the lower limit of the TRT would be predicted to yield 100% males. Average incubation temperatures between the pivotal temperature and the upper limit of the TRT would be predicted to produce a female-biased sex ratio, and average temperatures between the pivotal temperature and the lower limit of the TRT would be predicted to produce a male-biased sex ratio.

The availability of small data loggers for recording temperatures provides a practical means of directly recording incubation temperatures within sea turtle nests.[48,72] If direct measurement of incubation temperature is not feasible, then incubation duration provides an alternative method for predicting sex ratios for a given nest.[37,101,102] The total duration of incubation will reflect temperatures during the middle third of incubation, and assuming that temperatures are relatively constant throughout the entire incubation period, incubation duration should be a reliable predictor of sex ratio. In general, longer incubation durations are indicative of male-biased sex ratios, and shorter incubation durations are indicative of female-biased sex ratios.

It is significant that the accuracy of these indirect methods is dependent on validation data that are based on direct sexing techniques. For example, when incubation temperatures are used, the accuracy of the sex ratio predictions is dependent on the validity of the pivotal temperature and TRT data for that population.

4.3.7.3 Predicting an Overall Sex Ratio for a Nesting Beach

One of the primary reasons for developing the sexing techniques discussed above is to provide a means of examining the overall hatchling sex ratios produced from nesting beaches. Once a method has been chosen for estimating sex ratios from nests, you then need to develop an effective strategy for evaluating the sex ratio produced from the nesting beach of interest. This subject has been reviewed previously in detail.[21] Major points that should be considered include the spatial and temporal diversity of nests during a nesting season. In most situations, it is not possible to monitor temperature in all nests, so subsets of nests that accurately represent the diversity and abundance of nest locations on the nesting beach can be used (e.g., beach flat, dune, and vegetation zone). In addition, nests should be examined throughout the nesting season to evaluate any temporal variation in sex ratio during a nesting season. As indicated previously, seasonal variation in hatchling sex ratios has been detected in a number of studies. Weather parameters such as air temperature, rainfall, and cloud cover should be recorded throughout the nesting season for possible correlation with sex ratios. Studies have also shown annual variations, which suggest that multiyear data are critical to understanding sex ratio dynamics of sea turtle populations. Multiyear studies of sex ratios from a nesting beach should consistently address spatial and temporal diversity of nests during each of the nesting seasons to make year-to-year comparisons meaningful.

4.3.8 Manipulation of Hatchling Sex Ratios

Because sea turtles possess TSD, it is possible to artificially alter hatchling sex ratios. However, there has been considerable debate as to whether sex ratio manipulation is an appropriate conservation strategy considering our limited knowledge of TSD and the effects of sex ratios on reproductive ecology in sea turtles.[40,41,104,105] Historically, sex ratio manipulations began as an unintentional byproduct of conservation practices for protecting eggs. In some conservation programs, eggs were transferred to polystyrene boxes and incubated in egg hatcheries to prevent predation and poaching, and to enhance hatching success.[106] This practice can skew sex ratios, often by cooling the eggs and thus increasing the production of males.[24,106] Some conservation programs use egg hatcheries in which eggs are moved to protected locations and buried. This practice has the potential of altering incubation temperatures.[107] As an example, the majority of Kemp's ridley eggs have been moved to egg hatcheries for several decades.[108,109] Evaluation of hatchling sex ratios from those hatcheries in recent years has indicated a strong female bias in this recovering species.[76] It is not clear whether this bias reflects the natural sex ratio that would be produced if nests were left *in situ*, but the egg corrals most certainly do not contain the diversity of incubation temperatures that are present on the natural nesting beach. Regardless, it is clear that sex ratios can be manipulated, so the question is whether sex ratios should be intentionally manipulated to enhance the recovery of a population.

On the surface it seems logical that strong male biases would not be advantageous for the recovery of a population, and that the production of female biases could enhance recovery by increasing egg production.[40] However, several factors should be considered prior to such manipulations. First, it is not clear what proportion of males is needed for successful fertilization of most or all females in a population.[41,104] Even if a small minority of males could fertilize a large majority of females, the artificial skewing of sex ratios could alter ecological factors such as intra- and intersexual competition, sperm competition, and multiple paternity, which might affect the reproductive ecology and evolution of a species.[104] Thus, intentionally producing a female-biased sex ratio may have both advantages and disadvantages.

At the very least, nesting beach conservation programs should attempt to monitor hatchling sex ratios, especially if nests are translocated to hatcheries or to safer locations on the beach (e.g., above high tide line). If egg hatcheries are used, the resulting sex ratio should be considered when choosing the location of the hatchery.[107] It may be useful to have two hatcheries with different thermal environments that would allow the production of males or females, depending on the hatchery in which eggs are placed.[40,41] For egg hatcheries, a decision must be made about a desired sex ratio. One option would be to monitor incubation temperatures in natural nests on the nesting beach and attempt to duplicate the natural sex ratio. If the natural sex ratio is not known or cannot be predicted, an alternative is to choose a sex ratio based on its potential for enhancing the recovery of the population (discussed above) or based on sex ratios from other populations. As reviewed previously, a variety of sex ratios have been reported in sea turtle populations, with many having female biases. It is not clear, however, it those data represent a random sampling of all sea turtle nesting beaches. It is plausible that highly biased sex ratios are more often reported because there is an inherent interest to find and publish extreme sex ratios. As such, this issue is still speculative.

In summary, the manipulation of hatchling sex ratios can most certainly have an effect on the reproductive ecology and recovery of sea turtle populations. Unfortunately, we do not understand all the ecological and evolutionary ramifications of artificially skewing a sex ratio. Although it seems logical that female biases would be preferred over male biases in a conservation program for a recovering population, one should be cautious when manipulating sex ratios until more data are available on this subject.

4.4 PHYSIOLOGY OF TSD

The majority of studies of sex determination in sea turtles have focused on pivotal temperatures, TRTs, and sex ratios. Although some studies have addressed the physiology underlying sex determination in sea turtles, the majority of information regarding the physiology of TSD in turtles has been documented in freshwater turtles, many of which have a similar pattern of sex determination (i.e., MF pattern) and similar pivotal temperatures and TRTs.[110,111] As such, this section of the chapter will focus on sea turtles, but also include information from freshwater turtles when applicable.

4.4.1 Thermosensitive Period and Gonadal Differentiation

Studies of loggerhead[44] and olive ridley sea turtles[103] indicate that temperature affects sex determination during the approximate middle third of incubation. This is consistent with data reported for freshwater turtles.[45] Temperature appears to have both a cumulative and a quantitative effect on sex determination.[112] That is, an egg must be exposed to a male-producing or a female-producing temperature for an extended time period before sex is determined, so a short-duration spike of male-producing or female-producing temperature will not irreversibly determine sex. For turtles with the MF pattern of sex determination, the warmer the temperature, the more potent it is for producing females, and the cooler the temperature, the more potent it is for producing males.[112]

Gonadal differentiation has been described in the olive ridley sea turtle, and the first signs of sexual differentiation occur toward the end of the thermosensitive period.[14] Sexual differentiation at female-producing temperatures includes the thickening of cortical epithelial tissue (i.e., a proliferation of the gonad's cortex) and degeneration of medullary cords (i.e., regression of the gonad's medulla). In contrast, at male-producing temperatures, the cortex does not proliferate, and the medullary cords do not regress and eventually will develop into seminiferous tubules. By the time of hatching, an ovary can be histologically distinguished from a testis (Figure 4.2).

4.4.2 Estrogen Hypothesis

The physiological cascade underlying TSD is not well understood, but the leading hypothesis is that female-producing incubation temperatures stimulate the production of estrogen by the gonad, and the estrogen then stimulates the gonad to differentiate into an ovary.[110] This hypothesis is supported by a number of findings. Numerous studies of turtles have shown that exogenous estrogen injected into or applied topically to eggs can cause the production of females from eggs incubated at male-producing temperatures.[45] This includes a study of the olive ridley in which females were produced by treating eggs with estrogen.[103] Furthermore, treatment of turtle eggs with aromatase inhibitors (aromatase is the enzyme that produces estrogen from androgens) results in the masculinization of gonads.[113–116] In the European pond turtle, *Emys orbicularis*, higher estrogen levels and higher aromatase enzyme activity occur in the developing ovaries in comparison to the testes.[117,118] A study in the leatherback sea turtle also revealed higher aromatase enzyme activity in developing ovaries.[119] In a study of the diamondback terrapin, *Malaclemys terrapin*, higher levels of aromatase mRNA were recorded at female-producing temperatures, in comparison to male-producing temperatures, during the thermosensitive period of TSD.[120] Collectively, the findings from these studies support the hypothesis that estrogen production may play a pivotal role in the sex determination cascade of reptiles with TSD.

Although there is strong support for the involvement of estrogen in TSD, the results from studies of several reptiles do not support this hypothesis. Studies of the olive ridley sea turtle;[121] the red-eared slider turtle, *Trachemys scripta*;[122–124] the saltwater crocodile, *Crocodylus porosus*;[125,126] and the American alligator, *Alligator mississippiensis*[127,128] did not detect female-specific elevations in estrogen levels, aromatase levels, and/or aromatase enzyme activity in the gonads during the thermosensitive period of TSD. However, some of these studies did detect elevations after the thermosensitive period. Thus, although the estrogen hypothesis has substantial support, other studies suggest that elevated estrogen levels may be only a downstream event that occurs after sex determination in the ovary.

Alternatively, some findings suggest that the brain (rather than the gonads) may be a source of estrogen production during early development.[120,121] Consistent with the hypothesis of brain involvement in TSD, a study of the olive ridley indicates that the nervous system innervates the gonads prior to their sexual differentiation.[129]

4.4.3 Genetics of TSD

In contrast to birds and mammals, heteromorphic sex chromosomes have not been identified in reptiles with TSD, including the sea turtles.[84] In fact, it has been hypothesized that the XX/XY sex-determining system in mammals and the ZZ/ZW system in birds have evolved from autosomes of ancestral vertebrates with TSD.[130] A number of genes have been evaluated as potential factors in the sex determination–gonadal differentiation cascade of TSD in reptiles. The testis-determining gene in mammals (i.e., sex-determining region Y [SRY]) has not been detected in reptiles or birds, and is believed to have evolved well after the mammalian ancestors diverged from stem amniotic vertebrates.[130] In fact, SRY has not even been identified in primitive mammals; i.e., monotreme mammals.[130]

SOX9 is an SRY-related gene that appears to have a conserved role in testis differentiation.[131,132] Studies of the olive ridley indicate that SOX9 is expressed in gonads that are differentiating as testes, but its expression is downregulated in gonads developing as ovaries.[133–135] Although these findings suggest that SOX9 may be involved in testis differentiation, studies of TSD in the alligator indicate that SOX9 expression is a downstream event and is not the testis-determining gene.[136]

DMRT-1 is another gene that exhibits elevated expression in the differentiating testis of mammals[137] and birds,[138] and is required for testis differentiation in mammals.[139] DMRT-1 could potentially be involved in TSD because it has been shown to be expressed in the testis of the red-eared slider,[140] but it has not been studied in sea turtles.

Anti-müllerian hormone (AMH, also called müllerian inhibiting substance [MIS]) is produced by the differentiating testis and causes the müllerian ducts to degenerate in male vertebrates.[141] AMH has been identified in reptiles with TSD,[136,142] and it is expressed during the thermosensitive period in the alligator.[136] AMH has not been investigated in sea turtles.

Steroidogenic factor-1 (SF-1) appears to be a master regulator of steroidogenic genes, and it is required for the development of the gonads, the adrenal glands, and the ventromedial hypothalamus.[143–147] SF-1 also regulates the AMH gene.[148] SF-1 has been shown to have a sex-specific pattern of expression in mammals[149,150] and birds.[151] One of the steroidogenic genes regulated by SF-1 is the aromatase gene, so SF-1 could have a role in TSD if estrogen is involved (see Section 4.4.2). SF-1 has been identified in reptiles with TSD,[111,152] and has been shown to have a sex-specific expression pattern in the alligator and in a freshwater turtle with TSD.[152,153] SF-1 has not been examined in sea turtles.

The gene producing DAX1 has been implicated in mammalian sex determination because overexpression of DAX1 is associated with male-to-female sex reversal.[154,155] DAX1 has been identified in a reptile with TSD (the alligator), but no sex-specific pattern of expression was detected.[152] DAX1 has not been investigated in sea turtles.

The gene producing Wilms tumor 1 (WT1) is necessary for the proper development of the kidneys and gonads in mammals.[156] WT1 has also been hypothesized to act synergistically with SF-1 in regulating AMH, whereas DAX1 antagonizes this synergy.[157] WT1 has been identified in the alligator, but no sex-specific pattern of expression has been detected during TSD.[152] WT1 has not been investigated in sea turtles.

In summary, the genetics of TSD is not well understood, but a number of potential factors in the sex determination cascade have been identified. Although the testis-determining gene in mammals (SRY) does not appear to be present in reptiles, many other genes in the sex determination–sex differentiation cascade are conserved in amniotic vertebrates, and some could potentially be involved in TSD. In addition, as the specific functions of these genes are elucidated, the results will provide insight into the putative involvement of estrogen in TSD. Although only a few studies have addressed the genetics of sex determination in sea turtles, information from other reptiles with TSD can act as a template for designing studies that can efficiently evaluate the potential role of specific genes in the sex determination of sea turtles.

From a conservational viewpoint, some freshwater turtles may represent a more practical subject for TSD studies (in comparison to sea turtles) because the eggs of some species are available commercially in large numbers from captive breeding operations.[111]

ACKNOWLEDGMENTS

The research from our laboratory that was cited in this review was made possible through previous support by MS-AL Sea Grant Consortium, the Alabama Center for Estuarine Studies, the National Marine Fisheries Service, and the University of Alabama at Birmingham. The author would like to acknowledge Alyssa Geis and Chris Murdock for critiquing this manuscript.

REFERENCES

1. Charnier, M. Action de la temperature sur la sex-ratio chez l'embryon d'Agama agama (Agamidaie, Lacertilien). *C.R. Seanc. Soc. Biol.* 160, 620–622. 1966.
2. Bull, J.J. Sex determination in reptiles. *Q. Rev. Biol.* 55, 3–21. 1980.
3. Janzen, F. and Paukstis, G.L. Environmental sex determination in reptiles: ecology, evolution, and experimental design. *Q. Rev. Biol.* 66, 149–179. 1991.
4. Cree, A., Thompson, M.B., and Daugherty, C.H. Tuatara sex determination. *Nature* 375, 543. 1995.
5. Marshall Graves, J.A. and Shetty, S. Sex from W to Z: evolution of vertebrate sex chromosomes and sex determining genes. *J. Exp. Zool.* 290, 449–462. 2001.
6. Reinhold, K. Nest-site philopatry and selection for environmental sex determination. *Evol. Ecol.* 12, 245–250. 1998.
7. Shine, R. Why is sex determined by nest temperature in many reptiles? *Trends Ecol. Environ.* 14, 186–189. 1999.
8. Girondot, M. A fifth hypothesis for the evolution of TSD in reptiles. *Trends Ecol. Environ.* 14, 359–360. 1999.
9. Yntema, C.L. and Mrosovsky, N. Sexual differentiation in hatchling loggerheads (*Caretta caretta*) incubated at different controlled temperatures. *Herpetologica* 36, 33–36. 1980.
10. Miller, J.D. and Limpus, C.J. *Proceedings of the Melbourne Herpetological Symposium*, The Zoological Board of Victoria, Australia, 1981.
11. Morreale, S.J. et al. Temperature dependent sex determination: current practices threaten conservation of sea turtles. *Science* 216, 1245–1247. 1982.
12. Mrosovsky, N., Dutton, P.H., and Whitmore, C.P. Sex ratios of two species of sea turtles nesting in Suriname. *Can. J. Zool.* 62, 2227–2239. 1984.
13. McCoy, C.J., Vogt, R.C., and Censky, E.J. Temperature-controlled sex determination in the sea turtle *Lepidochelys olivacea*. *J. Herpetol.* 17, 404–406. 1983.
14. Merchant-Larios, H., Fierro, I.V., and Urruiza, B.C. Gonadal morphogenesis under controlled temperature in the sea turtle *Lepidochelys olivacea*. *Herpetol. Monogr.* 1989, 43–61. 1989.
15. Wibbels, T., Rostal, D.C., and Byles, R. High pivotal temperature in the sex determination of the olive ridley sea turtle from Playa Nancite, Costa Rica. *Copeia* 1998, 1086–1088. 1998.

16. Rimblot, F. et al. Sexual differentiation as a function of the incubation temperature of eggs in the sea turtle *Dermochelys coriacea* (Vandelli, 1761). *Amphibia-Reptilia* 6, 83–92. 1985.
17. Binckley, C.A. et al. Sex determination and sex ratios of Pacific leatherback turtles, *Dermochelys coriacea*. *Copeia* 2, 291–300. 1998.
18. Dalrymple, G.H., Hampp, J.C., and Wellins, J. Male biased sex ratio in a cold nest of a hawksbill sea turtle (*Eretmochelys imbricata*). *J. Herpetol.* 19, 158–159. 1985.
19. Horrocks, J.A. and Scott, N.M. Nest site location and nest success in the hawksbill turtle *Eretmochelys imbricata* in Barbados, West Indies. *Mar. Ecol. Prog. Ser.* 69, 1–8. 1991.
20. Mrosovsky, N. et al. Pivotal and beach temperatures for hawksbill turtles nesting in Antigua. *Can. J. Zool.* 70, 1920–1925. 1992.
21. Godfrey, M.H. and Mrosovsky, N. Estimating hatchling sex ratios. In: *Research and Management Techniques for the Conservation of Sea Turtles*, Eckert, K.L. et al., Eds., IUCN/SSC Marine Turtle Specialist Group Publication No. 41, 1999.
22. Aguilar, H.R. Influencia de la temperatura de incubacion sobre la determination del sexo y la duracion del periodoe incubacion en la tortuga lora (*Lepidochelys kempi*, Garman, 1880). Instituto Politecnico Nacional Mexico, D.F., 1987.
23. Shaver, D.J. et al. Styrofoam box and beach temperatures in relation to incubation and sex ratios of Kemp's ridley sea turtles. In: *Proceedings of the Eighth Annual Workshop on Sea Turtle Biology and Conservation*, Schroeder, B.A., Ed., 1988, pp. 103–108.
24. Wibbels, T. et al. Predicted sex ratios from the International Kemp's Ridley Sea Turtle Head Start Research Project. In: *Proceedings of the First International Symposium on Kemp's Ridley Sea Turtle Biology, Conservation and Management*, Caillouet, C.W., Jr. and Landry, A.M., Jr., Eds., Department of Commerce, 1989, pp. 82–89.
25. Diaz, A.C. Importancia de la temperatura de incubacion en la determinacion sexual de la tortuga negra, *Chelonia agassizii*, en la playa, de Colola, Mich. Mexico. In: *Escuela de Biologma*, Universidad Michoacana de San Nicolas de Hidalgo, 1986.
26. Hernandez, M.G. Temperatura de incubacion en huevos de tortuga negra (*Chelonia agassizii*) y su influencia en el sexo de las cmas en la playa de Colola, Mich. Mexico. In: *Facultad de Biologma*, Universidad Michoacana de San Nicolas de Hidalgo, 1995.
27. Andrade, R.G. Proporcion sexual natural en crmas de tortuga negra *Chelonia agassizii* (Bocourt, 1868) en Michoacan, Mexico. In: *Facultad de Biologma*, Universidad Michoacana de San Nicolas de Hidalgo, 2002.
28. Hewavisenthi, S. and Parmenter, C.J. Hydric environment and sex determination in the flatback turtle (*Natator depressus* Garman) (Chelonia; Cheloniidae). *Aust. J. Zool.* 48, 653–659. 2000.
29. Ewert, M.A., Jackson, D.R., and Nelson, C.E. Patterns of temperature-dependent sex determination in turtles. *J. Exp. Zool.* 270, 3–15. 1994.
30. Mrosovsky, N. and Pieau, C. Transitional range of temperature, pivotal temperature and thermosensitive stages for sex determination in reptiles. *Amphib. Reptilia* 12, 169–179. 1991.
31. Mrosovsky, N. Sex ratios of sea turtles. *J. Exp. Zool.* 270, 16–27. 1994.
32. Chevalier, J., Godfrey, M.H., and Girondot, M. Significant difference of temperature-dependent sex determination between French Guiana (Atlantic) and Playa Grande (Costa-Rica, Pacific) leatherbacks (*Dermochelys coriacea*). *Ann. Sci. Nat.* 20, 147–152. 1999.

33. Limpus, C.J., Reed, P.C., and Miller, J.D. Temperature dependent sex determination in Queensland sea turtles: intraspecific variation in *Caretta caretta*. In: *Biology of Australasian Frogs and Reptiles*, 1985.
34. Mrosovsky, N. Pivotal temperatures for loggerhead turtles (*Caretta caretta*) from northern and southern nesting beaches. *Can. J. Zool.* 66, 661–669. 1988.
35. Etchberger, C.R. et al. Effects of oxygen concentration and clutch on sex determination and physiology in red-eared slider turtles (*Trachemys scripta*). *J. Exp. Zool.* 258, 394–403. 1991.
36. Georges, A., Limpus, C.J., and Stoutjesdijk, R. Hatchling sex in the marine turtle *Caretta caretta* is determined by proportion of development at a temperature, not daily duration of exposure. *J. Exp. Zool.* 270, 432–444. 1994.
37. Marcovaldi, M.A., Godfrey, M.H., and Mrosovsky, N. Estimating sex ratios of loggerhead turtles in Brazil from pivotal incubation durations. *Can. J. Zool.* 75, 755–770. 1997.
38. Maxwell, J.A., Motara, M.A., and Frank, G.H. A micro-environmental study of the effect of temperature on the sex ratios of the loggerhead turtle, *Caretta caretta*, from Tongaland, Natal. *S. Afr. J. Zool.* 23, 342–350. 1988.
39. Rimblot-Baly, F. et al. Sensibilite a la temperature de la differenciation sexuelle chez la Tortue Luth, *Dermochelys coriacea* (Vandelli, 1971); application des donnees de l'incubation artificielle a l'etude de la sex-ratio dans la nature. *Ann. Sci. Nat. Zool. Paris* 8, 277–290. 1987.
40. Vogt, R.C. Temperature controlled sex determination as a tool for turtle conservation. *Chelonian Conserv. Biol.* 1, 159–162. 1994.
41. Mrosovsky, N. and Godfrey, M.H. Manipulating sex ratios: turtle speed ahead! *Chelonian Conserv. Biol.* 1, 238–240. 1995.
42. Hanson, J., Wibbels, T., and Martin, R.E. Use of miniature temperature data loggers to estimate sex ratios of hatchling loggerhead sea turtles. In: *Proceedings of the 18th International Sea Turtle Symposium*, NOAA Technical Publications, 1988.
43. Lang, J., Andrews, H.V., and Whitaker, R. Sex determination and sex ratios in *Crocodylus palustris*. *Am. Zool.* 29, 935–952. 1989.
44. Yntema, C.L. and Mrosovsky, N. Critical periods and pivotal temperatures for sexual differentiation in loggerhead turtles. *Can. J. Zool.* 60, 1012–1016. 1982.
45. Wibbels, T., Bull, J.J., and Crews, D. Temperature-dependent sex determination: a mechanistic approach. *J. Exp. Zool.* 270, 71–78. 1994.
46. Standora, E.A. and Spotila, J.R. Temperature dependent sex determination in sea turtles. *Copeia* 3, 711–722. 1985.
47. Spotila, J.R. et al. Temperature dependent sex determination in the green turtle (*Chelonia mydas*): effects on the sex ratio on a natural nesting beach. *Herpetologica* 43, 74–81. 1987.
48. Kaska, Y. et al. Natural temperature regimes for loggerhead and green turtle nests in the eastern Mediterranean. *Can. J. Zool.* 76, 723–729. 1998.
49. Godley, B.J. et al. Thermal conditions in nests of loggerhead turtles: further evidence suggesting female skewed sex ratios of hatchling production in the Mediterranean. *J. Exp. Mar. Biol. Ecol.* 263, 45–63. 2001.
50. Girondot, M. Statistical description of temperature-dependent sex determination using maximum likelihood. *Evol. Ecol. Res.* 1, 479–486. 1999.
51. Fisher, R.A. *The Genetical Theory of Natural Selection*, Clarendon Press, Oxford, 1930.
52. Charnov, E.L. and Bull, J.J. The primary sex ratio under environmental sex determination. *J. Theor. Biol.* 139, 431–436. 1989.

53. Godfrey, M.H., Barreto, R., and Mrosovsky, N. Estimating past and present sex ratios of sea turtles in Suriname. *Can. J. Zool.* 74, 267–277. 1996.
54. Mrosovsky, N. and Provancha, J. Sex ratio of hatchling loggerhead sea turtles: data and estimates from a 5-year study. *Can. J. Zool.* 70, 530–538. 1992.
55. Limpus, C.J., Reed, P.C., and Miller, J.D. Islands and turtles: The influence of choice of nesting beach on sex ratio. In: *Inaugural Great Barrier Reef Conference*, Baker, J.T. et al., Eds., JCU Press, 1983, pp. 397–402.
56. Hays, G.C., Godley, B.J., and Broderick, A.C. Long-term thermal conditions on the nesting beaches of green turtles on Ascension Island. *Mar. Ecol. Prog. Ser.* 185, 297–299. 1999.
57A. Hays, C.G. et al. The importance of sand albedo for the thermal conditions on sea turtle nesting beaches. *Oikos* 93, 87–95. 2001.
57B. Hendrickson, J.R. The green sea turtle, *Chelonia mydus*, in Malaysia and Sarawak, *Proc. Zool. Soc. London* 30, 455-535.
58. Broderick, A.C. et al. Incubation periods and sex ratios of green turtles: highly female biased hatchling production in the eastern Mediterranean. *Mar. Ecol. Prog. Ser.* 202, 273–281. 2000.
59. Godley, B.J. et al. Temperature-dependent sex determination of Ascension Island green turtles. *Mar. Ecol. Prog. Ser.* 226, 115–124. 2002.
60. Schulz, J.P. Sea turtles nesting in Surinam. *Zool. Verh. (Leiden)* 143, 1–143. 1975.
61. Mrosovsky, N., Lavin, C., and Godfrey, M.H. Thermal effects of condominiums on a turtle beach in Florida. *Biol. Conserv.* 74, 151–156. 1995.
62. Limpus, C.J. et al. Migration of green (*Chelonia mydas*) and loggerhead (*Caretta caretta*) turtles to and from eastern Australian rookeries. *Wildl. Res.* 19, 347–358. 1992.
63. Chaloupka, M. and Limpus, C. Trends in the abundance of sea turtles resident in southern Great Barrier Reef waters. *Biol. Conserv.* 102, 235–249. 2001.
64. Carr, A. and Giovannoli, L. The ecology and migrations of sea turtles. 2. Results of fieldwork in Costa Rica 1955. *Am. Mus. Novit.* 1957.
65. Caldwell, D.K. Carapace length–body weight relationship and size and sex ratio of the northeastern Pacific green turtle, *Chelonia mydas carrinegra. Contrib. Sci.* 62, 3–10. 1962.
66. Hirth, H.F. and Carr, A. The green turtle in the Gulf of Aden and the Seychelles Islands. *Ver. Kon. Ned. Adad Wet. Afd. Natur.* 58, 1970.
67. Mortimer, J.A. The feeding ecology of the west Caribbean green turtle (*Chelonia mydas*) in Nicaragua. *Biotropica* 13, 49–58. 1981.
68A. Ross, P.R. Adult sex ratio in the green sea turtle. *Copeia* 1984, 776–778. 1984.
68B. Stabenau, E.K., Sanley, K.S., and Landry, A.M. Jr. Sex ratios of sea turtles on the upper Texas coast. *J. Herpetol.* 30, 427–430, 1996.
69. Wibbels, T. et al. Female-biased sex ratio of immature loggerhead sea turtles inhabiting the Atlantic coastal waters of Florida. *Can. J. Zool.* 69, 2973–2977. 1991.
70. Wibbels, T. et al. Sexing techniques and sex ratios for immature loggerhead sea turtles captured along the Atlantic Coast of the United States. In: *Ecology of East Florida Sea Turtles*, U.S. Dept. Commerce NOAA Technical Report NMFS 53, 1987.
71. Shoop, C.R., Ruckdeschel, C.A., and Kenney, R.D. Female-biased sex ratio of juvenile loggerhead sea turtles in Georgia. *Chelonian Conserv. Biol.* 3 (1), 93–96. 1998.
72. Hanson, J., Wibbels, T., and Martin, R. Predicted female bias in sex ratios of hatchling loggerhead sea turtles from a Florida nesting beach. *Can. J. Zool.* 76, 1–12. 1998.
73. Mrosovsky, N., Hopkins-Murphy, S.R., and Richardson, J.I. Sex ratio of sea turtles: seasonal changes. *Science* 225, 739–741. 1984.

74. TEWG, T.E.W.G. Assessment update for the Kemp's ridley and loggerhead sea turtle populations in the Western North Atlantic, U.S. Dept. Commerce, NOAA Technical Memorandum, NMFS-SEFSC-444, 2000, 115 pp.
75. Henwood, T. Movements and seasonal changes in loggerhead turtle *Caretta caretta* aggregations in the vicinity of Cape Canaveral, Florida (1978–84). *Biol. Conserv.* 40, 191–202. 1987.
76. Geis, A. et al. Predicted sex ratios of hatchling Kemp's ridleys produced in egg corrals during the 1998, 1999, and 2000 nesting seasons. In: *Proceedings of the 21st International Sea Turtle Symposium*, 2001.
77. Gregory, L.F. and Schmid, J.R. Stress responses and sexing of wild Kemp's ridley sea turtles (*Lepidochelys kempii*) in the northeastern Gulf of Mexico. *Gen. Comp. Endocrinol.* 124, 66–74. 2001.
78. Geis, A., Barichivich, J., and Wibbels, T. Use of testosterone RIA to estimate sex ratio of juvenile Kemp's ridley turtles captured at Cedar Key, FL. *Amer. Zool.* 41, 1451. 2001.
79. Coyne, M.S. Population sex ratio of the Kemp's ridley sea turtle (*Lepidochelys kempii*): problems in population modeling, Texas A&M University, 2000, pp. 96–119.
80. Cannon, A.C. Gross necropsy results of sea turtles stranded on the upper Texas and western Louisiana coasts, 1 January–31 December 1994. In: *Characteristics and Causes of Texas Marine Strandings*. R. Zimmerman, Ed., U.S. Dept. Commerce, NOAA Technical Report NMFS 143, 1998, pp. 81–85.
81. Wibbels, T. et al. Sex ratio of immature green turtles inhabiting the Hawaiian archipelago. *J. Herpetol.* 27 (3), 327–329. 1993.
82. Koga, S.K. and Balazs, G.H. Sex ratios of green turtles stranded in the Hawaiian Islands. In: *Proceedings of the Fifteenth Annual Symposium on Sea Turtle Biology*, J.A. Keinath et al., Eds., pp. 148–151. NOAA Amphibia-Reptilia 12, 169–179. 1996.
83. Bowen, B.W. et al. Global population structure and natural history of the green turtle (*Chelonia mydas*) in terns of matriarchal phylogeny. *Evolution* 46, 865–881. 1992.
84. Wibbels, T., Owens, D.W., and Limpus, C.J. Sexing juvenile sea turtles: is there an accurate and practical method? *Chelonian Conserv. Biol.* 2000.
85. Bolten, A.B. et al. Sex ratio and sex-specific growth rates of immature green turtles, *Chelonia mydas*, in the southern Bahamas. *Copeia* 4, 1098–1103. 1992.
86. Lahanas, P.N. et al. Genetic composition of a green turtle (*Chelonia mydas*) feeding ground population: evidence for multiple origins. *Mar. Biol.* 130, 345–352. 1998.
87. Horikoshi, K. Sex ratio of green turtle hatchlings in Tortuguero, Costa Rica. In: *Proceedings of the Eleventh Annual Workshop on Sea Turtle Biology and Conservation*, NMFS-SEFSC-302, 1991, pp. 59–60.
88. Limpus, C.J. and Reed, P.C. The green turtle, *Chelonia mydas*, in Queensland: a preliminary description of the population structure in a coral reef feeding ground. In: *Biology of Australasian Frogs and Reptiles*, 1985, pp. 47–52.
89. Limpus, C.J., Fleay, A., and Guinea, M. Sea turtles of the Capricornia Section, Great Barrier Reef. In: *The Capricornia Section of the Great Barrier Reef Marine Park: Past, Present and Future*, Ward, W.T. and Saenger, P., Eds., Royal Society of Queensland and Australian Coral Reef Society, Brisbane, 1984, pp. 61–78.
90. Limpus, C.J., Couper, P.J., and Read, M.A. The green turtle, *Chelonia mydas*, in Queensland, population structure in warm temperate feeding area. *Mem. Queensland Mus.* 35, 139–154. 1994a.

91A. Limpus, C.J., Couper, P.J., and Read, M.A. The loggerhead turtle, *Caretta caretta*, in Queensland, population structure in a warm temperate feeding area. *Mem. Queensland Mus.* 37, 195–204. 1994b.

91B. Wibbels, T. Diagnosing the sex of sea turtles in foraging habitats. In: *Research and Management Techniques for the Conservation of Sea Turtles,* Eckert, K.L. et. al., Eds., IUCN/SSC Marine Turtle Specialist Group Publication No. 4, 1999, pp. 139–143.
92. Limpus, C.J. A study of the loggerhead sea turtle, *Caretta caretta*, in eastern Australia, Ph.D. dissertation, University of Queensland, Australia, 1985.
93. Wood, J.R. et al. Laparoscopy of the green sea turtle. *Br. J. Herpetol.* 6, 323–327. 1983.
94. Merchant-Larios, H. Determining hatchling sex. In: *Research and Management Techniques for the Conservation of Sea Turtles*. Eckert, K.L. et al., Eds., IUCN/SSC Marine Turtle Specialist Group Publication No. 4, 1999, pp. 130–135.
95. Mrosovsky, N. and Yntema, C.L. Temperature dependence of sexual differentiation in sea turtles: implications for conservation practices. *Biol. Conserv.* 18, 271–280. 1980.
96. van der Heiden, A.M., Briseno-Duenas, R., and Rios-Olmeda, D. A simplified method for determining sex in hatchling sea turtles. *Copeia* 1985, 779–782, 1985.
97. Mrosovsky, N. and Benabib, M. An assessment of two methods of sexing hatchling sea turtles. *Copeia* 1990, 589–591. 1990.
98. Gross, T.S. et al. Identification of sex in hatchling loggerhead turtles (*Caretta caretta*) by analysis of steroid concentrations in chorioallantoic/amniotic fluid. *Gen. Comp. Endocrinol.* 99, 204–210. 1995.
99. Wibbels, T. et al. Blood sampling techniques for hatchling cheloniid sea turtles. *Herpetol. Rev.* 29, 218–220. 1998.
100. Mrosovsky, N. and Provancha, J. Sex ratio of loggerhead sea turtles hatching on a Florida beach. *Can. J. Zool.* 67, 2533–2539. 1989.
101. Godfrey, M.H. et al. Pivotal temperature and predicted sex ratios for hatchling hawksbill turtles from Brazil. *Can. J. Zool.* 77, 1465–1473. 1999.
102. Mrosovsky, N., Baptistotte, C., and Godfrey, M.H. Validation of incubation duration as an index of the sex ratio of hatchling sea turtles. *Can. J. Zool.* 77, 831–835. 1999.
103. Merchant-Larios, H. et al. Correlation among thermosensitive period, estradiol response, and gonad differentiation in the sea turtle *Lepidochelys olivacea*. *Gen. Comp. Endocrinol.* 107, 373–385. 1997.
104. Lovich, J.E. Possible demographic and ecologic consequences of sex ratio manipulation in turtles. *Chelonian Conserv. Biol.* 2, 114–117. 1996.
105. Girondot, M., Fouillet, H., and Pieau, C. Feminizing turtle embryo as a conservation tool. *Conserv. Biol.* 12, 353–362. 1998.
106. Mrosovsky, N. *Conserving Sea Turtles*, The Zoological Society of London, London, 1983.
107. Mortimer, J.A. Reducing threats to eggs and hatchlings: hatcheries. In: *Research and Management Techniques for the Conservation of Sea Turtles*, Eckert, K.L. et al., Eds., IUCN/SSC Marine Turtle Specialist Group Publication No. 4, 1999.
108. Marquez, M.R., Villanueva, A., and Burchfield, P. Nesting population and production of hatchlings of Kemp's ridley sea turtle at Rancho Nuevo, Tamaulipas, Mexico. In: *Proceedings of the First International Symposium on Kemp's Ridley Sea Turtle Biology, Conservation, and Management*, Caillouet, C.W., Jr. and Landry, A.M., Jr., Eds., Department of Commerce, 1989, pp. 16–19.
109. Marquez, M.R. Synopsis of biological data on the Kemp's ridley sea turtle, *Lepidochelys kempi* (Garman, 1880). NOAA Technical Memorandum NMFS-SEFC-343, 1994, 91 pp.
110. Pieau, C. Temperature variation and sex determination in reptiles. *Bioessays* 18, 19–26. 1996.

111. Wibbels, T., Cowan, J., and LeBoeuf, R.D. Temperature-dependent sex determination in the red-eared slider turtle, *Trachemys scripta*. *J. Exp. Zool.* 281, 409–416. 1998.
112. Wibbels, T., Bull, J.J., and Crews, D. Chronology and morphology of temperature-dependent sex determination. *J. Exp. Zool.* 260, 71–381. 1991.
113. Crews, D. and Bergeron, J.M. Role of reductase and aromatase in sex determination in the red-eared slider turtle (*Trachemys scripta*), a turtle with temperature-dependent sex determination. *J. Endocrinol.* 148, 279–289. 1994.
114. Wibbels, T. and Crews, D. Putative aromatase inhibitor induces male sex determination in a female unisexual lizard and in a turtle with temperature-dependent sex determination. *J. Endocrinol.* 141, 295–299. 1994.
115. Richard-Mercier, N. et al. Endocrine sex reversal of gonads by the aromatase inhibitor letrozole (CGS 20267) in *Emys orbicularis*, a turtle with temperature-dependent sex determination. *Gen. Comp. Endocrinol.* 100, 314–326. 1995.
116. Belaid, B. et al. Sex reversal and aromatase in the European pond turtle: treatment with letrozole after the thermosensitive period for sex determination. *J. Exp. Zool.* 290, 490–497. 2001.
117. Dorizzi, M. et al. Involvement of estrogens in sexual differentiation of gonads as a function of temperature in turtles. *Differentiation* 47, 9–17. 1991.
118. Desvages, G. and Pieau, C. Aromatase activity in gonads of turtle embryos as a function of the incubation temperature of the egg. *J. Steroid Biochem. Mol. Biol.* 41, 851–853. 1992.
119. Desvages, G., Girondot, M., and Pieau, C. Sensitive stages for the effects of temperature on gonadal aromatase activity in embryos of the marine turtle *Dermochelys coriacea*. *Gen. Comp. Endocrinol.* 92, 54–61. 1993.
120. Jeyasuria, P. and Place, A. The brain-gonadal embryonic axis in sex determination of reptile: a role for cytochrome P450 Arom. *J. Exp. Zool.* 281, 428–449. 1998.
121. Salame-Mendez, A. et al. Response of diencephalons but not the gonad to female-promoting temperature with elevated estradiol levels in the sea turtle *Lepidochelys olivacea*. *J. Exp. Zool.* 280, 304–313. 1998.
122. White, R.B. and Thomas, P. Whole-body and plasma concentrations of steroids in the turtle, *Trachemys scripta*, before, during, and after the temperature-sensitive period for sex determination. *J. Exp. Zool.* 264, 159–166. 1992.
123. Willingham, E. et al. Aromatase activity during embryogenesis in the brain and adrenal-kidney-gonad of the red-eared slider turtle, a species with temperature-dependent sex determination. *Gen. Comp. Endocrinol.* 119, 202–207. 2000.
124. Murdock, C. and Wibbels, T. Cloning and expression of aromatase in a turtle with temperature-dependent sex determination. *Gen. Comp. Endocrinol.* submitted.
125. Smith, C. and Joss, J. Steroidogenic enzyme activity and ovarian differentiation in the saltwater crocodile, *Crocodylus porosus*. *Gen. Comp. Endocrinol.* 93, 232–245. 1994a.
126. Smith, C. and Joss, J. Uptake of 3H-estradiol by embryonic crocodile gonads during the period of sexual differentiation. *J. Exp. Zool.* 270, 219–224. 1994b.
127. Smith, C. et al. Aromatase enzyme activity during gonadal sex differentiation in alligator embryos. *Differentiation* 58, 281–290. 1995.
128. Gabriel, W. et al. Alligator aromatase cDNA sequence and its expression in embryos at male and female incubation temperatures. *J. Exp. Zool.* 290, 439–448. 2001.
129. Gutierrez-Ospina, G. et al. Acetylcholinesterase-positive innervation is present at undifferentiated stages of the sea turtle *Lepidocehlys olivacea* embryo gonads: implications for temperature-dependent sex determination. *J. Comp. Neurol.* 410, 90–98. 1999.

130. Marshall Graves, J.A. The rise and fall of SRY. *Trends Genet.* 18, 259–264. 2002.
131. da Silva, S.M. et al. SOX9 expression during gonadal development implies a conserved role for the gene in testis differentiation in mammals and birds. *Nat. Genet.* 14, 62–67. 1996.
132. Kent, J. et al. A male-specific role for SOX9 in vertebrate sex determination. *Development* 122, 2813–2822. 1996.
133. Moreno-Mendoza, N., Harley, V.R., and Merchant-Larios, H. Differential expression of SOX9 in gonads of the sea turtle *Lepidochelys olivacea* at male- or female-promoting temperatures. *J. Exp. Zool.* 284, 705–710. 1999.
134. Moreno-Mendoza, N., Harley, V.R., and Merchant-Larios, H. Temperature regulates SOX9 expression in cultured gonads of *Lepidochelys olivacea*, a species with temperature sex determination. *Dev. Biol.* 229, 319–326. 2001.
135. Torres-Maldonado, L. et al. Timing of SOX9 downregulation and female sex determination in gonads of the sea turtle *Lepidochelys olivacea*. *J. Exp. Zool.* 290, 498–503. 2001.
136. Western, P.S. et al. Temperature-dependent sex determination in the American alligator: AMH precedes SOX9 expression. *Dev. Dyn.* 216, 411–419. 1999.
137. Raymond, C. et al. A region of human chromosome 9p required for testis development contains two genes related to known sexual regulators. *Hum. Mol. Genet.* 8, 989–996. 1999.
138. Smith, C.A. and Sinclair, A.H. Sex determination in the chicken embryo. *J. Exp. Zool.* 290, 691–699. 2001.
139. Raymond, C.S. et al. DMRT1, a gene related to worm and fly sexual regulators, is required for mammalian testis differentiation. *Genes Dev.* 14, 2587–2595. 2000.
140. Kettlewell, J., Raymond, C., and Zarkower, D. Temperature-dependent expression of turtle Dmrt1 prior to sexual differentiation. *Genesis* 26, 174–178. 2000.
141. Cate, R.L. et al. Isolation of the bovine and human genes for müllerian inhibiting substance and expression of the human gene in animal cells. *Cell* 45, 685–698. 1986.
142. Wibbels, T. and LeBoeuf, R.D. Development of a müllerian-inhibiting hormone sexing technique in sea turtles. In: *Proceedings of the 17th Annual Sea Turtle Symposium*, NOAA Technical Publications, 1998.
143. Honda, S. et al. Ad4BP regulating steroidogenic P450 genes is a member of steroid hormone receptor superfamily. *J. Biol. Chem.* 268, 7494–7502. 1993.
144. Lynch, J.P. et al. Steroidogenic factor-1, an orphan nuclear receptor, regulates the expression of the rat aromatase gene in gonadal tissues. *Mol. Endocrinol.* 7, 776–786. 1993.
145. Luo, X., Ikeda, Y., and Parker, K.L. A cell-specific nuclear receptor is essential for adrenal and gonadal development and sexual differentiation. *Cell* 77, 481–490. 1994.
146. Ingraham, H.A. et al. The nuclear receptor SF-1 acts at multiple levels of the reproductive axis. *Genes Dev.* 8, 2302–2312. 1994.
147. Parker, K.L. and Schimmer, B.P. Steroidogenic factor 1: a key determinant of endocrine development and function. *Endocr. Rev.* 18, 361–377. 1997.
148. Shen, W. et al. Nuclear receptor steroidogenic factor-1 regulates the müllerian inhibiting substance gene: a link to the sex determination cascade. *Cell* 77, 651–661. 1994.
149. Hatano, O. et al. Sex-dependent expression of a transcription factor, Ad4BP, regulating steroidogenic P-450 genes in the gonads during prenatal and postnatal rat development. *Development* 120, 2787–2797. 1994.
150. Ikeda, Y. et al. Developmental expression of mouse steroidogenic factor-1, an essential regulator, of the steroid, hydroxylases. *Mol. Endocrinol.* 8, 654–662. 1994.

151. Smith, C.A., Smith, M.J., and Sinclair, A.H. Expression of chicken steroidogenic factor-1 during gonadal sex differentiation. *Gen. Comp. Endocrinol.* 113, 187–196. 1999.
152. Western, P.S. and Sinclair, A.H. Sex, genes, and heat: triggers of diversity. *J. Exp. Zool.* 290, 624–631. 2001.
153. Fleming, A. et al. Developmental expression of steroidogenic factor 1 in a turtle with temperature-dependent sex determination. *Gen. Comp. Endocrinol.* 116, 336–346. 1999.
154. Bardoni, B. et al. A dosage sensitive locus at chromosome Xp21 is involved in male to female sex reversal. *Nat. Genet.* 7, 497–500. 1994.
155. Muscatelli, F. et al. Mutations in the DAX-1 gene give rise to both Z-linked adrenal hyplasia congenita and hypogonadotropic hypogonadism. *Nature* 372, 672–676. 1994.
156. Pelletier, J. et al. WT1 mutations contribute to abnormal genital system development and hereditary Wilms' tumor. *Nature* 353, 431–434. 1991.
157. Nachtigal, M.W. et al. Wilm's Tumour 1 and Dax-1 modulate the orphan nuclear receptor SF-1 in sex specific gene expression. *Cell* 93, 445–454. 1998.

5 Reproductive Cycles of Males and Females

Mark Hamann, Colin J. Limpus, and David W. Owens

CONTENTS

0-8493-1123-3/03/$0.00+$1.50

5.1 INTRODUCTION

Reproductive biology and some aspects of endocrinology in sea turtles have been widely investigated and reviewed over the last two decades (Owens, 1980; 1982; Ehrhart, 1982; Owens and Morris, 1985; Miller, 1997; Owens, 1997; also see Kuchling, (1999) for a review on turtle reproduction). Similar to most ectotherms, sea turtles are seasonal breeders, although in some populations nesting occurs year round (Witzell, 1983; Marquez, 1994; Hirth, 1997). Most populations have reproductive cycles constrained by proximal environmental conditions, aiding both survival of the parents and offspring while allowing maximal reproductive effort (Miller, 1997). A percentage of males from at least some populations can breed annually in the wild (Limpus, 1993; Wibbels et al., 1990; FitzSimmons, 1997). This is not usually the case for most females, with the exception of both ridley species (*Lepidochelys olivacea* and *L. kempii*) (Miller, 1997) and captive *Chelonia mydas* (Wood and Wood, 1980). Female *C. mydas* appear to be incapable of breeding on annual cycles in nature (see reviews by Ehrhart, 1982; Miller, 1997), but a small percentage of female *Caretta caretta* and *Natator depressus* breed in consecutive years (Hughes, 1974; Limpus et al., 1984a; Parmenter and Limpus, 1995; Broderick and Godley, 1996). In at least one species (*C. mydas*) breeding rates are regulated to some extent by regional climatic events driven by El Niño southern oscillation (ENSO) (Limpus and Nichols, 1988; 2000), and it appears that levels of endogenous energy reserves may play a vital role in both intra- and interannual reproductive effort in both sexes.

Although significant breakthroughs in these areas have been and continue to be made, less attention has been given to developing an understanding of the mechanisms involved in gametogenesis, ovulation and egg production, and factors regulating the timing of reproductive cycles. These shortfalls in our understanding of sea turtle biology most probably reflect logistic difficulties in (1) the capture and study of turtles outside of the nesting season, (2) accurate identification of reproductive condition, and (3) an inability to distinguish successful from unsuccessful courtship events. In this chapter we have sought to do three things: (1) to review and summarize the available literature regarding reproductive cycles of sea turtles, (2) to identify gaps and controversial areas in the literature, and (3) to document the conservation implications of the compilation and extension of reproductive information.

5.2 GAMETOGENESIS

Reproductive cycles generally refer to the series of anatomical and physiological events that lead to the production of male and female gametes, fertilization, and production of offspring. In adults of both sexes, the process of gametogenesis involves primordial germ cells undergoing further mitotic and meiotic divisions within the gonads. These processes (termed spermatogenesis in males and oogenesis and vitellogenesis in females) are presumably controlled by proximal or ultimate events that switch on a cascade of physiological processes that act upon reproductive ducts and organs to facilitate the production of male and female gametes (spermatozoa and oocytes, respectively) (Licht et al., 1979; 1980; 1985; Owens and Morris, 1985; Wibbels et al., 1990; and for general reviews of seasonal reproduction in reptiles, refer to Licht, 1982; Whittier and Crews, 1987).

5.3 OBSERVATION OF REPRODUCTIVE ANATOMY

Identification of basic reproductive parameters such as gender, age class, and reproductive state are prerequisites for most studies on reproductive cycles and physiological systems. The characterization of these parameters is logistically difficult and often physically challenging for the researcher. Three methods are currently employed by sea turtle biologists to obtain such information: necropsy, laparoscopy, and ultrasonography. These definitive methods are preferred over the sole usage of external features such as body size, weight, body condition, tail length, and endocrine studies because these latter parameters do not permit definitive and quantifiable characterization of various reproductive stages (Limpus and Reed, 1985; Limpus, 1992; Wibbels et al., 2000).

When working with threatened or endangered wildlife, examination of euthanized specimens to obtain reproductive data is often impractical. However, *in situ* necropsies, or more detailed wet lab investigations on animals that die in markets or are found dead on beaches (from natural causes or misadventure), can reveal significant biological information such as gender, maturity, reproductive state, and the reproductive history of adult females. General anatomical data are limited for most species, as is information on developmental changes in gross and ultrastructural properties of reproductive organs and ducts (Limpus, 1992; Limpus and Limpus, 2002a).

Another method allowing direct observation of reproductive organs and ducts is laparoscopic surgery. The technical procedure, applications to sea turtle biology, and associated benefits and problems have been well described over the last two decades (Wood et al., 1983; Limpus, 1985; Limpus and Reed, 1985; Owens, 1999; Wibbels et al., 2000). To reiterate, the main benefit is that laparoscopic examinations allow direct and detailed color observation of reproductive organs and ducts in live animals. They can be used to determine gender, maturity, and reproductive status of individual turtles (Limpus and Reed, 1985; Wibbels et al., 1990; Limpus, 1992; Limpus et al., 1994a; 1994b; Wibbels et al., 2000). Some limitations of laparoscopic surgery are the high level of training required to conduct the surgery and interpret the resultant image, and if the procedure is not performed correctly, it may cause death of the turtle. Regardless, it still remains the most comprehensive nonlethal method for the examination of internal organs. It has been used widely in Queensland, Australia, and the southeastern U.S. to collect reproductive data from *C. caretta*, *Eretmochelys imbricata*, *N. depressus*, and *C. mydas* as an essential basis for several research projects. These include studies on annual reproductive cycles, population demographic studies, physiological systems, and determination of reproductive state for tracking studies (Wibbels et al., 1990; Limpus, 1992; FitzSimmons, 1997; Limpus and Chaloupka, 1997; Braun-McNeill et al., 1999; Jessop et al., 1999a; Chaloupka and Limpus, 2001; Limpus and Limpus, 2001; 2002b; Hamann et al., 2002).

Similarly, ultrasonography (see Rostal et al., 1990; Owens, 1999; Wibbels et al., 2000) has been used extensively in several sea turtle projects for quantitative analysis of follicle size, examination of intraoviducal egg development, characterization of reproductive condition, prediction of the likelihood of future reproductive events in breeding females, and the assessment of reproductive condition for tracking studies (Rostal et al., 1990; 1996; 1997; 1998; 2001; Plotkin et al., 1995). However, this noninvasive procedure is limited by its inability to image oviducts and ovarian features such as corpora lutea and corpora albicantia. Thus, it cannot be used to quantify reproductive maturity in nonbreeding females or past breeding history in adult females. In addition, its use is currently restricted to the examination of breeding females, and continuing work (unpublished) by both Owens and Limpus has shown that they were unable to obtain recognizable images of ovaries in nonbreeding females, or of testes or epididymis using ultrasonography. Regardless, the development of this technique over the last decade has been significant, and its usage promises to further enhance our understanding of the reproductive biology of adult female sea turtles.

5.4 MALES

5.4.1 Anatomy of the Male Reproductive System

Similar to most vertebrates, the male reproductive system in sea turtles is composed of simultaneously functioning paired testes and associated ducts (ducts epididymis, ductus [vas] deferens). In nonbreeding adult males the testes are cylindrical (Figure 5.1) (Limpus, 1992) and weigh around 50–100 g in *L. olivacea* and 200–400 g in

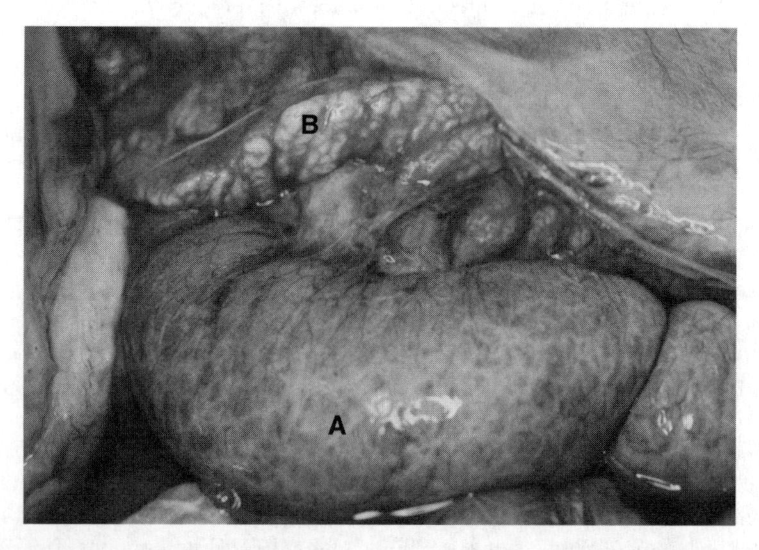

FIGURE 5.1 Testis (A) and epididymus (B) of a spermatogenic male *C. caretta* from the eastern Australian stock at courtship time. (Photo by Colin Limpus.)

C. mydas (Owens, 1980). The bulk of their volume is from seminiferous tubules. Within the seminiferous tubules is a population of epithelial cells, including a slowly dividing population of stem cells. In postpubescent males the epididymis (Figure 5.1) is pendulous and distinctly enlarged (Limpus and Reed, 1985). It is a convoluted duct extending from the ductuli efferentes, draining the testicular lobules to the ductus deferens, which conducts spermatozoa to the urethra. Urethral tissue is the site of spermatozoan accumulation and storage prior to ejaculation. The penis is an intromittent organ, >30 cm in length in *C. mydas*, and the hook at the end of the penis, adjacent to the sperm duct, presumably assists in intromission and sperm transfer (FitzSimmons, 1997; Miller, 1997).

Spermatozoa are neither motile nor capable of fertilizing ova until they have passed through the epididymis and undergo final maturation. The ultrastructure of spermatozoa has not been formally described in sea turtles; however, in a phylogenetic study using cladistic analysis, Jamieson and Healy (1992) found that turtles from a range of Cryptodire and Pleurodire genera formed a single primitive clade. Freshwater species of Cryptodire and Pleurodire turtles have spermatozoa that are 50–55 μm long and 0.9 μm wide with conspicuous spheroidal mitochondria in the midpiece (Hess et al., 1991; Healy and Jamieson, 1992). Several structures of *Chrysemys picta* spermatozoa are unique from those seen in mammals and other reptiles (Hess et al., 1991). The head is curved and pointed, 11–12 μm long by 0.9 μm wide, and contains a nucleus contiguous with intranuclear tubules. The middle section consists of proximal and distal centrioles surrounded by mitochondria. These mitochondria are speculated to maintain longevity of the sperm while in the oviduct (Hess et al., 1991). Sea turtle oviducts are very long (see below), and sperm competition may occur in some females (Owens, 1980; FitzSimmons, 1998). Thus, assessing whether these unique spermatozoa structures exist in sea turtles and developing an understanding of their function may provide a basis for gaining further insight into the movement of spermatozoa through the oviduct, potential longevity of turtle spermatozoa, and storage of spermatozoa within the oviduct.

5.4.2 Spermatogenesis

At puberty, the testes begin to secrete greatly increased amounts of the steroid hormone testosterone. This hormone has a multitude of effects including stimulation of secondary sex characteristics (such as tail elongation and softening of the plastron), the maturation of seminiferous tubules, and in adult turtles, the commencement of spermatogenesis (Wibbels et al., 1991; 1990; Licht et al., 1985). During spermatogenesis, testosterone influences Sertoli cells, which differentiate into seminiferous tubules. Previously dormant primordial germ cells divide by mitosis and differentiate into spermatogonia, eventually becoming primary spermatocytes and migrating to the lumen of the seminiferous tubule. Primary spermatocytes then undergo two meiotic divisions, developing first into secondary spermatocytes and eventually into spermatids. The spermatogenic cycle for sea turtles was first described by Wibbels et al. (1990) and has been reviewed by Owens (1997); we will not reiterate the same points here.

Histological analysis of sperm samples collected via testes biopsy suggests that the spermatogenic process lasts approximately 9 months in *C. caretta* (Wibbels et al., 1990), with primary and secondary spermatocytes present for 6 months and spermatids becoming abundant 2–3 months prior to maximal spermiogenesis (Wibbels et al., 1990; Rostal et al., 1998). Visual differentiation between the epididymis of spermatogenic and nonspermatogenic adult males is possible from late spermatogenic stage 2 (Wibbels et al., 1990) through early stage 8 (Figure 5.2; Limpus, unpublished data). The relative mass of gonads (gonadal somatic index [GSI]) collected from male *C. mydas* indicates that during active spermatogenesis the GSI increases from 1.33 to 3.08 g/kg (Licht et al., 1985). Among temperate zone reptiles the spermatogenic cycle can occur either pre- or postnuptial. Although detailed descriptions exist only for *C. caretta* (Wibbels et al., 1990) and *L. kempii* (Rostal et al., 1998), there is a general consensus that spermatogenesis in sea turtles occurs prenuptially, and is completed prior to the courtship period (Licht et al., 1985; Wibbels et al., 1990; Engstrom, 1994; Rostal et al., 1998). Because the testes become flaccid during this quiescent period, it is most likely that sperm in the epididymis is viable for only a few months. In annual breeding males it is therefore likely that only a short (2–3 month) quiescent period exists between maximal spermiogenesis during the courtship period and the beginning of the next spermatogenic cycle.

Recent correlative evidence suggests that breeding rates of male *C. mydas* in southern Queensland fluctuate synchronously with the numbers of females breeding annually (Limpus and Nicholls, 1988; 2000). Moreover, they appear to respond to ENSO on a similar time scale to that of females (Limpus and Nicholls, 2000). Males require lower levels of fat deposition for breeding than females (Kwan, 1994), and it appears that a high proportion of males in a particular foraging area prepare to breed each year. Indeed, annual baseline breeding rates of males from Shoalwater Bay in southern Queensland is approximately 15–20% (FitzSimmons, 1997). Furthermore, Licht et al. (1985) report that most "if not all" males in their captive *C. mydas* population showed annual signs of spermatogenesis and elevated testosterone. Although some males migrate considerable distances to courtship areas, a large proportion of males in the southern Great Barrier Reef (GBR) population appear to be resident in the vicinity of the courtship area year round (Limpus, 1993; FitzSimmons, 1997). Some males from this population have been followed for more than 10 years, and among them are several males that have been recorded in multiple breeding seasons, including some annual breeders (FitzSimmons, 1997). It is, however, unknown whether the resident group of males is breeding more frequently than males migrating into the area, or whether they have significantly lengthened breeding periods. Furthermore, data pertaining to breeding rates in other *C. mydas* populations and other species are lacking and present one of the challenges for future research. It would be interesting to know whether breeding rates differ among males from different foraging areas for the same genetic stock and between stocks within the same species. Similarly, the issue can be investigated from the perspective of whether smaller species (e.g., *Lepidochelys* spp.) breed more frequently than larger species (e.g., *C. mydas* or *Dermochelys coriacea*) or whether carnivores recover into the next breeding season sooner than herbivores.

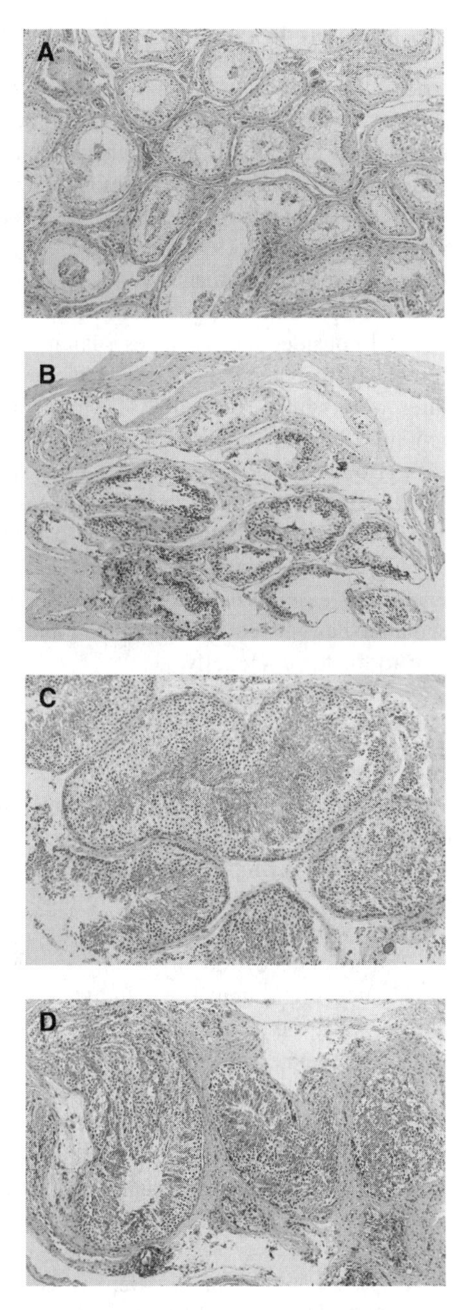

FIGURE 5.2 Micrographs (hematoxylin and eosin stain) of spermatogenic stages in adult male marine turtle testes. (A) *C. mydas*: stage 1. (B) *C. mydas*: stage 2. (C) *C. mydas*: stage 6. (D) *C. caretta*: stage 6. (Photos by Colin Limpus.)

5.4.3 Courtship and Scramble Polygamy

Male sea turtles are generally promiscuous seasonal breeders, and exhibit scramble mate-finding tactics (Ehrhart, 1982; Limpus, 1993; FitzSimmons, 1997; Jessop et al., 1999a). Similar to females, they are migratory and show strong site fidelity to both courtship and foraging areas (Limpus, 1993; FitzSimmons, 1997). Courtship appears to be confined to a distinct period just prior to the start of the nesting season (Ehrhart, 1982; Owens and Morris, 1985; Limpus, 1993), and male *C. mydas* appear to spend around 30 days searching for a mate (Wood and Wood, 1980; Limpus, 1993). In the most comprehensively studied population to date (*C. mydas* in the southern GBR, Australia) males may travel considerable distances searching for potential mates, and recapture distances are further afield in breeding as opposed to nonbreeding males (80% of recaptures were within 3650 and 1900 m of the initial capture site, respectively) (FitzSimmons, 1997). Competition between males has been recorded in many courtship areas (Booth and Peters, 1972; Balazs, 1980; Limpus, 1993; FitzSimmons, 1997; Miller, 1997). In some species and areas, aggressive male-to-male and male-to-female courtship activities have also been noted, one example being the black turtle (*Chelonia agassizi*) of the eastern Pacific (Alvarado and Figueroa, 1989). In general, male sea turtles show limited male-to-male aggression, and the number of attendant males with each mounted pair and the range of courtship damage on males appear to fluctuate annually.

5.4.4 Regulation of Courtship

Both male and female sea turtles are capital breeders, i.e., they store energy that can be later mobilized for reproduction (Stearns, 1989). Recently Jessop et al. (1999a) and Jessop (2000) proposed that the reproductive fitness of a particular male was likely to be status-dependent. Briefly, high-status males (those with higher somatic energy stores and elevated levels of testosterone) were most likely to have higher intensity mate-searching behavior and therefore be exposed to more females in a given amount of time. The associated tradeoff is almost certainly the increased energetic cost involved in such high-intensity scramble mating. Males exhibiting high-intensity courtship may reach their refractory period earlier and thus have a lesser period in which to find females. Alternatively, some males may adopt less energetic courtship strategies, and although these males may not search as large an area, they will be able to actively participate in mate searching and mate acquisition for longer. Courtship aggregations may show significant intra- and interannual variation in the density and ratio of breeding males and receptive females.

The courtship tactics used by males (high- or low-intensity scramble) may vary annually in their effectiveness at locating as many females as possible while maintaining metabolic homeostasis. In years of low-density courtship, high-intensity scramble behavior may result in higher reproductive success, whereas in high-density years, a lower (medium) scramble tactic may be the most appropriate (Jessop, 2000). From a metabolic viewpoint it also appears that the cessation of the courtship is marked by significant changes such as decreased body condition, identifiable as lowered plasma triglyceride levels and increased plasma protein levels (Hamann and Jessop, unpublished data); however, these relationships need further validation.

5.5 FEMALES

5.5.1 Anatomy of the Female Reproductive System

Female sea turtles have paired reproductive organs located abdominally. During puberty, hormonal changes increase the size and structure of both the ovary and oviduct. In comparison with immature or pubescent females, mature females typically have an ovary with an expanded stroma and a convoluted oviduct at least 1.5 cm in diameter (adjacent to the ovary) suspended in the body cavity. Oviducts of adults are very long, and lengths of 4–5 and >6 m have been recorded from *L. olivacea* and *C. mydas*, respectively (Owens, 1980; Hamann and Limpus, unpublished data). Other characteristics of an adult female may include (1) yellow vascularized vitellogenic follicles >0.3 cm in diameter (Figure 5.3), (2) presence of ovarian scars (corpora lutea or corpora albicantia; described below), (3) presence of atretic (regressing) follicles, and (4) presence of oviducal eggs (Limpus and Reed, 1985). Each characteristic is indicative of a particular stage of the reproductive cycle (Limpus and Reed, 1985; Limpus, 1992; Limpus and Limpus, 2002b).

5.5.2 Determination of Reproductive History

During ovulation, a complement of the mature follicles moves through the ovary wall into the oviduct (reviewed by Miller, 1997), although this has not been specifically described for sea turtles. It is expected that, similar to most reptiles, corpora lutea develop from hypertrophy of the empty follicle and/or the granulosa cells to form a luteal cell mass (Guraya, 1989). In sea turtles corpora lutea are approximately 1.5 cm in diameter (Limpus, 1985), and are characterized by a craterlike appearance

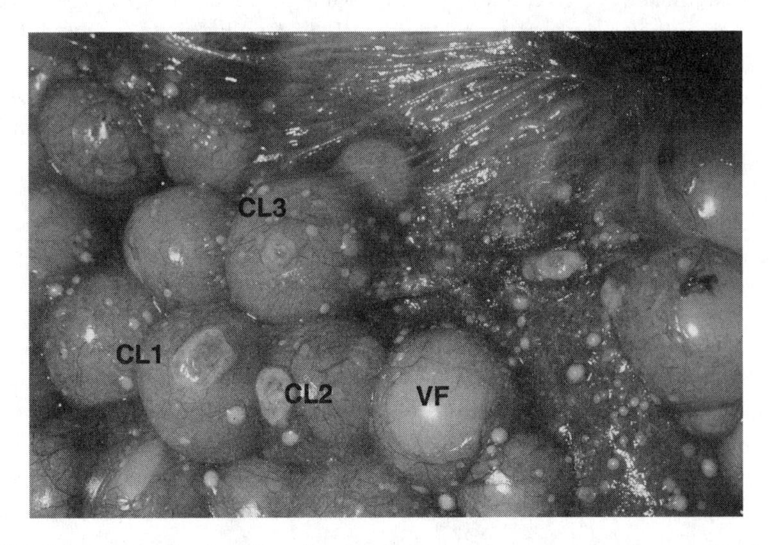

FIGURE 5.3 Ovary of a breeding female *C. caretta* (eastern Australian stock) that has ovulated three clutches within the current breeding season (three size classes of corpora lutea; CL1, CL2, and CL3) and has sufficient mature follicles (VF) for producing two more clutches. (Photo by Colin Limpus.)

(Figure 5.3). Corpora lutea act as steroid secretory glands, releasing progesterone in response to increased luteinizing hormone (Wibbels and Owens, unpublished data). Increased progesterone is thought to stimulate albumin production in post-ovulatory females (Owens, 1980; Owens and Morris, 1985). Corpora lutea regress during the nesting season such that at the end of the nesting season different size classes of corpora lutea may be evident on the surface of the ovary (Owens, 1980; Limpus, 1985) (Figure 5.3) Within a few months of the completion of the nesting season, healing corpora lutea are typically disk shaped. These scars further regress, and in females that have bred in the last season (i.e., <1 year ago), they are approximately 0.5 cm in diameter (termed corpora albicantia). Thereafter, they regress to small (approximately 0.1–0.2 cm) permanent scars on the ovary. Their presence indicates that the female has ovulated and presumably bred in a previous year (Limpus, 1985; 1992).

5.5.3 Vitellogenesis

Vitellogenesis is the process through which protein and lipid is progressively stored in the growing oocytes of oviparous animals, making up the yolk of the mature egg (Guraya, 1989). The process is remarkably similar in all reptiles studied to date (Guraya, 1989). However, little data are available on the physiological and biochemical processes that underlie vitellogenesis in sea turtles.

Vitellogenin (VTG), the main protein involved in vitellogenesis, is a relatively large (205 kDa) protein synthesized in the liver and transported to the ovary in plasma as part of a lipoprotein complex (Heck et al., 1997). As such, VTG carries lipid (predominantly triglyceride) to the growing oocytes. Estrogen production by the ovarian follicles appears to be the principal stimulus for the onset of VTG production in turtles (Ho, 1987) and increased estrogen has been linked to VTG secretion in *L. kempii* (Heck et al., 1997). Subsequently, Rostal et al. (1998) used polyacrylamide assays to monitor the presence or absence of VTG in annually breeding *L. kempii*. The protein band was visible in the postbreeding period persisting through until courtship around 7 months later (Rostal et al., 1998). More recently, Vargas (2000) has developed an enzyme-linked immunosorbent assay (ELISA) for sea turtle VTG in *L. kempii* using primary antibody derived from *Trachemys scripta*. This antibody has also been successfully tested in *C. mydas* using western blots (Hamann, unpublished data).

As yet, no research with sea turtles has focused on VTG receptors or patterns of synthesis in relation to oocyte growth. An understanding of these stages is important because they mediate key steps in oocyte maturation. It appears that in both birds and fish the uptake of yolk precursors including VTG is controlled by a 95-kDa protein receptor (George et al., 1987; Bujo et al., 1994; Davail et al., 1998). These receptors are presumed to lie in the plasma membrane of the growing oocyte, and their production is thought to precede yolk deposition. Moreover, they function as transport receptors for lipoproteins and regulatory protein for lipid deposition (Barber et al., 1991). A detailed understanding of VTG production, mobilization, and the biochemistry of vitellogenesis is needed for sea turtles.

Little is known about potential factors that may influence when a turtle enters vitellogenesis (or spermatogenesis). Similar to males, female sea turtles are capital breeders, and in at least some populations of *C. mydas* breeding rates are linked to climatic conditions at the foraging area (Limpus and Nicholls, 1988; 2000; Chaloupka, 2001). These climatic alterations may influence nutritional pathways (Limpus and Nicholls, 1988; 2000) by altering factors such as the abundance, quality, and distribution of food. In addition, climatic conditions may improve feeding rates or digestive efficiency among individual turtles. Presumably, each year a turtle (male or female) must make a choice whether to enter vitellogenesis (spermatogenesis) or to remain quiescent. The factors that influence this decision could be environmental cues such as temperature or ultimate cues such as a genetically determined energy threshold. If the conditions are favorable, the turtle will enter vitellogenesis (spermatogenesis) and breed in the following season; if not, then the individual will remain quiescent, at least until the following year.

Once an individual enters vitellogenesis, a series of physiological mechanisms are initiated that promote follicular growth. The first visible signs (increased follicle size) occur around 8–10 months prior to the breeding season (Wibbels et al., 1990; Rostal et al., 1997). In migratory birds, hyperphagia and increased lipolysis combine to ensure that adequate energy is accumulated and stored prior to breeding (Berthold, 1993; Guillemette, 2001). Although similar associations have not been investigated in sea turtles, vitellogenic females showed increases in plasma hormones (corticosterone, testosterone, estrogen, and epinephrine), triglyceride, and adipose tissue lipids. Moreover, turtles at the end of vitellogenesis (during courtship or in the early nesting season) showed decreased plasma VTG and estrogen, elevated plasma testosterone, corticosterone, epinephrine, triglyceride levels, and maximal follicle size (see Owens, 1997; Rostal et al., 1996; 1997; 1998; Hamann, 2002; Hamann et al., 2002a; Hamann et al., 2002b). In addition, total lipid in yolk follicles collected from courting females was similar to levels found in egg yolks during the early, middle, and late nesting season (Hamann et al., 2002b). These data suggest that lipid deposition and follicular development is completed prior to the nesting season.

There are significant gaps in our understanding of vitellogenesis and its regulating factors. Specifically, investigations could target ovarian synthesis of steroids, seasonal changes in VTG production, and exogenous and endogenous factors that may influence the timing of vitellogenesis and the regulation of body condition. It would be interesting to determine whether VTG production could be detected in females prior to the visual distinction of a developing follicle.

Another interesting area of research would be to investigate whether the hormone leptin, or an analogous hormone, is found in sea turtle adipose tissue. Leptin in mammals appears to signal nutritional status to several other physiological systems and modulates their function (Friedman and Halaas, 1998). More specifically, hyperleptinemia has been induced *in vivo* using hydrocortisone infusion (Askari et al., 2000), and has a profound effect on appetite and energy balance in humans (Maffei et al., 1995; Ahima and Flier, 2000). Recent experimental data have shown that exogenous leptin induced decreased feeding rates and weight loss in lizards (Niewiarowski et al., 2000). Indeed, Paolucci et al. (2001) found a seasonal pattern of leptin production in an oviparous, seasonally breeding lizard. These data suggest that leptin could be involved with the control of energy thresholds and important decision-making stages in reptiles. Expression of a similar "obesity gene" in sea turtles could be one signal that initiates or regulates vitellogenesis or metabolic homeostasis during the nesting (aphagia) season.

5.5.4 Follicular Atresia

The degeneration of ovarian follicles (atresia) is common in most vertebrates and can occur in follicles at various stages of development (Guraya, 1989). In this chapter, we will limit our discussion to atresia of mature preovulatory follicles. Atresia of these follicles has been reported in all species of sea turtle (Owens, 1980; Limpus, 1985; Rostal et al., 1996; 1997; Hamann et al., 2002b; Limpus, unpublished data). Our understanding of the mechanisms and functional role(s) of atresia is limited. However, the perceived benefit for the female of selecting a follicle for atresia is that the lipid can be resorbed, mobilized, and used for other metabolic needs (Kuchling and Bradshaw, 1993). It would be interesting to investigate whether females have the ability to compensate for decreased somatic energy by selecting follicles for atresia, or whether some females, especially those that migrate longer than average distances, have higher rates of follicular atresia to compensate for increased migratory costs.

5.5.5 Courtship and Clutch Preparation

Observations of courtship activity suggest that courtship generally occurs in the vicinity of the nesting beach (Booth and Peters, 1972; Owens and Morris, 1985; Limpus, 1993). Females may mate with several males, and average cumulative mating times are on the order of 25 h (Wood and Wood, 1980; Limpus, 1993; FitzSimmons, 1997). It is not yet possible from behavioral observations to distinguish successful from unsuccessful mating, or to determine whether insemination occurred (Wood and Wood, 1980; Limpus, 1993; FitzSimmons, 1997). Although spermatozoa have been found adjacent to the vagina and the junction between the magnum and aglandular zone, specialized sperm storage areas have not been identified in sea turtles (Solomon and Baird, 1979).

Although the courtship period appears to be well constrained temporally for the individual, arrival of turtles at the nesting beach is scattered over several months (Limpus, 1985; Dobbs et al., 1999; Godley et al., 2001; Limpus et al., 2001a). In captive *C. mydas* the average period from mating to nesting is 34.7 days (Wood and Wood, 1980). This period comprises two phases: the first period from insemination to ovulation, the second period from ovulation to oviposition. The latter has been extensively studied (Miller, 1997), but the former has never been investigated.

The control of ovulation and egg development has been linked to various endocrine pathways (see Owens 1980; 1997; Owens and Morris, 1985). Briefly, ovulation occurs approximately 36 h postoviposition and coincides with peaks in gonadotropins (luteinizing hormone and follicle-stimulating hormone) and a decrease in plasma testosterone (Licht et al., 1982; Wibbels et al., 1990). Albumin production and deposition coincide with a peak in progesterone, and shell formation is generally completed 9–10 days after ovulation (Owens, 1980; Owens and Morris, 1985; Miller, 1985; Solomon and Baird, 1979). Development of the embryo advances to middle gastrulation, when it is arrested until shortly after oviposition (Miller, 1985).

5.5.6 Oviposition

Although for some species frequent daytime nesting has been observed (e.g., *L. kempii*, *E. imbricata*, and *N. depressus*), for most sea turtle populations nesting usually occurs nocturnally (see Ehrhart, 1982; Miller, 1997). Although hormones such as prostaglandin, arginine vasotocin (AVT), and neurophysin have all been related to particular stages of oviposition (Figler et al., 1989; Guillette et al., 1991), little is known about the concert of physiological and mechanical events that occur to initiate a nesting emergence.

5.5.7 Reproductive Output

Female *C. mydas* from the southern GBR genetic stock have a mean life expectancy of around 55–60 years, including a reproductive period of around 19 years (Chaloupka and Limpus, in press). Tag recapture data of nesting females from this population indicated that the average remigration interval is greater than 5 years (5.8 and 5.9 years; Limpus et al., 1994c and Hamann, 2002, respectively), and females on average lay five clutches of 115 eggs (Bustard, 1972; Limpus et al., 1984b; Hamann, 2002). To summarize, they have an estimated lifetime reproductive output of approximately 2000 eggs. Even though these turtles have a high annual survivorship (Chaloupka and Limpus, in press), because they take decades to reach maturity, there will be a low probability of an individual's surviving to adulthood. Similarly, given the long interval between breeding seasons for adult females, a large proportion of individuals will not survive to breed a second season because of natural attrition of the breeding cohorts. Therefore, maximizing seasonal reproductive output (in terms of eggs laid) is an extremely important facet of sea turtle life history.

5.5.7.1 Ecological Variation in Reproductive Output

Differences in reproductive output may be dependent on numerous endogenous (e.g., genetics, age, body size, health and condition, and reproductive history) and exogenous (e.g., migratory distance, latitude of the foraging area, and foraging area quality) factors. Female turtles migrate to rookeries from foraging areas some tens to thousands of kilometers distant, and the foraging areas supporting a nesting population may cover a broad geographical range (Carr, 1965; Meylan, 1982; 1999; Mortimer and Carr, 1987; Limpus et al., 1992; Bowen and Karl, 1997; Miller et al., 1998; Mortimer and Balazs, 2000; Horricks et al., 2001). Furthermore, proximal cues (such as temperature and photoperiod) will undoubtedly differ in strength, intensity, and/or timing between foraging areas (especially along a latitudinal gradient). Consequently, some interesting questions arise. Are females from various locations responding to the same cues? Is there some plasticity in the way females respond to proximal cues? Are females that reside in optimal (both quantity and quality) foraging areas breeding more frequently and/or having higher reproductive output than those residing in less than optimal foraging areas? Is there a relationship between the average body size of individual nesting females (or reproductive history) with the number of females breeding for the year at a particular rookery or population?

Unfortunately, quantitative data do not exist to investigate these questions for most species. However, in Southern Queensland (Australia) female *C. caretta* foraging at Heron Island begin migration to nest at Mon Repos approximately 2 weeks earlier than those from the more distant and southern areas in Moreton Bay (Limpus, 1985). With regard to growth rates, density-dependent growth has been reported in Caribbean *C. mydas* populations (Bjorndal et al., 2000), and growth rates in southern Queensland populations appear to be related to ENSO climatic events (Limpus and Chaloupka, 1997). Specifically, female *C. mydas* at Moreton Bay have faster growth rates and obtain a larger size at maturity than females in Shoalwater Bay and Heron Reefs (Chaloupka et al., in press). It is presumed that resource (energy) acquisition is one factor that influences growth rates and breeding frequencies in presexually mature and mature turtles, respectively. Whether turtles with faster growth rates (presexual maturity) differ in their age and size at maturity or have different breeding rates is unknown. Limited data suggest that female *C. caretta* residing at Heron Island have a remigration interval 1.5 years longer than those females that nest at the same rookeries and reside in Moreton Bay, some 560 km into higher latitudes (Limpus, 1985). The continuation of long-term monitoring studies investigating reproductive cycles, in addition to the quantification of gender, age class, and reproductive output for these populations, may lead to definitive answers to these questions.

Seasonal reproductive output appears to be dependent on length of the breeding season and the breeding history of the individual. Although for *C. mydas*, *C. caretta*, and *N. depressus*, experienced breeders (remigrants) were larger than first-time breeders (neophytes) (Limpus, 1985; Parmenter and Limpus, 1995; Limpus et al., 2001a; Hamann, 2002), no significant difference in body size was found between experienced and first-time breeding *D. coriacea* (Tucker and Frazer, 1991). Reproductive output (number of clutches laid in a season) has been correlated with when a female first arrived at the nesting beach for the breeding season, with early arrivals laying more clutches (*C. mydas*: Limpus et al., 2001a). Interestingly, in some populations of *C. mydas* and *D. coriacea*, experienced breeders arrived earlier at the nesting beach than presumed first-time breeders (*C. mydas*: Hamann, 2002; *D. coreacea*: Tucker and Frazer, 1991). In contrast, in some populations the presumed first-time breeders appeared to arrive earlier in the season (*C. mydas*: Bjorndal and Carr, 1989). Additionally, experienced breeders have been recorded laying more clutches of eggs for the season than presumed first-time breeders (*C. mydas*: Carr et al., 1978; Bjorndal and Carr, 1989; Hamann, 2002; *C. caretta*: Limpus, 1985; *D. coriacea*: Tucker and Frazer, 1991).

Several hypotheses may account for the low reproductive output of first-time breeders. First, they may recruit fewer follicles for the first breeding season. Second, some animals, especially those that arrive late in the nesting season, may be interrupted by proximate environmental conditions such as a thermal constraint, or sporadic exogenous conditions that prevent continued nesting later in the season.

However, this is not supported by preliminary results of gonad examination of first-time breeding females: these data indicate that females with reduced clutch production ovulate approximately all available follicles similar to females that are more productive. Third, turtles living at different distances from the rookery can be expected to arrive for commencement of nesting at different times; in particular, turtles from very distant feeding grounds may arrive later in the season. First-time migrants may swim slower and navigate less precisely, thus consuming more energy on migration. However, the first-time breeding *C. caretta* tracked by satellite telemetry from her home foraging area in Moreton Bay did not display any evidence of slow swimming or less precise navigation (Limpus and Limpus, 2001). It is also possible that some first-time breeders may simply leave their home foraging area at a later date than the more experienced nesters, especially if hormone concentrations required to initiate migration are a function of the number of mature follicles in the ovary. Unfortunately, most studies addressing variability in egg production have focused on the turtles once they have arrived at the nesting beach. There has also been a deficiency in studies at the foraging areas to address the factors that might impact the number and size of mature follicles that a female develops in her ovaries prior to commencement of her breeding migration.

5.5.7.2 A Role for Hormones in Maximizing Reproductive Effort

Valverde et al. (1999), Jessop et al. (1999b; 2000), and Jessop (2001) recently investigated hormonal mechanisms that may act to facilitate maximal reproductive output. It now appears that at least four species of sea turtles (*L. olivacea*, *C. caretta*, *C. mydas*, and *E. imbricata*) have a physiological ability to downregulate or desensitize their corticosterone stress response (adrenocortical modulation) (Gregory et al., 1996; Valverde et al., 1999; Jessop et al., 1999b; 2000; Jessop, 2001). Moreover, this reduced stress response occurred in females prior to migration and persisted through the nesting season, regardless of the level of reproductive investment that remained (Jessop, 2001). Thus, the effects that ecological stressors (such as disturbance by conspecifics or competition for nesting space) may otherwise have on a female's ability to successfully oviposit are negated or temporarily set aside as physiological safeguards are set in place to maximize current reproductive output.

5.5.8 Regulation of a Nesting Season

Marine turtles show significant weight gains during the internesting period while preparing clutches for laying (*C. caretta:* Limpus, 1973; *D. coreacea*: Eckert et al., 1989; *E. imbricata*: Limpus et al., 1983). Limpus (1973) incorrectly attributed these weight gains to the female's feeding during the internesting period, whereas he ignored the possibility that water uptake could account for the weight changes. There was negligible food contained in the gastrointestinal (GI) tract of the internesting females examined in this latter study compared to the abundance of food in the GI tract of nonbreeding *C. caretta* that live within the same internesting habitat (Limpus et al., 2001b).

Although limited foraging has been recorded in gravid *C. mydas* females at Raine Island (Queensland, Australia) (Tucker and Read, 2001), diet studies, visual examination of GI tracts during laparoscopic examinations of gonads, and satellite telemetry studies generally indicate that *C. mydas* is primarily aphagic during the nesting season (Bjorndal, 1982; 1985; Balazs et al., 2000; Limpus, unpublished data). Thus, before they depart their foraging habitat on a breeding migration, females must allocate sufficient lipid reserves to allow for the entire season's reproductive output and the return migration, ideally without compromising metabolic processes. Nesting success (percent of nesting attempts that result in successful oviposition) varies among species and populations, and is often lower than 100% (Miller, 1997; Loop et al., 1995; Godley et al., 2001; Hamann, 2002). Correlative evidence suggests that *C. mydas* females nesting in a year characterized by high rates of unsuccessful nesting have higher rates of follicular atresia (Limpus et al., 1991; 1993). Plasma triglyceride levels are significantly lower in female *C. mydas* after prolonged periods (>3 days) of unsuccessful nesting (Hamann et al., 2002b). Moreover, total lipid values in adipose tissue in gravid *C. mydas* with atretic ovaries (Hamann et al., 2002b) are reduced and similar to levels in females after completion of a nesting season (Kwan, 1994). An interesting question thus arises: How many unsuccessful nesting attempts can an individual female sustain without its reducing her potential reproductive output (through depleting energy stores)?

The end of the nesting season could be triggered by insufficient mature ovarian follicles to produce another clutch. Alternatively, specific factors that signal the end of the nesting season may well be related to body condition or environmental conditions. In birds that undertake lengthy breeding seasons, or periods of aphagia, their behavior appears to be tightly regulated by a genetically determined energy threshold. Once body condition declines below this threshold and protein stores are put at risk, refeeding is initiated (Cherel et al., 1988; Gauthier-Clerc et al., 2001). From a physiological standpoint, several changes occur in sea turtles at the end of the nesting season: plasma hormone (testosterone, estrogen, and corticosterone) and plasma triglyceride levels typically decline to near basal levels, and total plasma protein levels have been observed to increase (Licht et al., 1979; 1980; Wibbels et al., 1990; Rostal et al., 1997; 1998; 2001; Whittier et al., 1997; Hamann et al., 2002b). There are few data available examining seasonal changes in body condition in sea turtles, and data from Limpus et al. (2001a) indicate that females nesting at Bramble Cay lost an average of 0.9 kg following each clutch. Moreover, the lack of a sharp increase in corticosterone at the end of the nesting season in female sea turtles suggests that body condition probably does not decline to critical levels as it does in birds (Whittier et al., 1997, Rostal et al., 2001; Hamann et al., 2002a). Therefore, despite slight evidence for a shift toward protein catabolism in *C. mydas* (Hamann et al., 2002b), potential metabolic signals in sea turtles are less clear.

5.5.9 Arribadas and Year-Round Nesting

Two important variations of the typical seasonal nesting pattern of sea turtles are the mass nesting behavior observed in some populations of the genus *Lepidochelys* and year round nesting seen in some populations of other species.

5.5.9.1 Arribadas

Unique to the *Lepidochelys* genus is a breeding event known commonly as an *arribada* (or *arribadazon*). Briefly, *L. kempii* and some populations of *L. olivacea* exhibit mass nesting events in which large groups of females emerge synchronously to lay eggs. Generally occurring at night in *L. olivacea* and during the day in *L. kempii*, mass nesting events occur over a period of 1–3 days and reoccur at intervals of approximately 30 days (see Miller, 1997). Anywhere from 100 to 10,000 or more females may be involved, and the emerging hypothesis is that this mass nesting behavior serves as a deterrent for nest predators through "predator satiation" (Eckrich and Owens, 1995). Although it has been thought of as "socially facilitated nesting" (Owens et al., 1982), it now appears that these groups of females aggregate for nesting but disperse randomly during the internesting period. The turtles respond independently to one or more proximal cues and commence a subsequent arribada some 30 days later (Plotkin et al., 1995; 1997). Pritchard (1969) hypothesized that ovulation and egg development would occur at approximately the same time for the whole arribada cohort, and that females were retaining eggs until suitable emergence cues arose. Although *Lepidochelys* spp. have internesting intervals significantly longer than other species, ovulation still occurs within 2–3 days postoviposition (Licht et al., 1982; Miller, 1997). Because arribadas are presumed to comprise several smaller groups of turtles (Plotkin et al., 1995), perhaps prolonged renesting intervals in arribada females acts to synchronize the final stages of egg development in as many females as possible. In addition, arribadas may serve to delay oviposition when conditions are unfavorable (Plotkin et al., 1997).

Although uncommon in other species, Limpus (1985) reports two instances of prolonged oviducal egg retention (41 and 42 days) in female *C. caretta* with disabled hind flippers. Dissection confirmed that in both cases the oviducal eggs were from the most recent ovulation. Moreover, eggs from one of these females were buried in an artificial nest and achieved emergence success of 76.2%, providing further evidence to support Pritchard's hypothesis that prolonged egg retention occurs to allow a female to wait for suitable nesting cues.

5.5.9.2 Year-Round Nesting

Nesting seasons for most populations are constrained temporally. However, for most species, uninterrupted year-round nesting has been recorded at some locations (Witzell, 1983; Marquez, 1994; Hirth, 1997; also see Miller, 1997), although in most of these cases the majority of nesting activity occurred in a peak period spread over several months. Bimodal nesting with small and large peak periods has been recorded for at least one population of *D. coriacea*; however, relationships between these two apparently separate cohorts are unknown (Chevalier et al., 1999). Unfortunately, comprehensive data sets detailing year-round nesting are not available, and thus we can only speculate reasons to support why and how they persist. Presumably, three factors control nesting seasonality: (1) the ability to find a mate and successfully copulate, (2) the suitability of a beach to successfully incubate sea turtle eggs, and (3) the suitability of the beach to allow efficient offshore dispersal of the hatchlings.

Some equatorial rookeries that support year-round nesting may receive females from foraging areas in both hemispheres; thus, regionally different climatic or oceanographic patterns may underlie some of the variation in nesting seasonality.

There are little data to confirm that some males are breeding outside of peak courtship times, and it would seem likely that peak nesting periods occur despite year-round activity because it makes sense to have some degree of synchronization and high density during courtship. However, in year-round nesting locations, if we assume that some males breed asynchronously, are the females that are nesting out of sync from the rest of the nesting cohort limiting their potential reproductive fitness by having limited mate selection? Are males limiting their fitness through the need for increased effort in searching for mates and limited mate choice?

5.6 REPRODUCTIVE CYCLES AND SEA TURTLE CONSERVATION

An increasing awareness exists of the role of sea turtles in the environment (Bjorndal, 1982; 1985; Rogers, 1989; Bouchard and Bjorndal, 2000); marine ecosystem effects resulting from ecological extinction (Jackson, 1997; Jackson et al., 2001); and anthropogenic and climatic impacts on our beaches, coasts, and seas (Davenport, 1997; Jackson, 1997; Jackson et al., 2001). We as managers, scientists, or conservation enthusiasts need to ask questions from the perspective of quickly changing ecosystems. What impact will alterations to the marine and coastal environments have on sea turtle life history characteristics? How has an altered environment affected sea turtle populations, and how will it do so in the future?

First, to reiterate points made by Owens (1980; 1997), we need to understand more about reproductive cycles, physiological processes, and how turtles respond to environmental cues. Second, we need to react to the challenges of Miller (1997). We need to collect quantitative data on breeding rates and reproductive, physiological, and environmental cycles to gain a more complete understanding of how (or whether) alterations to the environment are likely to influence sea turtle populations. At least two areas of probable concern are apparent for sea turtle managers: global warming and rising sea levels, and increased contamination of our oceans and beaches.

Global climate change is thought to influence the marine environments through increased temperatures and rising sea levels. Potential impacts include an increased frequency of coral bleaching, alterations to sea grass habitats, and habitat loss (see Hoegh-Guldberg, 1999; Short and Neckles, 1999; Daniels et al., 1993). The potential influence of increased temperatures on life history attributes and conservation of sea turtles is significant (see Mrosovsky et al., 1984; Davenport, 1989; 1997). In foraging habitats, increased temperatures may impact food sources and nutritional pathways. Decreased food abundance or quality could slow growth rates and affect breeding rates in sea turtles, particularly in *C. mydas* because these effects are likely to be greatest at lower trophic levels (Limpus and Nicholls, 1988; 2000; Limpus and Chaloupka, 1997; Short and Neckles, 1999; Bjorndal et al., 2000; Chaloupka and Limpus, 2001). As the rates of climatic change vary between tropical and temperate zones (Houghton et al., 1996), climatic conditions that may lead to the onset of vitellogenesis or migration may not lead to an adequate arrival date at the nesting beach (or both), or the optimum time to arrive at the nesting beach may vary over time (Both and Visser, 2001). At nesting beaches, increased temperatures may affect embryo development, and could lead to a significant shift toward female-producing temperatures (Davenport, 1997). Increased sea levels could substantially alter available nesting environments (Daniels et al., 1993) and factors controlling incubation, such as moisture, salinity, and gas exchange (see Ackerman, 1997).

It is probable that increased contaminant levels at both nesting beaches and foraging areas could affect physiological systems. For example, altered sex ratios and decreased fertility have been reported for some alligator populations that are exposed to a variety of xenobiotics (see Crain and Guillette, 1998). Although no direct cause has been identified, the reported incidence of fibropapilloma virus among sea turtles is highest in habitats in close proximity to large human population centers (see Davidson, 2001). Data from southern Queensland suggest that outbreaks of the toxic cyanobacteria *Lyngbya majuscula* are increasing in frequency and severity, and have the potential to alter sea grass quality and quantity, and thus potentially affect sea turtle distribution, growth, and breeding rates (Dennison et al., 1999; Osborne et al., 2001).

The metaphor of the environmental canary has been used when describing the decline of nesting *D. coriacea* populations in the eastern Pacific (Reina et al., 2000), and can be expanded to include what this decline insinuates about the quality or rate of change to conditions in foraging, migratory areas, and nesting beaches. Other early warning systems of population change may well be manifest in alterations to sex ratios of young recruits, growth rates, breeding rates, or a changing demographic within foraging areas. Thus, continued collection of baseline and experimental data across species and populations dealing with reproductive cycles, physiological control systems, and pertinent ecological parameters is of paramount importance. Otherwise, in times of rapidly changing environments, we will not have the necessary information to assess possible and probable impacts on sea turtle populations and apply early and appropriate management practices.

ACKNOWLEDGMENTS

The authors would like to thank Chloe Schäuble, Karen Arthur, and Tim Jessop for helpful comments on the manuscript.

REFERENCES

Ackerman, R.A., The nest environment and the embryonic development of sea turtles, in *The Biology of Sea Turtles*, Lutz, P.L. and Musick, J.A., Eds., CRC Publishing, Boca Raton, FL, 83–107, 1997.

Ahima, R.S. and Flier, J.S., Adipose tissue as an endocrine organ, *Trends Endocrinol. Metab.*, 11, 327, 2000.

Alvarado, J. and Figueroa, A., Breeding dynamics in the black turtle (*Chelonia agassizi*) in Michoacan, Mexico, in *Proceedings of the Ninth Annual Workshop on Sea Turtle Biology and Conservation*, Eckert, S., Eckert, K., and Richardson, T., compilers, NOAA Technical Memorandum NMFS-SEFC-232, Athens, GA, 1989.

Askari, H., Liu, J., and Dagogo, J.S., Hormonal regulation of human leptin in vivo: effects of hydrocortisone and insulin, *Int. J. Obesity*, 24, 1254, 2000.

Balazs, G.H., Synopsis of biological data on the green turtle in the Hawaiian Islands, National Oceanic and Atmospheric Administration Technical Memorandum, NMFS, Southwest Fisheries Centre, 1, 1980, Honolulu, Hawaii.

Balazs, G.H., Katahira, L.K., and Ellis, D.M., Satellite tracking of hawksbill turtles nesting in the Hawaiian Islands, in *Proceedings of the Eighteenth Annual Workshop on Sea Turtle Biology and Conservation*, Abreu-Grobois, F.A. et al., compilers, NOAA Technical Memorandum NMFS-SEFC-436, 2000, Mazatlan, Mexico.

Barber, D.L., et al., The receptor for yolk lipoprotein deposition in the chicken oocyte, *J. Biol. Chem.*, 266, 18761, 1991.

Berthold, P., *Bird Migration*, Oxford University Press, New York, 1993.

Bjorndal, K.A., The consequences of herbivory for the life history pattern of the Caribbean green turtle, *Chelonia mydas*, in *Biology and Conservation of Sea Turtles*, Bjorndal, K.A., Ed., Smithsonian Institution Press, Washington DC, 111, 1982.

Bjorndal, K.A., Nutritional ecology of sea turtles, *Copeia*, 1985, 736, 1985.

Bjorndal, K.A., Bolten, A.B., and Chaloupka, M.Y., Green turtle somatic growth model: evidence for density dependence, *Ecol. Appl.*, 10, 269, 2000.

Bjorndal, K.A. and Carr, A., Variation in clutch size and egg size in the green turtle nesting population at Tortuguero, Costa Rica. *Herpetologica*, 45, 181, 1989.

Booth, J. and Peters, J.A., Behavioural studies on the green turtle (*Chelonia mydas*) in the sea, *Anim. Behav.*, 20, 808, 1972.

Both, C. and Visser, M.E., Adjustment to climate change is constrained by arrival date in a long-distance migrant bird, *Nature*, 411, 296, 2001.

Bouchard, S.S. and Bjorndal, K.A., Sea turtles as biological transporters of nutrients and energy from marine to terrestrial ecosystems, *Ecology*, 81, 2305, 2000.

Bowen, B.W. and Karl, S.A., Population genetics, phylogeography, and molecular evolution, in *The Biology of Sea Turtles*, Lutz, P.L. and Musick, J.A., Eds., CRC Press, Boca Raton, FL, 29–50, 1997.

Braun-McNeil, J. et al., Sex ratios of immature sea turtles: does water temperature make a difference? in *Proceedings of the Nineteenth Annual Symposium on Sea Turtle Conservation and Biology*, Kalb, H. and Wibbels, T., compilers, Technical Memorandum NMFS-SEFSC-443, South Padre Island, TX, 1999.

Broderick, A.C. and Godley, B.J., Population and nesting ecology of the green turtle *Chelonia mydas*, and the loggerhead turtle, *Caretta caretta*, in northern Cyprus, *Zool. Middle East*, 13, 27, 1996.

Bujo, H., et al., Chicken oocyte growth is mediated by an eight ligand binding repeat member of the LDL receptor family, *EMBO J.*, 13, 5165, 1994.

Bustard, H.R., *Sea Turtles: Their Natural History and Conservation*, Collins, London, 1972.

Carr, A., The navigation of the green sea turtle, *Sci. Am.*, 212, 79, 1965.

Carr, A., Carr, M.H., and Meylan, A.B., The ecology and migrations of sea turtles. 7: The west Caribbean green turtle colony, *Bull. Am. Mus. Nat. Hist.*, 162, 1, 1978.

Chaloupka, M., Historical trends, seasonality and spatial synchrony in green sea turtle egg production, *Biol. Conserv.*, 101, 263, 2001.

Chaloupka, M. and Limpus, C., Trends in the abundance of sea turtles resident in southern Great Barrier Reef waters, *Biol. Conserv.*, 102, 235, 2001.

Chaloupka, M. and Limpus, C.J., Estimates of sex- and stage-specific survival probabilities for green sea turtles resident in southern Great Barrier Reef waters, *Can. J. Zool.*, in press.

Chaloupka, M., Limpus, C.J., and Miller, J., Sea turtle growth dynamics in a spatially disjunct metapopulation, *Can. J. Zool.*, in press.

Cherel, Y. et al., Fasting in king penguin. 1. Hormonal and metabolic changes during breeding, *Am. J. Physiol.*, 254, R170, 1988.

Chevalier, J. et al., Study of the bimodal nesting season for leatherback turtles (*Dermochelys coriacea*) in French Guiana, in *Proceedings of the Nineteenth Annual Symposium on Sea Turtle Conservation and Biology*, Kalb, H. and Wibbels, T., compilers, Technical Memorandum NMFS-SEFSC-443, South Padre Island, TX, 1999.

Crain, D.A. and Guillette, L.J., Jr., Reptiles as models of contaminant-induced endocrine disruption, *Animal Reproduction Science,* 53, 77, 1998.

Daniels, R.C., White, T.W., and Chapman, K.K., Sea-level rise: destruction of threatened and endangered species habitat in South Carolina, *Environ. Manage.*, 17, 373, 1993.

Davail, B. et al., Evolution of oogenesis: the receptor for vitellogenin from the rainbow trout, *J. Lipid Res.*, 39, 1929, 1998.

Davenport, J., Sea turtles and the greenhouse effect, *Br. Herpetol. Soc. Bull.*, 29, 11, 1989.

Davenport, J., Temperature and the life-history strategies of sea turtles, *J. Therm. Biol.*, 22, 479, 1997.

Davidson, O.G., *Fire in the Turtle House: The Green Sea Turtle and the Fate of the Ocean*, Public Affairs, New York, 2001.

Dennison, W.C. et al., Blooms of the cyanobacteria *Lyngbya majuscula* in coastal waters of Queensland, Australia, *Bull. Inst. Oceanogr. Monarco*, 19, 501, 1999.

Dobbs, K.A. et al., Hawksbill turtle, *Eretmochelys imbricata*, nesting at Milman Island, northern Great Barrier Reef, Australia, *Chelonian Conserv. Biol.*, 3, 344, 1999.

Eckert, S.A. et al., Diving and foraging behavior of leatherback sea turtles (*Dermochelys coreacea*), *Can. J. Zool.*, 67, 2834, 1989.

Eckrich, C.E. and Owens, D.W., Solitary versus arribada nesting in the olive ridley sea turtles (*Lepidochelys olivacae*): a test of the predator-satiation hypothesis, *Herpetologica*, 51, 349, 1995.

Ehrhart, L.M., A review of sea turtle reproduction, in *Biology and Conservation of Sea Turtles*, Bjorndal, K.A., Ed., Smithsonian Institution Press, Washington, DC, 1982.

Engstrom, T.N., Observations on the testicular cycle of the green turtle, *Chelonia mydas*, in the Caribbean, in *Proceedings of the Fourteenth Annual Symposium on Sea Turtle Biology and Conservation*, Bjorndal, K.A., Bolten, B.A., Johnson, D., and Elizar, P.L., compilers, NOAA Technical Memorandum NMFS-SEFSC-53, Miami, FL, 1994.

Figler, R.A. et al., Increased levels of arginine vasotocin and neurophysin during nesting in sea turtles, *Gen. Comp. Endocrinol.*, 73, 223, 1989.

FitzSimmons, N.N., Male marine turtles: gene flow, philopatry and mating systems of the green turtle *Chelonia mydas*, Ph.D. thesis, The University of Queensland, Brisbane, Australia, 241, 1997.

FitzSimmons, N.N., Single paternity of clutches and sperm storage in the promiscuous green turtle (*Chelonia mydas*), *Mol. Ecol.*, 7, 575, 1998.

Frazer, N.B., Sea turtle conservation and halfway technology, *Conserv. Biol.*, 6, 179, 1992.

Friedman, J.M. and Halaas, J.L., Leptin and the regulation of body weight in mammals, *Nature*, 395, 763, 1998.

Gauthier-Clerc, M. et al., State dependent decisions in long term fasting king penguins, *Aptenodytes patagonicus*, during courtship and incubation, *Anim. Behav.*, 62, 661, 2001.

George, R., Barber, D.L., and Schneider, W.J., Characterization of the chicken oocyte receptor for low and very low density lipoproteins, *J. Biol. Chem.*, 262, 16838, 1987.

Godley, B.J., Broderick, A.C., and Hays, G.C., Nesting of green turtles (*Chelonia mydas*) at Ascension Island, South Atlantic, *Biol. Conserv.*, 97, 151, 2001.

Gregory, L.F. et al., Plasma corticosterone concentrations associated with acute captivity stress in wild loggerhead sea turtles (*Caretta caretta*), *Gen. Comp. Endocrinol.*, 104, 312, 1996.

Guillemette, M., Foraging before spring migration and before breeding in Common Eiders: does hyperphagia occur, *Condor*, 103, 633, 2001.

Guillette, L.J. et al., Plasma estradiol-17-beta, progesterone, prostaglandin F, and prostaglandin E-2 concentrations during natural oviposition in the loggerhead turtle (*Caretta caretta*), *Gen. Comp. Endocrinol.*, 82, 121, 1991.

Guraya, S.S., *Ovarian Follicles in Reptiles and Birds*, Springer-Verlag, Germany, 1989.

Hamann, M., Reproductive cycles, interrenal gland function and lipid mobilisation in the green sea turtle (*Chelonia mydas*), Ph.D. thesis, The University of Queensland, Brisbane, Australia, submitted February 2002.

Hamann, M. et al., Interactions among endocrinology, annual reproductive cycles and the nesting biology of the female green sea turtle, *Mar. Biol.*, 40, 823, 2002a.

Hamann, M., Limpus, C.J., and Whittier, J.M., Patterns of lipid storage and mobilisation in female green sea turtles (*Chelonia mydas*), *J. Comp. Physiol. B*, 172, 485, 2002b.

Healy, J.M. and Jamieson, B.G.M., Ultrastructure of the spermatozoon of the tuatara (*Sphenodon punctatus*) and its relevance to the relationships of the sphenodontidae, *Proc. R. Soc. Lond.*, 335, 193, 1992.

Heck, J. et al., Estrogen induction of plasma vitellogenin in the Kemp's ridley sea turtle (*Lepidochelys kempii*), *Gen. Comp. Endocrinol.*, 107, 280, 1997.

Hess, R.A., Thurston, R.J., and Gist, D.H., Ultrastructure of the turtle spermatozoon, *Anat. Rec.*, 229, 473, 1991.

Hirth, H., Synopsis of biological data on the green turtle *Chelonia mydas* (Linnaeus 1758), U.S. Department of the Interior Biological Report 97 (1), 1, 1997.

Ho, S., Endocrinology of vitellogenesis, in *Hormones and Reproduction in Fishes, Amphibians and Reptiles*, Norris, D.O. and Jones, R.E., Eds., Plenum Press, New York, 1987, 145–169.

Hoegh-Guldberg, O., Climate change, coral bleaching and the future of the world's coral reefs, *Marine and Freshwater Research.*, 50, 839, 1999.

Horricks, J.A. et al., Migration routes and destination characteristics of post-nesting hawksbill turtles satellite tracked from Barbados, West Indies, *Chelonian Conserv. Biol.*, 4, 107, 2001.

Houghton, J.T., *Climate Change 1995*, Cambridge University Press, Cambridge, 1996.

Hughes, G.R., The sea turtles of South-East Africa. II, South African Association for Marine Biological Research Oceanographic Research Institute Durban Investigational Report 36, 1974.

Jackson, J.B.C., Reefs since Columbus, *Coral Reefs*, 16, s23, 1997.

Jackson, J.B.C. et al., Historical overfishing and the recent collapse of coastal ecosystems, *Science*, 293, 629, 2001.

Jamieson, B.G.M. and Healy, J.M., The phylogenetic position of the tuatara, *Sphenodon* (Sphenodontida, Amniota) as indicated by cladistic analysis of the ultrastructure of spermatozoa, *Proc. R. Soc. Lond. B*, 335, 207, 1992.

Jessop, T.S., Endocrinal ecology of the green turtle *Chelonia mydas*, Ph.D. thesis, The University of Queensland, Brisbane, Australia, 174, 2000.

Jessop, T.S., Modulation of the adrenocortical stress response in marine turtles: a hormonal tactic maximising maternal reproductive investment, *J. Zool.*, 251, 57, 2001.

Jessop, T.S. et al., Interactions between behavior and plasma steroids within the scramble mating system of the promiscuous green turtle, *Chelonia mydas*, *Horm. Behav.*, 36, 86, 1999a.

Jessop, T.S., Limpus, C.J., and Whittier, J.M., Plasma steroid interactions during high-density green turtle nesting and associated disturbance, *Gen. Comp. Endocrinol.*, 115, 90, 1999b.

Jessop, T.S. et al., Evidence for a hormonal tactic maximizing green turtle reproduction in response to a pervasive ecological stressor, *Gen. Comp. Endocrinol.*, 118, 407, 2000.

Kuchling, G., *The Reproductive Biology of the Chelonia*, Springer, Berlin, 1999.

Kuchling, G. and Bradshaw, S.D., Ovarian cycle and egg production of the western swamp tortoise *Pseudemydura umbrina* (Testudines: Chelidae) in the wild and in captivity, *J. Zool.*, 229, 405, 1993.

Kwan, D., Fat reserves and reproduction in the green turtle, *Chelonia mydas*, *Wildl. Res.*, 21, 257, 1994.

Licht, P., Endocrine patterns in the reproductive cycle of turtles, *Herpetologica*, 38, 51, 1982.

Licht, P. et al., Serum gonadotropins and steroids associated with breeding activities in the green sea turtle *Chelonia mydas*. I. Captive animals, *Gen. Comp. Endocrinol.*, 39, 274, 1979.

Licht, P., Rainey, W., and Clifton, K., Serum gonadotropin and steroids associated with breeding activities in the green sea turtle *Chelonia mydas*. II. Mating and nesting in natural populations, *Gen. Comp. Endocrinol.*, 40, 116, 1980.

Licht, P. et al., Changes in LH and progesterone associated with the nesting cycle and ovulation in the olive ridley sea turtle, *Lepidochelys olivacea*, *Gen. Comp. Endocrinol.*, 48, 247, 1982.

Licht, P., Wood, J.F., and Wood, F.E., Annual and diurnal cycles in plasma testosterone and thyroxine in the male green sea turtle *Chelonia mydas*, *Gen. Comp. Endocrinol.*, 57, 335, 1985.

Limpus, C.J., Loggerhead turtles (*Caretta caretta*) in Australia: food resources while nesting, *Herpetologica*, 29, 42, 1973.

Limpus, C.J., A study of the loggerhead turtle, *Caretta caretta,* in eastern Australia, Ph.D. thesis, The University of Queensland, Brisbane, Australia, 1985.

Limpus, C.J., The hawksbill turtle, *Eretmochelys imbricata*, in Queensland: population structure within a southern Great Barrier Reef feeding ground, *Wildl. Res.*, 19, 489, 1992.

Limpus, C.J., The green turtle, *Chelonia mydas*, in Queensland: breeding males in the southern Great Barrier Reef, *Wildl. Res.*, 20, 513, 1993.

Limpus, C.J., Carter, D., and Hamann, M., The green turtle, *Chelonia mydas*, in Queensland: The Bramble Cay rookery in the 1979–1980 breeding season, *Chelonian Conserv. Biol.*, 4, 34, 2001a.

Limpus, C. and Chaloupka, M., Nonparametric regression modelling of green sea turtle growth rates (southern Great Barrier Reef), *Mar. Ecol. Prog. Ser.*, 149, 23, 1997.

Limpus, C.J., Couper, P.J., and Read, M.A., The green turtle, *Chelonia mydas*, in Queensland: population structure in a warm temperate feeding area, *Mem. Queensland*, 35, 139, 1994a.

Limpus, C.J., Couper, P.J., and Read, M.A., The loggerhead turtle, *Caretta caretta*, in Queensland: population structure in a warm temperate feeding area, *Mem. Queensland*, 37, 195, 1994b.

Limpus, C.J., Eggler, P., and Miller, J.D., Long interval remigration in eastern Australian *Chelonia*, in *Proceedings of the Thirteenth Annual Symposium on Sea Turtle Biology and Conservation*, Schroeder, B.A., Witherington, B.E., compilers, NOAA Technical Memorandum NMFS-SEFSC-341, Miami, 1994c.

Limpus, C.J. et al., The loggerhead turtle, *Caretta caretta*, in Queensland: observations on feeding ecology in warm temperate waters, *Mem. Queensland*, 46, 631, 2001b.

Limpus, C.J. et al., The hawksbill turtle *Eretmochelys imbricata*, in north eastern Australia: the Campbell Island Rookery, *Aust. Wildl. Res.*, 10, 185, 1983.

Limpus, C.J. et al., Migration of green (*Chelonia mydas*) and loggerhead (*Caretta caretta*) turtles to and from eastern Australian rookeries, *Wildl. Res.*, 19, 347, 1992.

Limpus, C.J., Fleay, A., and Baker, V., The flatback turtle, *Chelonia depressus* in Queensland: reproductive periodicity, philopatry and recruitment, *Aust. Wildl. Res.*, 11, 579, 1984a.

Limpus, C.J., Fleay, A., and Guinea, M., Sea turtles of the Capricorn section, Great Barrier Reef, in *The Capricorn Section of the Great Barrier Reef: Past, Present and Future*, Ward, W.T. and Saenger, P., Eds., The Royal Society of Queensland, Brisbane, Australia, 61, 1984b.

Limpus, C.J. and Limpus, D.J., The loggerhead turtle, *Caretta caretta*, in Queensland: breeding migrations and fidelity to a warm temperate feeding area, *Chelonian Conserv. Biol.*, 4, 142, 2001.

Limpus, C.J. and Limpus, D.J., The loggerhead turtle, *Caretta caretta*, in the equatorial and southern Pacific Ocean: a species in decline, in *Biology and Conservation of Loggerhead Turtles*, Bolten, A. and Witherington, B., Eds., Smithsonian Institution Press, Washington, DC, 2002a.

Limpus, C.J. and Limpus, D.J., The biology of the loggerhead turtle, *Caretta caretta*, in the southwest Pacific Ocean foraging areas, in *Biology and Conservation of Loggerhead Turtles*, Bolten, A. and Witherington, B., Eds., Smithsonian Institution Press, Washington, DC, 2002b.

Limpus, C.J., Miller, J.D., and Parmenter, J., The green turtle, *Chelonia mydas* population of Raine Island and the northern Great Barrier Reef: 1843–1989, Queensland Parks and Wildlife Service, Brisbane, 1991.

Limpus, C.J., Miller, J.D., and Parmenter, C.J., The northern Great Barrier Reef green turtle, *Chelonia mydas*, breeding population, in *Raine Island and Environs, Great Barrier Reef: Quest to Preserve a Fragile Outpost of Nature*, Smyth, A.K., Zevering, K.H., and Zevering, C.E., Eds., Raine Island Corporation and Great Barrier Reef Marine Park Authority, Townsville, 47, 1993.

Limpus, C.J. and Nicholls, N., The southern oscillation regulates the annual numbers of green turtles (*Chelonia mydas*) breeding around northern Australia, *Aust. Wild. Res.*, 15, 157, 1988.

Limpus, C.J. and Nicholls, N., ENSO Regulation of Indo-Pacific green turtle populations, in *Applications of Seasonal Climate Forecasting in Agricultural and Natural Ecosystems*, Hammer, G.L., Nicholls, N., and Mitchell, C., Eds., Kluwer Academic Publishers, Dordrecht, Germany, 399, 2000.

Limpus, C.J. and Reed, P., The green turtle, *Chelonia mydas*, in Queensland: a preliminary description of the population structure in a coral reef feeding ground, in *Biology of Australasian Frogs and Reptiles*, Grigg, G., Shine, R., and Ehmann, H., Eds., Royal Zoological Society of New South Wales, Sydney, 47, 1985.

Loop, K.A., Miller J.D., and Limpus, C.J., Nesting by the hawksbill turtle (*Eretmochelys imbricata*) on Milman Island, Great Barrier Reef, Australia, *Wildl. Res.*, 22, 241, 1995.

Maffei, M., et al., Leptin levels in human and rodent: measurement of plasma leptin and ob RNA in obese and weight-reduced subjects, *Nat. Med.*, 1, 1155, 1995.

Marquez, M.R., Synopsis of the biological data on the Kemp's ridley turtle, *Lepidochelys kempii (Garman, 1880)*, NOAA Technical Memorandum NMFS-SEFSC-343, Miami, 91, 1994.

Meylan, A., Sea turtle migration — evidence from tag returns, in *Biology and Conservation of Sea Turtles*, Bjorndal, K.A., Ed., Smithsonian Institution Press, Washington, DC, 1982.

Meylan, A.B., International movements of immature and adult hawksbill turtles (*Eretmochelys imbricata*) in the Caribbean region, *Chelonian Conserv. Biol.*, 3, 189, 1999.

Miller, J.D., Embryology of marine turtles, in *Biology of the Reptilia*, Gans, C., Billett, F., and Maderson, P.F.A., Eds., John Wiley & Sons, New York, 269, 1985.

Miller, J.D., Reproduction in sea turtles, in *The Biology of Sea Turtles*, Lutz, P.L. and Musick, J.A., Eds., CRC Press, Boca Raton, FL, 51, 1997.

Miller, J.D. et al., Long-distance migrations by the hawksbill turtle, *Eretmochelys imbricata*, from north-eastern Australia, *Wildl. Res.*, 25, 89, 1998.

Mortimer, J.A. and Carr, A., Reproduction and migrations of the Ascension Island green turtle (*Chelonia mydas*), *Copeia*, 1987, 103, 1987.

Mortimer, J.A. and Balazs, G.H., Post-nesting migrations of hawksbill turtles in the granitic Seychelles and implications for conservation, NOAA Technical Memorandum, NMFS SEFSC, Miami, 443, 22, 2000.

Mrosovsky, N., Hopkins-Murphy, S.R., and Richardson, J.I., Sex ratio of sea turtles: seasonal changes, *Science*, 225, 739, 1984.

Niewiarowski, P.H., Balk, M.L., and Londraville, R.L., Phenotypic effects of leptin in an ectotherm: a new tool to study the evolution of life histories and endothermy?, *J. Exp. Biol.*, 203, 295, 2000.

Osborne, N.J.T., Webb, P.M., and Shaw, G.R., The toxins of *Lyngbya majuscula* and their human and ecological health effects, *Environ. Int.*, 27, 381, 2001.

Owens, D.W., The comparative reproductive physiology of sea turtles, *Am. Zool.*, 20, 549, 1980.

Owens, D., The role of reproductive physiology in the conservation of sea turtles, in *Biology and Conservation of Sea Turtles*, Bjorndal, K.A., Ed., Smithsonian Institution Press, Washington, DC, 1982.

Owens, D.W., Hormones in the life history of sea turtles, in *The Biology of Sea Turtles*, Lutz, P.L. and Musick, J.A., Eds., CRC Press, Boca Raton, FL, 315, 1997.

Owens, D., Reproductive cycles and endocrinology, in *Research and Management Techniques for the Conservation of Sea Turtles*, Eckert, K.L. et al., Eds., IUCN/SSC Marine Turtle Specialist Group, Publication No. 4, Pennsylvania, 1999.

Owens, D.W., Grassman, M.A., and Hendrickson, J.R., The imprinting hypothesis and sea turtle reproduction, *Herpetologica*, 38, 124, 1982.

Owens, D.W. and Morris, Y.A., The comparative endocrinology of sea turtles, *Copeia*, 1985, 723, 1985.

Paolucci, M., Rocco, M., and Varricchio, E., Leptin presence in plasma, liver and fat bodies in the lizard *Podarcis sicula* fluctuations throughout the reproductive cycle, *Life Sci.*, 69, 2399, 2001.

Parmenter, C.J. and Limpus, C.J., Female recruitment, reproductive longevity and inferred hatchling survivorship for the flatback turtle (*Natator depressus*) at a major eastern Australian rookery. *Copeia*, 1995, 474, 1995.

Plotkin, P.T., et al., Independent versus socially facilitated oceanic migrations of the olive ridley, *Lepidochelys olivacea*, *Mar. Biol.*, 122, 137, 1995.

Plotkin, P.T. et al., Reproductive and developmental synchrony in female *Lepidochelys olivacea*, *J. Herpetol.*, 31, 17, 1997.

Pritchard, P.C.H., Studies on the systematics and reproduction of the genus *Lepidochelys*, Ph.D. thesis, University of Florida, Gainesville, FL, 1969.

Reina, R.D. et al., Imminent extinction of Pacific leatherbacks and implications for marine biodiversity, in *Proceedings of the Twentieth Annual Symposium on Sea Turtle Biology and Conservation*, Technical Memorandum NMFS, Orlando, FL, 2000.

Rogers, R W., The influence of sea turtles on the terrestrial vegetation of Heron Island, Great Barrier Reef (Queensland, Australia), *Proc. R. Soc. Queensland*, 100, 67, 1989.

Rostal, D. et al., Ultrasound imaging of ovaries and eggs in Kemp's ridley sea turtles (*Lepidochelys kempii*), *J. Zoo Wildl. Med.*, 21, 27, 1990.

Rostal, D.C. et al., Reproductive physiology of nesting leatherback turtles (*Dermochelys coriacea*) at Las Baulas National Park, Costa Rica, *Chelonian Conserv. Biol.*, 2, 230, 1996.

Rostal, D.C. et al., Nesting physiology of Kemp's ridley sea turtles, *Lepidochelys kempii*, at Rancho Nuevo, Tamaulipas, Mexico, with observations on population estimates, *Chelonian Conserv. Biol.*, 2, 538, 1997.

Rostal, D.C. et al., Seasonal reproductive cycle of the Kemp's ridley sea turtle (*Lepidochelys kempii*), *Gen. Comp. Endocrinol.*, 109, 232, 1998.

Rostal, D.C. et al., Changes in gonadal and adrenal steroid levels in the leatherback sea turtle (*Dermochelys coriacea*) during the nesting cycle, *Gen. Comp. Endocrinol.*, 122, 139, 2001.

Short, F.T. and Neckles, H.A., The effects of global climate change on seagrasses, *Aquat. Bot.*, 63, 169, 1999.

Solomon, S.E. and Baird, T., Aspects of the biology of *Chelonia mydas* L., *Oceanogr. Mar. Biol.*, 17, 347, 1979.

Stearns, S.C., Trade-offs in life history evolution, *Funct. Ecol.*, 3, 259, 1989.

Tucker, A.D. and Frazer, N.B., Reproductive variation in leatherback turtles, *Dermochelys coreacea*, at Culebra National Wildlife Refuge, Puerto Rico, *Herpetologica*, 47, 115, 1991.

Tucker, A.D. and Read. M.A., Frequency of foraging by gravid green turtles (*Chelonia mydas*) at Raine Island, Great Barrier Reef, *J. Herp.*, 2001.

Valverde, R.A. et al., Basal and stress-induced corticosterone levels in olive ridley sea turtles (*Lepidochelys olivacea*) in relation to their mass nesting behavior, *J. Exp. Zool.*, 284, 652, 1999.

Vargas, P., Enzyme linked immunosorbent assay (ELISA) for the Kemps' ridley sea turtle (*Lepidochelys kempii*: Garman 1880), Vitellogenin. M.Sc. thesis, Texas A&M University, College Station, TX, 70, 2000.

Whittier, J.M. and Crews, D., Seasonal reproduction: patterns and control, in *Hormones and Reproduction in Fishes, Amphibians and Reptiles*, Norris, D.O. and Jones, R.E., Eds., Plenum, New York, 385, 1987.

Whittier, J.M., Corrie, F., and Limpus, C., Plasma steroid profiles in nesting loggerhead turtles (*Caretta caretta*) in Queensland, Australia: relationship to nesting episode and season, *Gen. Comp. Endocrinol.*, 106, 39, 1997.

Wibbels, T. et al., Seasonal changes in serum gonadal steroids associated with migration, mating, and nesting in the loggerhead sea turtle (*Caretta caretta*), *Gen. Comp. Endocrinol.*, 79, 154, 1990.

Wibbels, T., Owens, D., and Rostal, D., Soft plastra of adult male sea turtles: an apparent secondary sexual characteristic, *Herpetol. Rev.*, 22, 47, 1991.

Wibbels, T., Owens, D.W., and Limpus, C.J., Sexing juvenile sea turtles: is there an accurate and practical method?, *Chelonian Conserv. Biol.*, 3, 756, 2000.

Witzell, W.N., Synopsis of biological data on the hawksbill turtle, *Eretmochelys imbricata (Linnaeus, 1766)*, *FOA Fish. Synop.*, 137, 78, 1983.
Wood, J.R. and Wood, F.E., Reproductive biology of captive green sea turtles, *Chelonia mydas*, *Am. Zool.*, 20, 499, 1980.
Wood, J.R. et al., Laparoscopy of the green sea turtle, *Chelonia mydas*, *Brit. J. Herpetol.*, 6, 323, 1983.

6 Physiological and Genetic Responses to Environmental Stress

Sarah L. Milton and Peter L. Lutz

CONTENTS

6.1 WHAT IS STRESS?

Many people are uncomfortable with the term *stress* in animal biology. The root of the difficulty lies in the common usage of the word and its richness of meanings

0-8493-1123-3/03/$0.00+$1.50

that bedevil an exact scientific definition. In biology, the term embraces psychology to biomechanics, and it is only in the latter that it is used in the precise and quantitative terms of Hooke's law, where stress (the deforming force) is proportional to strain (the deformation). For the rest there is no agreement about whether stress refers to external or internal factors, what it consists of, or how it can be measured. Nevertheless, the fact that the concept is still widely used in biology, from the molecular to ecosystem level, indicates its utility and its necessity (Bonga, 1997). Perhaps the term should be used only in combination with the causal factor (i.e., crowding stress, temperature stress), with the concept that there is an (identified) tolerance range for the external factor within which the individual or community copes by means of adaptive responses, but that outside this range there is a quantitative or qualitative break in the (described) response.

The adaptive function of the stress response is to accommodate changes in the environment (stressors) by adjustments in behavior and/or changes in physiology. However, an excessive exposure to the stressor, in either intensity or duration, will result in dysfunctional debilitating responses. Environmental conditions to which an animal cannot adapt lead to both transient and relatively long-term physiological changes. Such changes often contribute to the development of disease, especially if the organism is exposed at the same time to potentially pathogenic stimuli. Various stressors, however, do not all produce the same outcomes; effects will depend on the quality, quantity, and duration of the stressor; the temporal relationship between the exposure to a stressor and the introduction of pathogenic stimuli; environmental conditions; and a variety of host factors (age, species, gender, etc.) (Ader and Cohen, 1993).

This chapter presents an overview of the relationship between sea turtles and some of the more important stressful aspects of their environment. Because stress is such a broad topic, many aspects of stress have been treated in previous chapters and elsewhere in this volume (see Lutcavage et al., 1997; George, 1997; Epperly, Chapter 13; and Herbst and Jacobson, Chapter 15, this volume). This chapter reviews a few environmental stressors of particular significance to sea turtles: temperature, chemical pollutants (organic and inorganic) and habitat degradation, and the sea turtle's physiological and potential genetic responses are discussed. Distinct environmental stressors affect the terrestrial nest and hatchlings, and are discussed separately from the other (oceanic) life stages.

6.2 WHY SEA TURTLES ARE AT SPECIAL RISK

Sea turtles naturally encounter a wide variety of stressors, both natural and anthropogenic, including environmental factors (salinity, pollution, temperature), physiological factors (hypoxia, acid–base imbalance, nutritional status), physical factors (trauma), and biological factors (toxic blooms, parasite burden, disease). Although they are physically robust and able to accommodate severe physical damage, sea turtles appear to be surprisingly susceptible to biological and chemical insults (Lutcavage and Lutz, 1997). For example, in the green sea turtle even a short exposure to crude oil shuts down the salt gland, produces dysplasia of the epidermal epithelium, and destroys the cellular organization of the skin layers, thus opening routes for infection (Lutcavage et al., 1995). The effects of many stressors, however, are likely to be less obvious, as in the (unknown) long-term effects of toxin exposure and bioaccumulation.

Because sea turtles are long-lived animals, the cumulative effect of various stressors is likely to be great. Because sea turtles spend discrete portions of their life in a variety of marine habitats, they are vulnerable at multiple life stages: as eggs on the beach, in the open ocean gyres, as juveniles in nearshore waters, and as adults migrating between feeding and nesting grounds. Thus, turtles may be exposed to a greater variety of environmental stressors than less migratory animals, with presumably different vulnerabilities at each stage. However, their exposure to a particular stressor may be limited by the length of that life history stage. For example, fibropapilloma disease appears to affect primarily juvenile green turtles of 40–90 cm carapace length (Ehrhart, 1991), but is rare in nesting adults. Exposure to weathered oil has significant health effects on swimming turtles (Lutcavage et al., 1995), but in one study demonstrated little impact on egg survival. Fresh oil, on the other hand, significantly affected egg survival (Fritts and McGehee, 1981). Vulnerability to certain stressors will also vary by ecological niche, i.e., polychlorobiphenyl (PCB) and dichlorodiphenyldichloroethylene (DDE) accumulations are consistently higher in loggerhead turtle tissues and eggs than in those of green turtles (George, 1997; Clark and Krynitsky, 1980), presumably because of dietary differences. Clark and Krynitsky (1980) also reported that DDE and PCB loads in both loggerhead and green turtle eggs were significantly lower than in bird eggs taken from the same location (Merritt Island, FL) and lower than contaminant levels in eggs from Everglades (FL) crocodiles. They speculated that adult turtles nesting on Merritt Island lived and fed in areas less contaminated than did the residential bird and Everglades crocodile populations.

Natural stressors include thermal stress (heat stress, cold stunning), seasonal or temperature-related changes in immune function, and the presence of disease, parasites, or epiphytes. Even these natural physiological stressors may, of course, be impacted or exaggerated by anthropogenic factors. For example, physiological responses to natural diving are significantly different from those produced by the forced submergence of trawl entanglement (Lutcavage et al., 1997), and animals with a depressed immune system related to pollutant levels would be more vulnerable to parasites and disease.

Anthropogenic stressors may have either direct or indirect impacts on sea turtle health. Direct impacts include such problems as oil spills, latex or plastic ingestion, fishing line entanglement, and the presence of persistent pesticides, hormone disrupting pollutants, and heavy metals. Indirect effects occur primarily through habitat degradation: eutrophication, the contribution of pollutants to toxic algal blooms, and collapse of the food web.

Inappropriate sea turtle behavior can put them at particular risk. For example, it appears that unlike marine mammals, adult sea turtles show no avoidance behavior when they encounter an oil slick (Odell and MacMurray, 1986); they also indiscriminately ingest tar balls and plastics (Lutz, 1990), and hatchlings congregate in ocean rift zones where floating debris concentrate. Their breathing pattern of large tidal volumes and rapid inhalation before diving will result in the most direct and effective exposure to petroleum vapors (the most toxic part of oil spills), as well as biotoxin aerosols resulting from dinoflagellate blooms.

Sea turtles are at particular risk from the stresses presented by degraded tropical coastal marine environments. Indeed, the high public awareness of sea turtles is such that they can serve as effective sentinels of tropical coastal marine ecosystem health (Aguirre and Lutz, in press).

6.3 STRESSORS

This review selects some of the most critical identified natural and anthropogenic stressors of sea turtle physiology, while omitting some (oil, nesting, capture stress) that have been previously reviewed (see Lutz and Musick, 1997).

6.3.1 Temperature

Both high and low temperatures are known to negatively impact sea turtle physiology, affecting feeding behavior, acid–base and ion balance, and stress hormone levels.

6.3.1.1 Hypothermia

Temperature has a marked effect on the feeding rates of sea turtles. At 20°C Kemp's ridley turtles decreased food consumption to 50% of control levels (at 26°C), and a similar reduction in food intake was found in green turtles at 15°C (Moon et al., 1997). Below 15°C both species ceased feeding. Interestingly, Moon et al. (1997) found that green and Kemp's ridley turtles' swimming behavior differed as temperatures decreased. When temperatures dropped below 20°C green turtles reduced swimming activity, but at these temperatures the ridleys became very agitated. Below 15°C both species became semidormant, hardly moving and only coming to the surface at intervals of up to 3 h to breathe. Field evidence supports these findings. During cold temperatures in winter, loggerhead turtles in Tunisian waters reduce overall activity even though they continue to forage (Laurent and Lescure, 1994).

Temperature also profoundly influences the physiology of sea turtles. In ridleys and greens, both venous blood partial pressure of oxygen (pO_2) and partial pressure of carbon dioxide (pCO_2) decreased with temperature (Moon et al., 1997), whereas venous blood pH increased. Similar temperature-dependent changes in blood pH, pCO_2, and pO_2 have been widely found in other reptiles, including loggerhead sea turtles (Lutz et al., 1989). Temperature-related adjustments of blood pH in the loggerhead appeared to be managed at both the lung and tissue (ion exchange) levels (Lutz et al., 1989). In both wild (Lutz and Dunbar-Cooper, 1987) and captive (Lutz et al., 1989) loggerheads, plasma potassium increased with temperature, which may be related to cellular-mediated adjustments in blood pH. Excessively low temperatures can also interfere with physiological functioning. For example, there was an abrupt failure in pH homeostasis and a sharp increase in blood lactate at temperatures below 15°C in the loggerhead (Lutz et al., 1989). At 10°C the loggerheads were lethargic and "floated" (Lutz, personal observation). Such positive buoyancy is probably due to cessation of intestinal mobility and the collection of ferment gases and is commonly observed in cold stunning.

Unlike certain freshwater turtles, which overwinter in frozen ponds and thus withstand months submerged in near-freezing water (Jackson, 2000), sea turtles (with the exception of leatherbacks) trapped in cold waters (below 8–10°C) may become lethargic and buoyant, floating at the surface. This condition is defined as *cold stunning* (Schwartz, 1978). Salt gland function may be impaired in cold-stunned animals, as evidenced by increased blood concentrations of sodium, potassium, chlorine, calcium, magnesium, and phosphorus (George, 1997; Carminati et al., 1994). Affected animals may not eat for days or even weeks prior to cold stunning, increasing overall physiological stress (Morreale et al., 1992). However, it is likely that it is the rate of cooling below 15°C that evokes cold stunning rather than the temperature per se. Satellite tracking studies of ocean migrating Kemp's ridley and loggerhead turtles indicate that they remain active in water temperatures as low as 6°C (Keinath, 1993). Sea turtles that overwinter in inshore waters are most susceptible to cold-stunning because temperature changes are most rapid in shallow water, especially in semienclosed areas such as lagoons (Witherington and Ehrhart, 1989). As temperatures drop below 5–6°C, death rates become significant, because the animals can no longer swim or dive, become vulnerable to predators, and may wash up onshore, where they are exposed to even colder temperatures.

As with other physiological stressors, cold stunning can affect *specific populations* of sea turtles more than others. For example, although cold-stunning events occur in Florida as well as in northern waters, the extended exposure to frigid waters experienced by turtles off New England or New York results in much higher mortality rates. Morreale et al. (1992) reported overall mortality rates as high as 94% over three winters in New York, whereas Witherington and Ehrhart (1989) reported only 10% mortality for cold-stunned turtles in a Florida estuary.

Habitat utilization is also a significant factor in differential mortality during cold-stun events. The waters off New York and New England appear to be an important habitat for juvenile Kemp's ridley turtles, with the result that a large percentage of identified cold-stunned animals are of this species (Figure 6.1). Of the 277 total sea turtles found on Cape Cod, MA, during the 1999–2000 winter season, 79% were Kemp's ridley turtles, 19% loggerheads, and 2% greens (Still et al., in press). During the 1985–1986 winter, 79% of the turtles retrieved on Long Island (NY) were Kemp's ridleys (Meylan and Sadove, 1986). Indeed, Kemp's ridleys have consistently made up more than 50% of the cold-stunned turtles found along Cape Cod for the past 20 winters, and 67–80% of cold-stunned turtles found off Long Island over a 3-year period were Kemp's ridleys (Morreale et al., 1992). By contrast, in five significant stunning events over a 9-year period in the Indian River Lagoon (FL), 73% of 467 recovered turtles were greens (Figure 6.1), 26% were loggerheads, but less than 1% (2 animals) were Kemp's ridleys (Witherington and Ehrhart, 1989).

Size is also an important factor in susceptibility to cold-stun events, because juveniles are the primary life history stage affected. The majority of Kemp's ridleys retrieved off Cape Cod in the 1999–2000 season were in the 25.0–29.9 cm curved carapace length (CCL) size class, as were many greens. Similarly, Morreale et al. (1992) reported a mean straight carapace length (SCL) of 29.4 cm for *Lepidochelys kempii* and 32.7 cm for *Chelonia mydas* for cold-stunned turtles collected off Long

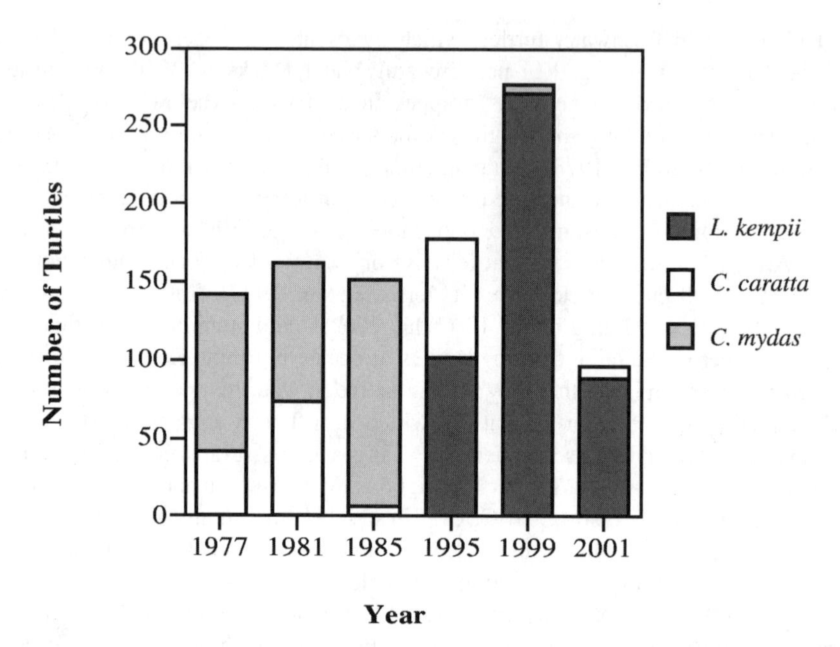

FIGURE 6.1 Species–habitat-specific susceptibility to cold-stun events at two different U.S. locations: the Indian River Lagoon, FL (south), and Cape Cod Bay, MA (north). Only large cold-stun events are shown: 1977–1985 data are from Florida (adapted from Witherington, B.E. and Ehrhart, L.M., Hypothermic stunning and mortality of marine turtles in the Indian River Lagoon system, Florida, *Copeia*, 1989, 696–703, 1989); 1995–2001 data are from Massachusetts (adapted from Still et al., 2000 and Still, B., Griffin, C., and Prescott, R., Factors affecting cold-stunning of juvenile sea turtles in Massachusetts, in: *Proceedings of the 22nd Annual Symposium on Sea Turtle Biology and Conservation*, J. Seminoff (compiler), U.S. Dept. Commerce NOAA Tech. Memo. NMFS-SEFSC, Miami, FL (in press). (With permission.)

Island between 1985 and 1987. It appears that larger Kemp's ridley turtles either do not make much use of this habitat (Morreale et al., 1992) or are more successful in emigrating from northern waters prior to the onset of lethal winter temperatures (Standora et al., 1992).

Smaller turtles also succumb more quickly than larger animals (Witherington and Ehrhart, 1989). In their study on cold-stunning events in the Indian River Lagoon, Witherington and Ehrhart (1989) noted that the smallest turtles were found on the first day of the cold snap, and largest turtles on the last day; over the 9 years of the study, nearly half of the green turtles recovered were in the 0–10 kg size class (SCL ranged from 24.6 to 75.4 cm).

It is also likely that there are *species differences* in susceptibility to hypothermia. Witherington and Ehrhart (1989) reported that the loggerhead cold-stunning death rate was less than that for green turtles, and suggested that this was because loggerheads are a more temperate zone species, whereas the Indian River Lagoon appears to be the northernmost limit of the green turtles' winter range. Leatherback turtles nest on tropical beaches, but are seen as far north as the waters off Newfoundland, in temperatures ranging from 0 to 15°C (Goff and Lien, 1988). Frair et al. (1972) reported a body temperature of 25.5°C for a leatherback held in 7.5°C water, which makes the idea of a cold-stunned adult leatherback unlikely!

In addition to migrating toward warmer waters at the onset of the cold season, larger turtles may physiologically avoid cold stunning by entering a hibernation-like state. There is evidence that both green (*Chelonia agassizi*) and loggerhead turtles bury themselves in bottom sediments for extended periods of time during winter (Felger et al., 1976; Carr et al., 1980–81).

The recommended treatment for cold stunning is fairly straightforward: hold the animals in warm water until their core temperature recovers (George, 1997). The success rate is high — of the turtles treated at the New England Aquarium during the 1999–2000 cold-stunning season, survival ranged from 66% (*C. mydas*) to 100% (*Caretta caretta*) (Still et al., in press). Holding the victims in fresh or brackish water until salt gland function recovers has also been recommended (George, 1997).

6.3.1.2 Hyperthermia

Excessive heat exposure is also a stress to poikilotherms, though for sea turtles hyperthermia would be a rare phenomenon when they are in the ocean. However, increased water temperatures may indirectly increase stress on sea turtles, in that increased surface temperatures increase the growth rates of both pathogens and toxic phytoplankton.

High temperatures can, however, be experienced while they are on land, basking or nesting.

In turtles basking at French Frigate Shoals (HI) carapace temperatures as high as 42.8°C have been recorded (Whittow and Balazs, 1982). Behavioral adaptations are used to moderate the ambient heat load. Surface temperatures can be reduced as much as 10°C by flipping sand onto flippers and the carapace, and basking turtles appear to choose cooler beaches (Whittow and Balazs, 1982).

Heat stress can be fatal for nesting females. Environmental temperatures above 40°C can result in stress for green sea turtles (see Spotila et al., 1997), whereas excessive heat exposure routinely results in a high mortality (tens of turtles per day) of postnesting females at the Raine Island (Australia) green turtle rookery (Jessop et al., 2000). In the Raine Island study, an increase in body temperature of females stranded on the beach from 28.2 to 40.7°C over 6 h resulted in a 16-fold mean increase in plasma corticosterone (a hormonal marker of stress), to levels comparable to those seen in animals subjected to 8 hr capture stress (Jessop et al., 2000). In the soft-shelled turtle, *Lissemys punctata punctata*, increases in adrenomedullary function were detected as temperatures increased from 30 to 35 and 38°C, resulting in increased levels of circulating epinephrine, norepinephrine, and glucose (Ray and Maiti, 2001).

6.3.2 Chemical Pollutants

Age, gender, and diet are all important factors in the potential for animals to be affected by or bioaccumulate persistent pollutants, as is the identity and effects of the specific contaminant. Manufactured chemicals released into the environment may act as endocrine-disrupting contaminants, affect tumor growth, depress immune function, or be acutely or chronically toxic. Two of the most significant groups of chemical stressors are the heavy metals and organopesticides.

6.3.2.1 Bioaccumulation

6.3.2.1.1 Heavy Metals

Despite the high toxicity of some compounds such as methylmercury, there is a relative paucity of data either for contaminated animals or for normal ranges (of trace elements) in tissues (for a review, see Pugh and Becker, 2001). In general, concentrations of heavy metals and trace elements appear to be lower in sea turtle tissues (by as much as one to two orders of magnitude) than values reported for marine birds and mammals, which may be a function of differences in their metabolic rates. Studies on liver concentrations of mercury indicate a correlation between diet and mercury accumulation, such as occurs in piscivorous marine mammals and seabirds, with mercury levels higher in the omnivorous loggerhead (Sakai, 1995; Storelli et al., 1998a; 1998b; Godley et al., 1999) than in herbivorous green and jellyfish-eating leatherback turtles (Godley et al., 1999; Davenport et al., 1990). Day et al. (2002) reported higher levels of mercury in loggerhead turtles residing near river mouths than those from farther away. One must be wary, however, of making assumptions based solely on trophic levels: Saeki et al. (2000) reported the surprising finding that arsenic levels were higher in hawksbill turtles (which consume primarily sponges) than in algae- and mollusk-eating green and loggerhead turtles. Changes in heavy metal accumulation with age (size) within a species have also been reported. For example, Sakai et al. (2000) found higher levels of copper in the livers of small green turtles than in larger ones; liver cadmium was also negatively correlated with size. They hypothesized a difference based on diet (i.e., life history stage), because cadmium levels are higher in the zooplankton diet of juvenile greens than in seagrasses. No data on heavy metal burdens are available for Kemp's or olive ridley turtles.

6.3.2.1.2 Pesticides

Reported levels of PCBs and other organic contaminants in sea turtle tissues are also generally an order of magnitude lower than those found in marine mammals (Becker et al., 1997). In particular, total dichlorodiphenyltrichloroethane (DDT) tissue concentrations in sea turtles are at the lowest end of the range reported for marine mammals and seabirds (Pugh and Becker, 2001). However, PCB contamination in sea turtles is widespread. One frequently detected congener, PCB 153, has been reported in the tissues of loggerheads and Kemp's ridleys along the East Coast of the U.S., in loggerheads and green turtles from the Mediterranean Sea, and in leatherbacks from the United Kingdom (Lake, 1994; Rybitski et al., 1995; Mckenzie et al., 1999). PCBs 153 and 138 were the dominant congeners detected in Hawaiian green turtle liver and adipose tissues, with detectable amounts of the more toxic congeners PCB 77, PCB 126, and PCB 169 (Miao et al., 2001). In these studies, levels were higher in loggerhead and Kemp's ridley turtles than in greens, most likely because these turtles are at a higher trophic level and thus more subject to bioaccumulation. Species-, gender-, or age-specific physiological differences clearly will play a role in the effects and accumulation of various chemicals; the "offloading" of pollutants to eggs, for example, is clearly not an option for male sea turtles as it is for the females. Unfortunately, most of such differences even in basic physiology are unknown (Milton et al., in press).

6.3.2.2 Effects

6.3.2.2.1 Toxicity

The toxicity of heavy metals and organopesticides is well established in other vertebrate groups (mammals and fish), with wide-ranging effects on the neurological, immunological, and reproductive systems. Although no long-term investigations in sea turtles have been reported, one might expect similar deleterious consequences.

For many compounds with potentially toxic effects, there are little or no data for sea turtles. Hexachlorobenzene (HCB), for example, is one of the most toxic and most persistent of the chlorobenzene compounds, which as a highly volatile compound is able to travel long distances in the atmosphere. No data on HCB, dioxin, or furan levels have been reported for sea turtle tissues or eggs. There is only one report of hexachlorocyclohexane and few for dieldrin, even though dieldrin is one of the most commonly detected and easily analyzed pesticides reported in marine biota (Pugh and Becker, 2001).

Although acutely toxic levels of xenochemicals have not been reported in sea turtles, even trace amounts may be of concern because of potential sublethal effects on health and normal physiology. Because of the difficulty of working with endangered animals, however, data are lacking on the normal physiology, immunology, and population biology of sea turtles, and it is difficult to determine chronic effects of pollutants. Such difficulties are compounded by the nature of the pollutants as well. For example, comparisons between studies on the harmful effects of organochlorines such as PCBs are difficult because of between-study variations in identification and quantification of congeners. Not all PCB congeners are metabolized at the same rate, and some are more toxic than others (Kannan et al., 1989). Despite these limitations, studies on other species indicate cause for concern. High organochlorines (such as PCBs and DDE) have been associated with uterine deformities and decreased pup production in seals (Baker, 1989; Reijnders, 1980); embryotoxicity and effects on the hypothalamus–pituitary–adrenal axis in herring gulls (*Larus argentatus*) (Fox et al., 1991; Lorenzen et al., 1999); decreased levels of circulating thyroid hormone and lesions of the thyroid gland in seals and rats (Byrne et al., 1987; Collins et al., 1977; Schumacher et al., 1993); decreased activity levels, feeding rates, and whole body corticosterone levels in tadpoles of the northern leopard frog (*Rana pipiens*) (Glennemeler and Denver, 2001); and decreased immune responsiveness in chicks (Andersson et al., 1991), rats (Smialowicz et al., 1989), primates (Tryphonas et al., 1989), mice (Thomas and Hinsdill, 1978), and beluga whales (De Guise et al., 1998). Beluga whales living in the highly contaminated St. Lawrence Seaway also have increased incidence of neoplasias (De Guise et al., 1995); PCBs apparently act as a tumor promoter as well as an immunosuppressant. PCB immunosuppression results in higher sensitivities of experimental animals to a wide variety of infectious agents, including bacteria (endotoxin), protozoa, and viruses (De Guise et al., 1998). Lahvis et al. (1995) found a direct correlation between suppressed immunological function *in vitro* and PCB load in bottlenose dolphins, whereas the PCB-linked impairment of immune function likely contributed to the recent mass mortalities in European harbor seals resulting from morbillivirus infections (Ross, 2000).

Similar patterns of accumulation, if not actual concentrations, are possible in some sea turtle species when compared to marine mammals because similar diets can lead to similar tissue lipid compositions (Guitart et al., 1999). In sea turtles, fibropapilloma is more prevalent in green turtles captured near densely populated, industrial regions than in animals from sparsely populated areas (Adnyana et al., 1997), although no correlation was detected between organochlorine, PCB, or organophosphate levels and green turtle fibropapilloma disease (GTFP) (Aguirre et al., 1994). However, the potential for chronic pollutants to decrease immune function either directly or indirectly (by increasing overall stress) could have significant impacts on sea turtle populations, because how they deal with physical stress (infection or trauma) is affected by environmental stress, and stress in general most likely depresses the turtle immune system (George, 1997).

In general, chronic illnesses, mass mortalities, and epidemics are being reported across a wide spectrum of taxonomic groups in increasing numbers, with novel occurrences of pathogens, invasive species, and illnesses affecting wildlife globally. Such disturbances impact multiple components of marine ecosystems, disrupt both functional and structural relationships between species, and affect the ability of ecosystems to recover from natural or anthropogenic perturbations (Sherman, 2000).

6.3.2.2.2 Endocrine Disruption

Hormone disrupters are insidious but high-impact disturbers of population fitness. It is now well established that some organopesticides released into the environment act as endocrine-disrupting contaminants, functioning as hormone agonists or antagonists to disrupt hormone synthesis, action, and/or metabolism. Laboratory studies provide strong evidence of organopesticides' causing endocrine disruption at environmentally realistic exposure levels (Vos et al., 2000). In the aquatic environment, effects have been observed in mammals, birds, reptiles, fish, and mollusks. Alligators living in environments contaminated with endocrine disrupters, for example, have suffered population declines because of the developmental and endocrine abnormalities effected by these contaminants on eggs, juveniles, and adults (Guillette, 2000). Endocrine-disrupting contaminants have also adversely affected a variety of fish species in freshwater systems, estuaries, and coastal areas, whereas marine invertebrates (snails and whelks) have suffered population declines in some areas because of the masculinization of females (Vos et al., 2000).

PCBs, which are widespread, low-level environmental contaminants, are strongly implicated as endocrine disrupters. There is evidence that PCBs are capable of disrupting reproductive and endocrine function in a variety of taxonomic groups, in addition to producing other adverse health effects such as immune suppression and teratogenicity. Bergeron et al. (1994) demonstrated that the estrogenic effect of some PCBs could cause a reversal of gonadal sex in freshwater turtles (*Trachemys scripta*), which, like sea turtles, have temperature-dependent sex determination. In some areas, sex-reversal in turtles is so prevalent that it can be utilized as a marker of environmental contamination.

The exposure of sea turtle eggs to such pollutants could be significant, because there is evidence that females offload contaminants to their eggs (Mckenzie et al., 1999). In one study, eggs sampled from 20 nests in northwest Florida had detectable amounts of polycyclic aromatic hydrocarbons (PAHs), dichlorodiphenyldichloroethane (DDD, a DDT metabolite), and PCBs (Alam and Brim, 2000). However, the effects of these compounds on sea turtles are not known. A direct application of DDE, another estrogen-like compound, to green turtle eggs did not alter normal sex ratios, incubation times, hatchling success or size, or number of deformities (Podreka et al., 1998).

6.3.3 Eutrophication and Algal Blooms

Eutrophication caused by excess nutrient pollution in coastal waters, particularly of nitrogen derived from sewage and agricultural fertilizers, affects sea turtles both directly and indirectly (Magnien et al., 1992; Burkholder, 1998). In particular, there is a growing link between harmful algal blooms (HABs) and eutrophication. Cyanobacteria blooms in Moreton Bay, Australia, for example, have been increasing in recent years in both size and severity, resulting in loss of seagrass beds, decreased fish catches, and increased levels of ammonia and toxins, including tumor promoters and immunosuppressants (Osborne et al., 2001). HABs thus may have many direct (toxic) and indirect harmful impacts on sea turtles and other marine fauna; in Moreton Bay, the cyanobacteria blooms affect green turtles by decreasing feeding directly (as well as indirectly through the loss of seagrasses) and through the ingestion of toxins (Arthur et al., 2002). A strong association has also been noted between the prevalence of a variety of diseases and coastal pollution in multiple taxonomic groups, such that the occurrence of the diseases derived from pathogens or algal-derived biotoxins often serve as indicators of declining ecological integrity in coastal areas (Epstein et al., 1998). Groups adversely affected by eutrophication-related diseases include humans, birds, marine mammals and turtles, fish, invertebrates, and seagrass beds (Epstein et al., 1998).

The most prevalent tropical–semitropical algal blooms are the so-called red tides (which may be any color or even be invisible), which are due primarily to dinoflagellate blooms and can lead to morbidity and mortality in many species. Immediate effects occur through aerosolized transport, and the sea turtle's mode of respiration (rapid inhalation to fill the lungs before a dive) puts the sea turtle at special risk here. Long-term effects may occur through the consumption of prey and toxin bioaccumulation.

Long-term exposure to biotoxins may exert more subtle, sublethal effects such as impaired feeding, physiological dysfunction, impaired immune function, and reduced growth and reproduction. Long-term effects often emerge as an increased susceptibility to disease (immunosuppression) and in the development of neoplasia (Epstein et al., 1998). Deaths are often attributed to viral factors as the immediate cause of mortality, whereas viral expression and host immunity have been affected by chronic biotoxin exposure. Such may be the case in GTFP, where oncogenic viruses and tumor-promoting toxins may be acting in concert (Landsberg, 1996), with particular effects on immunosuppressed animals (Bossart et al., 2002). Eutrophication may directly increase viral and bacterial loads as well, in addition to the increased severity and frequency of algal blooms (Herbst and Klein, 1995).

In sea turtles, there appears to be an association between the distribution of toxic dinoflagellates (*Prorocentrum* spp.) and the occurrence of fibropapilloma disease among Hawaiian green sea turtles (Landsberg et al., 1999). These benthic dinoflagellates are epiphytic on seagrasses and macroalgae, and would thus be consumed by foraging green turtles. *Prorocentrum* are of particular interest because this group produces the tumor-promoting toxin okadaic acid, also detected in the tissues of Hawaiian green turtles (*C. mydas*) with GTFP (Landsberg et al., 1999).

More direct, toxic effects of red tide blooms of *Gymnodinium* have been suggested, although a direct link has yet to be demonstrated between brevetoxin and large die-offs of turtles such as have recently occurred in Florida. Chronic brevetoxicosis has been suggested as the likely primary etiology for manatee deaths that occurred in the same time frame (Bossart et al., 1998); simultaneous epizootics for manatees, fish, and cormorants associated with *Gymnodinium* blooms have occurred in the past (O'Shea et al., 1991). Sea turtle strandings in Florida increased significantly during four recent red tide blooms of the dinoflagellate *Karenia brevis*, with live turtles displaying symptoms of neurological disorders (Redlow et al., 2002). In nonsurviving animals associated with these blooms, liver brevetoxins were often as high as or higher than those in manatees determined to have died of brevetoxin poisoning. Patterns of bioaccumulation or species-specific susceptibility were also detected: brevetoxins were highest in Kemp's ridley turtles, intermediate in loggerhead tissue (only 1 animal), and lowest in greens (Redlow et al., 2002). Such die-offs appear to primarily affect juvenile and subadult turtles that are residents of nearshore waters; however, effects on breeding populations could be significant should springtime HABs continue into the start of the nesting season.

A secondary but important effect of eutrophication is the general degradation of the marine environment, which can seriously devalue its use as turtle habitat. Even nontoxic algal blooms (brown tides) can result in the loss of seagrass beds at nutrient-rich locations (Havens et al., 2001), as can increased levels of turbidity or changes in salinity (Figure 6.2). Prolonged blooms can also add large amounts of decaying matter to the water, causing hypoxic or anoxic conditions and furthering the devastation (Epstein et al., 1998). Havens et al. (2001) reported that a dense lawn of macroalgae on the bottom of one Virginia estuary reduced sediment–water nitrogen exchange when the algae were actively growing, but resulted in high nitrogen release during algal senescence. Such significant impacts on invertebrates and seagrasses would be magnified up the food chain, potentially resulting in large areas of ocean "desert," which appear to be occurring with increasing frequency. In Hervey Bay, Australia, for example, more than 1000 km^2 of seagrass beds have been lost, resulting in significant mortality and migration of the dugong population and the reduction of commercial prawn and fish catches (Brodie, 1999). The effects of such large-scale eutrophication on resident sea turtle populations are completely unknown because in-water population studies are lacking in affected areas.

FIGURE 6.2 *Thalassia testudinum* in Florida Bay. Algal blooms and turbidity contribute to seagrass die-offs in turtle feeding grounds worldwide. (Photo courtesy of Dr. Michael Durako, University of North Carolina.)

6.3.4 Disease

Disease can be both a cause and a symptom of stress. Large numbers of leeches, for example, can lead to anemia and damage the dermis, thus opening routes for secondary infections, whereas barnacle loads increase stress by increasing drag (George, 1997). Models of swimming and drag suggest that a heavy barnacle load may increase drag up to tenfold and energetic requirements in swimming sea turtles by more than threefold (Gascoigne and Mansfield, 2002).

In general, bacterial infections are relatively rare in free-roaming sea turtles (although they occur more frequently in the crowded conditions of captivity); traumatic injury to the dermis and aspiration of seawater are the two primary routes by which bacteria enter (George, 1997; see also Chapter 15). Even infections that are less acutely toxic may have significant effects on sea turtle health that will increase overall stress on the animal. This is seen, for instance, in the buoyancy abnormalities associated with pneumonia reported by Jacobson et al. (1979). Health problems and diseases of sea turtles are reviewed extensively in the first volume of this series (George, 1997).

6.3.4.1 Trematodes

Among loggerhead turtles, the most damaging parasites are the spirorchid trematodes, which reside in the vascular system and affect up to 30% of the Atlantic loggerhead population (Wolke et al., 1982). Green turtles are also vulnerable. A histopathological examination of four dead green turtles by Raidal et al. (1998) revealed severe granulomatous vasculitis, with aggregations of spirorchid eggs and microabcesses in the intestines, kidney, liver, lung, and brain. This damage in turn permitted a variety of bacterial infections, including *Salmonella*, *Escherichia coli*, *Citrobacter*, and *Moraxella* spp. They concluded that Gram-negative bacterial infections caused systemic illness and death following the severe infestation by spirorchid cardiovascular flukes. Glazebrook and Campbell (1990) found cardiovascular flukes in green, loggerhead, and hawksbill turtles in the U.S., India, Pakistan, and Australia, as well as a variety of gastrointestinal (GI) flukes, barnacles, and mites. In that study, heart fluke infestations resulted in cases of bronchopneumonia and septicemia–toxemia, whereas all heavy infestations of cardiovascular flukes were associated with severe debilitation, generalized muscle wastage, and thickening and hardening of the walls of the major cardiac blood vessels.

6.3.4.2 GTFP

The epidemic of GTFP that has arisen over the last 15–20 years is of great concern. First recorded in the 1930s in the Florida Keys in a few green turtles, it appeared to increase in the 1960s and is now pandemic, with infection rates in some habitats of more than 70% (Aguirre and Lutz, in press). GTFP has been reported in every major ocean basin that is home to green sea turtles (Herbst, 1994). The rapid spread of this disease is exemplified by the record of its occurrence in the Indian River Lagoon on Florida's east coast. The first case in the Indian River was reported in 1982, and by late 1985 more than 50% of *C. mydas* captured in the lagoon had fibropapillomas (Herbst, 1994); current infection rates are approximately 67% (Hirama and Ehrhart, 2002). Although many turtles with GTFP will not die of the disease per se, the tumors, which may range up to more than 30 cm in diameter, interfere with normal functioning, cause physical weakening, and expose the carrier to other threats (Figure 6.3). Cutaneous tumors increase drag and may interfere with vision; large tumors could thus severely hamper the victim's ability to swim and dive; escape predation; and locate, capture, and swallow food. Internal tumors may affect organ function, digestion, buoyancy, cardiac function, and respiration (Herbst, 1994; Work and Balazs, 1999). Turtles with fibropapillomas are also more likely to become entangled in monofilament line or other debris (Witherington and Ehrhart, 1989). Turtles with advanced GTFP are chronically stressed. Those with large numbers of tumors are hypoferremic, anemic, and hypoproteinemic, and are in advanced stages of acidosis and calcium–phosphorus imbalance (Aguirre and Balazs, 2000). These symptoms, of course, may have additional effects on turtles: animals already in ion imbalance may be less able to handle additional osmotic stresses induced by cold stunning, for example, whereas anemic animals will have a lower oxygen-carrying capacity for diving, and would be more severely incapacitated if caught in a net or trawl. There is also likely to be a debilitating synergism between GTFP and spirorchidiasis; many animals suffer from both infections simultaneously, and many pathological outcomes are similar (Aguirre et al., 1998).

Although it initially appeared that GTFP was confined to green sea turtles, in which it is most prevalent, recent studies have found GTFP in loggerhead (Herbst, 1994), olive ridley (Aguirre et al., 1999), Kemp's ridley (Harshbarger, 1991), flatback (Limpus and Miller, 1994), and possibly leatherback turtles (Huerta et al., 2000).

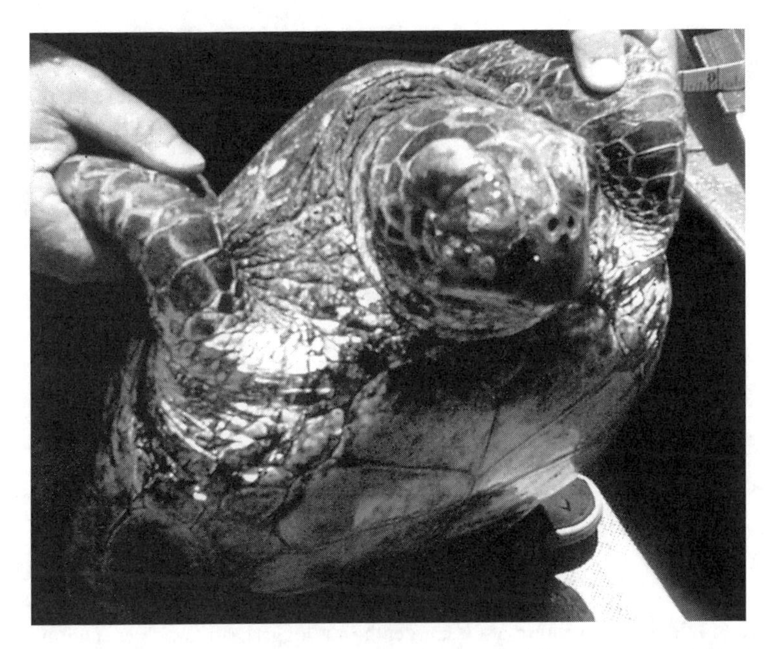

FIGURE 6.3 *Chelonia mydas* with fibropapillomatosis. (Photo courtesy of W. Teas.)

Although the precise etiology of GTFP is still under investigation, the disease has been linked to environmentally challenged habitats, and immunosuppression is strongly correlated with fibropapillomas in green turtles (Cray et al., 2001; Aguirre et al., 1994; Aguirre and Lutz, in press). Chronic stress, whether caused by environmental pollutants, parasites, or biotoxins, affects the immunological response of reptiles; thus, stressed sea turtles are likely to be less able to withstand the primary etiological factor for GTFP. There is convincing evidence of a virus as the transmissible causal factor for GTFP. Early work focused on papillomavirus (Jacobson et al., 1989), but recent work by Brown et al. (1999) failed to detect papillomavirus in freshly isolated tumor samples. More recently, a strong correlation has been detected between the presence of chelonian herpesvirus and papilloma (Lackovich et al., 1999), which has been supported by molecular (polymerase chain reaction) investigations (Lu et al., 2000; Quackenbush et al., 2001); papillomavirus was also detected.

6.3.5 Effects of Environmental Stressors on Hatchlings

Hatchlings must endure unique physiological stresses in emerging from the nest and swimming in the frenzy period away from shore to the open ocean gyres. Until hatching, the nest environment is controlled primarily by physical factors: the temperature, hydric environment, and gas exchange processes of the beach material (for a review, see Ackerman, 1997). As the embryos grow, they both consume more oxygen and produce more carbon dioxide, resulting in a hypoxic, hypercapnic nest environment. In addition, as the metabolic rate of the clutch increases with development, metabolic heat output increases as well (Figure 6.4), enough to raise nest

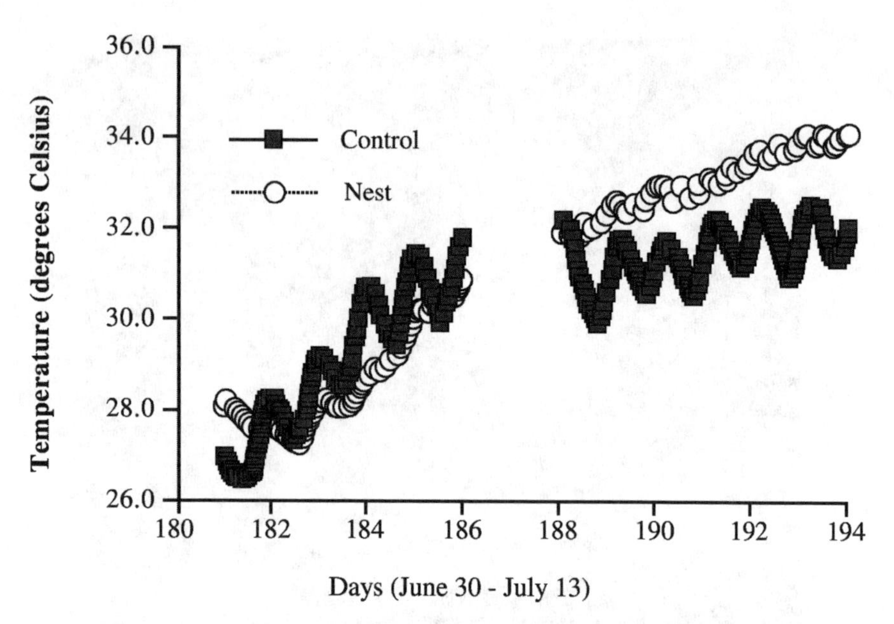

FIGURE 6.4 Mean temperature at 30 cm depth in a loggerhead turtle nest and in a control (sand, 4 m from nest) on a renourished Miami, FL, beach. During the final 2 weeks of incubation, metabolic heat raises nest temperatures above control. $N = 5$ nests. (Data adapted from Milton, S.L., Schulman, A.A., and Lutz, P.L., The effect of beach renourishment with aragonite versus silicate sand on beach temperature and loggerhead sea turtle nesting success. *J. Coast. Res.*, 13(3), 904–915, 1997. With permission.)

temperatures significantly over control (sand) temperatures by approximately 1–2°C (Milton et al., 1997). It is into this warm, low-oxygen environment that sea turtles hatch to dig their way to the surface, an energy-intensive effort that often exceeds the gas diffusion capacity of the environment as well as the aerobic capacity of the hatchlings such that anaerobic metabolism becomes necessary for successful nest emergence (Ackerman, 1977; Dial, 1987).

6.3.5.1 Emergence Stress and Lactate

Blood lactate levels in emerging green and loggerhead hatchlings increase significantly, with blood lactate concentrations in green turtle hatchlings approximately twice those of loggerhead hatchlings (Baldwin et al., 1989). Baldwin et al. (1989) suggested that emerging green turtles had higher lactate levels than loggerheads because they were digging from deeper nests, and were thus digging longer under possibly lower oxygen conditions. Recent work, however, indicates that the degree of lactate buildup, like many other stressors, is most significantly affected by interspecific differences. In a study by Giles et al. (in review), blood lactate concentrations in three species of hatchling sea turtles (*Dermochelys coriacea*, *C. caretta*, and *C. mydas*) were not significantly related to nest depth, oxygen levels, or temperature, but instead differed by species (Figure 6.5). Although lactate levels were highest in actively digging hatchlings of all three species (compared to those resting at the

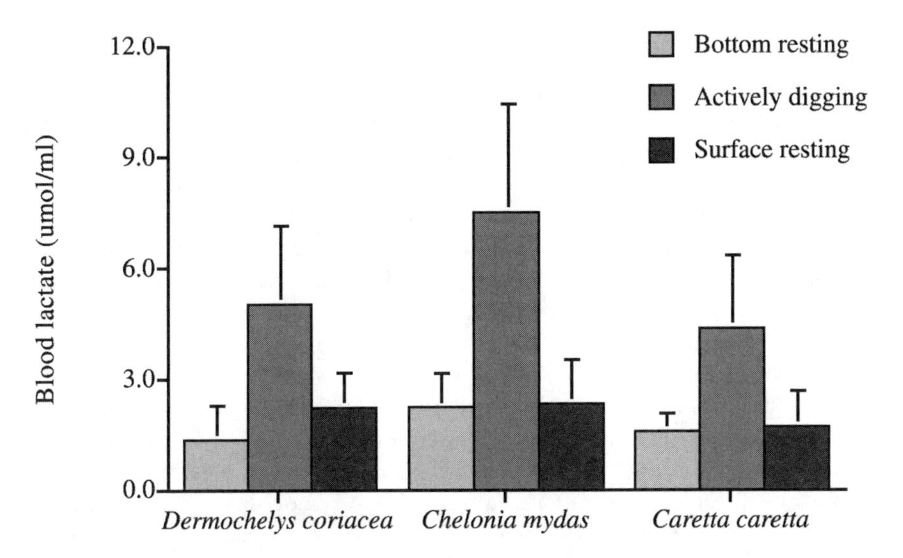

FIGURE 6.5 Mean blood lactate levels (1 ± SD) of hatchlings during emergence activities on a Florida beach. Lactate levels in actively digging hatchlings of all three species are significantly greater than for hatchlings of the same species resting at the surface or bottom of the nest. Mean nest depths were 60.5 ± 1.96 cm (*C. caretta*), 83.0 ± 8.06 cm (*C. mydas*), and 89.7 ± 873 cm (*D. coriacea*). There was no significant difference between lactate levels in hatchlings digging from the shallowest nests (*C. caretta*) and the deepest nests (*D. coriacea*). (Data are from Redfearn, 2000.)

bottom of the nest or at the sand surface), leatherback hatchlings, which emerge from the deepest nests, had the lowest blood lactate levels, whereas green turtle hatchlings emerging from shallower nests had the highest lactate levels (an average of 42% higher than in *D. coriacea* and 33% higher than *C. caretta*).

Low levels of lactate accumulation after exercise have also been reported in adult leatherback turtles (Paladino et al., 1996), a factor indicating that overall lactate production may reflect species-specific differences. Once emerged, hatchlings rest at or near the sand surface, which provides time for blood lactate levels to decline before the hatchlings begin their swimming frenzy, another energetically costly activity. It is not known, however, if the rest period is an adaptation to allow lactate levels to decrease or if this is a side effect of other inhibitory factors, such as sand temperature. High lactate levels are correlated with diminished behavioral capacities and lethargy in reptiles (Bennett, 1982), and would thus be an additional physiological (pH) and behavioral stress on swimming hatchlings, increasing the likelihood of predation (Stancyk, 1982; Witherington and Salmon, 1992). (Of course, resting at the sand surface also increases the likelihood of predation.) Crawling from the nest to the water also increases body lactate levels (Dial, 1987), and studies on loggerhead and green turtle hatchlings have shown that the hatchling frenzy is supported in part by anaerobic metabolism (Baldwin et al., 1989). Once hatchlings have successfully emerged, it may take as long as an hour for lactate levels to return to basal, resting levels (Baldwin et al., 1989; Giles, in review), after which hatchlings make their way down the beach and into the surf.

6.3.5.2 Temperature

High sand temperatures are an additional stress affecting hatchling behavior as well as nest success. Although thermal inhibition of movement most likely prevents daytime emergence, preventing additional thermal and dehydration stress and exposure to daytime predators (Mrosovsky, 1968; Gyuris, 1993), when temperatures are particularly high in nests, embryonic and hatchling deaths may result either as a direct result of crossing into the upper lethal temperature range or possibly as a result of behavioral (movement) inhibition to the point of nonemergence. Miller (1985) found that sea turtle eggs held at temperatures greater than 33°C for extended periods of time did not hatch, consistent with the thermal tolerance range for developing sea turtles proposed by Ackerman (1997) of between 25–27°C and 33–35°C; it was noted by both Cheeks (1997) and Fortuna and Hillis (1998) that higher than normal nest temperatures in the field decrease sea turtle nest success. In an *in situ* comparison between naturally or artificially shaded hawksbill turtle nests in St. Croix and those exposed to direct sunlight (after Hurricane Hugo removed shoreline vegetation), Fortuna and Hillis (1998) found that unshaded nests averaged 2.1°C warmer than shaded nests in the same location. Unshaded nests also had significantly lower mean hatch success and nearly three times as many full-term dead embryos, with an apparent exponential relationship between maximum nest temperature and the percentage of embryos that died late in development.

A similar correlation was noted between extreme temperatures (greater than 33°C, with some temperatures as high as 37.6°C) in loggerhead nests relocated to a Miami Beach, FL, hatchery and low emergence (but not hatching) success. Especially significant were high temperatures during the last 3 days of incubation and number of pipped dead hatchlings in the nest (Blair, 2001). A significant increase in the number of pipped dead occurred in nests experiencing maximum temperatures between 32 and 34°C. Although high temperatures may be directly lethal to developing embryos, it cannot be determined if hatchlings from nests with high hatching but low emergence success are affected directly by temperature or indirectly through temperature effects on behavior. Experiments on newly emerged individuals and small-group behavior at various temperatures have shown that crawling by newly emerged loggerhead hatchlings from the Miami Beach hatchery, even in a group, is significantly inhibited by temperatures above 33°C (Blair 2001), which may result in nonemergence of a nest despite high hatch success. Physiological and behavioral responses to increased temperatures include the well-described thermal inhibition that prevents hatchling emergence when sand temperatures are high (Witherington et al., 1990; Moran et al., 1999) as well as reduced swimming speeds at temperatures above 30°C and loss of coordinated muscle movement in loggerheads swimming at temperatures above 33°C (O'Hara, 1980).

6.3.5.3 Frenzy Swimming

Crawling and frenzy swimming are also metabolically costly; as in emergence, the hatchlings (*D. coriacea*, *C. caretta*, and *C. mydas*) again exceed their aerobic scope and blood lactate increases, though to a lesser extent than in digging hatchlings

(Wyneken and Milton, unpublished observations). Lactate levels are lower in swimming (nonfrenzy) hatchlings than in crawling, emerging, or frenzy-swimming animals, though species-specific differences exist. Only in leatherback hatchlings were there no significant differences in lactate levels induced by activity (crawling, resting, or frenzy or postfrenzy swimming). By contrast, green and loggerhead hatchlings appear to rely more heavily on anaerobic metabolism for burst activities: lactate levels were significantly higher in crawling and frenzy-swimming green and loggerhead hatchlings than in resting or swimming animals (Wyneken and Milton, unpublished observations). Swimming appears to be particularly efficient in leatherback hatchlings; recent work by Jones et al. (2002) shows that swimming 1- to 5-week-old leatherbacks have oxygen consumption rates comparable to resting metabolism; mass-specific oxygen consumption (VO_2) increases to only 96% over resting (in 5-week-old turtles), even when swimming at maximal rates, with positive correlations between breath rates and VO_2, and flipper stroke rates and VO_2.

Interspecific differences in the cost of locomotion are apparent when comparing olive ridley hatchlings to the leatherbacks (Figure 6.6). In olive ridley turtles, aerobic scope (oxygen consumption during exercise) was 370–400% of resting metabolism in 1- to 4-week-old hatchlings, whereas swimming in 4-week-old leatherback hatchlings is no more costly than resting. In the 4-week-old olive ridley hatchlings, VO_2 was also lower in maximally swimming animals than in freely swimming

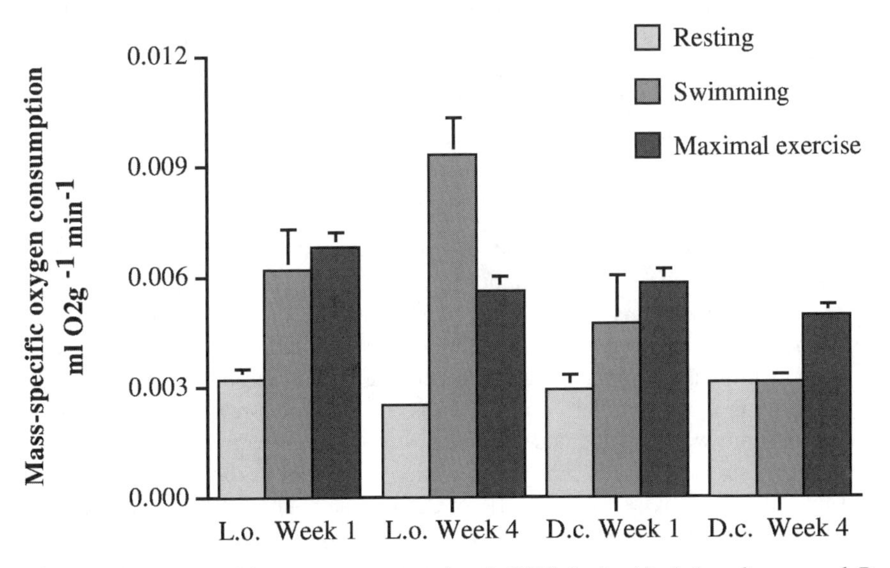

FIGURE 6.6 Mass-specific oxygen consumption (+SEM) in *Lepidochelys olivacea* and *D. coriacea* at 1 and 4 weeks of age during resting, swimming, and maximal (stimulated) swimming. (Data are adapted from Jones, T.T., Reina, R., and Lutz, P.L., A comparison of the ontogeny of oxygen consumption in leatherback sea turtle *Dermochelys coriacea* and olive ridley hatchlings, *Lepidochelys olivacea.* Different strokes for different life styles, in: *Proceedings of the 22nd Annual Symposium on Sea Turtle Biology and Conservation*, J. Seminoff (compiler), U.S. Dept. Commerce NOAA Tech. Memo. NMFS-SEFSC, Miami, FL (in press), 2002. With permission.)

hatchlings, indicating an increase in the anaerobic component (although VO_2 in both conditions was higher than in resting animals). Similarly, Wyneken (1992) reported that the cost of locomotion in leatherback hatchlings is as much as 20% lower during frenzy swimming than in green and loggerhead hatchlings, with leatherbacks having the slowest swimming speeds, stroke rates, and lowest metabolic rates. Because leatherback turtles of less than 110 cm CCL are not found in waters above 34° latitude (26°C) (Eckert, 2000), it has been suggested that leatherback hatchlings may become active, distance swimmers early in development, allowing them to forage in upwelling and convergence zones rather than being swept as passive feeders into the ocean gyres. Thus, although the physical requirements of emergence, crawling, and frenzy and postfrenzy swimming are common to all sea turtle species, the physiological stresses that these activities place on hatchlings again vary with interspecific metabolic differences.

6.4 RESPONSES TO STRESS

Stress responses may be expressed at multiple levels, from the immediate effects of acute stress on catecholamine levels to long-term effects such as immune suppression, changes in gene expression, and population effects, i.e., decreased reproductive rates. Harmful effects from both anthropogenic and natural insults include compromised physiology, impaired immune function, and an increase in the incidence of disease (Lutz, 1998). Immunosuppression is strongly correlated with GTFP in green turtles in Florida (Cray et al., in press; Sposato et al., 2002) and Hawaii (Aguirre et al., 1995), and it is likely that immunosuppressed turtles will suffer from other disease or parasite stressors as well.

6.4.1 Neuroendocrine Responses (Stress Hormones)

Selye (1936) proposed that different stresses produced a similar set of responses, which he called the general adaptation syndrome (GAS), i.e., alarm–resistance–exhaustion. In this widely adopted scheme, the primary response is at the neuroendocrine level, involving the hypothalamus–pituitary–adrenocorticoid axis. It is often identified as an increase in blood cortisol levels and has been taken as the stress-defining response (Nelson and Demas, 1996). Stress-related changes in corticosteroids are well documented in both freshwater and sea turtles.

Capture stress produces changes in corticosterone levels, but there are seasonal and size differences (Gregory et al., 1996; Gregory and Schmid, 2001). In examining acute captivity stress responses, Gregory et al. (1996) found that smaller turtles had higher levels of corticosterone in summer than did larger animals, whereas corticosterone levels were suppressed in both size classes in winter. It was suggested that the lower responses exhibited by large turtles in summer were related to reproductive condition, a finding supported by reduced adrenocortical function in heat-stressed breeding green turtles and in arribada olive ridleys exposed to turning stress (Jessop et al., 2000; Valverde et al., 1999). Similarly, male olive ridleys captured by hand and held in crowded conditions exhibited significantly higher corticosterone levels than females held under the same conditions (Schwantes, 1986). The stress of handling or capture in nets and trawls results in increased corticosterone in hatchling (Morris, 1982), juvenile (Morris, 1982; Wibbels et al., 1987), and adult sea turtles (Schwantes, 1986). Notably, forced submergence results in decreased corticosterone in freshwater turtles (Keiver et al., 1992).

In addition, corticosterone release is sensitive to temperature. Jessop et al. (2000) found that heat stress caused a 16-fold increase in circulating corticosterone in green sea turtles. In soft-shelled turtles adrenomedullary activity is stimulated by high temperatures and inhibited by low temperatures (Ray and Maita, 2001; Mahapatra et al., 1989).

Stress also results in increased blood levels of the catecholamine hormones epinephrine (EP), norepinephrine (NE), and dopamine, which, on an emergency basis, facilitate the fight or fight response by enhancing oxygen uptake and transfer, and the mobilization of energy substrates (Bonga, 1997). For example, forced submergence and acidosis greatly increases NE and EP levels in freshwater turtles (Wasser and Jackson, 1991). Hyperosmotic conditions deplete NE in soft-shells, whereas dehydration stress depletes EP but increases NE levels (Mahapatra et al., 1991). On the other hand, aldosterone and corticosterone levels were not affected by 4 days of freshwater exposure in Kemp's ridley turtles (Ortiz et al., 2000).

Although excessive or extended elevation of the stress hormones is immediately useful, it can have harmful effects by, for example, reallocating energy away from growth and reproduction, and suppressing immune functions (see Section 6.4.2) (Bonga, 1997). The experimental evidence for these effects is from species other than sea turtles, but it is undoubtedly a vertebrate-wide phenomenon. In the male common carp, prolonged elevation of cortisol levels inhibits testicular development and impairs the synthesis of the 11 oxygenated androgens (Consten et al., 2001); disease can also result in higher cortisol levels in fish (Mustafa et al., 2000; Sures et al., 2001).

There is some indirect evidence of such effects in sea turtles. Valverde et al. (1994) reported that olive ridley females restrained in the shade after nesting did not show the expected next-day progesterone peak indicative of ovulation, whereas unrestrained females captured in the water had ovulated (Valverde et al., 1992). Other work however, indicates that this response may be species-specific; postnesting loggerhead (Wibbels et al., 1992) and green turtles (Licht et al., 1980) subjected to severe handling stresses ovulated normally.

Increased levels of stress hormones have a variety of other harmful effects on turtles, including disturbed blood glucose levels (Keiver et al., 1992), impaired salt gland function (Reina and Cooper, 2000), and a compromised immune function (George, 1997).

Reina and Cooper (2000) found that both adrenaline and the cholinergic agonist methacholine inhibited salt gland activity in hatchling green sea turtles. Because the majority of salt excretion in sea turtles occurs through salt gland activity (Lutz, 1997), suppression of such activity could have significant effects on osmotic homeostasis in sea turtles, especially for hatchlings, which have an apparent requirement for seawater intake and concomitant high secretion rates (Bennett et al., 1986; Marshall and Cooper, 1988). Other potentially lethal ion imbalances may occur, for example, when salt gland function is inhibited during cold stunning.

Animals already subjected to physiological stresses (high or low temperatures, capture trauma, starvation) are likely to experience increased circulating glucocorticoid levels, which in turn depress immune function and accelerate catabolic processes. Recurrent environmental stressors may reduce survival if they result in persistent glucocorticoid secretion (Nelson and Demas, 1996); however, the potential links between environmental stressors, stress hormones, and immune function in sea turtles have not been investigated.

6.4.2 Immunological Responses

It is now commonly accepted that manipulation of neural and endocrine functions alters vertebrate immune responses, and the antigenic stimulation that generates an immune response results in changes in neural and endocrine functions; thus the immune status of an individual also has consequences for behavior (Ader and Cohen, 1993).

The suppression of the immune response by adrenocortical hormones, especially the glucocorticoids, is a well-described vertebrate response. Although most work describing the link between immunosuppression and elevated adrenocortical hormones has been done on mammals (Munck and Naray-Fejes-Toth, 1994), a few reptile studies have been performed. Saad and el Ridi (1988) observed significant lymphocytic destruction and the impairment of immune reactivity in the lizard *Chalcides ocellatus*, which they associated with sustained high levels of endogenous corticosteroid levels in the autumn and winter. By contrast, fully developed splenic lymphoid tissue and immune responses were coincident with low summer corticosteroid levels. The administration of exogenous corticosteroids to "summer" lizards depleted lymphoid elements and suppressed immune responses, whereas the pharmacological inhibition of corticosteroid synthesis in autumn ameliorated the natural winter-dependent immune depression (Saad and el Ridi, 1988).

The immune response of reptiles is of course affected by numerous factors. Steroid sex hormones, for example, also have significant effects on immunological activity in reptiles and other vertebrates, although again most studies in this area involve mammals. The reptile immune system is strongly affected by seasonal changes caused by both temperature changes and changes associated with the breeding cycle. Seasonal changes in thymic mass in turtles were first reported in 1912; Aime (1912) reported decreased thymic mass during winter estivation, and thymic regeneration in the spring. Androgens, like the glucocorticoids, appear to have immunocompromising properties. In poikilotherms, lymphoid mass and immunological activity is greatest in spring and summer, after breeding activities have been completed and testosterone levels decline. In the turtle *Mauremys caspica*, lymphocyte proliferation induced by mitogens showed high values in the spring and winter and decreased responses in summer and fall (Munoz et al., 2000; Munoz and De la Fuente, 2001), whereas a single injection of testosterone (200 μg/g body weight) produced thymic involution and intense lymphopenia in the spleen and peripheral blood compartment (Saad et al., 1991). Female mammals generally have higher immune activities by several indices than male conspecifics, whereas gonadectomized mice and rats treated with physiological or greater estrogen levels exhibited increased antibody responses to a variety of antigens (Nelson and Demas, 1996). It has been suggested that seasonal changes in immune responsiveness reflect seasonal changes in the neuroendocrine system, with a regular relationship between neuroendocrine and lymphoid systems (Zapata et al., 1992).

Two studies have also found seasonal patterns of immune responsiveness in sea turtles. McKinney and Bentley (1985) reported that lymphocyte blastogenic responses in *Chelonia mydas* to the mitogens phytohemagglutinin (PHA) and concanavalin A (ConA) varied between individuals but did not correlate with size–age, season, or temperature; however, responses to the mitogens pokeweed and lipopolysaccharide (LPS) were measurable only in spring. More recently, Keller et al. (2002) reported increases in both mitogen-induced lymphocyte proliferation and overall white blood cell counts during the summer months in loggerhead turtles. Differences in the seasonal patterns of immunological activity between other turtles and sea turtles may be due to differences in peak hormone levels, because some turtles breed immediately upon emerging from winter hibernation (Lee et al., 2002).

Although seasonal changes of the immune system have not been well described in sea turtles, seasonal cycles in testosterone levels have been well documented (Owens, 1997). The pattern is similar to other poikilotherms, with testosterone levels highest in the winter and early spring and decreasing as the mating season progresses (Wibbels et al., 1990). Because most species have seasonal fluctuations in reproductive activity, seasonal changes in immune function may be mediated by photoperiod effects on reproductive function and steroidal activity. Reptiles differ from other groups (mammals) in which laboratory studies show that decreasing photoperiods enhance immune function, whereas field studies report an increase in lymphatic tissue size and immune in winter (for a review, see Nelson and Demas, 1996). One example is the saltwater crocodile hatchling (*Crocodylus porosus*), in which suboptimal temperatures induced stress and immunosuppression with significant decreases in total white cell and lymphocyte counts (Turton et al., 1997).

The stress of coping with energetically demanding conditions can also indirectly cause illness and death by compromising immune function (Nelson and Demas, 1996). Although it has been assumed that low environmental temperatures and other stressors decrease immunoglobulin production and immune response in sea turtles, as they do in other reptiles (Zapata et al., 1992), these assumptions have not been examined. There has been no systematic examination of the relationships between acute and long-term stress on the immune function in sea turtles.

6.4.3 Gene Response, Molecular Biomarkers, and the Measurement of Stress: Potential Tools for the Future

In addition to short-term stress markers such as corticosterone levels, all organisms respond to environmental and physiological stress by altering gene expression (at the transcriptional and/or translational level) for a variety of compounds, including increasing synthesis of an evolutionarily conserved family of proteins known as the heat shock or stress proteins (HSPs). The HSP family is elicited by stressors as diverse as xenobiotics, heavy metals, heat, hypoxia, and osmotic stress.

These molecular stress responses have been studied mostly in organisms maintained under constant laboratory conditions; there is much less information on the regulation of stress responses in animals that are exposed to and tolerate large fluctuations in internal or external conditions (Rabergh et al., 2000). However, genetic changes such as increased HSP expression are becoming an important and powerful tool through which the direct effects of different stressors on organismal health and fitness can be measured by their effects on cellular and molecular processes. Most attempts to monitor the environmental status of an ecosystem rely on determining the abiotic components, such as contaminant analysis–loads, or assessing ecological responses to stressors (e.g., species richness, sex ratios, and indicator species fitness) (O'Connor, 1996). Such studies do not reveal the links between the stressor and its effects, and therefore we cannot predict how a species or ecosystem will respond to even one contaminant (Downs et al., 2001a), much less the more likely problem of a suite of stressors.

A number of different compensatory mechanisms may operate at multiple levels (cells, tissues, organ systems, and individual animal) to ameliorate stress before the fitness of an individual or its functional role in the community is altered (Allen and Starr, 1982), and thus stress affects higher levels of the biological hierarchy only when it overwhelms the homeostatic mechanisms of individual organisms. Rather than simply measuring stress responses, data regarding individual and population responses (especially for endangered species) would be far more useful if they could be used to forecast population changes. Forecasting stress responses in time to intervene and prevent population declines, however, requires linking changes at lower levels of biological organization with the fitness of individuals (and then accurately modeling the long-term demographic consequences).

The use of molecular biomarkers to assess organismal and ecosystem health is thus becoming a popular concept (Downs et al., 2001a). Although numerous studies, including many on sea turtles, examine a single or small set of physiological parameters to assess the overall physiological response to a stressor (Adams et al., 1992), and other studies support the validity of biomarker use as indicators of contaminant or stressor exposure (de Zwart et al., 1999; Adams and Ryon, 1994), very few attempt to integrate physiological status with multiple, specific biomarkers (Adams et al., 1992; Stegmann et al., 1992). A system to simultaneously assess multiple biomarkers to quantify known physiological responses to stressors would tell us: (1) whether an animal is physiologically stressed, (2) whether the animal is evolutionarily or physiologically adapted to a chronic stress, and (3) the physiological impact of the stress (Downs et al., 2001a). Such an integrated system using molecular biomarkers will allow for a diagnosis of an animal's physiological condition at the cellular level when challenged with a real or suspected stress.

With the development of molecular markers for specific individual or suites of stressors, such a system would become a powerful tool to identify environmental insults that are physiologically affecting an organism, providing a more accurate quantification of the health status of a population in response to a natural or anthropogenic stressor. Such a system, for example, has been developed for the intertidal eastern mud snail (*Ilyanassa obsoleta*), where biomarkers can differentiate between snails exposed to different stressors, including heat, cadmium, an herbicide and a pesticide, and a petroleum compound (Downs et al., 2001a). Other biomarker systems have been developed for species as diverse as cordgrass, estuarine fish, tadpoles, heat-stressed corals (Downs et al., 2000), and grass shrimp (Downs et al., 2001b). Representative stressors already used include elevated temperature, pesticides, heavy metals, and a pathogenic bacterium. Such a system could be extremely useful to measure the health status of marine turtle populations. Because turtles have nucleated red blood cells, a molecular biomarker system could theoretically be developed for diagnosis using blood samples relatively quickly, easily, and inexpensively.

A system for molecular diagnosis of stress might include biomarkers of general cellular integrity and oxidative stress (i.e., ubiquitin or malondialdehyde), HSPs such as hsp60 and hsp70, antioxidant enzymes (superoxide dismutases), enzymes that respond to pH stress (acid and alkaline phosphatases and dehydrogenases), and members of the P450 family (markers of xenochemical exposure). The synthesis of HSPs at normal physiological and at elevated temperatures, for example, has been correlated with the natural adaptation of nine lizard species to heat, in that animals adapted to desert conditions showed higher constitutive levels of hsp70 than lizards that inhabited cooler climates (Ulmasov et al., 1992). Lizards adapted to cooler climates also have a lower thermal threshold for HSP expression when exposed to heat shock than desert-adapted animals (Zatsepina et al., 2000). It has been suggested that increases in hsp70 mRNA levels in blood may serve as an early indicator of temperature stress in fish (Currie et al., 2000). The genetic response to stress in sea turtles is (naturally!) unknown, although changes in hsp70 and hsp60 have been detected in anoxic freshwater turtles (H. Prentice, personal communication).

REFERENCES

Ackerman, R.A., The respiratory gas exchange of sea turtle nests (*Chelonia, Caretta*), *Respir. Physiol.*, 31, 19–38, 1977.

Ackerman, R.A., The nest environment and the embryonic development of sea turtles, in: Lutz, P. and Musick, J. (eds.), *The Biology of Sea Turtles*, CRC Press, Boca Raton, FL, 1997, 432 pp.

Adams, S.M. and Ryon, M.G., A comparison of health assessment approaches for evaluating the effects of contaminant-related stress on fish populations, *J. Aquat. Ecosys. Health*, 3, 15–25, 1994.

Adams, S.M. et al., Relationships between physiological and fish population responses in a contaminated stream, *Environ. Toxicol. Chem.*, 11, 1549–1557, 1992.

Ader, R. and Cohen, N., Psychoneuroimmunology: conditioning and stress, *Ann. Rev. Psychol.*, 44, 53–85, 1993.

Adnyana, W., Ladds, P.W., and Blair, D., Observations of fibropapillomatosis in green turtles (*Chelonia mydas*) in Indonesia, *Aust. Vet. J.*, 75(10), 736–742, 1997.

Aguirre, A.A. and Balazs, G.H., Plasma biochemistry values of green turtles (*Chelonia mydas*) with and without fibropapillomas in the Hawaiian Islands, *Comp. Hematol. Int.*, 10(3), 132–137, 2000.

Aguirre, A.A. and Lutz, P.L., Marine turtles as sentinels of ecosystem health: is fibropapillomatosis an indicator?, *Ecosys. Health*, in press.

Aguirre, A.A., Organic contaminants and trace metals in the tissues of green turtles (*Chelonia mydas*) afflicted with fibropapillomas in the Hawaiian Islands, *Mar. Poll. Bull.*, 28, 109–114, 1994.

Aguirre, A.A. et al., Adrenal and hematological responses to stress in juvenile green turtles (*Chelonia mydas*) with and without fibropapillomas, *Physiol. Zool.*, 68(5), 831–854, 1995.

Aguirre, A.A. et al., Spirorchidiasis and fibropapillomatosis in green turtles from the Hawaiian Islands, *J. Wildl. Dis.*, 34(1), 91–98, 1998.

Aguirre, A.A. et al., Survey of fibropapillomatosis and other potential diseases in marine turtles from Moreton Bay, Queensland, Australia, in: *Proceedings of the 19th Annual Symposium on Sea Turtle Biology and Conservation*, South Padre Island, TX, 3–5 March, 1999, Miami, FL.

Aime, P., Note sure le thymus chez les cheloniens, Comptes rendus des Seances de la Societe de Biologie et de ses Filialses, 72, 889–890, 1912.

Alam, S.K. and Brim, M.S., Organochlorine, PCB, PAH, and metal concentrations in eggs of loggerhead sea turtles (*Caretta caretta*) from northwest Florida, USA, *J. Environ. Sci. Health B*, 35(6), 705–724, 2000.

Allen, T.F.H. and Starr, T., *Hierarchy: Perspectives for Ecological Complexity*, University of Chicago Press, Chicago, 1982.

Andersson, L. et al., Effects of polychlorinated biphenyls with Ah receptor affinity on lymphoid development in the thymus and bursa of Fabricius of chick embryos *in ovo*, *Toxicol. Appl. Pharmacol.*, 107, 183–188, 1991.

Arthur, K., Dennison, W., and Limpus, C., The effects of the toxic cyanobacteria *Lyngbya majuscula* on marine turtles, in: *Proceedings of the 22nd Annual Symposium on Sea Turtle Biology and Conservation,* J. Seminoff (compiler), U.S. Dept. Commerce NOAA Tech. Memo. NMFS-SEFSC, Miami, FL (in press).

Baker, J.R., Pollution-associated uterine lesions in grey seals from the Liverpool Bay of the Irish Sea, *Vet. Rec.*, 125, 303, 1989.

Baldwin, J. et al., Anaerobic metabolism during dispersal of green and loggerhead hatchlings, *Comp. Biochem. Physiol.*, 94A(4), 663–665, 1989.

Becker, P.R. et al., Concentrations of chlorinated hydrocarbons and trace elements in marine mammal tissues archived in the U.S. National Biomonitoring Specimen Bank, *Chemosphere*, 34(9/10), 2067–2098, 1997.

Bennett, A.F., The energetics of reptilian activity, in: *Biology of the Reptilia*, Vol. 13, Gans, C. and Pough, F.H. (eds.), Academic Press, New York, 1982.

Bennett, J.M., Taplin, L.E., and Grigg, G.C., Sea water drinking as a homeostatic response to dehydration in hatchling loggerhead turtles *Caretta caretta*, *Comp. Biochem. Physiol.*, 83A, 507–513, 1986.

Bergeron, J.M., Crews, P., and McLachlan, J.A., PCBs as environmental estrogens: turtle sex determination as a biomarker of environmental contamination, *Environ. Health Perspect.*, 102(9), 780–781, 1994.

Blair, K., High Sub-Lethal Temperature Effects on the Movement of Loggerhead Sea Turtle (*Caretta caretta*) Hatchlings, honors thesis, University of Miami, FL, 2001.

Bonga, S.E., The stress response in fish, *Physiol. Rev.*, 77, 591–625, 1997.

Bossart, G.D. et al., Brevetoxicosis in manatees (*Trichechus manatus latirostris*) from the 1996 epizootic: gross, histologic, and immunohistochemical features, *Toxicol. Pathol.*, 26(2), 276–282, 1998.

Bossart, G.D. et al., Viral papillomatosis in Florida manatees (*Trichchus manatus latirostris*), *Exp. Mol. Pathol.*, 72(1), 37–48, 2002.

Brodie, J., The problems of nutrients and eutrophication in the Australian marine environment, in: *State of the Marine Environment Report for Australia*, http://www.environment.gov.au/marine/publications/somer/somer_annex2/bro_txt.html, 1999.

Brown, D.R., Lackovich, J.K., and Klein, P.A., Further evidence for the absence of papillomavirus from sea turtle fibropapilloma, *Vet. Rec.*, 145, 616–617, 1999.

Burkholder, J.M., Implications of harmful microalgae and heterotrophic dinoflagellates in management of sustainable marine fisheries, *Ecol. Appl.*, 8, S37–S62, 1998.

Byrne, J.E., Carbone, J.P., and Hanson, E., Hypothyroidism and abnormalities on the kinetics of thyroid hormone metabolism in rats treated chronically with polychlorinated biphenyls and polybrominated biphenyls, *Endocrinology*, 121, 520–527, 1987.

Carminati, C.E. et al., Blood chemistry comparison of healthy vs. hypothermic juvenile Kemp's ridley sea turtles (*Lepidochelys kempii*), in: *Proceedings 14th Annual Workshop on Sea Turtles Conservation and Biology*, Bjorndal, K.A., Bolton, A.B., and Johnson, D.A. (compilers), NMFS Tech. Memo. NOAA-TM-NMFS-SEFSC-351, Miami, FL, 1994, p. 203.

Carr, A., Ogren, L., and McVea, C., Apparent hibernation by the Atlantic loggerhead turtle *Caretta caretta* off Cape Canaveral, Florida, *Biol. Conserv.*, 19, 7, 1980–1981.

Cheeks, R.C., Effects of Various Sand Types on Nest Temperature and Hatching Success in the Loggerhead (*Caretta caretta*) Sea Turtle, master's thesis, Florida Atlantic University, Boca Raton, FL, 1997, 54 pp.

Clark, D.R., Jr. and Krynitsky, A.J., Organochlorine residue in eggs of loggerhead and green sea turtles nesting at Merritt Island, Florida — July and August 1976, *Pestic. Monit. J.*, 14(1), 7–10, 1980.

Collins, W.T. et al., Effects of polychlorinated biphenyls (PCB) on the thyroid gland of rats, *Am. J. Pathol.*, 89, 119–136, 1977.

Comeau, S.G. and Hicks, J.W., Regulation of central vascular blood flow in the turtle, *Am. J. Physiol.*, 267 (2 Pt 2), R569–R578, 1994.

Consten, D., Lambert, J.G., and Goos, H.J., Cortisol affects testicular development in male common carp, *Cyprinus carpio* L., but not via an effect on LH secretion, *Comp. Biochem. Physiol. B Biochem. Mol. Biol.*, 129, 671–677, 2001.

Cray, C. et al., Altered *in vitro* immune responses in green turtles with fibropapillomatosis, *J. Zoo. Wildl. Med.*, 32(4), 436–440, 2001.

Currie S., Moyes, C.D., and Tufts B.L., The effects of heat shock and acclimation temperature on hsp70 and hsp30 mRNA expression in rainbow trout: *in vivo* and *in vitro* comparisons, *J. Fish Biol.*, 56, 398–408, 2000.

Davenport, J. et al., Metal and PCB concentration in the "Harlech" leatherback, *Mar. Turtle Newsl.*, 48, 1–6, 1990.

Day, R. et al., Mercury contamination in loggerheads of the southeastern U.S., in: *Proceedings of the 22nd Annual Symposium on Sea Turtle Biology and Conservation,* J. Seminoff (compiler), U.S. Dept. Commerce NOAA Tech. Memo. NMFS-SEFSC, Miami, FL (in press).

De Guise, S. et al., Possible mechanisms of action of environmental contaminants on St. Lawrence beluga whales (*Delphinapterus leaucas*), *Environ. Health Perspect.*, 103(Suppl. 4), 73–77, 1995.

De Guise, S. et al., Effects of *in vitro* exposure of beluga whale leukocytes to selected organochlorines, *J. Toxicol. Environ. Health*, Part A, 55, 479–493, 1998.

de Zwart, B.C., Fings-Dresen, M.H., and van Duivenbooden, J.C., Senior workers in the Dutch construction industry: a search for age-related work and health issues, *Exp. Aging Res.*, 25(4), 385–391, 1999.

Dial, B.E., Energetics and performance during nest emergence and the hatchling frenzy in loggerhead sea turtles (*Caretta caretta*), *Herpetology,* 43(3), 307–315, 1987.

Downs, C.A. et al., A molecular biomarker system for assessing the health of coral (*Montastraea faveolata*) during heat stress, *Mar. Biotechnol.*, 2, 533–544, 2000.

Downs, C.A. et al., A molecular biomarker system for assessing the health of gastropods (*Ilyanassa obsoleta*) exposed to natural and anthropogenic stressors, *J. Exp. Mar. Biol. Ecol.*, 259, 189–214, 2001a.

Downs, C.A., Fauth, J.E., and Woodley, C.E., Assessing the health of grass shrimp (*Palaeomonetes pugio*) exposed to natural and anthropogenic stressors: a molecular biomarker system, *Mar. Biotechnol.*, 3, 380–397, 2001b.

Eckert, S.A., Global distribution of juvenile leatherback sea turtles, Hubbs Sea World Research Institute Technical Report, San Diergo, CA, 99–294, 2000.

Ehrhart, L.M., Fibropapillomas in green turtles of the India River Lagoon, Florida: distribution over time and area, in: *Research Plan for Marine Turtle Fibropapilloma*, Balazs, G.H. and Pooley, S.G. (eds.), NMFS Tech. Memo. NOAA-TM-NMFS-SWFC-156, Honolulu, HI, 1991, p. 59.

Epstein, P.R. et al., *Marine Ecosystems — Emerging Diseases as Indicators of Change*, National Oceanic and Atmospheric Administration and National Aeronautics and Space Agency, The Center for Health and the Global Environment, Harvard Medical School, Boston, MA, 1998.

Felger, R.S., Cliffton, K., and Regal, P.J., Winter dormancy in sea turtles: independent discovery and exploitation in the Gulf of California by two local cultures, *Science*, 191, 283, 1976.

Fortuna, J.L. and Hillis, Z.M., Hurricanes, habitat loss, and high temperatures: implications for hawksbill hatch success at Buck Island Reef National Monument, in: *Proceedings of the Seventeenth Annual Sea Turtle Symposium*, Epperly, S.P. and Braun, J. (compilers), U.S. Dept. Commerce NOAA Tech. Memo. NMFS-SEFSC-415, 1998, 294 pp., Miami, FL.

Fox, G.A. et al., Reproductive outcomes in colonial fish-eating birds: a biomarker for developmental toxicants in Great Lakes food chains, *J. Great Lakes Res.*, 17(2), 153–157, 1991.

Frair, W., Ackerman, R.G., and Mrosovsky, N., Body temperature of *Dermochelys coriacea*: warm turtle from cold water, *Science*, 177, 791, 1972.

Fritts, T.H. and McGehee, M.A., Effects of Petroleum on the Development and Survival of Marine Turtle Embryos, Contract No. 14–16–0009–80–946, FWS/OBS-81/37, U.S. Fish and Wildlife Service, U.S. Department of the Interior, Washington, DC, 1981.

Gascoigne, J.C. and Mansfield, K.L., Barnacle drag and the energetics of sea turtle migration, in: *Proceedings of the 22nd Annual Symposium on Sea Turtle Biology and Conservation,* J. Seminoff (compiler), U.S. Dept. Commerce NOAA Tech. Memo. NMFS-SEFSC, Miami, FL (in press).

George, R.H., Health problems and diseases of sea turtles, in: *The Biology of Sea Turtles*, Lutz, P.L. and Musick, J.A. (eds.), CRC Press, Boca Raton, FL, 1997, pp. 363–385.

Giles, E.R., Wyneken, J., and Milton, S.L., Anaerobic metabolism and nest environment of loggerhead, green, and leatherback sea turtle hatchlings, *Can. J. Zool.*, in review.

Glazebrook, J.S. and Campbell, R.S.F., A survey of the diseases of marine turtles in northern Australia. II. Oceanarium-reared and wild turtles, *Dis. Aquat. Org.*, 9, 97–104, 1990.

Glennemeler, K.A. and Denver, R.J., Sublethal effects of chronic exposure to an organochlorine compound on northern leopard frog (*Rana pipiens*) tadpoles, *Environ. Toxicol.*, 16(4), 287–297, 2001.

Godley, B.J., Thompson, D.R., and Furness, R.W., Do heavy metal concentrations pose a threat to marine turtles from the Mediterranean Sea?, *Mar. Poll. Bull.*, 38, 497–502, 1999.

Goff, G.P. and Lien, J., Atlantic leatherback turtles, *Dermochelys coriacea*, in cold water off Newfoundland and Labrador, *Can. Field Nat.*, 102, 1, 1988.

Gregory, L.F. et al., Plasma corticosterone concentrations associated with acute captivity stress in wild loggerhead sea turtles (*Caretta caretta*), *Gen. Comp. Endocrinol.*, 104(3), 312–320, 1996.

Gregory, L.F. and Schmid, J.R., Stress responses and sexing of wild Kemp's ridley sea turtles (*Lepidochelys kempii*) in the northeastern Gulf of Mexico, *Gen. Comp. Endocrinol.*, 124(1), 66–74, 2001.

Guillette, L.J., Contaminant-induced endocrine disruption in wildlife, *Growth Horm. IGF Res.*, 10(Suppl. B), S45–S50, 2000.

Guitart, R. et al., Comparative study on the fatty acid composition of two marine vertebrates: striped dolphins and loggerhead turtles, *Comp. Biochem. Physiol. B Biochem. Mol. Biol.*, 124(4), 439–443, 1999.

Gyuris, E., Factors that control the emergence of green turtle hatchlings from the nest, *Wildl. Res.*, 20, 345–353, 1993.

Harshbarger, J.C., Sea turtle fibropapilloma cases in the registry of tumors in lower animals, in: *Research Plan for Marine Turtle Fibropapilloma*, Balazs, G.H. and Pooley, S.G. (eds.), NMFS Tech. Mem. NOAA-TM-NMFS-SWFC-156, Honolulu, HI, 1991, p. 63.

Havens, K.E. et al., Complex interactions between autotrophs in shallow marine and freshwater ecosystems: implications for community responses to nutrient stress, *Environ. Poll.* 113(1), 95–107, 2001.

Herbst, L.H., Fibropapillomatosis of marine turtles, *Annu. Rev. Fish Dis.*, 4, 389, 1994.

Herbst, L.H. and Klein, P.A., Green turtle fibropapillomatosis: challenges to assessing the role of environmental cofactors, *Environ. Health Perspect.*, 103(Suppl. 4), 27–30, 1995.

Hirama, S. and Ehrhart, L.M., Prevalence of green turtle fibropapillomatosis in three developmental habitats on the east coast of Florida, in: *Proceedings of the 22nd Annual Symposium on Sea Turtle Biology and Conservation*, J. Seminoff (compiler), U.S. Dept. Commerce NOAA Tech. Memo. NMFS-SEFSC, Miami, FL (in press), 2002.

Huerta, P. et al., First confirmed case of fibropapilloma in a leatherback turtle (*Dermochelys coriacea*), in: *Proceedings of the 20th Annual Symposium on Sea Turtle Biology and Conservation,* Orlando, FL, (in press), 29 February–4 March.

Jackson, D.C., Living without oxygen: lessons from the freshwater turtle, *Comp. Biochem. Physiol. A Mol. Integr. Physiol.*, 125(3), 299–315, 2000.

Jacobson, E.R. et al., Mycotic pneumonia in mariculture-reared green sea turtles, *J. Am. Vet. Med. Assoc.*, 175, 929, 1979.

Jacobson, E.R. et al., Cutaneous fibropapillomas of green turtles (*Chelonia mydas*), *J. Comp. Pathol.*, 101(1), 39–52, 1989.

Jessop, T.S. et al., Evidence for a hormonal tactic maximizing green turtle reproduction in response to a pervasive ecological stressor. *Gen. Comp. Endocrinol.*, 118(3), 407–417, 2000.

Jones, T.T., Reina, R., and Lutz, P.L., A comparison of the ontogeny of oxygen consumption in leatherback sea turtle *Dermochelys coriacea* and olive ridley hatchlings, *Lepidochelys olivacea.* Different strokes for different life styles, in: *Proceedings of the 22nd Annual Symposium on Sea Turtle Biology and Conservation*, J. Seminoff (compiler), U.S. Dept. Commerce NOAA Tech. Memo. NMFS-SEFSC, Miami, FL (in press).

Kannan, K. et al., Critical evaluation of polychlorinated biphenyl toxicity in terrestrial and marine mammals: increasing impact of non-ortho and mono-ortho coplanar polychlorinated biphenyls from land to ocean, *Arch. Environ. Contam. Toxicol.*, 18(6), 850–857, 1989.

Keinath, J.A., Movements and Behavior of Wild and Head-Started Sea Turtles, Ph.D. dissertation, College of William and Mary, Williamsburg, VA, 1993.

Keiver, K.M., Weinberg J., and Hochachka, P.W., The effect of anoxic submergence and recovery on circulating levels of catecholamines and corticosterone in the turtle, *Chrysemys picta*, *Gen. Comp. Endocrinol.*, 85(2), 308–315, 1992.

Keller, J.M. et al., Lymphocyte proliferation in loggerhead sea turtles: seasonal variations and contaminant effects, in: *Proceedings of the 22nd Annual Sea Turtle Symposium*, J. Seminoff (compiler), U.S. Dept. Commerce NOAA Tech. Memo. NMFS-SEFSC, Miami, FL (in press).

Lackovich, J.K. et al., Association of the herpesvirus with fibropapillomatosis of the green turtle *Chelonia mydas* and the loggerhead turtle *Caretta caretta* in Florida, *Dis. Aquat. Organ.*, 37(2), 89–97, 1999.

Lahvis, G.P. et al., Decreased lymphocyte responses in free-ranging bottlenose dolphins (*Tursiops truncatus*) are associated with increased concentrations of PCBs and DDT in peripheral blood, *Environ. Health Perspect.*, 103(Suppl. 4), 67–72, 1995.

Lake, J.L., PCBs and other chlorinated organic contaminants in tissues of juvenile Kemp's ridley turtles (*Lepidochelys kempii*), *Mar. Environ. Res.*, 38, 313–327, 1994.

Landsberg, J.H., Neoplasia and biotoxins in bivalves: is there a connection?, *J. Shellfish Res.*, 15, 203–230, 1996.

Landsberg, J.H. et al., The potential role of natural tumor promoters in marine turtle fibropapillomasis, *J. Aquat. Anim. Health*, 11, 199–210, 1999.

Laurent, L. and Lescure, J., L'Hivernage des tortues caouannes *Caretta caretta* (L.) dans le sud tunisien, *Rev. Ecol.* (*Terre Vie*), 49, 63, 1994.

Lee, A.M., Owens, D.W., and Roumillat, W.A., Reproductive biology and endocrine cycling of the diamondback terrapin, *Malaclemys terrapin*, in South Carolina estuaries, in: *Proceedings of the 22nd Annual Sea Turtle Symposium*, J. Seminoff (compiler), U.S. Dept. Commerce NOAA Tech. Memo. NMFS-SEFSC, Miami, FL (in press).

Licht, P., Rainey, W., and Cliffton, K., Serum gonadotropins and steroids associated with breeding activities in the green sea turtle, *Chelonia mydas*. II. Mating and nesting in natural populations, *Gen. Comp. Endocrinol.*, 40, 116, 1980.

Limpus, C.J. and Miller, J.D., The occurrence of cutaneous fibropapillomas in marine turtles in Queensland, in: *Proceedings of the Australian Marine Turtle Conservation Workshop*, 14–17 November 1990, Queensland Department of Environment and Heritage and The Australian Nature Conservation Agency, Brisbane, Australia, 1994, pp. 186–188.

Lorenzen, A. et al., Relationships between environmental organochlorine contaminant residues, plasma corticosterone concentrations, and intermediary metabolic enzyme activities in Great Lakes herring gull embryos, *Environ. Health Perspect.*, 107(3), 179–186, 1999.

Lu, Y. et al., Detection of herpesvirus sequences in tissues of green turtles with fibropapilloma by polymerase chain reaction, *Arch. Virol.*, 145(9), 1885–1893, 2000.

Lutcavage, M. et al., Physiologic and clinicopathologic effects of crude oil on loggerhead sea turtles, *Arch. Environ. Contam. Toxicol.*, 28, 417, 1995.

Lutcavage, M. et al., Human impacts on sea turtle survival, in: *The Biology of Sea Turtles*, Lutz, P.L. and Musick, J. (eds.), CRC Press, Boca Raton, FL, 1997, pp. 387–410.

Lutcavage, M. and Lutz, P.L., Diving physiology, in: *The Biology of Sea Turtles*, Lutz, P.L. and Musick, J.A. (eds.), CRC Press, Boca Raton, FL, 1997, pp. 277–296.

Lutz, P.L., Studies on the ingestion of plastic and latex by sea turtles, in *Proceedings 2nd International Conference on Marine Debris*, Shomura, R.S. and Godfrey, M.L., (eds.), NOAA Tech. Memo. NMFS-SWFS-154, Honolulu, HI, 1990.

Lutz, P.L., Salt, water, and pH balance in the sea turtle, in: *The Biology of Sea Turtles* Lutz, P.L. and Musick, J. (eds.), CRC Press, Boca Raton, FL, 1997, pp. 343–361.

Lutz, P.L., Health related sea turtle physiology, in: *Report of the Sea Turtle Health Assessment Workshop*, Fair, P. and Hansen, L.J. (eds.), NOAA Tech. Mem. NOS-NCCOS-CCE-HBR-0003, 1998, pp. 45–49.

Lutz, P.L., Bergey, A., and Bergey, M., The effect of temperature on respiration and acid-base balance in the sea turtle *Caretta caretta* at rest and during routine activity, *J. Exp. Biol.*, 144, 155–169, 1989.

Lutz, P.L. and Dunbar-Cooper, A., Variations in the blood chemistry of the loggerhead sea turtle, *Caretta caretta*, *Fish. Bull.*, 85, 37–43, 1987.

Magnien, R.E., Summers, R.M., and Sellner, K.G., External nutrient sources, internal nutrient pools, and phytoplankton production in Chesapeake Bay, *Estuaries*, 15, 497–516, 1992.

Mahapatra, M.S., Mahata, S.K., and Maiti, B.R., Effect of ambient temperature on serotonin, norepinephrine, and epinephrine contents in the pineal-paraphyseal complex of the soft-shelled turtle (*Lissemys punctata punctata*), *Gen. Comp. Endocrinol.*, 74(2), 215, 1989.

Mahapatra, M.S., Mahata, S.K., and Maiti, B., Effect of stress on serotonin, norepinephrine, epinephrine and corticosterone contents in the soft-shelled turtle, *Clin. Exp. Pharmacol. Physiol.*, 18(10), 719–724, 1991.

Marshall, A.T. and Cooper, P.D., Secretory capacity of the lachrymal salt gland of hatchling sea turtles, *Chelonia mydas*, *J. Comp. Physiol.*, 157B, 821–827, 1988.

Mckenzie, E. et al., Concentrations and patterns of organochlorine contaminants in marine turtles from Mediterranean and Atlantic waters, *Mar. Environ. Res.*, 47, 117–135, 1999.

McKinney, E.C. and Bentley, T.B., Cell-mediated response of *Chelonia mydas*, *Dev. Comp. Immunol.*, 9, 445–452, 1985.

Meylan, A.B. and Sadove, S., Cold-stunning in Long Island Sound, New York, *Mar. Turtle Newsl.*, 37, 7–8, 1986.

Miao, X.S. et al., Congener-specific profile and toxicity assessment of PCBs in green turtles (*Chelonia mydas*) from the Hawaiian Islands, *Sci. Total Environ.*, 281(1–3), 247–53, 2001.

Miller, J.D., Embryology of marine turtles, in: Gans, C., *Biology of Reptilia*, 14A, Billet, F. and Maderson, P.F.A. (eds.), Wiley-Interscience, New York, 1985, 269 pp.

Milton, S.L. et al., Health related sea turtle physiology, in: *Report of the 2nd Sea Turtle Health Assessment Workshop*, Fair, P. and Owens, D. (eds.), NOAA Tech. Mem., in press.

Milton, S.L., Schulman, A.A., and Lutz, P.L., The effect of beach renourishment with aragonite versus silicate sand on beach temperature and loggerhead sea turtle nesting success. *J. Coast. Res.*, 13(3), 904–915, 1997.

Moon, D.Y., MacKenzie, D.S., and Owens, D.W., Simulated hibernation of sea turtles in the laboratory: I. Feeding, breathing frequency, blood pH, and blood gases, *J. Exp. Zool.*, 278, 372–380, 1997.

Moran, K.L., Bjorndal, K.A., and Bolten, A.B., Effects of the thermal environment on the temporal pattern of emergence of hatchling loggerhead turtles *Caretta caretta*, *Mar. Ecol. Prog. Ser.*, 189, 251–261, 1999.

Morreale, S.J. et al., Annual occurrence and winter mortality of marine turtles in New York waters, *J. Herpetol.*, 26(3), 301–308, 1992.

Morris, Y.A., Steroid Dynamics in Immature Sea Turtles, M.S. thesis, Texas A&M University, College Station, TX, 1982, 92 pp.

Mrosovsky, N., Nocturnal emergence of hatchling sea turtles: control by thermal inhibition of activity, *Nature*, 220, 1338–1339, 1968.

Munck, A. and Naray-Fejes-Toth, A., Glucocorticoids and stress: permissive and suppressive actions, *Ann. N. Y. Acad. Sci.*, 746, 115–130, 1994.

Munoz, F.J. and De la Fuente, M., The immune response of thymic cells from the turtle *Mauremys caspica*, *J. Comp. Physiol. B*, 171(3), 195–200, 2001.

Munoz, F.J. et al., Seasonal changes in peripheral blood leukocyte functions of the turtle *Mauremys caspica* and their relationship with corticosterone, 17-beta-estradiol and testosterone serum levels, *Vet. Immunol. Immunopathol.*, 77(1–2), 27–42, 2000.

Mustafa, A. et al., Effects of sea lice (*Lepeophtheirus salmonis* Kroyer, 1837) infestation on macrophage functions in Atlantic salmon (*Salmo salar* L.), *Fish Shellfish Immunol.*, 10, 47–59, 2000.

Nelson, R.J. and Demas, G.E., Seasonal changes in immune function, *Q. Rev. Biol.*, 71(4), 511–548, 1996.

O'Connor, R.J., Toward the incorporation of spatiotemporal dynamics into ecotoxicology, in: *Population Dynamics in Ecological Space and Time*, Rhodes, O.E., Chesser, R.K., and Smith, M.H. (eds.), University of Chicago Press, Chicago, 281–317, 1996.

Odell, D.K. and MacMurray, C., Behavioral response to oil, in: *Final Report. Study of the Effect of Oil on Marine Turtles*, Vargo, S. et al. (eds.), Minerals Management Service Contract Number 14-12-0001-30063, Florida Institute of Oceanography, St. Petersburg, FL, 1986.

O'Hara, J., Thermal influences on swimming speed of loggerhead turtle hatchlings, *Copeia*, 1980, 773–780, 1980.

Ortiz, R.M. et al., Effects of acute fresh water exposure on water flux rates and osmotic responses in Kemp's ridley sea turtles (*Lepidochelys kempii*), *Comp. Biochem. Physiol. A Mol. Integr. Physiol.*, 127(1), 81–87, 2000.

Osborne, N.J.T., Webb, P.M., and Shaw, G.R., The toxins of *Lyngbya majuscula* and their human and ecological health effects, *Environ. Int.*, 27(5), 381–392, 2001.

O'Shea, T.J. et al., An epizootic of Florida manatees associated with a dinoflagellate bloom, *Mar. Mammal Sci.*, 7(2), 165–179, 1991.

Owens, D.W., Hormones in the life history of sea turtles, in: *The Biology of Sea Turtles,* Lutz, P.L. and Musick, J. (eds.), CRC Press, Boca Raton, FL, 1997, pp. 315–341.

Paladino, F.V. et al., Respiratory physiology of adult leatherback turtles (*Dermochelys coriacea*) while nesting on land, *Chelonian Conserv. Biol.*, 2(2), 223–229, 1996.

Podreka, S. et al., The environmental contaminant DDE fails to influence the outcome of sexual differentiation in the marine turtle *Chelonia mydas*, *Environ. Health Perspect.*, 106(4), 185–188, 1998.

Pugh, R.S. and Becker, P.R., *Sea Turtle Contaminants: A Review with Annotated Bibliography,* NISTIR 6700, Charleston, SC, 2001.

Quackenbush, S.L. et al., Quantitative analysis of herpesvirus sequences from normal tissue and fibropapillomas of marine turtles with real-time PCR, *Virology*, 287, 105–111, 2001.

Rabergh, C.M.I. et al., Tissue-specific expression of zebrafish (*Danio rerio*) heat shock factor 1 mRNAs in response to heat stress, *J. Exp. Biol.*, 203, 1817–1824, 2000.

Raidal, S.R. et al., Gram-negative bacterial infections and cardiovascular parasitism in green sea turtles (*Chelonia mydas*), *Aust. Vet. J.*, 76(6), 415–417, 1998.

Ray, P.P. and Maita, B.R., Adrenomedullary hormonal and glycemic responses to high ambient temperature in the soft-shelled turtle, *Lissemys punctata punctata*, *Gen. Comp. Endocrinol.*, 122(1), 17–22, 2001.

Redfearn, E., A comparative approach to understanding sea turtle hatchling metabolism during emergence, master's thesis, Florida Atlantic University, Boca Raton, FL, 2002. 105 pp.

Redlow, T., Foley, A., and Singel, K., Sea turtle mortality associated with red tide events in Florida, in: *Proceedings of the 22nd Annual Symposium on Sea Turtle Biology and Conservation*, J. Seminoff (compiler), U.S. Dept. Commerce NOAA Tech. Memo. NMFS-SEFSC, Miami, FL (in press).

Reijnders, P.H.J., Organochlorine and heavy metal residues in harbour seals from the Wadden Sea and their possible effects on reproduction, *Neth. J. Sea Res.*, 14, 30–65, 1980.

Reina, R.D. and Cooper, P.D., Control of salt gland activity in the hatchling green sea turtle, *Chelonia mydas*, *J. Comp. Physiol. B*, 170(1), 27–35, 2000.

Ross, P.S., Marine mammals as sentinels in ecological risk assessment, in: *Human Ecol. Risk Assess.*, 6(1), 29–46, 2000.

Rybitski, M.J., Hale, R.C., and Musick, J.A., Distribution of organochlorine pollutants in Atlantic sea turtles, *Copeia*, 2, 379–390, 1995.

Saad, A.H. et al., Testosterone induces lymphopenia in turtles, *Vet. Immunol. Immunopathol.*, 28(2), 173–180, 1991.

Saad, A.H. and el Ridi, R., Endogenous corticosteroids mediate seasonal cyclic changes in immunity of lizards, *Immunobiol.*, 177(4–5), 390–394, 1988.

Saeki, K. et al., Arsenic accumulation in three species of sea turtles, *Biometals*, 13(3), 241–250, 2000.

Sakai, H., Heavy metal monitoring in sea turtles using eggs, *Mar. Poll. Bull.*, 30, 347–353, 1995.

Sakai, H. et al., Growth-related changes in heavy metal accumulation in green turtle (*Chelonia mydas*) from Yaeyama Islands, Okinawa, Japan, *Arch. Environ. Contam. Toxicol.*, 39(3), 378–385, 2000.

Schumacher, U. et al., Histological investigations on the thyroid glands of marine mammals (*Phoca vitulina, Phocoena phocoena*) and the possible implications of marine pollution, *J. Wildl. Dis.*, 29, 103–108, 1993.

Schwantes, N., Aspects of Corticosterone Levels in Two Species of Sea Turtles *(Caretta caretta*) and (*Lepidochelys olivacea*), M.S. thesis, Texas A&M University, College Station, TX, 1986, 60 pp.

Schwartz, F.J., Behavioral and tolerance responses to cold water temperatures by three species of sea turtles (Reptilia, Cheloniidae) in North Carolina, *Florida Mar. Res. Pub.*, 33, 16–18, 1978.

Selye, H., A syndrome produced by diverse nocuous agents, *Nature*, 138, 32, 1936.

Sherman, B.H., Marine ecosystem health as an expression of morbidity, mortality and disease events, *Mar. Poll. Bull.*, 41, 232–54, 2000.

Smialowicz, R.J. et al., Evaluation of immunotoxicity of low level PCB exposure in the rat, *Toxicology*, 56, 197–211, 1989.

Sposato, P.L., Lutz, P.L., and Cray, C., Immunosuppression and fibropapilloma disease in wild green sea turtle populations (*Chelonia mydas*), in: *Proceedings of the 22nd Annual Symposium on Sea Turtle Biology and Conservation*, J. Seminoff (compiler), U.S. Dept. Commerce NOAA Tech. Memo. NMFS-SEFSC, Miami, FL (in press).

Spotila, J.R., O'Connor, M.P., and Paladino, F.V., Thermal biology, in: *The Biology of Sea Turtles*, Lutz, P.L. and Musick, J.A. (eds.), CRC Press, Boca Raton, FL, 1997, pp. 297–314.

Stancyk, S.E., Non-human predators of sea turtles and their control, in: *Biology and Conservation of Sea Turtles*, Bjorndal, K.A. (ed.), Smithsonian Institution Press, Washington, DC, 1982, pp. 139–152.

Standora, E.A., Morreale, S.J., and Burke, V.J., Application of recent advances in satellite transmitter microtechnology: integration with sonic and radio tracking of juvenile Kemp's ridleys from Long Island, NY, in: *Proceedings of the 11th Annual Symposium on Sea Turtle Biology and Conservation*, NOAA Tech. Memo. NMFS-SEFSC-302, 1992, pp. 111–113, Miami, FL.

Stegmann, J.J. et al., Molecular responses to environmental contamination: enzyme and protein systems as indicators of chemical exposure and effect, in: *Biomarkers: Biochemical, Physiological and Histological Markers of Anthropogenic Stress*, Huggert, R.J., Kimerle, R.A., Mehrle, P.M., Jr., and Bergman, H.L. (eds.), Lewis Publishers, Boca Raton, FL, 1992, pp. 235–335.

Still, B., Griffin, C., and Prescott, R., Factors affecting cold-stunning of juvenile sea turtles in Massachusetts, in: *Proceedings of the 22nd Annual Symposium on Sea Turtle Biology and Conservation*, J. Seminoff (compiler), U.S. Dept. Commerce NOAA Tech. Memo. NMFS-SEFSC, Miami, FL (in press).

Still, B., Tuxbury, K., Prescott, R., Ryder, C., Murley, D., Merigo, C., Smith, C., and Turnbull, E., A record cold stun season in Cape Cod Bay, Massachusetts, US, in: *Proceedings of the 21st Annual Symposium on Sea Turtle Biology and Conservation*, U.S. Dept. Commerce NOAA Tech. Memo. NMFS-SEFSC, Miami, FL(in press).

Storelli, M.M., Ceci, E., and Marcotrigiano, G.O., Distribution of heavy metal residues in some tissues of *Caretta caretta* (Linnaeus) specimens beached along the Adriatic Sea (Italy), *Bull. Environ. Contam. Toxicol.*, 60, 546–552, 1998a.

Storelli, M.M., Ceci, E., and Marcotrigiano, G.O., Comparison of total mercury, methylmercury, and selenium in muscle tissues and liver of *Stenella coeruleoalba* (Meyen) and *Caretta caretta* (Linnaeus), *Bull. Environ. Contam. Toxicol.*, 61, 541–547, 1998b.

Sures, B., Knopf, K., and Kloas, W., Induction of stress by the swimbladder nematode *Anguillicola crassus* in European eels *Anguilla anguilla*, after repeated experimental infection, *Parasitology*, 123(2), 179–184, 2001.

Thomas, P.T. and Hinsdill, R.D., Effect of polychlorinated biphenyls on the immune response of rhesus monkeys and mice, *Toxicol. Appl. Pharmacol.*, 44, 41–51, 1978.

Tryphonas, H. et al., Immuno-toxicity studies of PCB (Aroclor 1254) in the adult rhesus (*Macaca mulatta*) monkey — preliminary report, *Int. J. Immunopharmacol.*, 11, 199–206, 1989.

Turton, J.A. et al., Relationship of blood corticosterone, immunoglobulin and hematological values in young crocodiles (*Crocodylus porosus*) to water temperature, clutch of origin, and body weight, *Aust. Vet. J.*, 75(2), 114–119, 1997.

Ulmasov, K.A. et al., Heat shock proteins and thermoresistance in lizards, *Proc. Natl. Acad. Sci. U.S.A.*, 89(5), 1666–1670, 1992.

Valverde, R.A. et al., Adrenal responsiveness during nesting in the olive ridley sea turtle (*Lepidochelys olivacea*), *Am. Zool.*, 32, 21A, 1992.

Valverde R.A. et al., Hormone levels and ovulation correlates in the olive ridley sea turtle, in: *Proceedings 14th Annual Symposium on Sea Turtle Biology and Conservation*, Bjorndal, K.A. et al. (compilers), NOAA Tech. Memo. NMFS-SEFSC-351, Miami, FL, 1994, p. 156.

Valverde, R.A. et al., Basal and stress-induced corticosterone levels in olive ridley sea turtles (*Lepidochelys olivacea*) in relation to their mass nesting behavior, *J. Exp. Zool.*, 284(6), 652–662, 1999.

Vos, J.G. et al., Health effects of endocrine-disrupting chemicals on wildlife, with special reference to the European situation, *Crit. Rev. Toxicol.*, 30(1), 71–133, 2000.

Wasser, J.S. and Jackson, D.C., Effects of anoxia and graded acidosis on the levels of circulating catecholamines in turtles, *Respir. Physiol.*, 84(3), 363–377, 1991.

Whittow, G.C. and Balazs, G.H., Basking behavior of the Hawaiian green turtle (*Chelonia mydas*), *Pac. Sci.*, 36, 129, 1982.

Wibbels, T. et al., Sexing techniques and sex ratios for immature loggerhead sea turtles captured along the Atlantic coast of the U.S., in: *Ecology of East Florida Sea Turtles*, Witzel, E. (ed.), U.S. Department of Commerce, NOAA Tech. Rep. NMFS-53, Miami, FL, 1987, p. 65.

Wibbels, T. et al., Seasonal changes in gonadal steroid concentrations associated with migration, mating, and nesting in loggerhead sea turtles, *Gen. Comp. Endocrinol.*, 79, 154, 1990.

Wibbels, T. et al., Serum gonadotropins and gonadal steroids associated with ovulation and egg production in sea turtles, *Gen. Comp. Endocrinol.*, 87, 71, 1992.

Witherington, B.E., Bjorndal, K.A., and McCabe, C.M., Temporal pattern of nocturnal emergence of loggerhead turtle hatchlings from natural nests, *Copeia*, 4, 1165–1168, 1990.

Witherington, B.E. and Ehrhart, L.M., Hypothermic stunning and mortality of marine turtles in the Indian River Lagoon system, Florida, *Copeia*, 1989, 696–703, 1989.

Witherington, B.E. and Salmon, M., Predation on loggerhead turtle hatchlings after entering the sea, *J. Herpetol.*, 26(2), 226–228, 1992.

Wolke, R.E., Brooks, D.R., and George, A., Spirorchidiases in loggerhead sea turtles (*Caretta caretta*), *J. Wildl. Dis.*, 18(2), 175, 1982.

Work, T. and Balazs, G., Relating tumor score to hematology in green turtles with fibropapillomatosis in Hawaii, *J. Wildl. Dis.*, 35, 804–807, 1999.

Wyneken, J., Comparisons of oxygen utilization by hatchling loggerheads, greens, and leatherbacks during the swimming frenzy: sprinting vs. marathon strategies revisited, in: *Proceedings of the Eleventh Annual Workshop on Sea Turtle Biology and Conservation*, Salmon, M. and Wyneken, J. (compilers), NOAA Tech. Mem. NMFS-SEFSC-302, 1992, Miami, FL, 195 p.

Zapata, A.G., Varas, A., and Torroba, M., Seasonal variations in the immune system of lower vertebrates, *Immunol. Today*, 13, 142–147, 1992.

Zatsepina, O.G. et al., Thermotolerant desert lizards characteristically differ in terms of heat shock system regulation, *J. Exp. Biol.*, 203(6), 1017–1025, 2000.

7 Ontogeny of Marine Turtle Gonads

Jeffrey D. Miller and Colin J. Limpus

CONTENTS

7.1 INTRODUCTION

The ontogeny of the gonads of reptiles, including marine turtles, has been studied since the early 1800s, typically in association with the urinogenital system (Fox, 1977; Raynaud and Pieau, 1985). The early studies addressed such issues as embryogenesis, origin of germ cells, migration of germ cells, and development of the gonadal ridge as part of gaining an understanding of reptile development in the

0-8493-1123-3/03/$0.00+$1.50

context of what was known about amphibians, birds, and mammals (Wiedersheim, 1890a; 1890b; Wilson, 1896; 1900; Allen, 1906; Risley, 1933). Studies on the development of the cloaca and copulatory organs in turtles (Mitsukuri, 1896; Fleishmann and Hellmuth, 1902 in Fleishmann, 1902) and of the peritoneal canals (Moens, 1912) have contributed to the understanding of the morphology of the genital structures. Over the years, many specific aspects related to urinogenital development (i.e., germ cell movement, Jordan, 1917; kidney development, Burland, 1912; wolffian duct, Mitsukuri, 1888) have been examined; most of this work was descriptive and, in its day, theoretical. In 1977, Fox (1977) reviewed the ontogeny of the urinogenital system of reptiles. He traced the history of the kidney and associated ducts while comparing the descriptive morphology and conclusions drawn from the older studies dealing with the pronephros, mesonephros, and metanephros; he also dealt with the development of the gonads and associated ducts. More recently, the origin and development of oocytes (Hubert, 1985) and the embryonic development of the genital system in reptiles (Raynaud and Pieau, 1985) have been reexamined, with the emphasis on lizards and comments on turtles. At the same time, Ewert (1985) reviewed general embryology of turtles, and Miller (1985) described a series of developmental stages for marine turtles with emphasis on the Cheloniidae. Detailed descriptions of the development of *Dermochelys coriacea* were presented by Renous et al. (1989).

Although studies on marine turtles have contributed to the understanding of the development of the genital system in reptiles, no review has specifically addressed the ontogeny of marine turtle gonads. This chapter traces the ontogeny (morphology) of the gonads and associated ducts of marine turtles from embryogenesis through puberty to the adult, including changes linked with the reproductive cycle; lesser emphasis is placed on the development of the copulatory organs. The primary information is derived from studies on marine turtles; secondary support is obtained from studies on other turtles and, in turn, other reptiles. The approach focuses on marine turtles as much as possible rather than addressing the topic of gonadal ontogeny of reptiles (see Fox, 1977; Raynaud and Pieau, 1985). It should be remembered that morphological change, whether it be associated with development, growth, maturation, or reproduction, should not be viewed in isolation; the ontogenetic changes are driven by a complex suite of genetically coded interactive endocrine changes that occur in the context of endogenous and exogenous events throughout the life of the turtle (Owens, 1997).

7.2 EMBRYOGENESIS

During ovulation, multiple unfertilized ova are expelled from the ovary into the body cavity, from which they pass into the infundibulum of the oviduct; multiple ovulations for the entire clutch occur over a short time interval (Aitken et al., 1976; Owens, 1980). Fertilization occurs in the anterior of the oviduct (infundibulum or aglandular zone) before each ovum is surrounded by albumen (Miller, 1985). Cleavage and the formation of the gastrula occur as each ovum passes down the oviduct (Miller, 1985). Sequentially during this passage, each ovum is surrounded by albumen, the shell membrane, and shell. The eggs are ready to be oviposited after about 9 days following ovulation (Miller, 1985); however, the internesting interval is typically longer (Miller, 1985; 1997). Embryonic development is arrested at middle gastrulation until oviposition (stage 6, Miller, 1985).

At oviposition the blastodisc is composed of epiblast (presumptive ectoderm), and hypoblast (presumptive endoderm), and the area between is filling with migrating epiblastic cells (presumptive mesoderm) (Agassiz, 1857; Mitsukuri, 1894; Fujiwara, 1966; 1971). The dorsal expression of the chordamesodermal canal has the shape of an anteriorly opening, wide crescent (Mitsukuri, 1896–98; Miller, 1985). The canal has not broken through ventrally. During the early days of postovipositional development, the three germinal layers in the area opaca spread laterally and peripherally over the yolk mass that is contained in the follicular yolk membrane to form the extraembryonic splanchnopleure and the extraembryonic somatopleure (Agassiz, 1857; Mitsukuri, 1894; Fujiwara, 1966; 1971). These eventually give rise to the yolk sac and the allantoic membranes, and the amnion and chorion, respectively. The space between the yolk and the embryonic disk is filled by subgerminal fluid.

When oviposited into the nest chamber, the embryonic disk on the vitelline (follicular) membrane may land in any position relative to gravity. Within moments the vitelline membrane carrying the embryonic disk begins to rotate to the top pole of the yolk via the liquefaction of the surrounding albumen and the pull of gravity on the unevenly distributed, viscous yolk material contained within the vitelline membrane.

Over the next few hours, the albumen liquefies above the embryonic disk and passes through the margins of the embryonic area and vitelline membrane into the subgerminal area. This causes the vitelline membrane to distort to become more pear-shaped and causes the embryonic disk and yolk to rise toward the inner shell membrane. Simultaneously, the oviducal fluid that filled the microscopic canals among the aragonite crystals of the eggshell (Solomon and Baird, 1976; 1979; Solomon and Watt, 1985; Chan and Solomon, 1989) drains by capillary action down around the outer portion of the eggshell and/or inward to become part of the fluid layer just within the inner shell membrane. This action opens the pathway for gas exchange. The rising of the embryo on the distorted vitelline membrane reduces the distance over which gas exchange occurs. Together, these actions facilitate embryonic respiration (via diffusion) before development and vascularization of the extraembryonic membranes. Because the vitelline membrane is distorted and stretched, movement of the egg may cause it to rupture and the embryo to die (Limpus et al., 1979; Parmenter, 1980; Chan et al., 1985); the embryo remains subject to movement-induced mortality until it has established the extraembryonic membranes, about 25 days into incubation (Parmenter, 1980).

After oviposition, the development of the embryo is a continuous process. Once the egg has stabilized in the nest, the chordamesodermal canal breaks through ventrally. The neural plate forms above the notochord; the headfold becomes obvious. Somites begin to form just behind the neural folds and continue to form in pairs in a craniocaudal direction as the dorsal mesoderm subdivides into segments. Within 2.5 days at 30°C, the embryo reaches stage 10 (see Miller, 1985, for descriptions of embryonic stages).

7.2.1 Development of Kidneys and Gonadal Ridge

Within the abdominal cavity, the pronephros and mesonephros are first indicated in the anterior epithelium of the intermediary mesoderm as longitudinal ridges located on either side of the dorsal mesentery (Collins, 1990). In early development the nephrogenic cord, which is derived from intermediary mesoderm, segments to form nephrotomes. The functional pronephros in embryonic *Chelonia mydas* (12–13 mm, approximately stage 21–22) contains ciliated tubes that open into the coelom (Wiedersheim, 1890b). Large glomerulae develop external to the pronephros in *Dermochelys* and *Lepidochelys* (Fraser, 1950). The pronephros develops over a 13-day period during stages 12–24 in *Caretta caretta* incubated at 33°C. The pronephros and mesonephros are continuous in 8–13 mm embryos of *C. mydas* (Burland, 1912; Wiedersheim, 1890b). As each mesonephric vesicle develops and elongates, the nephrostomal canal degenerates. The distal end of the mesonephric vesicle inserts into the wolffian duct, and its proximal extension forms Bowman's capsule (Raynaud and Pieau, 1985). The kidneys of hatchling marine turtles are dorsoventrally flattened and lobate, and have many surface convolutions (*C. mydas*, DeRyke, 1926; *D. coriacea*, Burne, 1905).

The gonadal ridge develops from epithelial, mesothelial, and mesenchymal cells located between the base of the forming mesonephric tubules and the base of the dorsal mesentery ventral to the subcardinal veins during stage 17. Blood vessels invade the genital ridge together with a perforation of mesenchymal cells. The deposition of collagen fibers and the formation of the basal lamina separate the epithelial medullary cords from the loose mesenchymal cells of the stroma.

7.2.2 Origin and Migration of Germ Cells

In concert with the development of the gonadal ridge, the primordial germ cells develop in the extraembryonic hypoblast at the edge of the zona pellucida at the caudal end of the embryo (Allen, 1906; 1907; Jordan, 1917; Risley, 1933; Milaire, 1957). Primordial germ cells are distinguished from surrounding somatic cells because they are large cells with large, round nuclei (>16 μm) containing distinct nucleoli and large numbers of lipid droplets, yolk platelets, and glycogen particles in the cytoplasm; they stain positive with periodic acid-Schiff reagent (Fujimoto et al., 1979). They migrate by amoeboid action between the cells of the splanchnopleure to beneath the notochord and then enter the mesentery to reach the gonadal anlagen (Allen, 1906; Risley, 1933; Fujimoto et al., 1979). Primordial germ cells accumulate at the base of the gonadal ridge as the epithelium extends into the coelomic cavity over a period of approximately 10 days (Table 7.1).

Merchant-Larios et al. (1989) reported that primordial germ cells in *Lepidochelys* were first detected among the endodermal cells of the yolk sac as reported for *Caretta* (Fujimoto et al., 1979); this occurred at stages 15 and 16, about the tenth day of incubation, which was a bit later than reported for *Caretta*, but the conditions of incubation were not the same. By stage 18, primordial germ cells were in the hindgut epithelium and mesentery, and the genital ridges. The primordial germ cells were four to five times larger than surrounding somatic cells.

TABLE 7.1
Chronology of the Movement of Primordial Germ Cell (PGC) during Early Embryonic Development of Marine Turtle Gonads

Stage (Miller, 1985)	Day	Temperature	Description
12/13	4	30°C	PGCs not visibly separated from endoderm
13/14	5	30°C	PGCs visible in endoderm lateral to the midline
14/15	6	30°C	PGCs occur at the bilateral junctions between the splanchnic and somatic mesoderm
15/16	7	30°C	PGCs progressively accumulate in the root of the dorsal mesentery
16	8	30°C	
17	9	30°C	
17/18	10	30°C	PGCs are in the area of the presumptive genital ridge; the primitive gut and the dorsal mesentery have formed
18+	11	30°C	PGCs continue to accumulate in the root of the presumptive genital ridge
19	12	30°C	PGCs migrating to the genital ridge
20	13	30°C	
20+	14	30°C	Most PGCs have arrived in the genital ridge
21	15	30°C	

Source: Based on Fujimoto, T. et al. 1979. Observations of primordial germ cells in the turtle embryo (*Caretta caretta*): light and electron microscopic studies. *Dev. Growth Differ.* 21:3–10. With permission.

7.2.3 Gonadal Morphogenesis

Before sexual differentiation, the structure of the gonadal primordia is similar in both presumptive sexes. The gonadal primordia has two regions: the cortex (outer), which is characterized by a single layer of cuboidal epithelial cells with embedded germinal primordial cells, and the medulla (inner), which is derived from mesenchyme cells within the middle of the undifferentiated gonad. As sexual differentiation occurs (during stages 24–29, Merchant-Larios et al., 1989), the cells of the gonadal primordia propagate into two distinct and opposite patterns. At this time, changes in both the cortex and medulla are visible. Most of the primordial germ cells are situated in the cortex; both nerve and blood vessels penetrate the mass of mesenchyme cells in the medullary area.

In presumptive females, the ovary results from the simultaneous proliferation of cells in the cortex and regressive modification in the medulla. The cells of the cortex become more columnar in shape and the layer becomes thicker; primordial germinal cells are interspersed among the cells of the cortex. The medullary area shows some differentiation of cells to form primitive medullary (sex) cords, but for the most part, these regress and the area remains a thick mass of undifferentiated cells that will be penetrated by blood vessels.

In presumptive males, the testis results from the simultaneous regression of the cortex and the differentiation of seminiferous tubules in the medulla. In the cortex, cells regress to become a flattened epithelium, whereas cells in the medulla differentiate to form hollow sex cords that twist and anastomose to form the seminiferous tubules, which eventually connect to the rete testis and efferent tubules. The primordial germ cells migrate to reside among the cells lining the tubules. Mesenchyme cells condense to form the tunica albuginea that is situated between the cortex and the medulla of the gonad; it eventually becomes a thin, vascular, connective tissue sheath surrounding the testis and covered by a thin epithelium.

In marine turtles, the sex of the turtle is determined in the thermosensitive period (stages 22–27, Miller, 1985) of incubation rather than at fertilization (*C. caretta*: Yntema and Mrosovsky, 1980; 1982; Wibbels et al., 1991; *C. mydas:* Miller and Limpus, 1981; *Lepidochelys olivacea*: Mohanty-Hejmadi and Dimond, 1986; McCoy et al., 1983; Merchant-Larios and Villalpando, 1990; Merchant-Larios et al., 1989; 1997; *L. kempii*: Shaver et al., 1988; *Eretmochelys imbricata*: Wibbels et al., 1999a; *Natator depressus*: Hewavisenthi and Parmenter, 2000; *D. coriacea*: Desvages et al., 1993; Rimblot et al., 1985). Although much remains to be elucidated about the impact on embryonic development of variations in environmental temperature, gas concentration, and moisture availability during incubation (Ackerman, 1997), the clearest picture of the impact of incubation temperature on the differentiation of the gonads of marine turtles occurs toward the extremes of embryonic tolerance (approximately 23–33°C). In general, female hatchlings result from eggs incubated above 30°C and males result from eggs incubated below 28°C. The theoretical point at which a 1:1 sex ratio would be produced has been termed the pivotal temperature (Mrosovsky and Yntema, 1980), the critical temperature (Pieau, 1973), and the threshold temperature (Bull, 1980). The theoretical temperature can be estimated in the manner of lethal dose 50 (LD_{50}) calculations to provide an estimate of the 95% confidence limits (Limpus et al., 1983), which is useful for comparing among populations. From both the population function and conservation management points of view, the differences among populations are important.

7.2.4 Genital Ducts

The genital ducts provide the means by which the sex cells (sperm, ova) are passed to the exterior of the body. Within the testis, as the medulla differentiates, the seminiferous tubules become connected with the tubules of the rete testis that in turn connect to the tubules of the anterior mesonephros (termed vasa efferentia). The vasa efferentia combine with the anterior portion of the vas deferens to become the epididymis. The remainder of the mesonephric duct (Wolffian duct) becomes the vasa deferentia and, although originally derived from the mesonephros, carries only sperm in the adult. As testicular development continues, the paramesonephric duct begins to degenerate (Raynaud et al., 1970; Raynaud and Pieau, 1985).

In the female, the paramesonephric (müllerian) ducts develop from coelomic epithelium that thickens and forms a craniocaudal groove, the edges of which fuse to form a tube (Raynaud and Pieau, 1985; Wibbels et al., 1999b). The cephalic end of each tube opens as a funnel-shaped ostium tubae into the coelomic cavity. The ostium tubae becomes the infundibulum in the mature oviduct. The caudal end of the paramesonephric duct extends into the retroperitoneal connective tissue; as it develops, it parallels the mesonephric duct and eventually forms a connection with the cloaca (Raynaud and Pieau, 1985). In *C. mydas*, oviducts are indicated by stage 23 (13 mm body length), and the anterior end is formed by stage 25 (21 mm, Weidersheim, 1890b; Wilson, 1900).

Paramesonephric ducts develop in both sexes but function as oviducts only in females. Remnants may be retained in functional males. For example, in loggerhead turtles and less frequently in green turtles examined via necropsy and laparoscopy, the paramesonephric duct may persist and remain visible adjacent to the testis but terminate within the mesentery posteriorly and not connect to the cloaca (Limpus et al., 1982). The less degenerated paramesonephric ducts in males structurally resemble the female oviduct except that the lumen is lined with a squamous rather than a columnar epithelium. The function of the persistent paramesonephric duct in adult males is unresolved.

The cloaca is a "common sewer" that receives the output of the urogenital ducts and intestinal waste and passes to the exterior of the body. The cloaca originates from a diverticulum of the hindgut, fusing with the inward pocketing proctodeum (Raynaud and Pieau, 1985). The result is a tube that is separated from the intestine by a sphincter and has openings from the oviducts (female) or vas deferens (male), contains the penis or clitoris, and receives fluid from the bladder (derived from the allantois).

During development, the urinogenital prominence is first visible in stage 18 (Miller, 1985); by stage 23, the bulge extends to about the posterior edge of the hind digital plate. The undifferentiated urinogenital papilla is not visible until stage 24; the papilla is withdrawn by stage 28. In the female, it forms the clitoris; it is assumed, by homology, to follow the same pattern of differentiation as the penis, albeit on a smaller scale, although there are no direct studies. The penis is attached anteriorly on the floor of the male cloaca. It is formed from a pair of longitudinal vascular, spongy ridges of tissue (corpora cavernosa) that converge distally to form the glans penis with a medial groove (seminal furrow) (Raynaud and Pieau, 1985). The erected, adult penis results from the corpora cavernosa being engorged with blood to seal the medial seminal furrow into a tube (Figure 7.1). The engorgement of the corpora cavernosa extends the penis from the cloaca during copulation. No observations have been published that indicate whether the penis erects as it is inserted into the female or before. However, during our numerous courtship observations of males mounted on females, none have been observed with an erect penis outside the female.

FIGURE 7.1 Erect penis of an adult *C. mydas*. Note medial seminal furrow.

7.3 HATCHLING GONADS

Attempts have been made to visually distinguish between testis and ovary of hatchling marine turtles, but the technique has not proven to be useful (Whitmore et al., 1985; Mrosovsky and Benabib, 1990). At present, only two methods can be used to reliably determine the sex of a hatchling marine turtle: (1) histological examination and (2) serological typing (blood sampling) (Merchant-Larios, 1999). The former requires that the hatchling be killed and the gonads removed and, after preparation, examined under a microscope. The method is useful in the experimental context to establish the pivotal temperature for a population under management. The use of blood does not require the killing of hatchlings, but hormonal analysis (via radioimmunoassay) is expensive and works best on slightly older post-hatchlings (curved carapace length [CCL] > 30 cm) (Owens and Ruiz, 1980; Wibbels et al., 1998) because of the volumes of blood needed. Regardless of the method used, the pivotal temperature is useful in planning for management of the population and understanding of the sex ratio of hatchlings leaving the beach (Godfrey et al., 1996).

The gonads of hatchling marine turtles are morphologically defined by the time the turtles reach the beach surface (Table 7.2). At emergence the gonad appears as a whitish, elongate structure on the ventral surface of the kidney (Figure 7.2); it cannot be distinguished by eye as an ovary or a testis.

7.3.1 OVARY

Histologically, the cortex of the ovary is differentiated into columnar shaped cells with germinal cells spaced among them near the basal membrane (Figure 7.3). The medulla is dense with occasional small strings of cells among blood vessels and other cells; no tubules with open lumen are present.

TABLE 7.2
Characteristics of Hatchling Gonads

Region	Female	Male
Gonad		
Medulla	Dense, without tubules	Tubular
Cortex	Columnar epithelium	Squamous epithelium
Paramesonephric duct		
	Developed with lumen	Undeveloped, without lumen
	Inner epithelium columnar	Inner epithelium absent
	With long, thin shaft	Without long, thin stalk

Sources: From Miller, J.D. and Limpus, C.J. 1981. Incubation period and sexual differentiation in the green turtle, *Chelonia mydas* L. Pp. 66–73. In: *Proceedings of the Melbourne Herpetological Symposium* (C.B. Banks and A.A. Martin, eds.). Zoological Board of Victoria, Parkville, Victoria, Australia; and Merchant-Larios, H., Villalpando, I., and Centeno, B. 1989. Gonadal morphogenesis under controlled temperature in the sea turtle *Lepidochelys olivacea*. *Herpetol. Monogr.* 3:128–157. With permission.

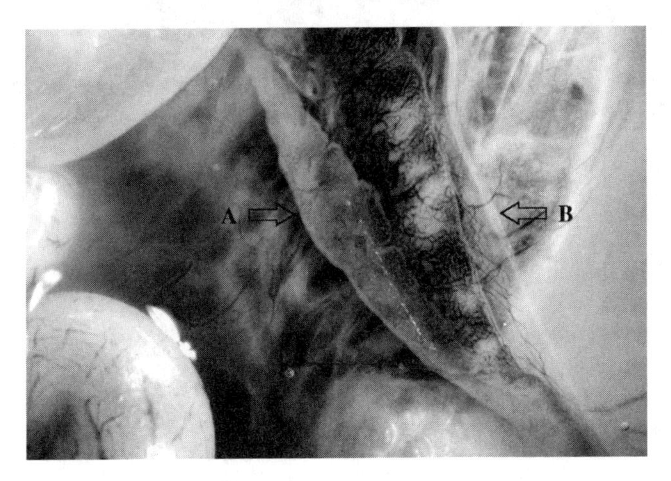

FIGURE 7.2 Gonad and paramesonephric duct in hatchling *C. caretta*. A. Gonad, B. Paramesonephric duct.

7.3.2 Testis

In histological cross section, the cortex of the testis is differentiated into flattened, squamous cells positioned next to the tunica albuginea (Figure 7.4). The medulla is filled with seminiferous tubules with germinal cells situated near the inner membrane. The space contained in the lumen of the tubules appears about the same as that occupied by the surrounding tissue.

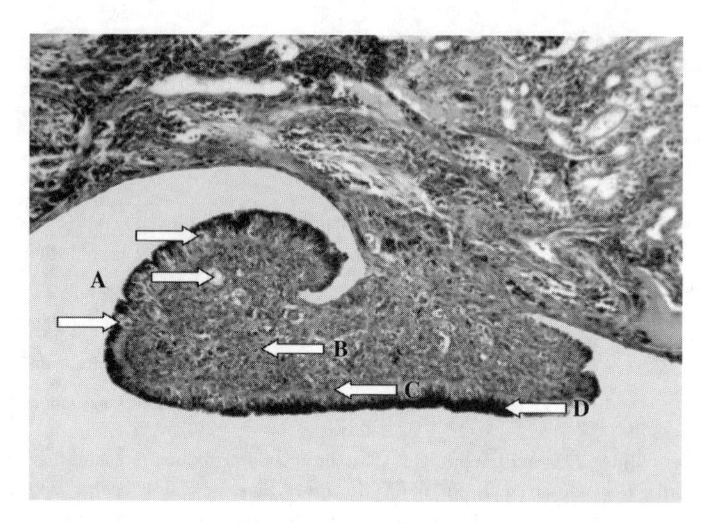

FIGURE 7.3 Hatchling ovary of *C. mydas*. (A) Germinal cells; (B) medulla; (C) tunica albuginea; (D) cortex.

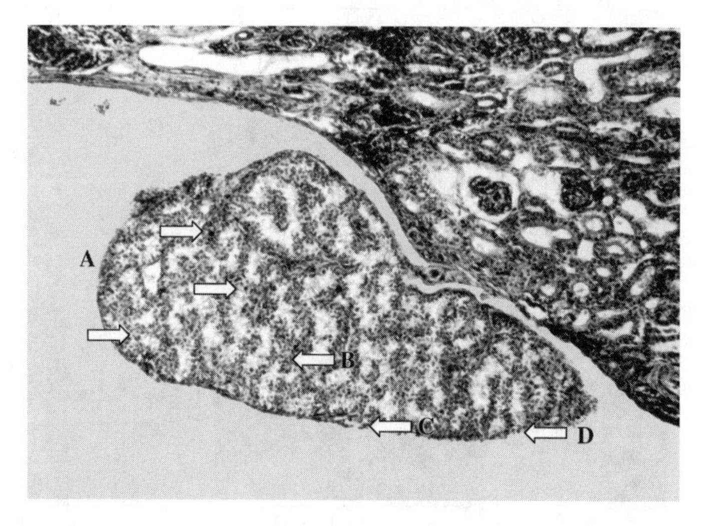

FIGURE 7.4 Hatchling testis *C. mydas*. (A) Germinal cells; (B) medulla containing seminiferous tubules; (C) tunica albuginea; (D) cortex.

7.3.3 Genital Ducts

The paramesonephric (müllerian duct, oviduct) in the female is white and straight, and can be seen traversing the posterior of the kidney. In histological cross section, the lumen is open and lined with columnar cells that sit on a distinct basement membrane (Figure 7.5A). The surrounding tissues are penetrated by blood vessels. The suspending mesentery is long and thin. Histologically in the male, in most

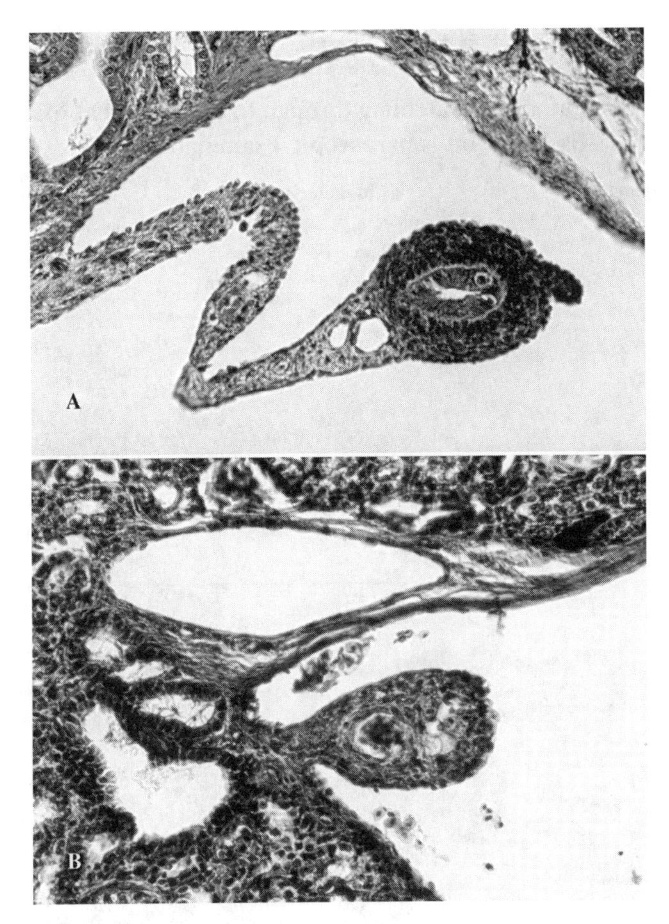

FIGURE 7.5 Paramesonephric ducts of (A) female and (B) male hatchling marine turtles (*C. caretta*). See text for descriptions.

individuals, the paramesonephric duct has virtually lost its embryonic lumen (Figure 7.5B); the cells have begun to degrade by hatching. The surrounding cells are dense and disorganized, and the stalk is short and thick.

The vas deferens is a white tube that is barely visible in the mesentery lateral to the kidney. In histological cross section, the vas deferens has a lumen lined with columnar cells positioned on a basement membrane. The tube is supported by connective tissue and embedded in mesentery. The epididymis is not visible in the mesentery.

7.4 PREPUBERTY

During the years before the turtle begins puberty, morphological change in the gonads is the result of growth. As both the ovary and the testis increase in size, their morphological differences become increasingly more visible (Table 7.3).

The ovary assumes a pale-yellowish tinge, and the previtellogenetic follicles appear as tiny spheres situated within the compact ovarian stroma (Color Figure

TABLE 7.3
Characteristics of a Post-Hatchling through to Prepubescent Marine Turtle Ovary and Testis Based on Laparoscopic Examination

Structure	Characteristic	Interpretation
Ovary	Stroma compact; previtellogenetic follicles small (<0.1 mm) and uniform in size	Post-hatchling, prepubescent female
Oviduct	White and straight	Post-hatchling, prepubescent female
Testis	Solid structure with tiny tubules visible via magnification	Post-hatchling, prepubescent male
Epididymis	Not expanded from body wall	Post-hatchling, prepubescent male

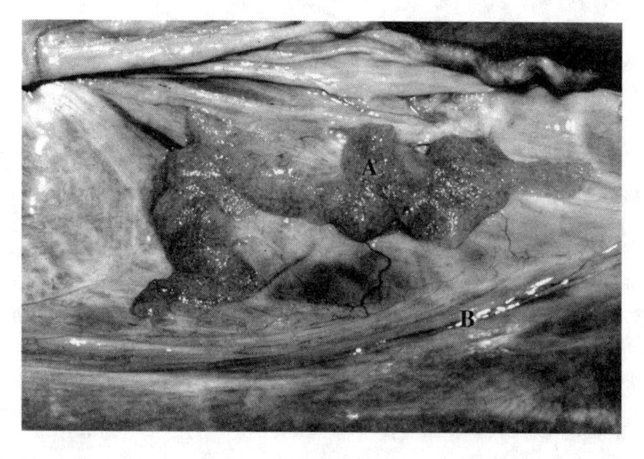

FIGURE 7.6 Prepubescent ovary (A) and oviduct (B), *C. caretta*, CCL = 69 cm.

7.6*). The oviduct appears as a white and straight duct contained in the mesentery adjacent to the gonad. The testis enlarges through the years to appear as a pale salmon-colored, solid structure (Color Figure 7.7); magnification with a hand lens (×10) reveals tiny convoluted tubules within the smooth epithelium. The epididymis is typically visible in the body wall adjacent to the testis as a thickened mass of tissue; it becomes more obvious as the turtle grows but does not extend from the body wall until the turtle enters puberty. The epididymis in the larger, prepubescent males can be seen through the peritoneum as a loosely convoluted, white duct.

* Color figures follow page 210.

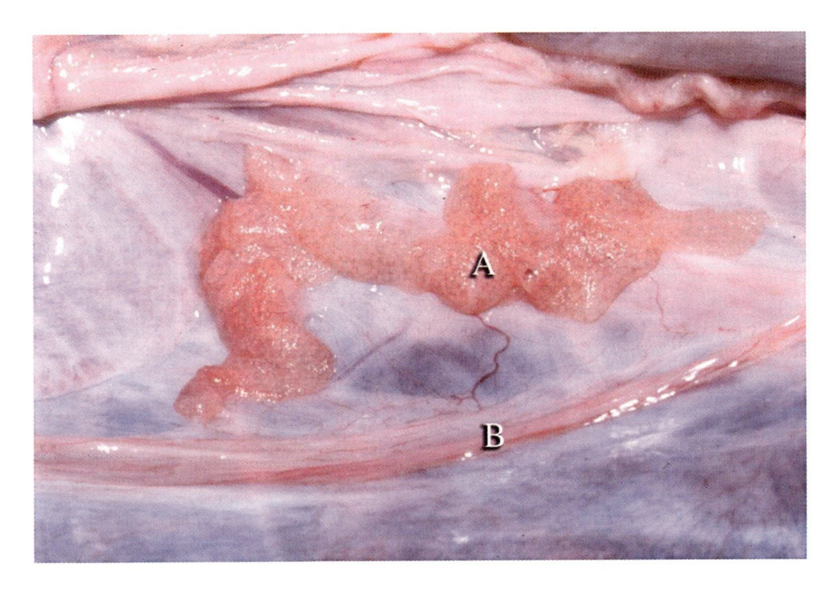

FIGURE 7.6 Prepubescent (A) ovary and (B) oviduct, *Chelonia mydas*, CCL = 69 cm.

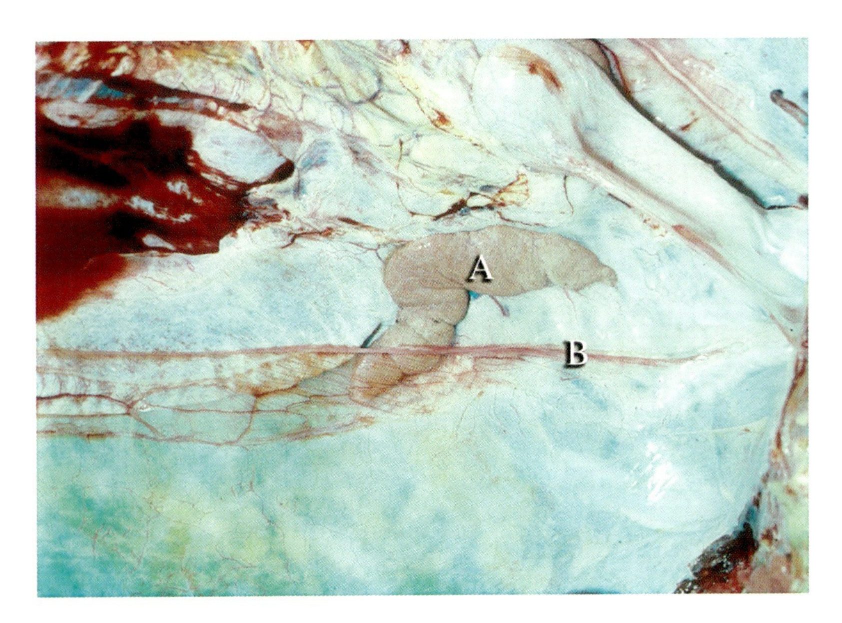

FIGURE 7. 7 Prepubescent testis (A), *Caretta caretta*, CCL = 89.5 cm, tail length beyond carapace = 11 cm. Note paramesonephric duct (B) supported by a distinct mesentery.

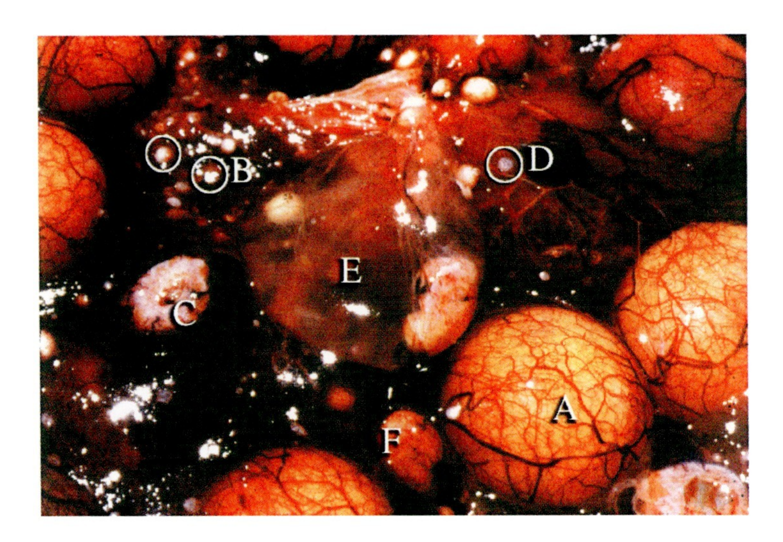

FIGURE. 7.8 Internesting adult *Caretta caretta* ovary with (A) mature preovulatory follicle, (B) previtellogenic follicles, (C) corpus luteum, (D) corpus albicans, (E) fluid- filled vesicle, and (F) atretic follicle.

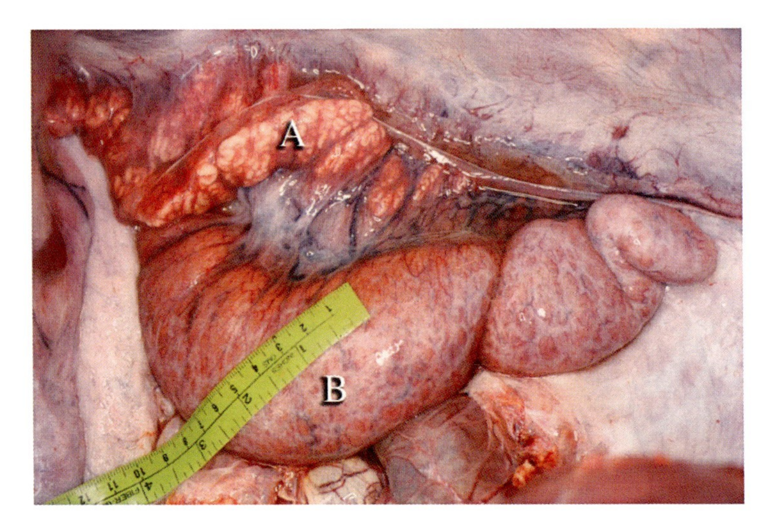

FIGURE 7. 9 Adult testis and adjacent epididymus of a breeding male *Caretta caretta*; (A) epididymis, (B) testis.

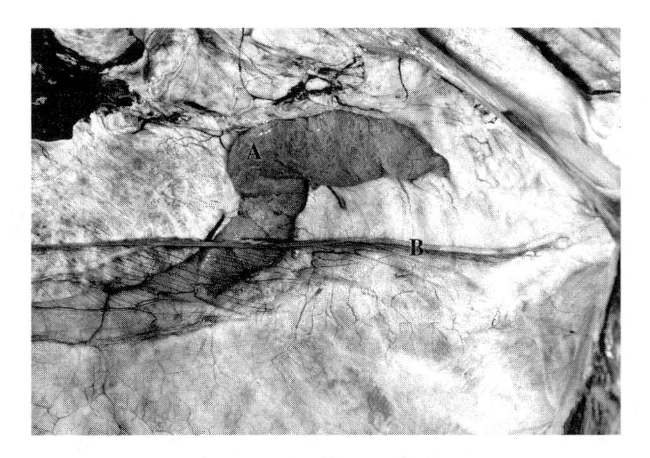

FIGURE 7.7 Prepubescent testis (A), *C. caretta*, CCL = 89.5 cm; tail length beyond carapace = 11 cm. Note paramesonephric duct (B) supported by a distinct mesentery.

7.5 PUBERTY

The changes that occur in the gonads and associated ducts of marine turtles during puberty have been studied only in the loggerhead turtle, *C. caretta* (Limpus, 1990; Limpus and Limpus, in press). The process of puberty, as demonstrated by sequential laparoscopic examination, required about 10 years from initiation to completion (Limpus, 1990).

During puberty, the female reproductive system undergoes a series of morphological changes, including the enlargement of the stroma of the ovary to become partly expanded. The previtellogenetic follicles are mostly of uniform size, but a few scattered atretic follicles may occur in the stroma.

Change occurs in the size and appearance of the oviduct assessed adjacent to the ovary before changes in the ovary become apparent (Table 7.4). The oviduct changes from being immature (white, straight, approximately cylindrical, and <2 mm in diameter) to the adult form (pink, convoluted, flattened, and >15 mm in diameter) over a period of about 4 years in loggerhead turtles (Limpus, 1990) in eastern Australia. Presumably, other species require about the same interval for their oviducts to reach adult size, although there may be differences in rates of maturation in different populations, particularly if there are differences in food abundance or temperatures. During puberty the oviduct typically is partly convoluted and <15 mm in flattened diameter.

In the loggerhead study (Limpus, 1990; Limpus and Limpus, in press), enlargement of the oviducts was followed by ovarian activity in the form of nonreproductive vitellogenesis. The ovarian stroma expanded from the compact form it exhibited previously. The distance between previtellogenetic follicles increased as the stroma expanded until the ovary had assumed the appearance of a curtain gathered at the top (Owens, 1980). For a small group of loggerhead turtles, the first nonreproductive vitellogenesis occurred in the second to fourth year following the changes in the oviduct. Although they departed from the foraging area and

TABLE 7.4
Characteristics of Pubescent Marine Turtle Ovaries and Testis Based on Laparoscopic Examination

Structure	Characteristic	Interpretation
Ovary	Stroma partly expanded	Previtellogenetic follicles mostly of uniform size, but a few scattered atretic follicles may occur
Oviduct	Partly convoluted	Pubescent female
Testis	Ellipsoidal	Pubescent male
Epididymis	Distinct ridge, obviously raised from body wall	Pubescent male

Source: From Limpus, C.J. 1990. Puberty and first breeding in *Caretta caretta*. Pp. 81–84. In: *Proceedings of the 10th Annual Workshop on Sea Turtle Biology and Conservation*. NOAA Tech. Memo. NMFS-SEFSC-278; Limpus, C.J. 1992. The hawksbill turtle, *Eretmochelys imbricata*, in Queensland: population structure within a southern Great Barrier Reef feeding ground. *Wildl. Res.* 19:489–506; and Limpus, C.J. and Limpus, D.J. (in press). The biology of the loggerhead turtle, *Caretta caretta,* in Southwest Pacific Ocean foraging areas. In: *The Biology of he Loggerhead Turtle,* Caretta caretta. (A. Bolten and B. Witherington, eds.). Smithsonian Institution Press, Washington. With permission.

subsequently returned to the foraging area (although there is no evidence that they migrated to the nesting area), most (seven of nine) did not ovulate during the interval. When examined via laparoscopy back in their foraging area, the seven were absorbing the yolky follicles (forming atretic follicles). The ovaries of these turtles completed vitellogenesis but did not ovulate; they contained atretic follicles but did not contain corpora albicantia, which form only following ovulation. Four of the seven were observed in their second nonreproductive vitellogenesis; three of these four turtles ovulated that season (2–3 years following the first). This nonreproductive vitellogenesis prior to the first successful breeding season has also been observed in green turtles.

A group of males was examined during puberty. The testis increased in size and changed shape to become ellipsoidal (Table 7.4). However, the change in the epididymis provided a more reliable indication of puberty. The epididymis became a distinct ridge that was obviously raised from the body wall.

7.6 ADULT

Marine turtles require decades to reach maturity (Chaloupka and Limpus, 1997; Limpus and Chaloupka, 1997; Limpus and Limpus, in press), and not all mature at the same size (Limpus et al., 1994a; 1994b). The new recruit female does not join the breeding population at the minimum breeding size (Limpus, 1992; Limpus and Limpus, in press). New recruits have a carapace length that is slightly smaller than the average for the breeding population, but greater than the minimum breeding size (Limpus, 1990; 1992; Limpus and Reed, 1985).

Several investigators have described the adult reproductive structures (*C. mydas*: Aitken et al., 1976; Owens, 1980, personal observations; *L. olivacea*: Owens, 1980; *Caretta* and *Natator*: personal observations). Various combinations of the size and shape of the ovary and oviduct, or the testis and epididymis, indicate the reproductive state of the turtle (Tables 7.5and 7.6).

7.6.1 Ovary

The mature ovary is attached to the dorsal body wall by a relatively narrow neck of tissue; the stroma is curtainlike and free-hanging in the body cavity. The stroma carries thousands (uncounted) of previtellogenetic follicles, each measuring 1–3 mm in diameter, on both sides (Hughes, 1974). Depending on the reproductive state of

TABLE 7.5
Characteristics of an Adult Marine Turtle Ovary and Testis Based on Laparoscopic Examination

Structure	Characteristic	Interpretation
Oviduct	Pink, very convoluted	Adult
Ovary	Stroma expanded forming a curtain with imbedded follicles of different diameters	Adult
Ovary with	Developing follicles (4–25 mm)	Preparing to breed
	Corpora albicantia (1–5 mm)	Has bred in past
	No corpora albicantia	New recruit
Testis	Cylindrical, seminiferous tubules may be visible within the testis	Male
Epididymis	Pendulous, distinct from body wall	Adult male
	White coils obvious within epididymis	Breeding during the next reproductive season
	No enlarged coils obvious within	Not breeding during the next reproductive season

Sources: Limpus, C.J. 1992. The hawksbill turtle, *Eretmochelys imbricata*, in Queensland: population structure within a southern Great Barrier Reef feeding ground. *Wildl. Res.* 19:489–506; and Limpus, C.J. and Limpus, D.J. (in press). The biology of the loggerhead turtle, *Caretta caretta,* in Southwest Pacific Ocean foraging areas. In: *The Biology of the Loggerhead Turtle,* Caretta caretta. (A. Bolten and B. Witherington, eds.). Smithsonian Institution Press, Washington. With permission.

TABLE 7.6
Characteristics of the Ovary of Adult Marine Turtles that Allow Interpretation of the Breeding History of the Individual

Structure	Characteristic	Interpretation
Oviduct	Oviducal eggs present	Currently breeding
Ovary with	Vesicular stroma Mature follicles (>25 mm) Large atretic follicles (pink) Corpora lutea (>7 mm)	Breeding current season
	If corpora lutea 1–5 mm	Has bred in past
	If no corpora albicantia present	New recruit
Ovary with	Vesicular stroma Corpora albicantia 4–7 mm Atretic follicles (pink, granular)	Bred last season
Ovary with	Nonvesicular stroma Corpora albicantia <> 3 mm with radiating white folds	Bred season before last
Ovary with	Small (<3mm) corpora albicantia	Adult, has bred in past
Ovary with	None of the above	Adult, has not bred

the turtle, the ovary may also contain mature follicles, corpora hemorrhagia, corpora albicantia, corpora lutea, and/or atretic follicles (Color Figure 7.8).

Each corpus luteum results from the stromal tissue that surrounded the oocyte before it was ovulated. Immediately after ovulation the ovarian tissue collapses to form a crater that is maintained during the preparation and gestation of the eggs prior to oviposition, then progressively degenerates. Progesterone is produced by the corpus luteum and is modulated through the hypothalamus. It inhibits follicular development and ovulation until about 40 h following oviposition (Owens, 1997). Eventually, the fluid-filled vesicle is absorbed and each corpus luteum appears to be drawn nearly closed, leaving white lines radiating outward from the small crater. Atretic follicles (corpora atretica) are follicles that were not ovulated. They are eventually invaded by stromal tissue and their contents are absorbed, leaving a scar in the ovary that becomes progressively smaller over several years.

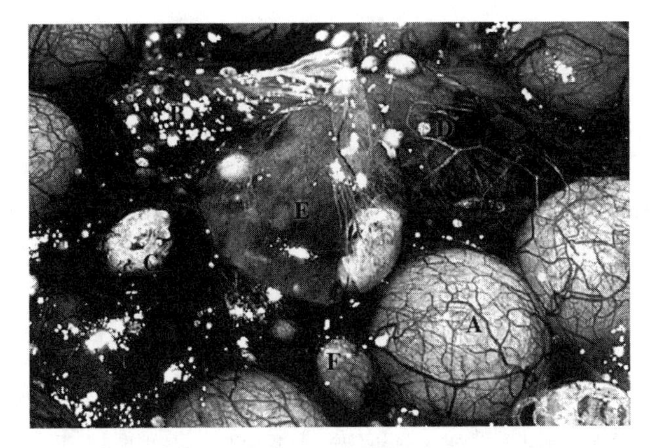

FIGURE 7.8 Internesting adult *C. caretta* ovary with (A) mature preovulatory follicle; (B) previtellogenic follicles; (C) corpus luteum; (D) corpus albucans; (E) fluid-filled vesicle; and (F) atretic follicle.

FIGURE 7.9 Adult testis and adjacent epididymis of a breeding male *C. caretta*. (A) Epididymis; (B) testis.

7.6.2 Testes

The mature testes are cylindrical, and seminiferous tubules are visible within (Color Figure 7.9). Histologically, the cortex of the testis is composed of a flattened epithelium supported by the tunica albuginea. In the medullary area, the seminiferous tubules are surrounded by loose connective tissue. Imbedded in the connective tissue between the testicular tubules are interstitial cells (Leydig's cells) that occur singly or in small groups near blood vessels. Interstitial cells secrete testosterone and lipoidal substances (Pellegrini, 1925a; 1925b). Inside the basement membrane of the seminiferous tubules are spermatogonia and, during the breeding period, primary spermatocytes. Sertoli's cells (sustentacular cells) are located against the basement membrane inside the seminiferous tubules (Risley, 1938a; 1938b; Altland, 1951). Sertoli cells link between adenohypophysial stimulation and tubular spermatogenesis (Lofts,

1968). In addition to being involved in hormone regulation, Sertoli's cells have a

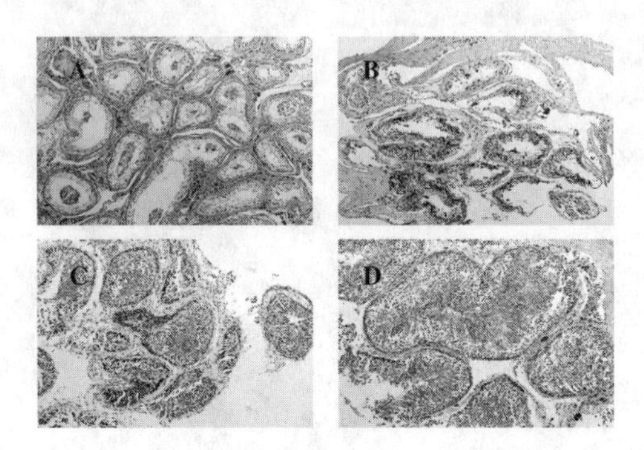

FIGURE 7.10 Micrographs of spermatogenic stages in adult male *C. mydas* marine turtle testes, hematoxylin and eosin stain. See Wibbels et al. (1990) for definition of the stages in the spermatogenetic cycle. (A) Stage 1, only spermatogonia present; (B) stage 2, primary spermatocytes and spermatogonia present; (C) stage 4, spermatids becoming spermatozoa; (D) stage 6, maximum spermatogenesis.

phagocytic role, eliminating sperm after spermatogenesis is finished (Fox, 1952).

The spermatogenesis within the testis of mature loggerhead turtles stands as the model for the other species (Figure 7.10) (Wibbels et al., 1990). In stage 1, the seminiferous tubules are involuted with only spermatogonia and possibly some spermatozoa in the lumen. In stage 2, the primary spermatocytes are present and spermatogonia become abundant. In stage 3, the secondary spermatocytes and early spermatids are abundant. In stage 4, spermatids are transforming and some spermatozoa are present. In stage 6, spermatogenesis reaches a maximum. In stage 7, spermatozoa remain abundant but spermatids and spermatocytes are reduced in number. In stage 8, the number of spermatozoa has reduced, the number of spermatids and spermatocytes may be absent or very low, and spermatozoa may be abundant in the lumen.

7.6.3 Genital Ducts

The oviducts in adult marine turtles are long (4–6 m, Owens, 1980). They are approximately 2 cm in flattened diameter and very concertinaed when not holding eggs (Figure 7.11). The oviduct has five sections: the infundibulum, aglandular zone, magnum, shell-forming zone, and vagina. Only the magnum and the shell-forming zone contribute to formation of the albumen and shell (Solomon and Baird, 1979). Ciliated and secretory cells line the magnum; the former assist the passage of sperm and the latter produce only albumen, which is homogenous. The shell-forming zone has the function of both producing the inner shell membrane and secreting the crystal portion of the shell (Solomon and Baird, 1979). The epithelium of the shell-forming zone contains mucus-secreting, ciliated, and nonciliated cells. The detailed description of the ultrastructure of the oviduct of *C. mydas* (Aitken and Solomon, 1976) is assumed to be representative of all species.

Sperm ascend the oviduct with the aid of ciliary tracts (Solomon and Baird, 1979). They found nests of sperm in the glands at the junction of the shell-forming

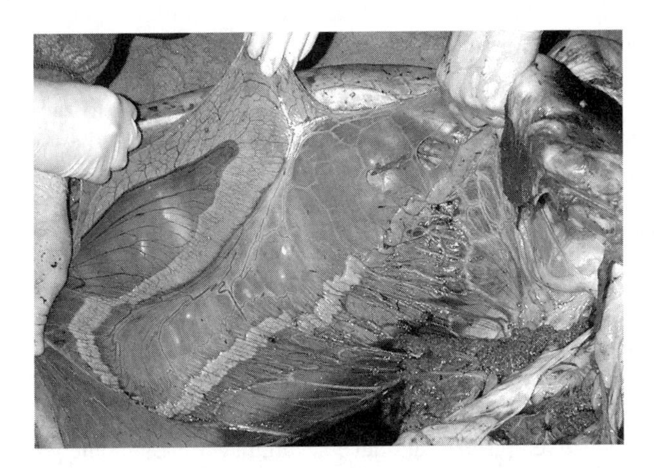

FIGURE 7.11 Oviduct of adult female *C. mydas* on mesentery.

zone and the vagina, near the base of the oviduct, but no spermatheca, which would indicate storage. They also reported quantities of sperm in the aglandular zone. This area is closer to the ovary than the magnum and is the last place during descent of the oocyte through the oviduct where sperm could fertilize the oocyte without having to penetrate albumen. Fertilization must occur either in the aglandular zone or in the infundibulum.

The epididymis of a mature male is pendulous and extends from the body wall. The seminiferous tubules are obvious within the testes, even in nonbreeding males. In breeding males, the seminiferous tubules are distended and appear white because they are filled with sperm. The entire epididymis is enlarged compared to that of a nonbreeding male. The vas deferens is a whitish tube traversing the body wall from the epididymis to the vicinity of the cloaca, where it becomes obscured by other tissue.

For turtles in their foraging area, the adult reproductive system is in one of three general states: quiescent, active, or regressive. The morphological expression of these states is obvious in the female but is only indicated in the male. Because marine turtles exhibit an iteroparous reproductive pattern, they do not all breed on the same cycle (Hirth, 1980; NRC, 1990). Individual females exhibit variable intervals between reproductive episodes (Limpus and Limpus, in press); males do not necessarily breed on the same cycle as females (Limpus, 1993).

7.6.4 Quiescence

Female turtles that are in a quiescent reproductive period did not breed in the immediately past season and will not breed in the next two seasons. They are recovering the energy reserves used during their previous reproductive cycle, growing, and building the reserves in general preparation for their next reproductive episode. The ovary of a reproductively quiescent turtle will contain no enlarged yolk-filled follicles, but it will contain numerous postovulatory scars (corpora albicantia) and scattered atretic follicles. The oviduct of a quiescent female is flaccid in appearance, very convoluted, and pink adjacent to the ovary.

In quiescent males, seminiferous tubules are pale white to salmon. The epididymis appears pendulous but not turgid and tubules will not be visible; its color is pale white.

7.6.5 Breeding Condition

A variable proportion of the adult female and male turtles in the foraging areas prepare for reproduction in any one year. For most green, loggerhead, and hawksbill adult female turtles, this preparation requires in excess of a year, on the basis of repeated examination via laparoscopy (Limpus, unpublished data). Before commencing vitellogenesis, the turtle must be sexually mature, have recovered from any previous reproductive episode, and have accumulated enough energy (fat) reserves to support vitellogenesis. Vitellogenesis requires the mobilization of stored energy (fat) and its modification via the liver under the control of the endocrine system with deposition into the previtellogenetic follicles. To prepare multiple clutches of follicles to support a breeding season requires at least 8 months. The quality and quantity of food are likely to play a major role in the timing of reproduction (Limpus and Nicholls, 2000).

The ovary of a turtle preparing to breed contains several hundred externally vascularized, yolk-filled follicles that are half a centimeter or more in diameter. It may contain postovulatory scars (corpora albicantia) and scattered atretic follicles. When the female commences her breeding migration from her foraging area, she carries a full complement of mature-sized yolked follicles and numerous other less-than-mature-sized follicles in each ovary. Once breeding has commenced after migration and copulation, the internesting female's ovary also contains increasing numbers of corpora lutea and decreasing numbers of mature follicles as each clutch is ovulated. The less-than-mature-sized follicles at the commencement of migration do not mature during the breeding season and are resorbed. The oviduct of a reproducing female is flaccid, having been stretched by the passage of eggs, or it may contain eggs.

In the testis of breeding males, the seminiferous tubules are distended and white. The coils of the epididymis appear turgid and enlarged, and the tubules are white and externally visible. Wibbels et al. (1990) described timing of the spermatogenic cycle in adult male loggerhead turtles preparing to breed. Histologically, the testis can be identified as developing out of the quiescent phase about 8 months prior to the courtship season, and has peak sperm production at about the time that migration to the breeding area begins.

7.6.6 Regression

Following a reproductive episode, each male and female turtle must return to its foraging area to recuperate. Each must recover the nutrient and energy reserves depleted during the breeding migration as well as fulfill its requirements for physiological maintenance before it can begin to accumulate the reserves required to initiate another reproductive episode. For the female, this may require a year (*L. olivacea*: Plotkin et al., 1994), a few years, or a decade or more (Limpus et al., 1992; Limpus and Limpus, in press). The ovary of a turtle that has just returned from a breeding episode may contain a few mature follicles, possibly some corpora hemorrhagia and numerous corpora albicantia from previous clutches in the season still supported on fluid-filled vesicles, and atretic follicles. Small (<4 mm) corpora albicantia from previous breeding seasons may be present. These characteristics become less obvious as the interval from the oviposition of the last clutch increases. The corpora albicantia continue to heal, decreasing to about 2 mm diameter over 2 years; the atretic follicles are fully resorbed during about a year. The small regressed corpora albicantia remain as permanent scars in the surface of the ovary. The oviduct of a regressive female remains flaccid, and the convolutions may not be as closely aligned as in a quiescent turtle. At times, the oviduct may appear thicker and/or contain visible blood vessels, more like an oviduct in a reproducing turtle.

In the male, seminiferous tubules of the testis are pale white to salmon. Histologically, remaining spermatozoa are resorbed and seminiferous tubules become involuted and lined with only spermatogonia (Wibbels et al., 1990). The epididymis loses the turgid appearance, and the color changes to pale white.

7.7 SUMMARY

Although the general pattern of ontogeny of the gonads of marine turtles can be described, there is much to be learned. Further descriptions of the changes in the structure and ultrastructure of individual parts of the genital system should be made among several species. The application of high-resolution laparoscopy, magnetic resonance imaging, and high-resolution ultrasound (among other techniques) to the study of the progression of development would maximize the information obtained from the fewest embryos. The illustrations of Agassiz (1857), Mitsukuri (1888; 1894; 1896–98), and others could be incorporated into modern computer graphic studies of ontogeny of marine turtles.

ACKNOWLEDGMENTS

This review is based on the pioneering work of literally dozens of researchers whose efforts before 1980 are often ignored in computer-based literature searches. Their descriptive prose and detailed drawings provide a wealth of information that should not be overlooked. It is more than good science to review the older literature; it is also good conservation because it helps to avoid unnecessary duplication and allows better formulation of research questions.

REFERENCES

Ackerman, R.A. 1997. The nest environment and the embryonic development of sea turtles. Pp. 83–106. In: *The Biology of Sea Turtles* (P. Lutz and J. Musick, eds.). CRC Press, Boca Raton, FL.

Agassiz, L. 1857. *Contributions to the Natural History of the United States of America.* First monograph, Vol. II. Little, Brown and Co., Boston.

Aitken, R.N. and Solomon, S.E. 1976. Observations on the ultrastructure of the oviduct of the Costa Rican green turtle (*Chelonia mydas* L.), *J. Exp. Mar. Biol. Ecol.* 21:75–90.

Aitken, R.N.C., Solomon, E.E., and Amoroso, E.C. 1976. Observations on the histology of the ovary of the Costa Rican green turtle, *Chelonia mydas* L. *J. Exp. Mar. Biol. Ecol.* 24:189–204.

Allen, B.M. 1907. A statistical study on the sex cells of *Chrysemys marginata. Anat. Anz.* 30:301–399.

Allen, B.M. 1906. The origin of the sex cells. *Anat. Anz.* 29:217–236.

Altland, P.D. 1951. Observations on the structure of the reproductive organs of the box turtle. *J. Morphol.* 89:599–621.

Bull, J.J. 1980. Sex determination in reptiles. *Q. Rev. Biol.* 55:3–21.

Burland, T.H. 1912. Observations on the development of the kidney in *Chelonia. Anat. Anz.* 41:497–511.

Burne, R.H. 1905. Notes on the muscular and visceral anatomy of the leathery turtle (*Dermochelys coriacea*). *Proc. Zool. Soc. Lond.* 1:291–324.

Chaloupka, M.Y. and Limpus, C.J. 1997. Robust statistical modelling of hawksbill sea-turtle growth rates (southern Great Barrier Reef). *Mar. Ecol. Prog. Ser.* 146:1–8.

Chan, E. and Solomon, S. 1989. The structure and formation of the eggshell of the leatherback turtle (*Dermochelys coriacea*) from Malaysia, with notes on attached fungal forms. *Anim. Technol.* 4:91–102.

Chan, E.H., Salleh, H.U., and Liew, H.C. 1985. Effects of handling on hatchability of eggs of the leatherback turtle, *Dermochelys coriacea* (L.). *Pertanika* 8:265–271.

Collins, P. 1990. The origin, development, and degeneration of the pronephros in *Caretta caretta. J. Morphol.* 205:297–305.

DeRyke, W. 1926. The vascular structure of the kidney of *Chrysemys marginata belli* (Gay) and *Chelydra serpentina* (L.). *Anat. Rec.* 33:163–177.

Desvages, G., Girondot, M., and Pieau, C. 1993. Sensitive stages for the effects of temperature on gonadal aromatase activity in embryos of the marine turtle *Dermochelys coriacea. Gen. Comp. Endocrinol.* 92:54–61.

Ewert, M. 1985. Embryology of turtles. Pp. 75–267. In: *Biology of the Reptilia*. Vol. 14 (C. Gans, F. Billet, and P.F.A. Madderson, eds.). John Wiley & Sons, New York.

Fleishmann, A. 1902. Morphologische studien über Kloake und Phallus der Amnioten: I. Dei Eidechsen und Schlangen, by P. Unterhössel; II. Die Schildkröten und Krokodile, by K. Hellmuth; III. Die Vögel, by C. Pomayer; IV. Die Säugethiere; and V. Die Stilistik des Urodäum und Phallus bei den Amnioten, by A. Fleishmann. *Morph. Jb.* 30:539–589.

Fox, H. 1977. The urogenital system in reptiles. Pp. 1–157. In: *Biology of the Reptilia*. Vol. 6 (C. Gans and T.S. Parsons, eds.). Academic Press, New York.

Fox, W. 1952. Seasonal variation in the male reproductive system of Pacific coast garter snakes. *J. Morphol.* 90:481–553.

Fraser, E.A. 1950. The development of the vertebrate excretory system. *Biol. Rev.* 25:150–187.

Fujimoto, T. et al. 1979. Observations of primordial germ cells in the turtle embryo (*Caretta caretta*): light and electron microscopic studies. *Dev. Growth Differ.* 21:3–10.

Fujiwara, M. 1966. The early development of the marine turtle with special reference to the formation of the germ layers in amniota. *Bull. Tokyo Gakugai Univ.* 18 (Ser. IV):45–60.

Fujiwara, M. 1971. Perforation of the chordomesodermal canal in the Pacific roggerhead [sic] turtle. *Acta. Herpetol. Jap.* 4:16–17.

Godfrey, M.H., Barreto, R., and Mrosovsky, N. 1996. Estimating past and present sex ratios of sea turtles in Surinam. *Can. J. Zool.* 74:267–277.

Hewavisenthi, S. and Parmenter, C.J. 2000. Hydric environment and sex determination in the flatback turtle (*Natator depressus* Garman) (Chelonia:Cheloniidae). *Aust. J. Zool.* 48:653–659.

Hirth, H.F. 1980. Some aspects of the nesting behavior and reproductive biology of sea turtles. *Am. Zool.* 20 (3):507–524.

Hubert, J. 1985. Embryology of the squamata. Pp. 1–23. In: *Biology of the Reptilia, Vol. 15, Development B.* (C. Gans and F. Billett, eds.)

Hughes, G. 1974. The sea turtles of South-East Africa. II. The biology of the Tongaland loggerhead turtles *Caretta caretta* L. with comments on the leatherback turtle *Dermochelys coriacea* L. and the green turtle *Chelonia mydas* L. in the study region. *S. Afr. Assoc. Mar. Biol. Res. Oceanogr. Res. Inst. Invest. Rep.* 36:1–96.

Jordan, H.E. 1917. Embryonic history of the germ cells of the loggerhead turtle (*Caretta caretta*). *Publ. Carnegie Inst. Contrib. Embryol.* 251:313–344.

Limpus, C.J. 1990. Puberty and first breeding in *Caretta caretta*. Pp. 81–84. In: *Proceedings of the 10th Annual Workshop on Sea Turtle Biology and Conservation*. NOAA Tech. Memo. NMFS-SEFSC-278, Miami, FL.

Limpus, C.J. 1992. The hawksbill turtle, *Eretmochelys imbricata*, in Queensland: population structure within a southern Great Barrier Reef feeding ground. *Wildl. Res.* 19:489–506.

Limpus, C.J. 1993. The green turtle, *Chelonia mydas*, in Queensland: breeding males in the southern Great Barrier Reef. *Wildl. Res.* 20:513–523.

Limpus, C.J., Baker, V., and Miller, J.D. 1979. Movement induced mortality of loggerhead eggs. *Herpetologica* 35:335–338.

Limpus, C.J. and Chaloupka, M. 1997. Nonparametric regression modelling of green sea turtle growth rates (southern Great Barrier Reef). *Mar. Ecol. Prog. Ser.* 149:23–34.

Limpus, C.J., Couper, P.J., and Read, M.A. 1994a. The green turtle, *Chelonia mydas*, in Queensland: population structure in a warm temperate feeding area. *Mem. Queensland Mus.* 35:139–154.

Limpus, C.J., Couper, P.J., and Read, M.A. 1994b. The loggerhead turtle, *Caretta caretta*, in Queensland: population structure in a warm temperate feeding area. *Mem. Queensland Mus.* 37:195–204.

Limpus, C.J. et al. 1992. Migration of green (*Chelonia mydas*) and loggerhead (*Caretta caretta*) turtles to and from eastern Australian rookeries. *Aust. Wildl. Res.* 19:347–358.

Limpus, C.J. and Limpus, D.J. (in press). The biology of the loggerhead turtle, *Caretta caretta,* in Southwest Pacific Ocean foraging areas. In: *The Biology of the Loggerhead Turtle,* Caretta caretta (A. Bolten and B. Witherington, eds.). Smithsonian Institution Press, Washington, D.C.

Limpus, C.J., Miller, J.D., and Reed, P. 1982. Intersexuality in a loggerhead sea turtle, *Caretta caretta. Herpetol. Rev.* 13:32–33.

Limpus, C. and Nicholls, N. 2000. ENSO regulation of Indo-Pacific green turtle populations. In: *Applications of Seasonal Climate Forecasting in Agricultural and Natural Ecosystems*. Pp. 399–408 (G. Hammer, N. Nicholls, and C. Mitchell, eds.). Kluwer Academic Publishers, Dordrecht, Germany.

Limpus, C.J. and Reed, P. 1985. The green turtle in Queensland: population structure in a coral reef feeding ground. Pp. 47–52. In: *Biology of Australasian Frogs and Reptiles*. (G. Grigg, R. Shine, and H. Ehmann, eds.). Surrey Beatty and Sons, Sydney.

Limpus, C.J., Reed, P., and Miller, J.D. 1983. Islands and turtles. The influence of choice of nesting beach on sex ratio. Pp. 397–402. In: *Proceedings of the Inaugural Great Barrier Reef Conference* (J.T. Baker et al., eds.). James Cook University Press, Townsville, Australia.

Lofts, B. 1968. The sertoli cell. *Gen. Comp. Endocrinol.* Suppl. 3:636–648.

McCoy, C., Vogt, R.C., and Censky, J.E. 1983. Temperature controlled sex determination in the sea turtle *Lepidochelys olivacea. J. Herpetol.* 17:404–406.

Merchant-Larios, H. 1999. Determining hatchling sex. Pp. 130–135. In: *Research and Management Techniques for the Conservation of Sea Turtles* (K.L. Eckert et al., eds.). IUCN/SSC Marine Turtle Specialist Group Publication No. 4. Washington, D.C.

Merchant-Larios, H. et al. 1997. Correlation among thermosensitive period, estradiol response and gonad differentiation in the sea turtle *Lepidochelys olivacea. Gen. Comp. Endocrinol.* 107:373–385. Washington, D.C.

Merchant-Larios, H. and Villalpando, I. 1990. Effect of temperature on gonadal sex differentiation in the sea turtle *Lepidochelys olivacea*: an organ culture study. *J. Exp. Zool.* 254:327–331. Washington, D.C.

Merchant-Larios, H., Villalpando, I., and Centeno, B. 1989. Gonadal morphogenesis under controlled temperature in the sea turtle *Lepidochelys olivacea. Herpetol. Monogr.* 3:128–157.

Milaire, J. 1957. Contribution à la connaissance morphologique et cytologique des bourgeons de members chez quelques reptiles. *Arch. Biol. Paris* 68:429–512.

Miller, J.D. 1985. Embryology of marine turtles. Pp 270–328. In: *Biology of the Reptilia*. Vol. 14 (C. Gans, F. Billet, and P.F.A. Madderson, eds.). John Wiley & Sons, New York.

Miller, J.D. 1997. Reproduction in sea turtles. Pp. 51–81. In: *The Biology of Sea Turtles* (P. Lutz and J.A. Musick, eds.). CRC Press, Boca Raton, FL.

Miller, J.D. and Limpus, C.J. 1981. Incubation period and sexual differentiation in the green turtle, *Chelonia mydas* L. Pp. 66–73. In: *Proceedings of the Melbourne Herpetological Symposium* (C.B. Banks and A.A. Martin, eds.). Zoological Board of Victoria, Parkville, Victoria, Australia.

Mitsukuri, K. 1888. The ectoblastic origin of the Wolffian duct in Chelonia. *Zool. Anz.* 11:1–111.

Mitsukuri, K. 1894. On the process of gastrulation in Chelonia. *J. Coll. Sci. Univ. Tokyo* 6:227–277.

Mitsukuri, K. 1896–98. On the fate of the blastopore. The relations of the primitive streaks and the formation of the posterior end of the embryo in Chelonia, together with remarks on the nature of meroblastic ova in vertebrates. *J. Coll. Sci. Univ. Tokyo* 10:1–116.

Moens, N.L. 1912. Die Peritonealkanale de Schildkröten und Krokodile. *Morph. Jb.* 44:1–80.

Mohanty-Hejmadi, P. and Dimond, M.T. 1986. Temperature dependent sex determination in the olive ridley turtle. Pp. 159–162. In: *Progress in Developmental Biology, Part A* (H.C. Slavkin, ed.). Alan R. Liss, New York.

Mrosovsky, N. and Benabib, M. 1990. An assessment of two methods of sexing hatchling sea turtles. *Copeia* 1990:589–581.

Mrosovsky, N. and Yntema, C.L. 1980. Temperature dependence of sexual differentiation in sea turtles: implications for conservation practices. *Biol. Conserv.* 18:271–280.

NRC (National Research Council). 1990. *The Decline of the Sea Turtles.* P. 259. National Academy of Science Press, Washington, DC.

Owens, D. 1980. The comparative reproductive physiology of sea turtles. *Am. Zoologist* 20:549–564.

Owens, D. 1997. Hormones in the life history of sea turtles. Pp. 315–341. In: *The Biology of Sea Turtles* (P. Lutz and J. Musick, eds.). CRC Press, Boca Raton, FL.

Owens, D. and Ruiz, G. 1980. New methods of obtaining blood and cerebrospinal fluid from marine turtles. *Herpetologica* 36:17–20.

Parmenter, C.J. 1980. Incubation of green sea turtles (*Chelonia mydas*) in Torres Strait, Australia: the effect on hatchability. *Aust. Wildl. Res.* 7:487–491.

Pellegrini, G. 1925a. Sur la Cellule interstitielle du testicule. *C. R. Assoc. Anat.* 20:314–316.

Pellegrini, G. 1925b. Sulle modificazione degli elementi interstiziale del testicolo negli animali ad attività sessuale periodica. *Arch. Ital. Anat. Embriol.* 22:550–585.

Pieau, C. 1973. Nouvelles données expérimentales concernant les effets de la température sur la différenciation sexuelle chez les embryons de chéloniens. *C. R. Hebd. Séanc. Acad. Sci., Paris, Ser. D* 277:2789–2792.

Plotkin, P., Byles, R., and Owens, D. 1994. Migratory and reproductive behavior of *Lepidochelys olivacea* in the eastern Pacific. Pp. 138. In: *Proceedings of the Thirteenth Annual Symposium on Sea Turtle Conservation and Biology* (B.A. Schroeder and B. Witherington, compilers). NOAA Tech. Memo. NMFS-SEFSC-341, Miami, FL.

Raynaud, A. and Pieau, C. 1985. Embryonic development of the genital system. Pp. 149–300. In: *Biology of the Reptilia.* Vol. 15. (C. Gans and F. Billet, eds.). John Wiley & Sons, New York.

Raynaud, A., Pieau, C., and Raynaud, J.J. 1970. Étude histologique comparative de l'allongement des canaux de Müller, de l'arrêt de leur progression en direction caudale et de leur destruction, chez les embryons males de diverses espèces de reptiles. *Ann. Embriol. Morphogen.* 3:21–47.

Renous, S. et al. 1989. Caractéristiques du développement embryonnaire de la tortue luth, *Dermochelys coriacea* (Vandelli 1761). *Ann. Sci. Nat. Zool. Paris. Ser. 13,* 10:197–229.

Rimblot, F.J.J. et al. 1985. Sexual differentiation as a function of the incubation temperature of eggs in the sea turtle *Dermochelys coriacea* (Vandelli 1761). *Amphibia-Reptilia* 6:83–92.

Risley, P.L. 1933. Contribution on the development of the reproductive system in *Sternotherus oderatus* (Latreille). I. The embryonic origin and migration of the primordial germ cells. *Z. Zellforsch. Mikrosk. Anat.* 18:458–492.

Risley, P.L. 1938a. Development of embryonic gonad grafts in female diamond-back terrapins. *Anat. Rec.* 72 (Suppl.):101.

Risley, P.L. 1938b. Seasonal changes in the testis of the musk turtle, *Sternotherus odoratus* L. *J. Morphol.* 63:301–317.

Shaver, D. et al. 1988. Styrofoam box and beach temperatures in relation to incubation and sex ratios of Kemp's ridley sea turtles. Pp. 103–108. In: *Proceedings of the Eighth Annual Workshop on Sea Turtle Conservation and Biology* (B.A. Schroeder, ed.). NOAA Tech. Memo. NMFS-SEFC-214, Miami, FL.

Solomon, S. and Baird, T. 1976. Studies on the eggshell (oviducal and oviposited) of *Chelonia mydas* L. *J. Exp. Mar. Biol. Ecol.* 22:145–160.

Solomon, S. and Baird, T. 1979. Aspects of the biology of *Chelonia mydas* L. *Oceangr. Mar. Biol. Ann. Rev.* 17:347–361.

Solomon, S. and Watt, J. 1985. The structure of the eggshell of the leatherback turtle (*Dermochelys coriacea*). *Anim. Technol.* 36:19–28.

Whitmore, C., Dutton, P., and Mrosovsky, N. 1985. Sexing of hatchling sea turtles: gross appearance versus histology. *J. Herpetol.* 19:430–431.

Wibbels, T. et al. 1990. Seasonal changes in serum gonadal steroids associated with migration, mating and nesting in loggerhead sea turtle (*Caretta caretta*). *Gen. Comp. Endocrinol.* 79:154–64.

Wibbels, T. et al. 1991. Female-biased sex ratio of immature loggerhead sea turtles inhabiting the Atlantic coastal waters of Florida. *Can. J. Zool.* 69:2973–2977.

Wibbels, T. et al. 1998. Blood sampling techniques for hatchling Cheloniid sea turtles. *Herpetol. Rev.* 29:218–220.

Wibbels, T., Hillis-Starr, Z., and Phillips, B. 1999a. Female-biased sex ratio of hatchling hawksbill turtles from a Caribbean nesting beach. *J. Herpetol.* 33:142–144.

Wibbels, T., Wilson, C., and Crews, D. 1999b. Müllerian duct development and regression in a turtle with temperature-dependent sex determination. *J. Herpetol.* 33:149–152.

Wiedersheim, R. 1890a. Über die Entwicklung des Urogenitalalapparates bei Krokodilen und Schildkrõten. *Anat. Anz.* 5:337–344.

Wiedersheim, R. 1890b. Uber die Entwicklung des Urogenitalalapparates bei Krokodilen und Schildkrõten. *Arch. Mikrosk. Anat. EntwMech Org.* 36:410–468.

Wilson, G. 1896. The development of the *ostium abdominale tubae* in the crocodile. *Anat. Anz.* 12:79–85.

Wilson, G. 1900. The development of the Müllerian ducts of reptiles. *Trans. Roy. Soc. Edinburgh* 39:613–621.

Yntema, C.L. and Mrosovsky, N. 1980. Sexual differentiation in hatching loggerheads (*Caretta caretta*) incubated at different controlled temperatures. *Herpetologica* 36:33–36.

Yntema, C.L. and Mrosovsky, N. 1982. Critical periods and pivotal temperatures for sexual differentiation in loggerhead sea turtles. *Can. J. Zool.* 60:1012–1014.

8 Adult Migrations and Habitat Use

Pamela Plotkin

CONTENTS

8.1 INTRODUCTION

Adult sea turtles are among the largest living reptiles and the only reptiles that exhibit long-distance migrations that rival those of terrestrial and avian vertebrates. Many details of these large-scale movements are poorly understood because sea turtles swim over vast areas. Data accumulated from several decades of mark–recapture and telemetry studies demonstrate that adult sea turtle migrations are resource-driven, with migrants traveling hundreds to thousands of kilometers between established feeding and breeding areas at regular or seasonal intervals. For some species, however, resources are not always predictable in time and space. For example, food resources can vary spatially and temporally, and critical breeding habitats may be ephemeral. Thus, some sea turtles have evolved special migratory behaviors to compensate for environmental variability and unpredictability.

The mechanisms that adult sea turtles employ as they travel through seemingly featureless ocean have been an enigma since Archie Carr first described the amazing trans-Atlantic journey of female green turtles to nesting beaches on Ascension Island (Carr, 1965). Results from laboratory studies using hatchling sea turtles have been extrapolated to explain the environmental cues used by adults

0-8493-1123-3/03/$0.00+$1.50

during migration. Hatchlings can perceive and respond to several environmental cues, including magnetic field intensity (Lohmann and Lohmann, 1996), magnetic inclination angle (Lohmann and Lohmann, 1994), visual cues (Mrosovsky and Shettleworth, 1968), water temperature gradients (Owens, 1980a), wave direction (Lohmann et al., 1990; Wyneken et al., 1990), and chemicals in the water (Grassman et al., 1984). Despite this wealth of information, it is quite tenable that hatchlings rely on different cues than do adult turtles.

In the last decade, we have gained a much better understanding of the navigational abilities of adult sea turtles, but the mechanisms used to guide them during migration remain speculative (Papi and Luschi, 1996; Papi et al., 2000). Movements of sea turtles to specific sites (Papi et al., 1995; 1997) and their return to these areas even after displacement (Luschi et al., 1996; Papi et al., 1997) confirm that sea turtles do indeed navigate. Recent studies have suggested or demonstrated that these navigational feats may be guided by biological compasses (Papi and Luschi, 1996; Luschi et al., 1998), currents (Morreale et al., 1996; Papi et al., 2000), waterborne chemicals (Luschi et al., 1998; Papi et al., 2000), windborne information (Luschi et al., 2001), bathymetric features (Morreale et al., 1994), and water temperature (Plotkin, 1994).

Sea turtles evolved distinct migratory strategies during their evolutionary history as they adapted to different ocean habitats (Hendrickson, 1980). These adaptations are illustrated by the migratory and habitat use patterns that are beginning to emerge following several decades of research. Interspecific and intraspecific variation in migratory behavior exists among contemporary sea turtles, and it is probable that considerable variation exists in navigational mechanisms used among and within species as well.

The remainder of this chapter provides a summary of the current state of knowledge of adult sea turtle migration patterns and habitat use by each species and, where available, the potential navigational mechanisms employed. Most of our knowledge comes from studies conducted on postnesting females because these turtles are easy to capture, mark, and tag. Overall, very little is known about adult male sea turtles.

8.2 LEATHERBACK, *DERMOCHELYS CORIACEA*

The leatherback is widely distributed throughout the world's oceans from boreal to tropical waters. Leatherbacks inhabit the oceanic zone, are highly migratory (Pritchard, 1973; 1976; Morreale et al., 1996; Hughes et al., 1998), and are capable of transoceanic migrations (Eckert, 1998) and diving to great depths (Eckert et al., 1989). Much of the details of leatherback migrations remain elusive, in part because the turtles occur far from land and travel such great distances; however, recent and ongoing studies will soon provide more specific information regarding the migratory behavior of this ocean traveler (Eckert and Sarti, 1997; Eckert, 1998; Lutcavage et al., in press).

Little is known of the prereproductive migrations of leatherbacks and the location of breeding grounds; it is believed that they conform to the generalized model for sea turtle reproduction (Owens, 1980b). Females migrate to nearshore waters of tropical beaches several weeks prior to the nesting season. Most female leatherbacks undertake reproductive migration to nesting beaches every 2–3 years, where they oviposit on average five to six clutches at 9-day intervals (Boulon et al., 1996; Steyermark et al., 1996). Estimating fecundity for leatherbacks is challenging because females do not display strong beach fidelity. Females may travel among adjacent (Steyermark et al., 1996) or distant beaches (Keinath and Musick, 1993) within a nesting season.

After the nesting season, females migrate long distances across deep oceanic waters (Morreale et al., 1994; 1996; Eckert and Sarti, 1997; Eckert, 1998; Hughes et al., 1998; Lutcavage et al., in press) and in some instances across ocean basins (Eckert, 1998). In some regions, migratory corridors along deepwater bathymetric contours have been described, with multiple postnesting females from the same beach migrating through these areas in subsequent years (Morreale et al., 1994; 1996). However, in other regions no such corridors have been detected for postnesting female cohorts (Eckert, 1998). Leatherbacks do not migrate to resident feeding grounds, as has been well described for some species. Instead, leatherbacks appear to swim continuously (Eckert and Sarti, 1997; Eckert, 1998), possibly to areas of high food concentration (Grant et al., 1996; Eckert and Sarti, 1997), where they appear to feed on organisms associated with the deep scattering layer (Eckert et al., 1989).

Navigational cues used by leatherbacks during migration are not known, but potentially important cues suggested thus far include ocean currents, ocean fronts, bathymetric features, and magnetic cues (Morreale et al., 1994; 1996; Lutcavage, 1996).

8.3 OLIVE RIDLEY, *LEPIDOCHELYS OLIVACEA*

The olive ridley has a circumtropical distribution, occurring in the Atlantic, Pacific, and Indian Oceans (Pritchard, 1969). Knowledge of olive ridley migrations is fragmentary throughout most of its range, with the exception of the eastern Pacific and the northern Indian Ocean. The olive ridley is highly migratory and spends most of its nonbreeding life cycle in the oceanic zone (Cornelius and Robinson, 1986; Pitman, 1990; 1993; Arenas and Hall, 1992; Plotkin, 1994; Plotkin et al., 1994; 1995; Beavers, 1996; Beavers and Cassano, 1996).

Olive ridleys occupy the neritic zone during the breeding season. Reproductively active males and females migrate toward the coast and aggregate at nearshore breeding grounds located near beaches where mass nesting emergences (commonly known as arribadas) also occur (Pritchard, 1969; Hughes and Richard, 1974; Cornelius, 1986; Dash and Kar, 1990; Plotkin et al., 1991; 1996; Kalb et al., 1995; 1997; Pandav et al., 2000). A significant proportion of the breeding also takes place far from shore (Pitman, 1990; Kopitsky et al., 2000), and some males and females may not migrate to nearshore breeding aggregations. Some males appear to remain in oceanic waters, are nonaggregated, and mate opportunistically as they intercept females *en route* to nearshore breeding grounds and nesting beaches (Plotkin, 1994; Plotkin et al., 1994; 1996; Kopitsky et al., 2000).

After mating, females remain nearshore for several weeks to several months. Solitary nesters emerge onto beaches to lay eggs individually throughout much of the species' range. Solitary nesters have weak site fidelity (Kalb, 1999), lay two clutches annually at 14-day intervals (Pritchard, 1969; Kalb, 1999), and may use multiple, geographically distant beaches within a nesting season (Kalb, 1999). Arribada nesting females emerge onto beaches to lay eggs *en masse* at a few select beaches in the Atlantic, Pacific, and Indian Oceans. The arribada nesters have strong site fidelity (Plotkin et al., 1995; Kalb, 1999), lay two clutches approximately every 28 days (Pritchard, 1969; Kalb, 1999), and may delay nesting for 6–8 weeks when environmental conditions are unfavorable (Plotkin et al., 1997). Once mating and nesting is completed, olive ridleys quickly migrate back to oceanic waters.

The postreproductive migrations of olive ridleys are unique and complex. Their migratory pathways vary annually (Plotkin, 1994), there is no spatial and temporal overlap in migratory pathways among groups or cohorts of turtles (Plotkin et al., 1994; 1995), and no apparent migration corridors exist. Unlike other marine turtles that migrate from a breeding ground to a single feeding area, where they reside until the next breeding season, olive ridleys are nomadic migrants that swim hundreds to thousands of kilometers over vast oceanographic stretches (Plotkin, 1994; Plotkin et al., 1994; 1995).

Despite the multitude of cues that may be used in long-distance navigation, operation of a specific cue has not been demonstrated. However, Plotkin (1994) suggested that water temperature might be the predominant cue used during post-reproductive migrations to oceanic feeding areas in the eastern Pacific because of the spatial and temporal correspondence between turtle movements and the locations of divergence and convergence zones, thermal fronts, and cool water masses.

8.4 KEMP'S RIDLEY, *LEPIDOCHELYS KEMPII*

The Kemp's ridley has a relatively restricted range, occurring in the neritic zone of the Gulf of Mexico and western Atlantic (Marquez, 1994). Evidence accumulated from several decades of tag returns and telemetry studies has demonstrated that Kemp's ridley is a neritic migrant that swims along the U.S. and Mexican coasts, nearshore in continental shelf waters (Byles, 1989; Byles and Plotkin, 1994; Marquez, 1994; Renaud, 1995; Shaver, 1999; 2001). Narrow migratory corridors extend along the entire U.S. and Mexican gulf coast (Byles and Plotkin, 1994).

Reproductively mature females undertake annual migrations from the western Atlantic and Gulf of Mexico (Renaud et al., 1996) to their principal nesting beach, Rancho Nuevo, located near the central Mexican gulf coast in the state of Tamaulipas. Females aggregate nearshore Rancho Nuevo in advance of the nesting season, and mating takes place approximately 30 days prior to first oviposition for the season (Chavez et al., 1967; Pritchard, 1969; Mendonca and Pritchard, 1986; Rostal, 1991). Mating also occurs elsewhere in coastal and inshore waters from south Texas to areas south of Rancho Nuevo in Tamaulipas and Veracruz, Mexico (Shaver, 1992).

Nesting begins in late April and may last until mid-August (Marquez, 1994). The vast majority of females emerge *en masse* to nest at Rancho Nuevo during the May, June, and July arribadas. The arribadas at Rancho Nuevo typically occur every 28 days (Pritchard and Marquez, 1973). Females lay approximately three nests per season (Rostal et al., 1997) and remain relatively close to the nesting beach during the internesting period between clutches (Mendonca and Pritchard, 1986).

Solitary nesting also occurs at Rancho Nuevo (Rostal et al., 1997) and other beaches. A small number of females regularly nest on the Texas coast (Shaver, 1998), and very rarely on western Atlantic beaches such as Florida (Fletemeyer, 1990; Meylan et al., 1990a; Libert, 1998; Johnson et al., 2000), North Carolina (T. Conant, unpublished data cited in Bowen et al., 1994), and South Carolina. Nesting has been documented as far south as Colombia (Chavez and Kaufmann, 1974); however, Kemp's ridleys rarely nest south of the tip of the Yucatan Peninsula (Marquez, 1994).

After the last clutch is oviposited, females begin postnesting migrations away from their nesting beach, traveling north or south along the coast (Mysing and Vanselous, 1989; Byles, 1989; Shaver, 1999; 2001). Postnesting migrations have been recorded as far south as Colombia (Marquez, 1994) and as far north as Virginia; however, most Kemp's ridleys migrate to areas concentrated between north Texas coastal waters and Campeche, Mexico (Chavez, 1969; Pritchard and Marquez, 1973; Byles, 1989; Shaver, 1999; 2001). These long-distance migrations encompass hundreds of kilometers (Byles, 1989) and occur primarily in shallow waters less than 50 m deep (Byles, 1989; Renaud, 1995; Renaud et al., 1996; Shaver, 1999; 2001). Females may establish relatively circumscribed ranges in coastal waters for several months (Byles, 1989; Byles and Plotkin, 1994), suggesting that resident feeding areas exist.

In contrast to the females, adult males appear to be nonmigratory. Shaver et al. (in press) tracked 11 adult male Kemp's ridleys, and most of them remained resident in coastal waters near Rancho Nuevo for several months after the nesting season. Only one male migrated away from the breeding grounds; he migrated to the north Texas coast near Galveston. This is quite different from the generalized pattern that has been described for male sea turtles (Rostal, 1991). Most males depart the breeding grounds by the time the greatest number of females emerge to lay eggs (i.e., when most females have already copulated) (Hendrickson, 1958; Booth and Peters, 1972; Ehrhart, 1982; Frazier, 1985) and migrate to distant feeding grounds (Plotkin et al., 1995; 1997; Hays et al., 2001b).

8.5 HAWKSBILL, *ERETMOCHELYS IMBRICATA*

Hawksbills are distributed in tropical waters throughout much of the Atlantic, Pacific, and Indian Oceans (Witzell, 1983). Hawksbills live in close association with hard-substrate communities such as coral reefs, where they forage primarily on sponges (Meylan, 1988), and may also occur in coastal lagoons and bays. Hawksbills were once believed to be nonmigratory residents of reefs adjacent to their respective nesting beaches (Hendrickson, 1980; Witzell, 1983; Frazier, 1985), but postreproductive tagging, telemetry, and genetic studies have revealed that hawksbills do indeed migrate and that many are highly migratory, traveling hundreds to thousands of kilometers between nesting beaches and foraging areas (Meylan, 1982; Parmenter, 1983; Broderick et al., 1994; Byles and Swimmer, 1994; Groshens and Vaughan, 1994; Miller et al., 1998; Meylan, 1999; Prieto et al., 2001). Data from one adult male hawksbill marked and later recaptured indicate that males are also highly migratory (Nietschmann, 1981).

Very little is known about hawksbill reproductive migrations from foraging areas to breeding grounds. Females migrate to nest at their natal beaches (Bass, 1999) every 2–3 years (Witzell, 1983). Females inhabiting the same foraging area do not all migrate to the same nesting beach (Miller et al., 1998). Mating is not well documented for this species, but it has been observed in the shallow waters adjacent to nesting beaches (Witzell, 1983) and probably occurs approximately 30 days prior to first nesting (Owens, 1980b). Hawksbills are solitary nesters that lay four to seven clutches every 14–16 days (Witzell, 1983). Females remain nearshore the nesting beach during the internesting period (Starbird, 1993; Starbird et al., 1999). As soon as the last nest is oviposited, females begin postnesting migrations back to foraging areas (Starbird, 1993; Starbird et al., 1999; Mortimer and Balazs, 2000; Horrocks et al., 2001).

Postnesting hawksbills migrate to specific foraging areas within short range (25–200 km) (Ellis et al., 2000; Hillis-Starr et al., 2000; Mortimer and Balazs, 2000; Horrocks et al., 2001; Lageux et al., in press) and long range (200 km or more) (Byles and Swimmer, 1994; Miller et al., 1998; Horrocks et al., 2001; Prieto et al., 2001; Lageux et al., in press) of their nesting beaches. Such variation in migratory behavior is found among females nesting at the same beaches (Miller et al., 1998; Horrocks et al., 2001; Prieto et al., 2001). No apparent patterns have emerged to explain why some females migrate short distances while others bypass reefs close to their nesting beaches and migrate greater distances.

Both short-distance and long-distance migrations appear to be relatively quick, directed movements that may occur across deep oceanic waters or channels (Ellis et al., 2000; Horrocks et al., 2001) or shallow coastal waters (Ellis et al., 2000). Once a female reaches her foraging ground, she remains resident there (Ellis et al., 2000; Mortimer and Balazs, 2000), presumably until her next reproductive migration.

8.6 FLATBACK, *NATATOR DEPRESSUS*

The flatback has the most restricted migratory range of all sea turtles. It is endemic to the tropical waters of the Australian continental shelf (Limpus et al., 1981), occurring in shallow, turbid waters and bays (Limpus et al., 1983; 1989). Flatbacks were once characterized as nonmigratory (Hendrickson, 1980), but tagging studies have confirmed that they do undertake long-distance migrations between foraging and breeding areas (Limpus et al., 1981; 1983). Flatbacks nest on mainland beaches, continental island beaches, and sand cays within Australian territorial waters on the northeast coast (Limpus, 1971; Limpus et al., 1981; 1989), north coast (Limpus et al., 1983; Guinea et al., 1991; Guinea, 1994), and west coast (Prince, 1994). Foraging areas extend just beyond the Australian territorial waters into adjacent waters of the Indonesian archipelago and Papuan coast (Parmenter, 1994).

Females migrate from foraging areas to nesting beaches on average every 1–3 years (Limpus et al., 1984; Parmenter, 1994). Females show strong site fidelity to their nesting beaches (Limpus et al., 1984; Parmenter, 1994). Mating occurs in the vicinity of the nesting beach approximately 1 month prior to the start of the nesting season (Limpus et al., 1989; 1993). Nesting occurs year-round at some beaches (Limpus et al., 1983; 1989) and seasonally at others (Limpus, 1971; Limpus et al.,

1984; 1989; Guinea, 1994). Flatbacks are solitary nesters that lay an average of three clutches per season at approximately 16-day intervals (Limpus et al., 1984). Females presumably remain nearshore during the internesting period and return to foraging grounds after the last clutch has been oviposited; no published data exist to support this assumption.

Postnesting flatbacks migrate hundreds to thousands of kilometers to their foraging grounds, located primarily in turbid, shallow, inshore waters of northern Queensland and along the north Australian coast (Limpus et al., 1983; Parmenter, 1994) and possibly northward to the Irian Jaya coast (Limpus et al., 1993).

8.7 LOGGERHEAD, *CARETTA CARETTA*

Loggerheads occur in subtropical and temperate waters across continental shelves and estuarine areas in the Atlantic, Pacific, and Indian Oceans (Dodd, 1988). Throughout this range, loggerheads spend most of their time in nearshore and inshore waters, sometimes associated with reefs and other natural and artificial hard substrates (Dodd, 1988). Loggerheads are highly migratory, capable of traveling hundreds to thousands of kilometers between foraging and breeding areas (Caldwell et al., 1959; Bell and Richardson, 1978; Timko and Kolz, 1982; Meylan et al., 1983; Limpus et al., 1992; Papi et al., 1997; Plotkin and Spotila, 2002). Female loggerheads do not appear to migrate to just one foraging area. Rather, they move continuously and thus appear to forage at a series of coastal areas (Timko and Kolz, 1982; Papi et al., 1997; Plotkin and Spotila, 2002).

Females migrate to nest at their natal beaches (Schierwater and Schroth, 1996) about every 3 years (Limpus, 1985; Dodd, 1988). Both females and males migrate asynchronously from foraging areas to breeding areas several weeks to months prior to the nesting season (Limpus, 1985). Males arrive a few weeks in advance of the females (Henwood, 1987). Some males appear to be nonmigratory and may reside in breeding areas throughout the year (Henwood, 1987). Mating occurs during or immediately after migration to breeding areas located nearshore nesting beaches (Caldwell et al., 1959; Limpus, 1985; Wibbels et al., 1987).

Females lay an average of four clutches approximately every 2 weeks (Dodd, 1988). During the internesting period, females remain nearshore (Hopkins and Murphy, 1981; Stoneburner, 1982; Sakamoto et al., 1990; Hays et al., 1991; Tucker et al., 1996). Females begin postnesting migrations as soon as their last clutch is oviposited (Stoneburner, 1982; Tucker et al., 1996; Plotkin and Spotila, 2002). Females typically migrate nearshore, moving north or south of their nesting beach (Papi et al., 1997; Plotkin and Spotila, 2002), but may also make brief offshore movements after the nesting season into deep oceanic waters (Byles and Dodd, 1989).

8.8 GREEN, *CHELONIA MYDAS*

Green turtles occur in tropical and subtropical waters of the Atlantic, Pacific, and Indian Oceans. They inhabit the neritic zone, occurring in nearshore and inshore waters where they forage primarily on sea grasses and algae (Mortimer, 1982), and temporarily inhabit the oceanic zone during migrations from foraging areas to breeding areas and back. Some of these long-distance reproductive migrations are spectacular feats, with turtles swimming thousands of kilometers across the open ocean directly to beaches located on small, isolated oceanic islands (Carr, 1965; Luschi et al., 1998).

Female green turtles migrate from foraging areas to their natal beaches (Meylan et al., 1990b) every 2–4 years and show a high degree of nest site fidelity (Miller, 1997). Mating may occur *en route* to the nesting beach (Meylan et al., 1992), far from the nesting beach at distant mating grounds (Limpus, 1993), or nearshore the nesting beach (Carr and Ogren, 1960; Booth and Peters, 1972; Broderick and Godley, 1997; Godley et al., 2002). Females oviposit an average of three clutches at 10- to 17-day intervals (Miller, 1997) and remain near the nesting beach during the internesting period (Carr et al., 1974; Dizon and Balazs, 1982).

Postnesting females migrate hundreds to thousands of kilometers from their nesting beach to resident coastal foraging areas (Balazs, 1994; Balazs et al., 1994; 2000; Papi et al., 1995; Schroeder et al., 1996; Cheng and Balazs, 1998; Luschi et al., 1998; Papi et al., 2000; Luschi et al., 2001). Postbreeding males also migrate long distances from breeding areas to foraging grounds at the end of the mating season (Hays et al., 2001a) or may remain in the vicinity of the nesting beach (Garduno et al., 2000). In general, these migrations can be characterized as relatively fast, directed movements toward specific locations (Schroeder et al., 1996; Luschi et al., 1998), which may occur nearshore (Schroeder et al., 1996) or in deep oceanic water (Balazs, 1994; Luschi et al., 1998), with cohorts traveling along similar pathways during part of the migration (Luschi et al., 2001).

The navigational mechanisms used by green turtles migrating from Ascension Island to coastal foraging grounds in Brazil have provided insights into the navigational abilities of adult sea turtles. These studies have demonstrated that green turtles are able to maintain straight courses over long distances in the open ocean (Luschi et al., 1998), can correct their course during the migration according to environmental information (Luschi et al., 1998), may be guided in part by currents (Luschi et al., 1998) or windborne information (Luschi et al., 2001), do not rely on sea surface temperatures (Hays et al., 2001b), and can navigate in the absence of magnetic cues (Papi et al., 2000).

8.9 EAST PACIFIC GREEN, *CHELONIA AGASSIZI*

East Pacific green turtles are restricted to the coastal waters, lagoons, and bays along the west coast of America from Baja California and the Gulf of California to southern Peru and the Galapagos Islands (Alvarado and Figueroa, 1998), where they feed primarily on sea grasses and algae (Seminoff et al., 2000).

Females migrate from foraging areas to nesting beaches on average every 3–4 years (Alvarado et al., 2000). The primary nesting beaches are located in Michoacan, Mexico; however, sporadic nesting also occurs elsewhere along the Mexican and Central American coast. Mating takes place nearshore the nesting beach (Alvarado and Figueroa, 1989). Females oviposit between one and seven clutches per season at 11- to 13-day intervals (Alvarado et al., 2000). Females remain nearshore the nesting beach during the internesting period and return to foraging grounds after the last clutch has been oviposited (Byles et al., 1995). Postnesting females migrate north and south of their nesting beaches to foraging areas in the Gulf of California and to coastal areas in Central and South America (Byles et al., 1995).

8.10 CONCLUSIONS

Distinct migratory patterns exist among extant sea turtles, and these patterns are best understood in the context of the locations and sizes of their foraging areas. The first pattern is exhibited by leatherbacks and east Pacific olive ridleys. These turtles migrate to oceanic waters where they forage over very broad areas, seeking out highly productive waters such as fronts and convergence zones. These foraging areas vary spatially and temporally, and are frequently unpredictable. The second pattern is exhibited by Kemp's ridleys, loggerheads, and flatbacks. These species migrate to highly productive neritic foraging areas located on continental shelves. Many forage over broad areas, typically swimming along a coastline; however, some establish small, circumscribed foraging areas. Their foraging areas are fairly predictable in space and time, however small-scale variations are possible. The third pattern is exhibited by green and hawksbill turtles. These species migrate to well-established, fixed foraging areas located nearshore. Their foraging range is relatively small and virtually no spatial or temporal variation exists.

In the past decade we've made great advances in describing sea turtle migratory patterns and pathways but there still remains much to learn. Describing and understanding the migratory behavior of and navigational mechanisms used by sea turtles remains to be one of the most exciting challenges ahead.

REFERENCES

Alvarado Diaz, J., Delgado Trejo, C., and Figueroa Lopez, A., Reproductive biology of the black turtle in Michoacan, Mexico, in *Proceedings of the Eighteenth International Sea Turtle Symposium*, Abreu-Grobois, F.A. et al., Compilers, NOAA Tech. Memo., NMFS-SEFSC-436, 2000, 159, Miami, FL.

Alvarado, J. and Figueroa, A., Breeding dynamics of the black turtle (*Chelonia agassizi*) in Michoacan, Mexico, in *Proceedings of the Ninth Annual Workshop on Sea Turtle Conservation and Biology*, Eckert, S.A., Eckert, K.L., and Richardson, T.H., Compilers, NOAA Tech. Memo., NMFS-SEFC-232, 1989, 5, Miami, FL.

Alvarado, J. and Figueroa, A., East Pacific green sea turtle, *Chelonia mydas*, in *National Marine Fisheries Service and U.S. Fish and Wildlife Service Status Reviews for Sea Turtles Listed under the Endangered Species Act of 1973*, Plotkin, P.T., Editor, National Marine Fisheries Service, Silver Spring, MD, 1998, 24.

Arenas, P. and Hall, M., The association of sea turtles and other pelagic fauna with floating objects in the eastern tropical Pacific Ocean, in *Proceedings of the Eleventh Annual Workshop on Sea Turtle Biology and Conservation*, Salmon, M. and Wyneken, J., Compilers, NOAA Tech. Memo., NMFS-SEFC-302, 1992, 7, Miami, FL.

Balazs, G.H., Homeward bound: satellite tracking of Hawaiian green turtles from nesting beaches to foraging pastures, in *Proceedings of the Thirteenth Annual Symposium on Sea Turtle Biology and Conservation*, Schroeder, B.A. and Witherington, B.E., Compilers, NOAA Tech. Memo., NMFS-SEFSC-341, 1994, 205, Miami, FL.

Balazs, G.H. et al., Satellite telemetry of green turtles nesting at French Frigate Shoals, Hawaii, and Rose Atoll, American Samoa, in *Proceedings of the Fourteenth Annual Symposium on Sea Turtle Biology and Conservation*, Bjorndal, K.A. et al., Compilers, NOAA Tech. Memo., NMFS-SEFSC-351, 1994, 184, Miami, FL.

Balazs, G.H., Katahira, L.K., and Ellis, D.M., Satellite tracking of hawksbill turtles nesting in the Hawaiian Islands, in *Proceedings of the Eighteenth International Sea Turtle Symposium*, Abreu-Grobois, F.A. et al., Compilers, NOAA Tech. Memo., NMFS-SEFSC-436, 2000, 279, Miami, FL.

Bass, A.L., Genetic analysis to elucidate the natural history and behavior of hawksbill turtles (*Eretmochelys imbricata*) in the wider Caribbean: a review and re-analysis, *Chelonian Conserv. Biol.*, 3, 195, 1999.

Beavers, S.C., The ecology of the olive ridley turtle (*Lepidochelys olivacea*) in the eastern tropical Pacific Ocean during the breeding/nesting season, M. Sc. thesis, Oregon State University, Corvallis, OR, 1996.

Beavers, S.C. and Cassano, E.R., Movements and dive behavior of a male sea turtle (*Lepidochelys olivacea*) in the eastern tropical Pacific, *J. Herpetol.*, 30, 97, 1996.

Bell, R. and Richardson, J.I., An analysis of tag recoveries from loggerhead sea turtles (*Caretta caretta*) nesting on Little Cumberland Island, Georgia, *Fla. Marine Res. Pub.*, 33, 20, 1978.

Booth, J. and Peters, J., Behavioural studies on the green turtle (*Chelonia mydas*) in the sea, *Anim. Behav.*, 20, 808, 1972.

Boulon, R.H., Jr., Dutton, P.H., and McDonald, D.L., Leatherback turtles (*Dermochelys coriacea*) on St. Croix, U.S. Virgin Islands: fifteen years of conservation, *Chelonian Conserv. Biol.*, 2, 141, 1996.

Bowen, B.W., Conant, T.A., and Hopkins-Murphy, S.R., Where are they now? The Kemp's Ridley Headstart Project, *Conserv. Biol.*, 8, 853, 1994.

Broderick, A.C. and Godley, B.J., Observations of reproductive behaviour of male green turtles (*Chelonia mydas*) at a nesting beach in Cyprus, *Chelonian Conserv. Biol.*, 2, 615, 1997.

Broderick, D. et al., Genetic studies of the hawksbill turtle *Eretmochelys imbricata*: evidence for multiple stocks in Australian waters, *Pacific Conserv. Biol.*, 1, 123, 1994.

Byles, R.A., Satellite telemetry of Kemp's ridley sea turtle, *Lepidochelys kempii*, in the Gulf of Mexico, in *Proceedings of the Ninth Annual Workshop on Sea Turtle Conservation and Biology*, Eckert, S.A., Eckert, K.L., and Richardson, T.H., Compilers, NOAA Tech. Memo., NMFS-SEFC-232, 1989, 25, Miami, FL.

Byles, R., Alvarado, J., and Rostal, D., Preliminary analysis of post-nesting movements of the black turtle (*Chelonia agassizi*) from Michoacan, Mexico, in *Proceedings of the Twelfth Annual Workshop on Sea Turtle Biology and Conservation*, Richardson, J.I. and Richardson, T.H., Compilers, NOAA Tech. Memo., NMFS-SEFSC-361, 1995, 12.

Byles, R.A. and Dodd, C.K., Satellite biotelemetry of a loggerhead sea turtle (*Caretta caretta*) from the east coast of Florida, in *Proceedings of the Ninth Annual Workshop on Sea Turtle Biology and Conservation*, Eckert, S.A., Eckert, K.L., and Richardson, T.H., Editors, NOAA Tech. Memo., NMFS SEFSC-232, 1989, 215, Miami, FL.

Byles, R.A. and Plotkin, P.T., Comparison of the migratory behavior of the congeneric sea turtles *Lepidochelys olivacea* and *L. kempii*, in *Proceedings of the Thirteenth Annual Symposium on Sea Turtle Biology and Conservation*, Schroeder, B.A. and Witherington, B.E., Compilers, NOAA Tech. Memo., NMFS-SEFSC-341, 1994, Miami, FL.

Byles, R.A. and Swimmer, J.Y.B., Post-nesting migration of *Eretmochelys imbricata* in the Yucatan Peninsula, in *Proceedings of the Fourteenth Annual Symposium on Sea Turtle Biology and Conservation*, Bjorndal, K.A. et al., Compilers, NOAA Tech. Memo., NMFS-SEFSC-351, 1994, 202, Miami, FL.

Caldwell, D.K., Carr, A., and Ogren, L.H., The Atlantic loggerhead sea turtle, *Caretta caretta* (*L.*), in America, I. Nesting and migration of the Atlantic loggerhead turtle, *Bull. Fla. State Mus.*, 4, 295, 1959.

Carr, A.F., The navigation of the green turtle, *Sci. Am.*, 212, 78, 1965.

Carr, A. and Ogren, L., The ecology and migrations of sea turtles, 4. The green turtle in the Caribbean Sea, *Bull. Am. Mus. Nat. Hist.*, 121(1), 1, 1960.

Carr, A., Ross, P., and Carr, S., Internesting behaviour of the green turtle *Chelonia mydas* at a mid ocean island breeding round, *Copeia*, 3, 703, 1974.

Chavez, H., Tagging and recapture of the lora turtle (*Lepidochelys kempii*), *Int. Turtle Tort. Soc. J.*, 3, 14019, 32, 1969.

Chavez, H., Contreras, G.M., and Hernandez, D.T.P.E., Aspectos biologicos y protection de la tortuga lora, *Lepidochelys kempii* (*Garman*) en la costa de Tamaulipas, Mexico, *Inst. Nac. Invest. Biol. Pesqueras,* 1967, 17, Mexico.

Chavez, H. and Kaufmann, R., Informacion sobre la tortuga marina *Lepidochelys kempii* (Garman), con referencia a un ejemplar marcado en Mexico y observado en Colombia, *Bull. Mar. Sci.*, 24, 372, 1974.

Cheng, I.J. and Balazs, G.H., The post-nesting long range migration of the green turtles that nest at Wan-An Island, Penghu Archipelago, Taiwan, in *Proceedings of the Seventeenth Annual Sea Turtle Symposium*, Epperly, S.P. and Braun, J., Compilers, NOAA Tech. Memo., NMFS-SEFSC-415, 1998, 29, Miami, FL.

Cornelius, S.E., *The Sea Turtles of Santa Rosa National Park*, Fundacion de Parques Nacionales, Costa Rica, 1986.

Cornelius, S.E. and Robinson, D.C., Post-nesting movements of female olive ridley turtles tagged in Costa Rica, *Vida Silvestre Neotropical*, 1, 12, 1986.

Dash, M.C. and Kar, C.S., *The Turtle Paradise: Gahirmatha (An Ecological Analysis and Conservation Strategy)*, Interprint, New Delhi, 1990.

Dizon, A.E. and Balazs, G.H., Radio telemetry of Hawaiian green turtles at their breeding colony, *Mar. Fish. Rev.*, 44, 13, 1982.

Dodd, C.K., Jr., Synopsis of the biological data on the loggerhead sea turtle *Caretta caretta* (Linnaeus 1758), U.S. Fish Wildl. Serv., Biol. Report 88–14, 1988, Washington, D.C.

Eckert, S.A., Perspectives on the use of satellite telemetry and other electronic technologies for the study of marine turtles, with reference to the first year long tracking of leatherback sea turtles, in *Proceedings of the Seventeenth Annual Sea Turtle Symposium*, Epperly, S.P. and Braun, J., Compilers, NOAA Tech. Memo., NMFS-SEFSC-415, 1998, 44, Miami, FL.

Eckert, S.A. et al., Diving and foraging behavior of leatherback sea turtles (*Dermochelys coriacea*), *Can. J. Zool.*, 67, 2834, 1989.

Eckert, S.A. and Sarti, M.L., Distant fisheries implicated in the loss of the world's largest leatherback nesting population, *Mar. Turtle Newsl.*, 78, 2, 1997.

Ehrhart, L.M., A review of sea turtle reproduction, in *Biology and Conservation of Sea Turtles*, Bjorndal, K.A., Editor, Smithsonian Institution Press, Washington, DC, 1982, 29.

Ellis, D.M. et al., Short-range reproductive migrations of hawksbill turtles in the Hawaiian Islands as determined by satellite telemetry, in *Proceedings of the Eighteenth International Sea Turtle Symposium*, Abreu-Grobois, F.A. et al., Compilers, NOAA Tech. Memo., NMFS-SEFSC-436, 2000, 252, Miami, FL.

Fletemeyer, J., Kemp's ridley sea turtle nests in Palm Beach, *Fla. Nat.*, 63, 5, 1990.

Frazier, J., *Marine turtles in the Comoro Archipelago*, Royal Netherlands Academy of Arts and Sciences, Amsterdam, The Netherlands, 1985.

Garduno, M. et al., Satellite tracking of an adult male and female green turtle from Yucatan in the Gulf of Mexico, in *Proceedings of the Nineteenth Annual Symposium on Sea Turtle Biology and Conservation*, Kalb, H.J. and Wibbels, T., Compilers, NOAA Tech. Memo., NMFS-SEFSC-443, 2000, 158, Miami, FL.

Godley, B.J. et al., Reproductive seasonality and sexual dimorphism in green turtles, *Mar. Ecol. Prog. Ser.*, 226, 125, 2002.

Grant, G.S., Malpass, H., and Beasley, J., Correlation of leatherback turtle and jellyfish occurrence, *Herpetol. Rev.*, 27, 123, 1996.

Grassman, M.A. et al., Olfactory-based orientation in artificially imprinted sea turtles, *Science*, 224, 83, 1984.

Groshens, E.B. and Vaughan, M.R., Post-nesting movements of hawksbill sea turtles from Buck Island Reef National Monument, St. Croix, USVI, in *Proceedings of the Thirteenth Annual Symposium on Sea Turtle Biology and Conservation*, Schroeder, B.A. and Witherington, B.E., Compilers, NOAA Tech. Memo., NMFS-SEFSC-341, 1994, 69, Miami, FL.

Guinea, M.L., Nesting seasonality of the flatback sea turtle *Natator depressus* (*Garman*) at Fog Bay, Northern Territory, in *Proceedings of the Australian Marine Turtle Conservation Workshop*, James, R., Compiler, Queensland Dept. of Environment and Heritage and Australian Nature Conservation Agency, 1994, 150, Queensland, Australia.

Guinea, M.L. et al., Nesting seasonality of the flatback turtle at Bare Sand Island, Northern Territory Australia, *Mar. Turtle Newsl.*, 52, 4, 1991.

Hays, G.C. et al., Satellite tracking of a loggerhead turtle (*Caretta caretta*) in the Mediterranean, *J. Mar. Biol. Assoc. UK*, 71, 743, 1991.

Hays, G.C. et al., The movements and submergence behaviour of male green turtles at Ascension Island, *Mar. Biol.*, 139, 395, 2001a.

Hays, G.C. et al., Movements of migrating green turtles in relation to AVHRR derived sea surface temperature, *Int. J. Remote Sensing*, 22, 1403, 2001b.

Hendrickson, J.R., The green sea turtles *Chelonia mydas* (*Linn.*) in Malaya and Sarawak, *Proc. Zool. Soc. Lond.*, 130, 455, 1958.

Hendrickson, J.R., The ecological strategies of sea turtles, *Am. Zool.*, 20, 597, 1980.

Henwood, T.A., Movements and seasonal changes in loggerhead turtle *Caretta caretta* aggregations in the vicinity of Cape Canaveral, FL, 1978–1984, *Biol. Conserv.*, 40, 191, 1987.

Hillis-Starr, Z., Coyne, M., and Monaco, M., Buck Island and back — hawksbill turtles make their move, in *Proceedings of the Nineteenth Annual Symposium on Sea Turtle Biology and Conservation*, Kalb, H.J. and Wibbels, T., Compilers, NOAA Tech. Memo., NMFS-SEFSC-443, 2000, 159, Miami, FL.

Hopkins, S.R. and Murphy, T.M., Reproductive ecology of *Caretta caretta* in South Carolina, study completion report, South Carolina Wildl. Mar. Res. Dept., E-1, Study No. V1-A-1, 1981, Charleston, S.C.

Horrocks, J.A. et al., Migration routes and destination characteristics of post-nesting hawksbill turtles satellite-tracked from Barbados, West Indies, *Chelonian Conserv. Biol.*, 4, 107, 2001.

Hughes, D.A. and Richard, J.D., The nesting of the Pacific ridley turtle (*Lepidochelys olivacea*) on Playa Nancite, Costa Rica, *Mar. Biol.*, 24, 97, 1974.

Hughes, G.R. et al., The 7000-km oceanic journey of a leatherback turtle tracked by satellite, *J. Exp. Mar. Biol. Ecol.*, 229, 209, 1998.

Johnson, S.A. et al., Kemp's ridley (*Lepidochelys kempii*) nesting in Florida, USA, in *Proceedings of the Nineteenth Annual Symposium on Sea Turtle Biology and Conservation*, Kalb, H.J. and Wibbels, T., Compilers, NOAA Tech. Memo., NMFS-SEFSC-443, 2000, 283, Miami, FL.

Kalb, H.J., Behavior and physiology of solitary and arribada nesting olive ridley sea turtles (*Lepidochelys olivacea*) during the internesting period, Ph.D. thesis, Texas A&M University, College Station, TX, 1999.

Kalb, H., Valverde, R.A., and Owens, D., What is the reproductive patch of the olive ridley sea turtle?, in *Proceedings of the Twelfth Annual Workshop on Sea Turtle Biology and Conservation*, Richardson, J.I. and Richardson, T.H., Compilers, NOAA Tech. Memo., NMFS-SEFSC-361, 1995, 57, Miami, FL.

Keinath, J.A. and Musick, J.A., Movements and diving behavior of a leatherback turtle, *Dermochelys coriacea*, *Copeia*, 4, 1010, 1993.

Kopitsky, K., Pitman, R.L., and Plotkin, P., Investigations on at-sea mating and reproductive status of olive ridleys, *Lepidochelys olivacea*, captured in the eastern tropical Pacific, in *Proceedings of the Nineteenth Annual Symposium on Sea Turtle Biology and Conservation*, Kalb, H.J. and Wibbels, T., Compilers, NOAA Tech. Memo., NMFS-SEFSC-443, 2000, 160, Miami, FL.

Lageux, C.J. et al., Migration routes and dive patterns of post-nesting hawksbills from the Pearly Cays, Nicaragua, in *Proceedings of the Twenty-Second Annual Symposium on Sea Turtle Biology and Conservation*, in press.

Libert, B., Kemp's ridley nesting in Volusia County, in *Proceedings of the Seventeenth Annual Sea Turtle Symposium*, Epperly, S.P. and Braun, J., Compilers, NOAA Tech Memo., NMFS-SEFSC-415, 1998, 219, Miami, FL.

Limpus, C.J., The flatback turtle *Chelonia depressa* (*Garman*) in southeast Queensland, Australia, *Herpetologica*, 27(4), 431, 1971.

Limpus, C.J., A study of the loggerhead sea turtle, *Caretta caretta* in eastern Australia, Ph.D. thesis, University of Queensland, Brisbane, Australia, 1985.

Limpus, C.J., The green turtle, *Chelonia mydas*, in Queensland: breeding males in the Southern Great Barrier Reef, *Wildlife Res.*, 20, 4, 513, 1993.

Limpus, C.J., Couper, P.J., and Couper, K.L.D., Crab Island revisited: reassessment of the world's largest flatback turtle rookery after twelve years, *Mem. Queensland Mus.*, 33, 277, 1993.

Limpus, C.J. et al., The flatback turtle *Chelonia depressa* in Queensland: the Peak Island rookery, *Herpetofauna*, 13, 14, 1981.

Limpus, C.J. et al., The flatback turtle, *Chelonia depressa*, in Queensland: Post-nesting migration and feeding ground distribution, *Aust. Wildl. Res.*, 10, 557, 1983.

Limpus, C.J. et al., Sea-turtle rookeries in north-western Torres Strait, *Aust. Wildl. Res.*, 16, 517, 1989.

Limpus, C.J. et al., Migration of green (*Chelonia mydas*) and loggerhead (*Caretta caretta*) turtles to and from eastern Australian rookeries, *Wildl. Res.*, 19, 347, 1992.

Limpus, C.J., Fleay, A., and Baker, V., The flatback turtle *Chelonia depressa* in Queensland Australia: reproductive periodicity, philopatry and recruitment, *Aust. Wildl. Res.*, 11, 579, 1984.

Lohmann, K.J. and Lohmann, C.M.F., Detection of magnetic inclination angle by sea turtles: a possible mechanism for determining latitude, *J. Exp. Biol.*, 194, 23, 1994.

Lohmann, K.J. and Lohmann, C.M.F., Detection of magnetic field intensity by sea turtles, *Nature*, 380, 59, 1996.

Lohmann, K.J., Salmon, M., and Wyneken, J., Functional autonomy of land and sea orientation systems in sea turtle hatchlings, *Biol. Bull.*, 179, 214, 1990.

Luschi, P. et al., Long-distance migration and homing after displacement in the green turtle (*Chelonia mydas*): a satellite tracking study, *J. Comp. Physiol. A*, 178, 447, 1996.

Luschi, P. et al., The navigational feats of green sea turtles migrating from Ascension Island investigated by satellite telemetry, *Proc. R. Soc. Lond. B*, 265, 2279, 1998.

Luschi, P. et al., Testing the navigational abilities of ocean migrants: displacement experiments on green sea turtles (*Chelonia mydas*), *Behav. Ecol. Sociobiol.*, 50, 528, 2001.

Lutcavage, M.E., Planning your next meal: leatherback travel routes and ocean fronts, in *Proceedings of the Fifteenth Annual Symposium on Sea Turtle Biology and Conservation*, Keinath, J.A. et al., Compilers, NOAA Tech. Memo., NMFS-SEFSC-387, 1996, 174.

Lutcavage, M.E. et al., Post nesting movements of leatherback turtles tracked from Culebra and Fajardo, PR, with pop-up archival and TDR satellite tags, in *Proceedings of the Twenty-Second Annual Symposium on Sea Turtle Biology and Conservation*, NOAA Tech. Memo, Miami, FL. in press.

Marquez-M.R., Synopsis of biological data on the Kemp's ridley turtle, *Lepidochelys kempii* (*Garman*, 1880), NOAA Tech. Memo., NMFS-SEFSC-343, 1994.

Mendonca, M.T. and Pritchard, P.C.H., Offshore movements of post-nesting Kemp's ridley sea turtles (*Lepidochelys kempii*), *Herpetologica*, 42(3), 373, 1986.

Meylan, A., Sea turtle migration — evidence from tag returns, in *Biology and Conservation of Sea Turtles*, Bjorndal, K.A., Editor, Smithsonian Institution Press, Washington, DC, 1982, 91.

Meylan, A., Spongivory in hawksbill turtles: a diet of glass, *Science*, 239, 393, 1988.

Meylan, A.B., International movements of immature and adult hawksbill turtles (*Eretmochelys imbricata*) in the Caribbean region, *Chelonian Conserv. Biol.*, 3, 189, 1999.

Meylan, A. et al., First recorded nesting by Kemp's ridley in Florida, USA, *Mar. Turtle Newsl.*, 48, 8–9, 1990a.

Meylan, A.B., Bjorndal, K.A., and Turner, B.J., Sea turtles nesting at Melbourne Beach, Florida. II. Post-nesting movements of *Caretta caretta*, *Biol. Conserv.*, 26, 79, 1983.

Meylan, A.B., Bowen, B.W., and Avise, J.C., A genetic test of the natal homing versus social facilitation models for green turtle migration, *Science*, 248, 724, 1990b.

Meylan, P.A., Meylan, A.B., and Yeomans, R., Interception of Tortuguero-bound green turtles at Bocas Del Toro Province, Panama, in *Proceedings of the Eleventh Annual Workshop on Sea Turtle Biology and Conservation*, Salmon, M. and Wyneken, J., Compilers, NOAA Tech. Memo., NMFS-SEFC-302, 1992, 74, Miami, FL.

Miller, J.D., Reproduction in sea turtles, in *The Biology of Sea Turtles*, Lutz, P.L. and Musick, J.A., Editors, CRC Press, Inc., Boca Raton, FL, 1997, 51.

Miller, J.D. et al., Long-distance migrations by the hawksbill turtle, *Eretmochelys imbricata*, from north-eastern Australia, *Wildl. Res.*, 25(1), 89, 1998.

Morreale, S.J. et al., Leatherback migrations along deepwater bathymetric contours, in *Proceedings of the Thirteenth Annual Symposium on Sea Turtle Biology and Conservation*, Schroeder, B.A. and Witherington, B.E., Compilers, NOAA Tech. Memo., NMFS-SEFSC-341, 1994, 109, Miami, FL.

Morreale, S.J. et al., Migration corridor for sea turtles, *Nature*, 384, 319, 1996.

Mortimer, J.A., Feeding ecology of sea turtles, in *Biology and Conservation of Sea Turtles*, Bjorndal, K.A., Editor, Smithsonian Institution Press, Washington, DC, 1982, 103.

Mortimer, J.A. and Balazs, G.H., Post-nesting migrations of hawksbill turtles in the granitic Seychelles and implications for conservation, in *Proceedings of the Nineteenth Annual Symposium on Sea Turtle Biology and Conservation*, Kalb, H.J. and Wibbels, T., Compilers, NOAA Tech. Memo., NMFS-SEFSC-443, 2000, 22, Miami, FL.

Mrosovsky, N. and Shettleworth, S.J., Wavelength preferences and brightness cues in the water finding behavior of sea turtles, *Behaviour*, 32, 211, 1968.

Mysing, J.O. and Vanselous, T.M., Status of satellite tracking of Kemp's ridley sea turtles, in *Proceedings of the First International Symposium on Kemp's Ridley Sea Turtle Biology, Conservation, and Management*, Caillouet, C.W., Jr., and Landry, A.M., Jr., Editors, Texas A&M University Sea Grant College Publication, TAMU-SG-89–105, 1989, 112, College Station, TX.

Nietschmann, B., Following the underwater trail of a vanishing species — the hawksbill turtle, *Natl. Geogr. Soc. Res. Rep.*, 3, 459, 1981.

Owens, D.W., Studies of behavioral thermoregulation in hatchling sea turtles, *Am. Zool.*, 20, 763, 1980a.

Owens, D.W., The comparative reproductive physiology of sea turtles, *Am. Zool.*, 20, 549, 1980b.

Pandav, B. et al., Fidelity of male olive ridley sea turtles to a breeding ground, *Mar. Turtle Newsl.* 87, 9, 2000.

Papi, F. et al., Long-range migratory travel of a green turtle tracked by satellite: evidence for navigational ability in the open ocean, *Mar. Biol.*, 122, 171, 1995.

Papi, F. et al., Satellite tracking experiments on the navigational ability and migratory behavior of the loggerhead turtle *Caretta caretta*, *Mar. Biol.*, 129, 215, 1997.

Papi, F. et al., Open-sea migration of magnetically disturbed sea turtles, *J. Exp. Biol.*, 203, 3435, 2000.

Papi, F. and Luschi, P., Pinpointing "Isla Meta": the case of sea turtles and albatrosses, *J. Exp. Biol.*, 199, 65, 1996.

Parmenter, C.J., Reproductive migration in the hawksbill turtle (*Eretmochelys imbricata*), *Copeia*, 1, 271, 1983.

Parmenter, C.J., Species review: the flatback turtle — *Natator depressa*, in *Proceedings of the Australian Marine Turtle Conservation Workshop*, James, R., Compiler, Queensland Dept. of Environment and Heritage, and Australian Nature Conservation Agency, 1994, 60, Queensland, Australia.

Pitman, R.L., Pelagic distribution and biology of sea turtles in the eastern tropical Pacific, in *Proceedings of the Tenth Annual Workshop on Sea Turtle Biology and Conservation*, Richardson, T.H., Richardson, J.I., and Donnelly, M., Compilers, NOAA Tech. Memo., NMFS-SEFC-278, 1990, 143, Miami, FL.

Pitman, R.L., Seabird associations with marine turtles in the eastern Pacific Ocean, *Colonial Waterbirds*, 16, 194, 1993.

Plotkin, P.T., Migratory and reproductive behavior of the olive ridley turtle, *Lepidochelys olivacea* (*Eschscholtz*, 1829), in the eastern Pacific Ocean, Ph.D. thesis, Texas A&M University, College Station, TX, 1994.

Plotkin, P., Polak, M., and Owens, D.W., Observations on olive ridley sea turtle behavior prior to an arribada at Playa Nancite, Costa Rica, *Mar. Turtle Newsl.*, 53, 9, 1991.

Plotkin, P.T., Byles, R.A., and Owens, D.W., Post-breeding movements of male olive ridley sea turtles *Lepidochelys olivacea* from a nearshore breeding area, in *Proceedings of the Fourteenth Annual Symposium on Sea Turtle Biology and Conservation*, Bjorndal, K.A. et al., Compilers, NOAA Tech. Memo., NMFS-SEFSC-351, 1994, 119, Miami, FL.

Plotkin, P.T. et al., Independent vs. socially facilitated migrations of the olive ridley, *Lepidochelys olivacea*, *Mar. Biol.*, 122, 137, 1995.

Plotkin, P.T. et al., Departure of male olive ridley turtles (*Lepidochelys olivacea*) from a nearshore breeding area, *Herpetologica*, 52, 1, 1996.

Plotkin, P.T. et al., Reproductive and developmental synchrony in female *Lepidochelys olivacea*, *J. Herpetol.*, 31, 17, 1997.

Plotkin, P.T. and Spotila, J.R., Post-nesting migrations of loggerhead turtles *Caretta caretta* from Georgia, U.S.A.: conservation implications for a genetically distinct sub-population, *Oryx*, 36, 4, 2002.

Prieto, A., et al. Informe de la República de Cuba. Primera Reunión de Diálogo CITES Sobre la Tortuga Carey del Gran Caribe. Cuidad de México, 15–17, Mayo, 2001.

Prince, R.I.T., The flatback turtle (*Natator depressus*) in Western Australia: new information from the Western Australian Marine Turtle Project, in *Proceedings of the Australian Marine Turtle Conservation Workshop*, James, R., Compiler, Queensland Dept. of Environment and Heritage, and Australian Nature Conservation Agency, 1994, 146.

Pritchard, P.C.H., Studies of the systematics and reproductive cycles of the genus *Lepidochelys*, Ph.D. thesis, University of Florida, Gainesville, FL, 1969.

Pritchard, P.C.H., International migrations of South American sea turtles (Cheloniidae and Dermochelyidae), *Anim. Behav.*, 21, 18, 1973, Gland Switzerland.

Pritchard, P.C.H., Post nesting movements of marine turtles (Cheloniidae and Dermochelyidae) tagged in the Guianas, *Copeia*, 4, 749, 1976.

Pritchard, P.C.H. and Marquez, R., Kemp's ridley or the Atlantic ridley, *Lepidochelys kempii*, IUCN, Monograph (Marine Turtle Series), 1973.

Renaud, M.L., Movements and submergence patterns of Kemp's ridley turtles (*Lepidochelys kempii*), *J. Herpetol.*, 29, 370, 1995.

Renaud, M.L. et al., Kemp's ridley sea turtle (*Lepidochelys kempii*) tracked by satellite telemetry from Louisiana to nesting beach at Rancho Nuevo, Tamaulipas, Mexico, *Chelonian Conserv. Biol.*, 2, 108, 1996.

Rostal, D.C., The reproductive behavior and physiology of the Kemp's ridley sea turtle, *Lepidochelys kempii* (*Garman*, 1880), Ph.D. thesis, Texas A&M University, College Station, TX, 1991.

Rostal, D.C. et al., Nesting physiology of Kemp's ridley sea turtles, *Lepidochelys kempii*, at Rancho Nuevo, Tamaulipas, Mexico, with observations on population estimates, *Chelonian Conserv. Biol.*, 2, 538, 1997.

Sakamoto, W. et al., Circadian rhythm on diving motion of the loggerhead turtle, *Caretta caretta* during inter-nesting and its fluctuations induced by oceanic environmental events, *Nippon Suisan Gakkaishi*, 56, 263, 1990.

Schierwater, B. and Schroth, W., Molecular evidence for precise natal homing in loggerhead sea-turtles, *Bull. Ecol. Soc. Am.*, 77(3), Part 2, 393, 1996.

Schroeder, B.A., Ehrhart, L.M., and Balazs, G.H., Post-nesting movements of Florida green turtles: preliminary results from satellite telemetry, in *Proceedings of the Fifteenth Annual Symposium on Sea Turtle Biology and Conservation*, Keinath, J.A. et al., Compilers, NOAA Tech. Memo., NMFS-SEFSC-387, 1996, 289, Miami, FL.

Seminoff, J.A. et al., Using carapace-mounted submersible cameras to study foraging in black sea turtles, in *Proceedings of the Nineteenth Annual Symposium on Sea Turtle Biology and Conservation*, Kalb, H.J. and Wibbels, T., Compilers, NOAA Tech. Memo., NMFS-SEFSC-443, 2000, 185, Miami, FL.

Shaver, D.J., *Lepidochelys kempii* (Kemp's ridley sea turtle): reproduction, *Herpetol. Rev.*, 23, 59, 1992.

Shaver, D.J., Kemp's ridley turtle nesting on the Texas coast, 1979–1996, in *Proceedings of the Seventeenth Annual Sea Turtle Symposium*, Epperly, S.P. and Braun, J., Compilers, NOAA Tech. Memo., NMFS-SEFSC-415, 1998, 91, Miami, FL.

Shaver, D.J., Kemp's ridley sea turtle project at Padre Island National Seashore, Texas, in *Proceedings of the 17th Annual Gulf of Mexico Information Transfer Meeting*, McKay, M. and Nides, J., Editors, U.S. Dept. of the Interior, Minerals Management Service, Gulf of Mexico OCS Region, MMS 99–0042, 1999, 342, New Orleans, LA.

Shaver, D.J., U.S. Geological Survey/National Park Service Kemp's ridley sea turtle research and monitoring programs in Texas, in *Proceedings of the Gulf of Mexico Marine Protected Species Workshop*, 2001, New Orleans, LA.

Shaver, D.J. et al., Movements of adult male Kemp's ridley sea turtles (*Lepidochelys kempii*) in the Gulf of Mexico investigated by satellite telemetry, *Chelonian Conserv. Biol.*, in press.

Starbird, C.H., Internesting movements and behavior of hawksbill sea turtles (*Eretmochelys imbricata*) around Buck Island Reef National Monument, St. Croix, United States Virgin Islands, M. Sc. thesis, San Jose State University, San Jose, CA, 1993.

Starbird, C.H. et al., Internesting movements and behavior of hawksbill turtles (*Eretmochelys imbricata*) around Buck Island Reef National Monument, St. Croix, U.S. Virgin Islands, *Chelonian Conserv. Biol.*, 3, 237, 1999.

Steyermark, A.C., et al., Nesting leatherback turtles at Las Baulas National Park, Costa Rica, *Chelonian Conserv. Biol.*, 2, 173, 1996.

Stoneburner, D.L., Satellite telemetry of loggerhead sea turtle movement in the Georgia bight USA, *Copeia*, 2, 400, 1982.

Timko, R.E. and Kolz, A.L., Satellite sea turtle tracking, *Mar. Fish. Rev.*, 44, 19, 1982.

Tucker, A.D., Fitzsimmons, N.N., and Limpus, C.J., Conservation implications of internesting habitat use by loggerhead turtles, *Caretta caretta*, in Woongarra Marine Park, Queensland, Australia, *Pac. Conserv. Biol.*, 2, 157, 1996.

Wibbels, T., Owens, D.W., and Amoss, M.S., Seasonal changes in the serum testosterone titers of loggerhead sea turtles captured along the Atlantic coast of the USA, in *Proceedings of the Cape Canaveral, Florida Sea Turtle Workshop*, Witzell, W.N., Editor, NOAA Tech. Rep., NMFS 53, 1987, 59, Miami, FL.

Witzell, W.N., Synopsis of biological data on the hawksbill turtle *Eretmochelys imbricata* (*Linnaeus*, 1766), FAO Fisheries Synopsis No. 137, FAO, Rome, Italy, 1983.

Wyneken, J., Salmon, M., and Lohmann, K.J., Orientation by hatchling loggerhead sea turtles *Caretta caretta L.* in a wave tank, *J. Exp. Mar. Biol. Ecol.*, 139, 43, 1990.

9 Variation in Sea Turtle Life History Patterns: Neritic vs. Oceanic Developmental Stages

Alan B. Bolten

CONTENTS

9.1 INTRODUCTION

Sea turtles are slow growing and long lived. Their complex life history patterns encompass a diversity of ecosystems from terrestrial habitats where oviposition and embryonic development occur to developmental and foraging habitats in coastal waters (neritic zone) as well as in the open ocean (oceanic zone). Of all the sea turtle life stages, the biology of post-hatchling and early juvenile stages is the least understood (i.e., the "mystery of the lost year" [Carr, 1986; Bolten and Balazs, 1995]). For most sea turtle species, not even the location or duration of the early juvenile stage is known.

0-8493-1123-3/03/$0.00+$1.50

Except for the loggerhead (*Caretta caretta*), little progress has been made in our understanding of the early juvenile stage beyond what was summarized in Musick and Limpus (1997) and Bjorndal (1997). Recent studies on the early life stages of the loggerhead sea turtle have improved our knowledge of the biology of the oceanic juvenile stage (for a review, see Bolten, in press; Bjorndal et al., 2000a, in review).

In this chapter, three generalized sea turtle life history patterns are identified and evaluated with respect to phylogenetic relationships and reproductive traits. Characteristics of the developmental stages (oceanic vs. neritic) and adult foraging stage (oceanic vs. neritic) are the primary differences that distinguish the three patterns. These variations are reviewed, the consequences of oceanic vs. neritic developmental stages are discussed, and finally, speculation about how these differences may have evolved is presented. The dramatic decline in sea turtle populations and the extensive degradation of their ecosystems make it difficult to determine the functional roles of sea turtles in their ecosystems (see Chapter 10), and therefore, it is difficult to evaluate the selective factors that led to the present-day sea turtle life history patterns. It is particularly difficult, in the context of massive faunal declines and food web alterations, to speculate on the relative importance of the evolutionary pressures from competition for resources and predation that may have resulted in these observed life history patterns.

9.2 TERMINOLOGY

The terminology used to describe the life histories of sea turtles has been inconsistent for both the oceanographic terms and the developmental stages (Bolten, in press). To be consistent with standard oceanographic terminology, the following terms should be used (see Lalli and Parsons [1993] for review):

- The neritic zone describes the inshore marine environment (from the surface to the sea floor) where water depths do not exceed 200 m. The neritic zone generally includes the continental shelf, but in areas where the continental shelf is very narrow or nonexistent, the neritic zone conventionally extends to areas where water depths are less than 200 m.
- The oceanic zone is the vast open ocean environment (from the surface to the sea floor) where water depths are greater than 200 m.
- Organisms are pelagic if they occupy the water column, but not the sea floor, in either the neritic zone or oceanic zone. Organisms are epipelagic if they occupy the upper 200 m in the oceanic zone.
- Organisms on the sea floor in either the neritic zone or oceanic zone are described as benthic or demersal.

Organisms can therefore be pelagic in shallow neritic waters or in the deep oceanic waters. Similarly, organisms can be benthic in shallow neritic waters as well as in the deep ocean. We should describe sea turtle life stages by the oceanic realm that they inhabit. Therefore, the early juvenile stage found in the open ocean should be described as the oceanic stage, not the pelagic stage, and the juvenile stage found in coastal waters as the neritic stage, not the benthic stage.

9.3 SEA TURTLE LIFE HISTORY PATTERNS

Although there are only seven extant species, sea turtles exhibit a surprising diversity of life history traits that make them good subjects for comparative life history studies. Aspects of this diversity are illustrated by reproductive extremes from arribadas to solitary nesting, dietary specializations from seagrasses to sponges to jelly organisms, and metabolic adaptations from hibernation to endothermy (*sensu lato*). Despite this high diversity in life history traits among sea turtle species, there have been few comparative analyses or syntheses of their life history patterns. Hendrickson (1980) was the first to attempt to summarize the ecological strategies of sea turtles. However, his conclusions were limited by the lack of information at the time of his synthesis in both life history characteristics of the different sea turtle species and their taxonomic relationships, which are now better understood through molecular techniques. In another analysis, Van Buskirk and Crowder (1994) developed a dendrogram to compare the reproductive characteristics among the seven species. This chapter focuses on an analysis of the variation in the juvenile developmental stages.

Once hatchlings emerge from their nests, crawl down the beach, and enter the sea, post-hatchlings embark upon one of three basic developmental life history patterns:

- Complete development in the neritic zone (Type 1, Figure 9.1, top panel)
- Early juvenile development in the oceanic zone and later juvenile development in the neritic zone (Type 2, Figure 9.1, middle panel)
- Complete development in the oceanic zone (Type 3, Figure 9.1, bottom panel)

9.3.1 TYPE 1: THE NERITIC DEVELOPMENTAL PATTERN

The Type 1 life history pattern is characterized by developmental and adult stages occurring completely in the neritic zone (Figure 9.1, top panel). The Australian flatback turtle (*Natator depressus*) apparently has a completely neritic developmental pattern (Walker and Parmenter, 1990; Walker, 1994) and is the only extant example of the Type 1 life history pattern. Walker (1994) suggests there may be increased food resources in the neritic zone, but with a tradeoff of increased predation. Flatback hatchlings are larger than those of other cheloniid sea turtles (60 vs. 41–50 mm carapace length [Van Buskirk and Crowder, 1994]; 39 vs. 15–25 g hatchling mass [Miller, 1997]). As would be expected from the evolutionary tradeoff between the size and number of offspring, clutch size is smaller in flatbacks (53 vs. 100–182 [Hirth, 1980; Van Buskirk and Crowder, 1994], although East Pacific green turtles [*Chelonia mydas*] have small clutch sizes from 65 to 90 [Hirth, 1997]). Flatbacks thus produce fewer, larger progeny, a pattern typical of marine species with shorter dispersal distances relative to species that disperse more widely, such as Type 2 and 3 sea turtles.

Researchers have speculated that the larger hatchling size of flatback turtles may reduce predation in the neritic zone (Hirth, 1980; Walker and Parmenter, 1990;

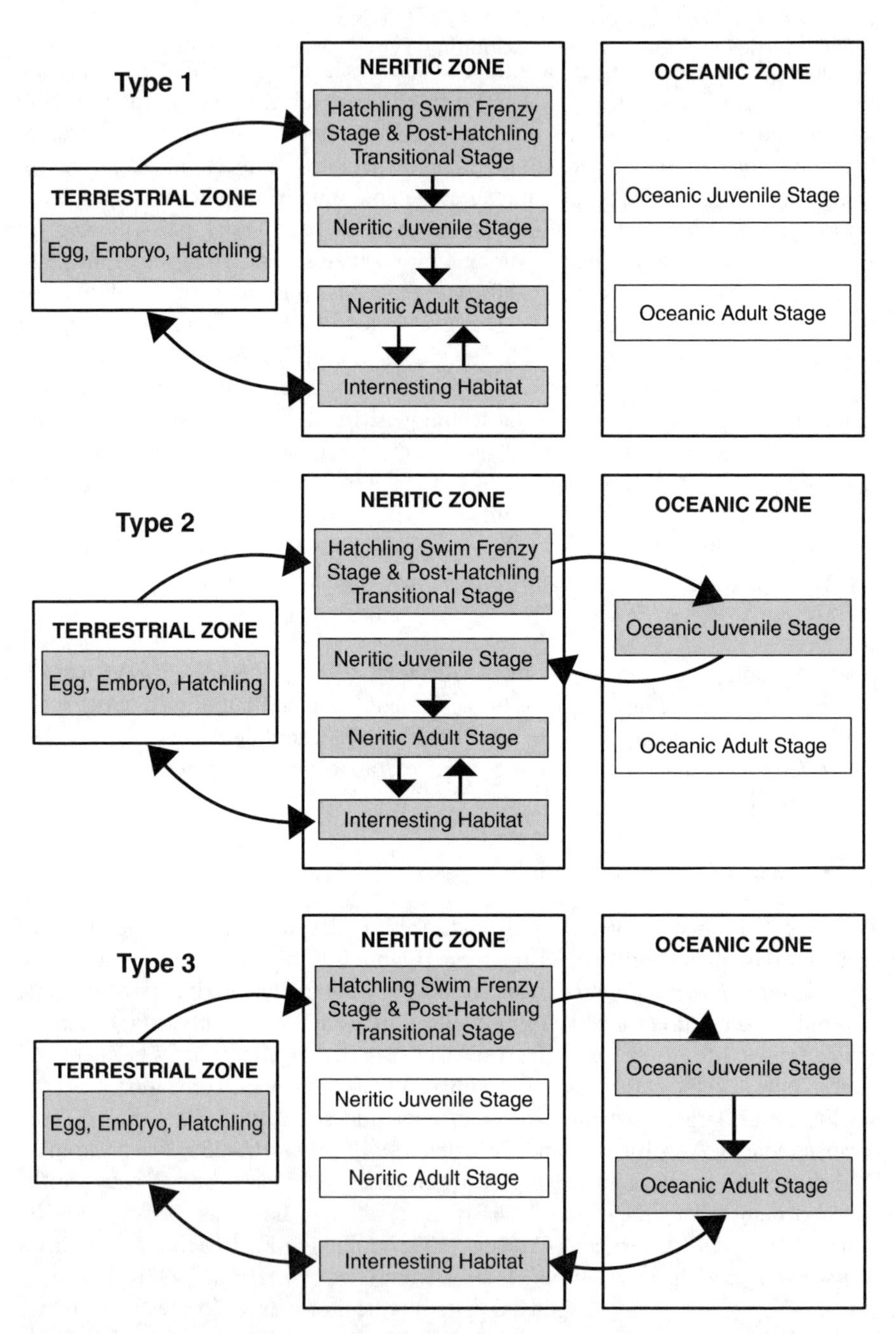

FIGURE 9.1 Three distinct sea turtle life history patterns illustrating the sequence of ecosystems inhabited. See text for a description of each type.

Walker, 1994; Musick and Limpus, 1997). However, the size difference between flatback turtles and those species exhibiting Type 2 or 3 patterns does not seem to be great enough to make a significant difference with respect to marine predators. The larger hatchling size of the flatback may be a response to terrestrial predators and the need to survive the predator gauntlet from the nest to the water. The larger size of flatback hatchlings allows them to escape some bird and crab predators on Australian beaches that prey upon the green turtle (*C. mydas*) and loggerhead hatchlings (Limpus, 1971). The fact that other sea turtle species have not also responded to terrestrial predators by increasing hatchling body size suggests that the selective pressures leading to the increased size of flatback hatchlings are not fully understood. Knowledge of the selective advantages of large hatchling size may be critical for understanding the success of the Type 1 life history pattern.

The early juvenile stage of Type 1 species (including post-hatchlings) probably feeds on the surface and within the water column, and may later develop a benthic feeding strategy once the turtle has gained buoyancy control and can dive to the sea floor. In the shallow waters that Type 1 species inhabit, their foraging behavior may be a mix of pelagic and benthic feeding throughout life. Support for this mixed foraging strategy is seen in the diet of both small and large flatback turtles (Limpus et al., 1988; Zangerl et al., 1988). However, the number of samples that has been evaluated is too small to be conclusive; more studies on the diet of flatbacks are needed.

9.3.2 Type 2: The Oceanic–Neritic Developmental Pattern

The Type 2 life history pattern is characterized by early development in the oceanic zone followed by later development in the neritic zone (Figure 9.1, middle panel). The best-known example of this life history pattern is that of the loggerhead turtle (for review, see Bolten, in press). Use of genetic markers has confirmed the relationships between oceanic foraging grounds and rookeries (Bowen et al., 1995; Bolten et al., 1998) that had been hypothesized based on length–frequency distributions (Carr, 1986; Bolten et al., 1993) and tag returns (Bolten, in press). Although based on rather few data (summarized in Carr, 1987a), this life history pattern is thought to be the pattern for the green turtle, hawksbill (*Eretmochelys imbricata*), and Kemp's ridley (*Lepidochelys kempii* [Collard and Ogren, 1990]). Little is known about the ecology of juvenile olive ridleys (*Lepidochelys olivacea*); differences among populations from different ocean basins suggest that this species exhibits either a Type 2 or a Type 3 life history pattern, perhaps in response to differences in resource availability. In the West Atlantic (Pritchard, 1976; Reichart, 1993; Bolten and Bjorndal, unpublished data) and Australia (Harris, 1994), olive ridleys appear to exhibit a Type 2 life history pattern, whereas East Pacific populations (Pitman, 1990) appear to exhibit a Type 3 life history pattern.

Following the hatchling swim-frenzy stage (Wyneken and Salmon, 1992), loggerheads have a transition period when the post-hatchling begins to feed and moves from the neritic zone into the oceanic zone (Bolten, in press). The duration, movements, and distribution of the post-hatchlings during this transition have been reviewed by Witherington (2002, in review a). This transition is relatively passive in that oceanographic and meteorological factors (e.g., currents and winds) have the greatest influence on the movements and distribution patterns of these turtles, although the post-hatchlings may actively position themselves using magnetic orientation cues to maximize the likelihood of successful transport (Lohmann and Lohmann, in press). After a developmental period in the oceanic zone lasting from 7 to 11.5 years, when the turtles reach a size of 46–64 cm curved carapace length (Bjorndal et al., 2000a; Bjorndal et al., in review), juvenile loggerheads in the Atlantic leave the oceanic zone and complete their development in the neritic zone (Musick and Limpus, 1997; Bjorndal et al., 2001; Bolten, in press).

Other Type 2 species recruit to neritic habitats at smaller sizes. Green turtles and hawksbills appear in neritic foraging grounds at about 20–35 cm carapace length (CL) and Kemp's ridleys at 20–25 cm CL (Bjorndal, 1997; Musick and Limpus, 1997). On the basis of size at recruitment to neritic habitats, the durations of the oceanic stages in these species may be shorter than that of Atlantic loggerheads. Size at recruitment is apparently not a function of size at maturity. Adult size of loggerheads falls between those of Kemp's ridleys and green turtles (Miller, 1997).

Recruitment from the oceanic, where the turtles are primarily epipelagic, to the neritic, where they are primarily benthic, may involve another transition period before the juvenile turtles become fully neritic (Kamezaki and Matsui, 1997; Laurent et al., 1998; Bolten, in press; Tiwari et al., in press). Adults of Type 2 species may leave neritic habitats during their reproductive migrations, which may involve oceanic migration corridors between the adult foraging areas (neritic) and internesting habitat (also neritic). Figure 9.2 presents the details of the loggerhead life history from the North Atlantic.

9.3.3 Type 3: The Oceanic Developmental Pattern

The Type 3 life history pattern is characterized by both developmental and adult stages occurring completely in the oceanic zone (Figure 9.1, bottom panel). Of course, post-hatchlings, once they leave the nesting beach, must traverse the neritic zone to reach the oceanic zone, and adults must return to the neritic zone for reproduction. The leatherback (*Dermochelys coriacea*) and olive ridley (East Pacific populations) are believed to exhibit this life history pattern. Very little is known about the biology (e.g., oceanic distribution, diet, or growth rates) of early developmental stages for olive ridleys and leatherbacks, but it is assumed that the juvenile stages occur in the oceanic zone.

Leatherbacks and olive ridleys are very different in many aspects, such as body size, thermal regulation, and foraging behavior. The leatherback is the largest sea turtle species, with a mean adult size of 149 cm CL, whereas the olive ridley is one of the smallest species, with a mean adult CL of 66 cm (Van Buskirk and Crowder, 1994). Leatherbacks are able to maintain a body temperature 15°C above ambient and forage in waters with temperatures as low as 0–15°C (Spotila et al., 1997). The olive ridley maintains a body temperature at most a few degrees above ambient (Spotila et al., 1997) and appears to be limited to warmer, tropical waters (Pitman, 1990; 1993; Polovina et al., in review). Leatherbacks are active predators, and may not undergo a substantial diet shift as they grow because the ability to capture and

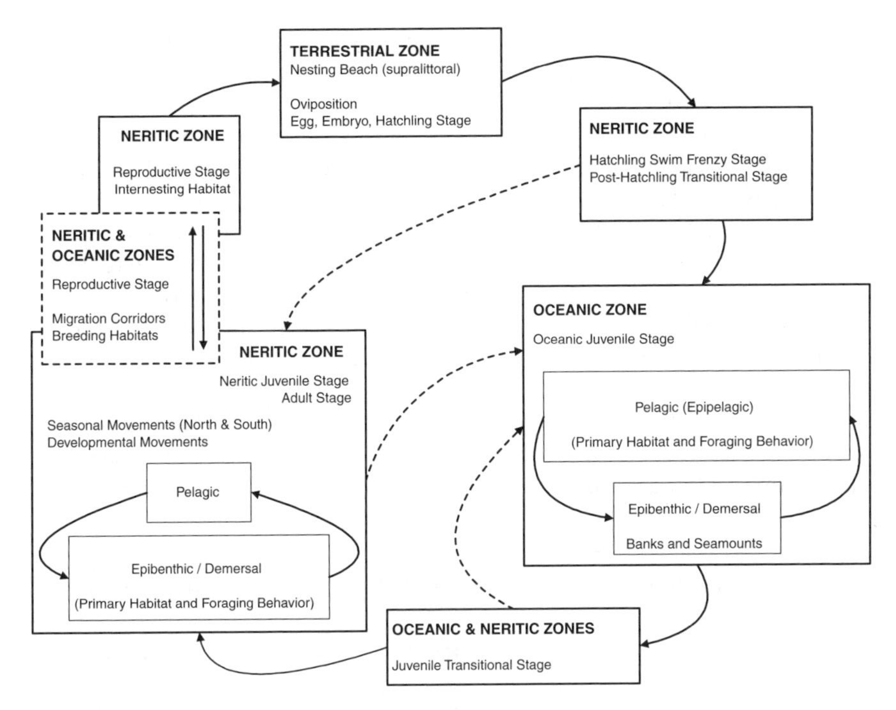

FIGURE 9.2 Life history diagram of the Atlantic loggerhead sea turtle. Boxes represent life stages and the corresponding ecosystems. Solid lines represent movements between life stages and ecosystems; dotted lines are speculative. (Modified from Bolten, A.B. In press. Active swimmers — passive drifters: the oceanic juvenile stage of loggerheads in the Atlantic system. In A.B. Bolten and B.E. Witherington, editors. *Biology and Conservation of the Loggerhead Sea Turtle*. Smithsonian Institution Press, Washington, DC. With permission.)

consume gelatinous prey species is not size dependent (Bjorndal, 1997). Olive ridleys apparently spend more time at the surface than leatherbacks (Pitman, 1993) and may exhibit a "float and wait" foraging strategy, although Polovina et al. (in review) report that olive ridleys only spend 20% of their time on the surface and 40% of their time diving deeper than 40 m.

9.4 RELATIONSHIPS OF THE THREE LIFE HISTORY PATTERNS TO PHYLOGENETIC PATTERNS AND REPRODUCTIVE TRAITS

9.4.1 Phylogenetic Patterns

A comparison of the life history patterns (Type 1, 2, or 3) of the seven species of extant sea turtles with their phylogenetic patterns is presented in Figure 9.3, left dendrogram (Bowen and Karl, 1997). Sea turtles are generally recognized as a monophyletic group (Bowen and Karl, 1997). The ancestor of sea turtles was probably a resident of coastal salt marshes, estuaries, and tidal creeks. Once these ancestors committed fully to the sea, a completely neritic developmental and adult life history (Type 1) would be the expected pattern because they would probably have stayed close to shore. The Type 1 pattern observed in the Australian flatback is secondarily derived on the basis of the phylogenetic position of the flatback (Figure 9.3, left dendrogram). A change to the Type 2 pattern may have resulted from selective pressures to exploit new food resources with fewer competitors in the oceanic zone or to avoid the higher predation risks in the neritic zone.

A change from the Type 2 to Type 3 pattern would be a natural outgrowth of the Type 2 pattern. Once the transition has been made for early development to occur in the oceanic zone, it would be reasonable to continue development in that habitat. The phylogenetic pattern suggests that the olive ridley (Lo in Figure 9.3) has recently derived the Type 3 pattern from the Type 2 pattern. The hypothesis that the Type 3 pattern in olive ridleys is recent is supported by both Type 2 (West Atlantic, Australia) and Type 3 (East Pacific) life history patterns being expressed in this species. In contrast, the phylogenetic pattern suggests that the leatherback (Dc in Figure 9.3)

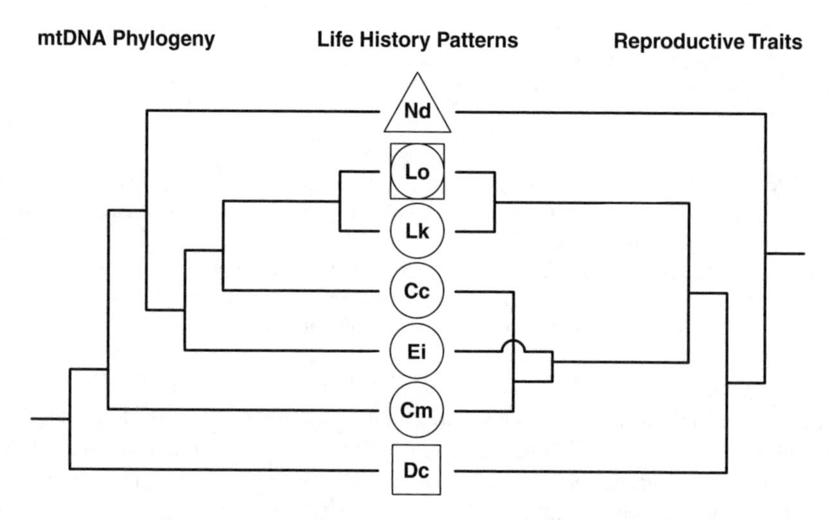

FIGURE 9.3 The relationships of the three types of life history patterns (see Figure 9.1) with a phylogeny based on mtDNA and a dendrogram of reproductive traits. The branching of the phylogenetic and reproductive trait dendrograms is not to scale. The triangle indicates the Type 1 life history pattern, the circles indicate those species exhibiting the Type 2 pattern, and the squares indicate those species exhibiting the Type 3 pattern. The olive ridley (Lo) is diagrammed with both a circle (Type 2) to represent West Atlantic and Australian populations and a square (Type 3) to represent East Pacific populations (see text for discussion). Cc = *Caretta caretta*; Cm = *Chelonia mydas*; Dc = *Dermochelys coriacea*; Ei = *Eretmochelys imbricata*; Lk = *Lepidochelys kempii*; Lo = *Lepidochelys olivacea*; and Nd = *Natator depressus*. (Left dendrogram of figure was modified from Bowen, B.W. and S.A. Karl. 1997. Population genetics, phylogeography, and molecular evolution. Pages 29–50 in P.L. Lutz and J.A. Musick, editors. *The Biology of Sea Turtles*. CRC Press, Boca Raton, FL; right side of figure was modified from Van Buskirk, J. and L.B. Crowder. 1994. Life-history variation in marine turtles. *Copeia* 1994:66–81.)

has probably exhibited the Type 3 pattern for a long time, with resultant physiological adaptations that have allowed leatherbacks to exploit both tropical and temperate oceanic realms.

9.4.2 Reproductive Traits

Van Buskirk and Crowder (1994) evaluated a number of reproductive traits (e.g., female size, clutch size, egg volume, hatchling size, clutch frequency, and remigration interval) and developed a dendrogram illustrating the relationships among the species (Figure 9.3, right dendrogram). The Van Buskirk and Crowder (1994) dendrogram is not congruent with the phylogeny of Bowen and Karl (1997), but has a greater similarity to the three life history patterns presented in Figure 9.1. This better fit would be expected between life history patterns (Figure 9.3, center) and reproductive behavior/demographic traits (Figure 9.3, right dendrogram). The two species with the greatest difference in reproductive traits (leatherbacks and flatbacks) are also the two species with the greatest difference in life history patterns.

9.5 A Closer Look at the Type 2 Pattern: Ontogenetic Habitat Shifts

Species exhibiting either the Type 1 or Type 3 pattern commit to either the neritic or oceanic zone, respectively, for their entire developmental stages as well as for the adult foraging stage. Only turtles with the Type 2 pattern have a major habitat change during their development. The Type 2 pattern is the most successful pattern if success is defined by the number of species with this life history pattern (five of the seven extant species exhibit the Type 2 pattern). As presented above (Section 9.4.1), the Type 1 pattern is hypothesized to be the ancestral pattern that still exists today (although presumably secondarily derived) in the Australian flatback. Why post-hatchling turtles leave the neritic zone for the oceanic, and why, after an extended development period in the oceanic zone, the turtles return to the neritic to complete their development, are two intriguing questions.

The existence of an early developmental stage in the oceanic habitat may be a result of higher predator pressure in neritic habitats and/or intra- and interspecific competition for food in neritic habitats. Such competition may not be apparent now because of depleted sea turtle populations, but evidence for density-dependent effects on growth rates has been reported for a population of green turtles in a neritic foraging habitat (Bjorndal et al., 2000b).

Even more puzzling is the shift from oceanic to neritic habitats. Why do juvenile turtles leave the oceanic zone where they have spent the first years of their lives successfully finding food, growing, and surviving? When they leave the oceanic zone for the neritic zone, they enter a new habitat with which they are unfamiliar, and must learn to find new food sources and avoid a new suite of predators. A current hypothesis to explain why ontogenetic habitat shifts occur is that a species shifts habitats to maximize growth rates (Werner and Gilliam, 1984). Bolten (in press) presents evidence for the Atlantic loggerhead population that supports the Werner and Gilliam hypothesis. The extrapolation of the size-specific growth function for the oceanic stage intersects

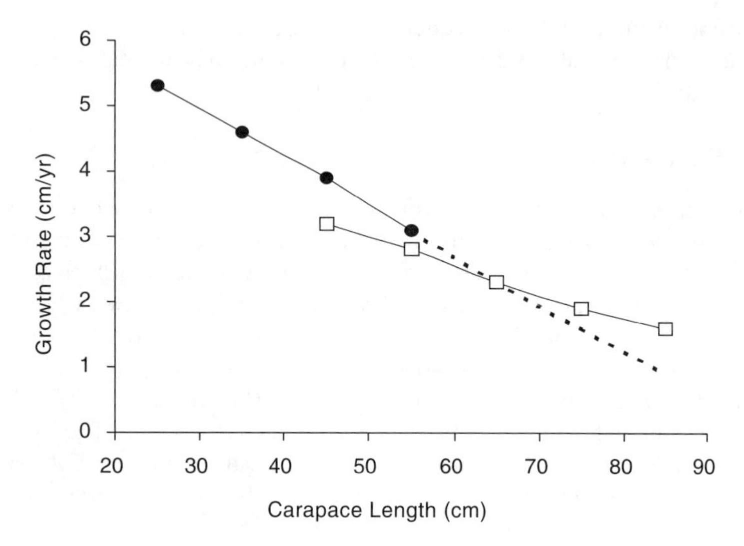

FIGURE 9.4 Size-specific growth functions of oceanic-stage (solid circles) and neritic-stage (open boxes) loggerheads based on length–frequency analyses. Dashed line is an extrapolation of the growth function for oceanic-stage loggerheads. The slopes of the lines are significantly different ($p < 0.001$). (Data from Bjorndal, K.A. et al. 2001. Somatic growth function for immature loggerhead sea turtles in southeastern U.S. waters. *Fish. Bull.* 99:240–246; and Bjorndal, K.A., A.B. Bolten, and H.R. Martins. 2000a. Somatic growth model of juvenile loggerhead sea turtles *Caretta caretta*: duration of pelagic stage. *Mar. Ecol. Prog. Ser.* 202:265–272. Modified from Bolten, A.B. In press. Active swimmers — passive drifters: the oceanic juvenile stage of loggerheads in the Atlantic system. In A.B. Bolten and B.E. Witherington, editors. *Biology and Conservation of the Loggerhead Sea Turtle.* Smithsonian Institution Press, Washington, DC. With permission.)

the size-specific growth function for the neritic stage (Figure 9.4). Therefore, for a given CL greater than approximately 64 cm (a size by which almost all of the loggerheads have left the oceanic zone [Bjorndal et al., 2000a]), growth rates will be greater in the neritic zone than in the oceanic zone. Snover et al. (2000) present data from a skeletochronology study that also demonstrate increased growth rates of turtles that have left the oceanic zone and entered the neritic zone.

There is substantial variation among the species that have the Type 2 pattern, suggesting a fair amount of variation in lifestyles. Morphological differences in oceanic-stage turtles include variation in countershading (suggesting different predator-avoidance behaviors) and front flipper length relative to body length (suggesting differences in swimming activity and resulting feeding behavior). The swimming behavior of post-hatchling green turtles appears to be different than that of loggerheads (Wyneken, 1997). Another source of variation is the duration of the oceanic developmental stage that may be significantly different for the different species based on the size at which they recruit to neritic habitats (see Section 9.3.2). In addition, resource partitioning along temperature gradients and among foraging strategies probably occurs among Type 2 species, but data are lacking. For example, loggerheads (Type 2 species) and olive ridleys (Type 3 species) apparently partition resources in the Pacific by water temperature; loggerheads inhabit cooler waters than do olive ridleys (Pitman, 1993; Polovina et al., in review). Thus, as more is learned about this life stage, further divisions in the Type 2 pattern may be required.

9.6 ANTHROPOGENIC IMPACTS ON THE EARLY JUVENILE STAGE

The future is bleak. Early juvenile-stage sea turtles face a myriad of obstacles irrespective of whether they are in the neritic or oceanic zone. Directed take of very small turtles for food is not common. However, directed take for the souvenir trade in polished shells or whole stuffed turtles, such as the once-popular but now illegal tourist trade in Madeira, Portugal (Brongersma, 1982), still exists in some regions.

Indirect take in fisheries, whether it is the high seas drift nets, longlines, or coastal trawlers, is a very serious problem for juvenile turtles (National Research Council, 1990; Wetherall et al., 1993; Balazs and Pooley, 1994; Witzell, 1999; Bolten et al., 2000). Throughout the world's oceans, the size distribution of loggerhead turtles caught in longline fisheries is the largest size class for the oceanic development stage (Bolten et al., 1994; Ferreira et al., 2001; Bolten, in press), which has significant demographic consequences (Crouse et al., 1987; Heppell et al., in press).

The lethal and sublethal effects of debris ingestion and entanglement are also major concerns (Balazs, 1985; Carr, 1987b; McCauley and Bjorndal, 1999; Witherington, in review b). Habitat loss, particularly in coastal areas, has been documented; habitat degradation in the oceanic zone is more difficult to document but nonetheless acute when the effects of pollution are considered (Lutcavage et al., 1997). Both oceanic and neritic ecosystems are changing as a result of overfishing and pollution. Changes to the suite of species interactions and food webs in these ecosystems are undoubtedly having a major negative impact on sea turtles.

9.7 CONCLUSIONS: RESEARCH DIRECTIONS

Our ability to solve the "mystery of the lost year" for those species for which the early juvenile stages remain unknown has been improved by the development of new research tools. Biotechnology is providing molecular tags to identify populations and track movements; biotelemetry is allowing researchers to evaluate movement and distribution patterns. Stable isotopes may provide clues about where to look for early juvenile stages and also provide information on trophic relationships. For researchers to make rapid progress in the study of early juvenile stages, multidisciplinary teams should be developed with expertise in the fields of physical and biological oceanography, population genetics, statistical modeling, demography, nutrition, and ecosystem analyses.

ACKNOWLEDGMENTS

My research on the oceanic juvenile stage has focused on the loggerhead population in the Atlantic and has been conducted in collaboration with Karen Bjorndal, whose inspiration and enthusiasm has made this project possible. Our research has been funded by the U.S. National Marine Fisheries Service and the Disney Wildlife Conservation Fund. Colleagues at the Department of Oceanography and Fisheries at the University of the Azores, especially Helen Martins and "equipa tartaruga," have made our research not only possible but most enjoyable. I would also like to thank Karen Bjorndal, Jeffrey Seminoff, and Jack Musick for their comments on earlier drafts of this manuscript.

REFERENCES

Balazs, G.H. 1985. Impact of ocean debris on marine turtles: entanglement and ingestion. Pages 387–429 in R.S. Shomura and H.O. Yoshida, editors. *Proceedings of the Workshop on the Fate and Impact of Marine Debris*. NOAA Technical Memorandum NOAA-TM-NMFS-SWFS-54, Honolulu, HI.

Balazs, G.H. and S.G. Pooley, editors. 1994. Research plan to assess marine turtle hooking mortality: results of an expert workshop held in Honolulu, Hawaii, November 16–18, 1993. NOAA Technical Memorandum NMFS-SWFSC-201, Honolulu, HI.

Bjorndal, K.A. 1997. Foraging ecology and nutrition of sea turtles. Pages 199–231 in P.L. Lutz and J.A. Musick, editors. *The Biology of Sea Turtles*. CRC Press, Boca Raton, FL.

Bjorndal, K.A., A.B. Bolten, and M.Y. Chaloupka. 2000b. Green turtle somatic growth model: evidence for density dependence. *Ecol. Appl.* 10:269–282.

Bjorndal, K.A. et al. In review. Compensatory growth in oceanic loggerhead sea turtles: response to a stochastic environment.

Bjorndal, K.A. et al. 2001. Somatic growth function for immature loggerhead sea turtles in southeastern U.S. waters. *Fish. Bull.* 99:240–246.

Bjorndal, K.A., A.B. Bolten, and H.R. Martins. 2000a. Somatic growth model of juvenile loggerhead sea turtles *Caretta caretta*: duration of pelagic stage. *Mar. Ecol. Prog. Ser.* 202:265–272.

Bolten, A.B. In press. Active swimmers — passive drifters: the oceanic juvenile stage of loggerheads in the Atlantic system. In A.B. Bolten and B.E. Witherington, editors. *Biology and Conservation of the Loggerhead Sea Turtle*. Smithsonian Institution Press, Washington, DC.

Bolten, A.B. and G.H. Balazs. 1995. Biology of the early pelagic stage — the "lost year." Pages 579–581 in K.A. Bjorndal, editor. *Biology and Conservation of Sea Turtles*. Revised edition. Smithsonian Institution Press, Washington, DC.

Bolten, A.B., K.A. Bjorndal, and H.R. Martins. 1994. Life history model for the loggerhead sea turtle (*Caretta caretta*) population in the Atlantic: potential impacts of a longline fishery. Pages 48–54 in G.H. Balazs and S.G. Pooley, editors. Research plan to assess marine turtle hooking mortality: results of an expert workshop held in Honolulu, Hawaii, November 16–18, 1993. NOAA Technical Memorandum NMFS-SWFSC-201, Honolulu, HI.

Bolten, A.B. et al. 1993. Size distribution of pelagic-stage loggerhead sea turtles (*Caretta caretta*) in the waters around the Azores and Madeira. *Arquipélago* 11A:49–54.

Bolten, A.B. et al. 1998. Transatlantic developmental migrations of loggerhead sea turtles demonstrated by mtDNA sequence analysis. *Ecol. Appl.* 8:1–7.

Bolten, A.B., H.R. Martins, and K.A. Bjorndal, editors. 2000. Proceedings of a workshop to design an experiment to determine the effects of longline gear modification on sea turtle bycatch rates [Workshop para a elaboração de uma experiência que possa diminuir as capturas acidentais de tartarugas marinhas nos Açores], 2–4 September 1998, Horta, Azores, Portugal. NOAA Technical Memorandum NMFS-OPR-19, Silver Spring, MD.

Bowen, B.W. et al. 1995. Trans-Pacific migrations of the loggerhead turtle (*Caretta caretta*) demonstrated with mitochondrial DNA markers. *Proc. Natl. Acad. Sci. U.S.A.* 92:3731–3734.

Bowen, B.W. and S.A. Karl. 1997. Population genetics, phylogeography, and molecular evolution. Pages 29–50 in P.L. Lutz and J.A. Musick, editors. *The Biology of Sea Turtles*. CRC Press, Boca Raton, FL.

Brongersma, L.D. 1982. Marine turtles of the eastern Atlantic Ocean. Pages 407–416 in K.A. Bjorndal, editor. *Biology and Conservation of Sea Turtles*. Smithsonian Institution Press, Washington, D.C.

Carr, A.F. 1986. Rips, FADS, and little loggerheads. *Bioscience* 36:92–100.

Carr, A.F. 1987a. New perspectives on the pelagic stage of sea turtle development. *Conserv. Biol.* 1:103–121.

Carr, A.F. 1987b. Impact of nondegradable marine debris on the ecology and survival outlook of sea turtles. *Mar. Pollut. Bull.* 18:352–356.

Collard, S.B. and L.H. Ogren. 1990. Dispersal scenarios for pelagic post-hatchling sea turtles. *Bull. Mar. Sci.* 47:233–243.

Crouse, D.T., L.B. Crowder, and H. Caswell. 1987. A stage-based population model for loggerhead sea turtles and implications for conservation. *Ecology* 68:1412–1423.

Ferreira, R.L. et al. 2001. Impact of swordfish fisheries on sea turtles in the Azores. *Arquipélago* 18A:75–79.

Harris, A. 1994. Species review: the olive ridley. Pages 63–67 in R. James, compiler. *Proceedings of the Australian Marine Turtle Conservation Workshop*, Sea World Nara Resort, Gold Coast, 14–17 November 1990. Queensland Department of Environment and Heritage, and Australian Nature Conservation Agency, Canberra, Australia.

Hendrickson, J.R. 1980. The ecological strategies of sea turtles. *Am. Zool.* 20:597–608.

Heppell, S.S. et al. In press. Population models for Atlantic loggerheads: past, present and future. In A.B. Bolten and B.E. Witherington, editors. *Biology and Conservation of the Loggerhead Sea Turtle*. Smithsonian Institution Press, Washington, D.C.

Hirth, H.F. 1980. Some aspects of the nesting behavior and reproductive biology of sea turtles. *Am. Zool.* 20:507–523.

Hirth, H.F. 1997. Synopsis of the biological data on the green turtle *Chelonia mydas* (Linneaus 1758). U.S. Fish and Wildlife Service Biological Report 97(1), Washington, D.C. 120 pages.

Kamezaki, N. and M. Matsui. 1997. Allometry in the loggerhead turtle, *Caretta caretta*. *Chelonian Conserv. Biol.* 2:421–425.

Lalli, C.M. and T.R. Parsons. 1993. *Biological Oceanography: An Introduction*. Pergamon Press, New York.

Laurent, L. et al. 1998. Molecular resolution of marine turtle stock composition in fishery bycatch: a case study in the Mediterranean. *Mol. Ecol.* 7:1529–1542.

Limpus, C.J. 1971. The flatback turtle, *Chelonia depressa* Garman in southeast Queensland, Australia. *Herpetologica* 27:431–446.

Limpus, C.J., E. Gyuris, and J.D. Miller. 1988. Reassessment of the taxonomic status of the sea turtle genus *Natator* McCulloch, 1908, with a redescription of the genus and species. *Trans. R. Soc. South Aust.* 112:1–9.

Lohmann, K.J. and C.M.F. Lohmann. In press. Orientation mechanisms of hatchling loggerheads. In A.B. Bolten and B.E. Witherington, editors. *Biology and Conservation of the Loggerhead Sea Turtle*. Smithsonian Institution Press, Washington, DC.

Lutcavage, M.E. et al. 1997. Human impacts on sea turtle survival. Pages 387–409 in P.L. Lutz and J.A. Musick, editors. *The Biology of Sea Turtles*. CRC Press, Boca Raton, FL.

McCauley, S.J. and K.A. Bjorndal. 1999. Conservation implications of dietary dilution from debris ingestion: sublethal effects in post-hatchling loggerhead sea turtles. *Conserv. Biol.* 13:925–929.

Miller, J.D. 1997. Reproduction in sea turtles. Pages 51–81 in P.L. Lutz and J.A. Musick, editors. *The Biology of Sea Turtles*. CRC Press, Boca Raton, FL.

Musick, J.A. and C.J. Limpus. 1997. Habitat utilization and migration in juvenile sea turtles. Pages 137–164 in P.L. Lutz and J.A. Musick, editors. *The Biology of Sea Turtles*. CRC Press, Boca Raton, FL.

National Research Council. 1990. *Decline of the Sea Turtles: Causes and Prevention*. National Academy Press, Washington, D.C.

Pitman, R.L. 1990. Pelagic distribution and biology of sea turtles in the eastern tropical Pacific. Pages 143–148 in T.H. Richardson, J.I. Richardson, and M. Donnelly, compilers. *Proceedings of the Tenth Annual Workshop on Sea Turtle Biology and Conservation*. NOAA Technical Memorandum NMFS-SEFC-278, Miami, FL.

Pitman, R.L. 1993. Seabird associations with marine turtles in the eastern Pacific Ocean. *Colonial Waterbirds* 16:194–201.

Polovina, J.J. et al. In review. Forage and migration habitat of loggerhead (*Caretta caretta*) and olive ridley (*Lepidochelys olivacea*) sea turtles in the Central North Pacific Ocean.

Pritchard, P.C.H. 1976. Post-nesting movements of marine turtles (Cheloniidae and Dermochelyidae) tagged in the Guianas. *Copeia* 1976:749–754.

Reichart, H.A. 1993. Synopsis of biological data on the olive ridley sea turtle *Lepidochelys olivacea* (Eschscholtz 1829) in the western Atlantic. NOAA Technical Memorandum NMFS-SEFSC-336, Miami, FL.

Snover, M.L., A.A. Horn, and S.A. Macko. 2000. Detecting the precise time at settlement from pelagic to benthic habitats in the loggerhead sea turtle. Page 174 in H. Kalb and T. Wibbels, compilers. *Proceedings of the Nineteenth International Symposium on Sea Turtle Biology and Conservation*. NOAA Technical Memorandum NMFS-SEFSC-443, Miami, FL.

Spotila, J. R., M.P. O'Connor, and F.V. Paladino. 1997. Thermal biology. Pages 297–314 in P.L. Lutz and J.A. Musick, editors. *The Biology of Sea Turtles*. CRC Press, Boca Raton, FL.

Tiwari, M. et al. In press. Morocco and Western Sahara: sites of an early neritic stage in the life history of loggerheads? In *Twentieth Annual International Symposium on Sea Turtle Biology and Conservation*. NOAA Technical Memorandum, Miami, FL.

Van Buskirk, J. and L.B. Crowder. 1994. Life-history variation in marine turtles. *Copeia* 1994:66–81.

Walker, T.A. 1994. Post-hatchling dispersal of sea turtles. Pages 79–94 in R. James, compiler. *Proceedings of the Australian Marine Turtle Conservation Workshop* held at Sea World Nara Resort, Gold Coast, 14–17 November 1990. Queensland Department of Environment and Heritage, and Australian Nature Conservation Agency, Canberra, Australia.

Walker, T.A. and C.J. Parmenter. 1990. Absence of a pelagic phase in the life cycle of the flatback turtle, *Natator depressa* (Garman). *J. Biogeogr.* 17:275–278.

Werner, E.E. and J.F. Gilliam. 1984. The ontogenetic niche and species interactions in size-structured populations. *Annu. Rev. Ecol. Syst.* 15:393–425.

Wetherall, J.A. et al. 1993. Bycatch of marine turtles in North Pacific high-seas driftnet fisheries and impacts on the stocks. *Int. North Pac. Fish. Comm. Bull.* 53:519–38.

Witherington, B.E. 2002. Ecology of neonate loggerhead turtles inhabiting lines of downwelling near a Gulf Stream front. *Mar. Biol.* 140:843–853.

Witherington, B.E. In review. A test of a "smart drifter" hypothesis describing the distribution of neonate loggerhead turtles in the open ocean.

Witherington, B.E. In review b. Frequency of tar and plastics ingestion by neonate loggerhead turtles captured from the western Gulf Stream off Florida, USA.

Witzell, W.N. 1999. Distribution and relative abundance of sea turtles caught incidentally by the U.S. pelagic longline fleet in the western North Atlantic Ocean, 1992–1995. *Fish. Bull.* 97:200–211.

Wyneken, J. 1997. Sea turtle locomotion: mechanisms, behavior, and energetics. Pages 165–198 in P.L. Lutz and J.A. Musick, editors. *The Biology of Sea Turtles*. CRC Press, Boca Raton, FL.

Wyneken, J. and M. Salmon. 1992. Frenzy and postfrenzy swimming activity in loggerhead, green, and leatherback hatchling sea turtles. *Copeia* 1992:478–484.

Zangerl, R., L.P. Hendrickson, and J.R. Hendrickson. 1988. A redescription of the Australian flatback sea turtle, *Natator depressus*. *Bishop Mus. Bull. Zool. I.* Bishop Museum Press, Honolulu, HI.

10 Roles of Sea Turtles in Marine Ecosystems: Reconstructing the Past

Karen A. Bjorndal and Jeremy B.C. Jackson

CONTENTS

10.1 INTRODUCTION

Populations of sea turtles have been drastically reduced since interactions between humans and sea turtles began. Although Caribbean sea turtle populations generally have been considered to be pristine when Columbus arrived in 1492, archeological research is now revealing that some sea turtle nesting aggregations in the Caribbean were extirpated or significantly reduced by Amerindians (Carlson, 1999; O'Day, 2001). Therefore, the roles that sea turtles played in the functioning of ecosystems in the Caribbean may have been substantially affected before European contact. Initially a result of directed harvest, population declines have more recently been driven by factors in addition to direct harvest, such as incidental capture in commercial fisheries, habitat degradation, introduction of feral predators on nesting beaches, and marine pollution (Eckert, 1995; Lutcavage et al., 1997; Witherington, in press). These population declines have produced a corresponding decline in the extent to which sea turtles fulfill their roles in maintaining the structure and function of marine ecosystems.

Because the massive proportions of the declines occurred so long ago, sea turtles are now viewed by many as charming anachronisms or quaint archaic relics. Their past roles as major marine consumers in many marine ecosystems from

0-8493-1123-3/03/$0.00+$1.50

tropical to subarctic waters have been forgotten. Thus, sea turtles are victims of the "shifting baseline syndrome" (Pauly, 1995; Sheppard, 1995). This pervasive syndrome is the use of inappropriate baselines to assess population change or stability. Referring to fisheries management, Pauly (1995) first described the syndrome as the tendency of scientists to use population levels at the beginning of their careers as the baseline against which to measure population change. Pauly stressed the importance of incorporating historical anecdotes of fish abundance into population models of commercial fish species. For sea turtles, we do not have the proper perspective, or a reliable baseline, against which to assess population trends. For example, hawksbills (*Eretmochelys imbricata*) have been extensively exploited for centuries for the keratinized scutes covering their shells, which are the source of tortoiseshell or *bekko* (Parsons, 1972; Groombridge and Luxmoore, 1989; Meylan, 1999). Because populations were already greatly reduced or extirpated before they were recorded, we have been unable to quantify past populations of hawksbills and their ecological function.

Why is an understanding of the ecological roles of sea turtles important? We propose three reasons.

1. Ecosystem function: To discover what we have lost in terms of ecosystem structure and function. The far-reaching effects of removing consumers from marine ecosystems have been demonstrated during the past decade in a series of studies (Dayton et al., 1995; 1998; Jackson, 1997; 2001; Pauly et al., 1998; Jackson et al., 2001; Pitcher, 2001). The fact that humans have been "fishing down food webs" (Pauly et al., 1998) with resulting widespread effects or trophic cascades is well documented (Jackson, 2001; Pitcher, 2001). Several studies have emphasized that current problems — collapse of marine ecosystems and commercial fisheries — are not only the result of recent events, but originate in prehistoric times (Jackson, 1997; 2001; Jackson et al., 2001). These studies have generated a new appreciation of the need to explore the characteristics of marine ecosystems before human intervention. Paleoecological, archaeological, and historical data are needed to reconstruct how marine ecosystems once functioned (Jackson, 2001). The historical perspective gained from these reconstructions provides essential guidance for restoring marine ecosystems and ensuring sustainable fisheries (Jackson et al., 2001; Pitcher, 2001). Restoring consumer populations to an abundance necessary to be ecologically functional is still possible because most of these species still exist, at much reduced levels (Jackson et al., 2001), with a few exceptions such as the extinct Caribbean monk seal, *Monachus tropicalis* (LeBoeuf et al., 1986).
2. Better understanding of environmental effects on remnant populations of sea turtles: To understand how environmental changes today — either natural or human-induced — may affect sea turtle populations. This understanding would greatly enhance our ability to make informed management decisions. What effect would changes in the designation of allowable use in zones of the Great Barrier Reef have on sea turtles there? What would

be the effect of developing a commercial harvest of jellyfish in the Gulf of Mexico — a major food resource for several sea turtle species? What is the effect of the depletion of shark populations — major predators on sea turtles around the world?

3. More meaningful goals for management and conservation of sea turtles: To define goals for sea turtle recovery programs that allow sea turtles to be ecologically functional in marine ecosystems. The mission of the Marine Turtle Specialist Group of the World Conservation Union (IUCN) is to "promote the restoration and survival of healthy marine turtle populations that fulfill their ecological roles" (Marine Turtle Specialist Group, 1995). Such goals coincide with the current emphasis on ecosystem management rather than single-species management. Sea turtles cannot be conserved without restoring and competently managing the marine systems they inhabit. The recovery plans for sea turtle species developed by the U.S. Fish and Wildlife Service and the National Marine Fisheries Service contain specific recovery goals, as required under the U.S. Endangered Species Act. None of these plans has set a recovery goal to restore sea turtle populations to their ecological roles (e.g., National Marine Fisheries Service and U.S. Fish and Wildlife Service, 1991a, b). Our lack of knowledge hinders setting such goals: How many sea turtles would be required for a population to be ecologically functional?

10.2 ECOLOGICAL ROLES OF SEA TURTLES

Sea turtles range widely over the Earth. They occur in oceanic and neritic habitats from the tropics to subarctic waters and venture onto terrestrial habitats to nest or bask in tropical and temperate latitudes (Table 10.1). Before sea turtle populations were depleted by humans, sea turtles occurred in massive numbers that are now difficult to imagine (King, 1982; Ross, 1982; Jackson, 1997; Jackson et al., 2001). At those high population levels, sea turtles had substantial effects on the marine systems they inhabited as consumers, prey, and competitors; as hosts for parasites and pathogens; as substrates for epibionts; as nutrient transporters; and as modifiers of the landscape.

Bjorndal (in press) summarized the current state of our knowledge of the ecological roles of loggerheads (*Caretta caretta*). Although our understanding of the ecological role of the loggerhead is extremely limited, it is the best-studied sea turtle species in this regard. Loggerheads prey upon a large number of species and, particularly at small sizes, are preyed upon by a wide range of predators (Bjorndal, in press). Sea turtles serve as substrate and transport for a diverse array of epibionts. Loggerheads nesting in Georgia had 100 species of epibionts from 13 phyla (Frick et al., 1998), and loggerheads nesting at Xcacel, Mexico, carried 37 taxa of algae in total, with up to 12 species on an individual turtle (Senties et al., 1999). Sea turtles can transfer substantial quantities of nutrients and energy from nutrient-rich foraging grounds to nutrient-poor nesting beaches. Less than one third of the energy and nitrogen contained in eggs deposited by loggerheads in Melbourne Beach, FL, returned to the ocean in the form of hatchlings (Bouchard and Bjorndal, 2000). Loggerheads can modify the physical structure of their habitat in a number of ways, including digging trenches through soft substrates in search of infauna prey (Preen, 1996).

The roles of sea turtles as consumers are the best known, but information is largely limited to lists of prey species. The diets of most species have been evaluated (Table 10.1), although there are considerable gaps for early life stages and some geographic areas (Bjorndal, 1997). Knowledge of selective feeding and rates of consumption, which is critical for quantitatively evaluating the ecological function of sea turtles as consumers, is generally lacking.

For the remainder of this chapter, we will present two case studies to illustrate how the ecological role of sea turtles as consumers can be quantified, as indicated by the amount of prey consumed. We selected the Caribbean green turtle (*Chelonia mydas*) and the Caribbean hawksbill because of the availability of data and the difference in diets: The green turtle is an herbivore that feeds primarily on seagrasses in the Caribbean, and the hawksbill is a carnivore that feeds largely on sponges.

In the two case studies, we have had to assume that diet and intake (rate of consumption) will not change with changes in population density. We realize that these assumptions may not be true. As populations become denser, diet species may change as preferred prey become less abundant and less-favored species must be consumed. Intake may decrease as intraspecific competition for food increases or may change with diet quality. Evidence for such density-dependent effects was observed for a population of immature green turtles for which somatic growth rates and condition index (mass/length^3) declined as population density increased, apparently in response to lower food resources (Bjorndal et al., 2000).

10.3 CASE STUDY: CARIBBEAN GREEN TURTLE

The decline of green turtles in the Caribbean during historic times is well recognized (Parsons, 1962). The example of the extirpation of the Cayman Islands green turtle nesting colony is relatively well recorded in historical documents. The Cayman Islands were apparently never inhabited and their resources were never utilized by Amerindians (Stokes and Keegan, 1996; Scudder and Quitmyer, 1998). Columbus sighted the islands of Cayman Brac and Little Cayman during his last voyage in 1503, and named them Las Tortugas because of the great number of turtles on the land and in the surrounding waters (Hirst, 1910). After that time, the Cayman Islands, which were not permanently settled by humans until 1734 (Williams, 1970), were visited by ships of many nations to take on green turtles and their eggs (Lewis, 1940). Consistent exploitation of Cayman green turtles by ships from Jamaica was initiated in 1655 when the English took Jamaica from Spain (Lewis, 1940). In 1684, when French and Spanish corsairs chased English turtling vessels out of Cayman and Cuban waters, Colonel Hender Molesworth reported to Britain that Jamaica would suffer because green turtle "is what masters of ships chiefly feed their men in port, and I believe that nearly 2000 people, black and white, feed on it daily at this point, to say nothing of what is sent inland. Altogether it cannot be easily imagined how prejudiced is this interruption of the turtle trade" (Smith, 2000). With safe access to the Caymans restored, Jamaican ships carried 13,000 turtles each year from the Caymans between

TABLE 10.1
General Summary of Distribution, Habitats, and Diets of Sea Turtle Species

Species	Distribution[a]	Habitats[b]	Diet[b]
Loggerhead (*Caretta caretta*)	Global, usually temperate and subtropical; sometimes tropical	SJ: epipelagic in oceanic LJ and A: demersal in neritic	SJ: epipelagic invertebrates, primarily jelly organisms LJ and A: invertebrates, primarily sessile or slow moving
Green turtle (*Chelonia mydas*)	Global, usually tropical and subtropical; sometimes temperate	SJ: unknown, believed to be epipelagic in oceanic LJ and A: demersal in neritic	SJ: unknown, believed to be carnivorous or omnivorous LJ and A: primarily herbivorous, seagrasses and algae; some invertebrates
Hawksbill (*Eretmochelys imbricata*)	Global, tropical	SJ: unknown, believed to be epipelagic in oceanic LJ and A: demersal in neritic	SJ: unknown, believed to be carnivorous or omnivorous LJ and A: invertebrates, primarily sponges in the Atlantic, perhaps more omnivorous in the Pacific
Olive ridley (*Lepidochelys olivacea*)	Pacific, Indian, and South Atlantic oceans, tropical	SJ: unknown, believed to be epipelagic in oceanic LJ and A: commonly epipelagic in oceanic, but also demersal in neritic	SJ: unknown, believed to be carnivorous or omnivorous LJ and A: invertebrates, primarily jelly organisms and crabs
Kemp's ridley (*Lepidochelys kempi*)	Gulf of Mexico, eastern U.S., and occasionally western Europe	SJ: unknown, believed to be epipelagic in oceanic LJ and A: demersal in neritic	SJ: poorly known, believed to be carnivorous or omnivorous LJ and A: invertebrates, primarily crabs
Flatback (*Natator depressus*)	Tropical Australia and possibly southern New Guinea	Neritic throughout life; SJ: apparently epipelagic; LJ and A: demersal	SJ: poorly known, pelagic snails and jelly organisms LJ and A: soft-bodied invertebrates
Leatherback (*Dermochelys coriacea*)	Global, tropical to subarctic	Pelagic throughout life, primarily in oceanic; also in neritic	SJ: unknown, believed to be jelly organisms LJ and A: jelly organisms

Notes: SJ = small juvenile; LJ = large juvenile; A = adult. For more detailed descriptions, the reader is referred to the cited references.

[a] From Pritchard, P.C.H. and J.A. Mortimer. 1999. Taxonomy, external morphology, and species identification. Pages 21–38 in K.L. Eckert, K.A. Bjorndal, F.A. Abreu-Grobois, and M. Donnelly, editors. Research and management techniques for the conservation of sea turtles. IUCN/SSC Marine Turtle Specialist Group Publication No. 4.

[b] From Bjorndal, K.A. 1997. Foraging ecology and nutrition of sea turtles. Pages 199–231 in P.L. Lutz and J.A. Musick, editors. *The Biology of Sea Turtles*. CRC Press, Boca Raton, FL.

1688 and 1730 (Jackson, 1997). By 1790, green turtles had become scarce in Cayman waters and soon could not support a fishery, so Cayman turtlers went to the waters of southern Cuba (Williams, 1970; Smith, 2000). By 1830, green turtles off south Cuba had diminished, so Cayman turtlers went to the Miskito Cays, off the coasts of Nicaragua and Honduras (Williams, 1970). By 1890, concerns were expressed over the growing scarcity of turtles in the Miskito Cays (Hirst, 1910). In 1901, Duerden (1901) urged the government of Jamaica to establish artificial hatching and rearing facilities for green turtles and hawksbills because of "the diminution in the supply [from the Miskito Cays] which is now being felt."

Although the Cayman green turtle story is the best known, it is far from being the only extirpation of green turtle populations in the Caribbean. Early historical accounts report "vast quantities" of sea turtles in areas where few, if any, sea turtles exist today. For example, the pirate John Esquemeling, in his account of the activities of buccaneers in America, described turtles that "resort in huge multitudes at certain seasons of the year, there to lay their eggs" on the Isle of Savona off the coast of Hispaniola, as well as on the west coast of mainland Hispaniola (Esquemeling, 1684). Neither area supports such sea turtle nesting today.

The pattern of overexploitation of green turtles is clear from these accounts. However, how many green turtles lived in the Caribbean before humans began harvesting them? Jackson (1997) used the Jamaican exploitation records described above to estimate the preexploitation number of adult green turtles in the Caribbean. Jackson's estimates ranged from 33 to 39 million adult green turtles.

If preexploitation green turtle populations were regulated by food limitations, the carrying capacity (K) of Caribbean seagrass beds for the green turtle would be a maximum estimate of population size. The seagrass *Thalassia testudinum* is the primary diet of green turtles in the Caribbean (Bjorndal, 1997), and the green turtle is one of the few species that consumes Caribbean seagrasses as a major part of its diet (Thayer et al., 1984) after the extinction of the diverse dugongid fauna before the Pleistocene (Domning, 2001). Populations of large herbivores are often "bottom-up" regulated by food limitation rather than "top-down" by predators (Sinclair, 1995; Jackson, 1997), so green turtle populations in the greater Caribbean may well have been controlled by food limitation (Bjorndal, 1982; Jackson, 1997), and density-dependent effects would have regulated productivity of green turtles (Bjorndal et al., 2000). Jackson (1997) used an estimate of the carrying capacity of the seagrass *T. testudinum* for green turtles from Bjorndal (1982) and generated an estimate of 660 million adult green turtles in the Caribbean. Bjorndal et al. (2000) estimated a range of carrying capacities of *T. testudinum* for green turtles based on three estimates of intake and two estimates of *T. testudinum* productivity (Table 10.2). The estimates ranged from 122 to 4439 kg of green turtle per hectare (ha) of *T. testudinum*, or 16–586 million 50-kg green turtles. This range nearly encompasses the range of 33–660 million adult green turtles of Jackson (1997). The estimates of K vary by an order of magnitude based on the two productivity levels of *T. testudinum* measured in areas heavily grazed and more moderately grazed by green turtles (Table 10.2). This variation is not surprising. The biomass, rate of production, and quality of seagrasses are all affected by grazing (Thayer et al., 1984). In grazing systems, highest plant productivity is often associated with light to moderate grazing

(McNaughton, 1985). A study now underway (Moran and Bjorndal, unpublished data) on the effects of green turtle grazing on *T. testudinum* productivity should greatly improve our estimates of *K*.

Under such heavy grazing regimes, seagrass ecosystems in the Caribbean would have had very different structures and dynamics than they do today. The current green turtle population in the Caribbean has been estimated to represent 3–7% of preexploitation population levels (Jackson et al., 2001). Major changes in biodiversity, productivity, and structure of *T. testudinum* pastures would be expected between grazed pastures with blade lengths of 2–4 cm and the essentially ungrazed pastures of today with blade lengths of up to 30 cm or more (Zieman, 1982). Dampier (1729) observed that blades of *T. testudinum* were only "six inches long" (15 cm) at a time when green turtles were much more abundant in the Caribbean. Grazing by green turtles significantly shortens nutrient cycling times in *T. testudinum* pastures (Thayer et al., 1982). Reduced blade life in grazed stands and thus reduced time for epibiont colonization would affect the epibionts that cover *T. testudinum* blades in some areas. Shorter blade lengths in grazed stands would decrease the baffling effect and thus the entrapment of particles and deposition of substrate and would substantially change the physical structure of these ecosystems that are important nursery areas for many species of fish and invertebrates. This change in structure may have contributed to the mass mortality of Florida seagrasses in the 1980s (Jackson, 2001). Seagrass mortality was positively density dependent and was correlated with high temperatures and salinities, sulfide toxicity, self-shading, hypoxia, and infection by the slime mold *Labyrinthula* spp. (Robblee et al., 1991; Harvell et al., 1999; Zieman et al., 1999). All of these factors, except temperature and salinity, are greatly increased in ungrazed seagrass pastures (Jackson, 2001). Again, the study now underway (Moran and Bjorndal, unpublished data) on the effects of green turtle grazing on *T. testudinum* productivity and structure should provide quantitative estimates of some of these effects.

We can conclude that natural populations of green turtles consumed a tremendous amount of *T. testudinum*. A population of 100 million green turtles with an average mass of 50 kg (a relatively modest population estimate from Jackson [1997] and Bjorndal et al. [2000]) with an average annual intake of 1.23 kg *T. testudinum* dry mass per kg turtle (Table 10.2) would consume 6.2×10^9 kg *T. testudinum* dry mass each year. That value is approximately half of the estimated total annual production of 1.2×10^{10} kg *T. testudinum* dry mass in the Caribbean (6,600,000 ha *T. testudinum* in the Caribbean [Jackson, 1997] $\times$ 1750 kg *T. testudinum* dry mass produced annually per ha [Table 10.2]).

10.4 CASE STUDY: CARIBBEAN HAWKSBILL

As stated above, Caribbean hawksbills have been extensively exploited for centuries for tortoiseshell, the keratinized scutes that cover their shells (Parsons, 1972; Groombridge and Luxmoore, 1989; Meylan, 1999). The current number of adult female hawksbills that nest each year in the Caribbean is estimated at 5000, on the basis of a thorough review by Meylan (1999). Because each female nests at an average interval of 2.7 years (Richardson et al., 1999), the estimate of adult female hawksbills

TABLE 10.2
Carrying Capacities for Green Turtles on *Thalassia testudinum* Pastures in the Caribbean

	Intake kg DM *Thalassia* • (kg Green Turtle)$^{-1}$ • year^{-1}					
	0.74[a]		1.17[b]		1.77[c]	
***Thalassia* productivity kg DM • ha^{-1} year^{-1}**	**kg Turtle • ha^{-1}**	**Number of Turtles in Caribbean[d]**	**kg Turtle • ha^{-1}**	**Number of Turtles in Caribbean[d]**	**kg Turtle • ha^{-1}**	**Number of Turtles in Caribbean[d]**
Heavy grazing[e]	292	38,544,000	185	24,420,000	122	16,104,000
Moderate grazing[f]	4,439	585,948,000	2,808	370,656,000	1,856	244,992,000

Notes: Calculations are based on three levels of intake estimated by three different methods and on two levels of *T. testudinum* productivity. DM = dry mass.

Sources: From Bjorndal, K.A., A.B. Bolten, and M.Y. Chaloupka. 2000. Green turtle somatic growth model: evidence for density dependence. *Ecol. Appl.* 10:269–282. With permission.

[a] From Bjorndal, K.A. 1982. The consequences of herbivory for the life history pattern of the Caribbean green turtle, *Chelonia mydas*. Pages 111–116 in K.A. Bjorndal, editor. *Biology and Conservation of Sea Turtles*. Smithsonian Institution Press, Washington, DC; based on calculation of energy budget for adult female.

[b] From Bjorndal, K.A. 1980. Nutrition and grazing behavior of the green turtle, *Chelonia mydas*. *Mar. Biol.* 56:147–154; based on indigestible lignin ratio and daily feces production.

[c] From Williams, S.L. 1988. *Thalassia testudinum* productivity and grazing by green turtles in a highly disturbed seagrass bed. *Mar. Biol.* 98:447–455, based on estimates of daily bite counts and bite size.

[d] Based on 6,600,000 ha *Thalassia* in the Caribbean (from Jackson, J.B.C. 1997. Reefs since Columbus. *Coral Reefs* 16:S23–S33) and turtle size = 50 kg.

[e] 216 kg DM • ha^{-1} • year^{-1}. (Recalculated from Williams, S.L. 1988. *Thalassia testudinum* productivity and grazing by green turtles in a highly disturbed seagrass bed. *Mar. Biol.* 98:447–455, Table 4.)

[f] 3,285 kg DM • ha^{-1} • year^{-1}. (From Zieman, J.C., R.L. Iverson, and J.C. Ogden. 1984. Herbivory effects on *Thalassia testudinum* leaf growth and nitrogen content. Marine Ecology Progress Series 15:151–158.)

in the Caribbean is 13,500. Estimates of sex ratio have ranged from male biased to female-biased (León and Diez, 1999), so if we assume a 1:1 sex ratio, the estimated total number of adult hawksbills in the Caribbean today is 27,000.

Sponges are abundant on modern Caribbean coral reefs, where their biomass and diversity often exceed that of corals (Goreau and Hartman, 1963; Rützler, 1978; Suchanek et al., 1983; Targett and Schmahl, 1984). Hawksbills in the Caribbean feed primarily on a relatively few species of sponges, although they also consume other invertebrates (Bjorndal, 1997; León and Bjorndal, in press). As the largest sponge predator, how much sponge biomass would an adult hawksbill consume annually? Unfortunately, there are no data on intake or digestion of sponges in hawksbills. We can derive a rough estimate, however, if we assume that the digestible energy intake of hawksbills would lie between those of the green turtle (an herbivore) and the loggerhead (a carnivore that feeds on invertebrates with fewer antiquality components than sponges) (Bjorndal, 1997).

The energy intake of green turtles feeding on *T. testudinum* can be estimated by multiplying the average annual intake from Table 10.2 (1.23 kg *T. testudinum* dry mass per kg turtle) by the energy content of grazed *T. testudinum* blades (14,000 kJ/kg dry mass [Bjorndal, 1980]), which equals 17,220 kJ/kg turtle each year. To estimate digestible energy intake, this value is multiplied by the energy digestibility coefficient for a diet of *T. testudinum* (60% for adults [Bjorndal, 1980]), which yields an estimate of 10,332 kJ/kg green turtle each year. For loggerheads, an annual energy intake of a highly digestible, balanced diet was estimated to be 13,140 kJ/kg turtle (Bjorndal, in press). With an estimate of 90% energy digestibility for the high-quality diet, our estimate of annual digestible energy intake for loggerheads is 11,826 kJ/kg turtle.

Therefore, a very rough estimate of annual digestible energy intake for a hawksbill would be 11,000 kJ/kg. To convert this estimate to the biomass of sponges consumed annually by an adult hawksbill, we will use the sponge *Chondrilla nucula* as the prey species because it is the best studied of the sponges in terms of composition and digestibility, and is a major prey species of hawksbills. *Chondrilla nucula* was consumed by hawksbills in seven of the eight studies of hawksbill diet in the Caribbean and, in most cases, made a major contribution to the diet (summarized in León and Bjorndal, in press). In the only study of selective feeding in hawksbills, there was strong selection for *C. nucula* (León and Bjorndal, in press). Because *C. nucula* has high energy, organic matter, and nitrogen content relative to most sponge species consumed by hawksbills (León and Bjorndal, in press), intake values for hawksbills estimated for a diet of *C. nucula* will be conservative. The average mass of an adult hawksbill is 70 kg (Witzell, 1983), the energy content for *C. nucula* is 15,900 kJ/kg dry mass (Bjorndal, 1990), and we will use a range of energy digestibility coefficients of 43–90%. The low value in this range is based on a value of 43.4% energy digestibility of *C. nucula* measured in green turtles (Bjorndal, 1990). Digestibility should be higher in hawksbills because they feed primarily on sponges. The upper estimate (90%) is near the upper limit of digestibilities of animal tissue measured in reptiles (Zimmerman and Tracy, 1989). The resulting estimate of sponge consumed by an adult hawksbill each year is 54–113 kg dry mass [$(11{,}000 \times 70)/(15{,}900 \times 0.90)$ or $(11{,}000 \times 70)/(15{,}900 \times 0.43)$].

Because the dry mass of *C. nucula* is about 15% of wet mass (León and Bjorndal, in press), these values are equivalent to 360–753 kg wet mass. The population of 27,000 adult hawksbills would consume from 1.5 to 3.1 million kg of sponge dry mass or 10–21 million kg of sponge wet mass each year.

On first consideration, 10–21 million kg of sponge wet mass seems a large quantity. We must consider that number, however, from the perspective of the quantity of sponges that hawksbill populations once consumed in the Caribbean. As noted above, hawksbills have been harvested in the Caribbean since prehistoric times primarily for their scutes, but also for their meat and eggs (Meylan, 1999). On the basis of a thorough review of available data, Meylan and Donnelly (1999) documented declines in hawksbill populations in the Caribbean ranging from 75 to 98% over the last 100 years or less. Given the historic records of annual harvests of thousands of hawksbills in the Caribbean during the eighteenth and nineteenth centuries (summarized in Meylan and Donnelly, 1999), an estimate of an overall decline of 95% in hawksbills from preexploitation to the present is conservative. If adult hawksbills consumed only sponges when population densities were at preexploitation levels, then we estimate that 540,000 adult hawksbills (27,000/0.05) consumed from 200 to 420 million kg of sponge wet mass each year. We consider the estimate of 540,000 adult hawksbills in preexploitation populations to be very conservative — perhaps underestimating the true value by an order of magnitude. This estimate does not include the amount of sponge consumed by the large number of immature hawksbills in the population.

The effect of this massive increase in the consumption of Caribbean sponges in the past would go beyond the direct effect of decreasing sponge populations. Hawksbills can also affect reef diversity and succession by influencing space competition. Scleractinian corals and sponges commonly compete for space on reefs with up to 12 interactions per square meter, and sponges are more often the superior competitor (references in León and Bjorndal, in press). Competition for space also exists among sponge species, and predation by hawksbills is believed to have a major role in maintaining sponge species diversity (van Dam and Diez, 1997).

The diet preference for *C. nucula* emphasizes the past role of hawksbills in space competition on coral reefs because *C. nucula* is a very aggressive competitor for space with reef corals. *C. nucula* is now a very common Caribbean demosponge. As summarized in León and Bjorndal (in press), *C. nucula* was the dominant sponge at 13% of shallow reef sites off Cuba (Alcolado, 1994), occupied up to 12% of the area on some Puerto Rican reefs (Corredor et al., 1988), and was one of the dominant sponges in the Exuma Cays, Bahamas (Sluka et al., 1996). *C. nucula* was involved in nearly half of all scleractinian coral competitive interactions on a reef in Puerto Rico (Vicente, 1990), caused >70% of all coral overgrowths in a study in the Florida Keys (Hill, 1998), and was considered one of the major threats to corals in a reef in Belize (Antonius and Ballesteros, 1998). Hill (1998) excluded sponge predators from coral–sponge interactions and found that *C. nucula* would rapidly overgrow the majority of corals with which it interacted. Hill (1998) concluded that spongivory might have substantial community-level effects in coral reefs.

Acroporid coral cover in the Caribbean during the first half of the twentieth century had declined dramatically from the Pleistocene (Jackson et al., 2001).

How much of this decline was a result of decreased hawksbill predation on sponges? The relatively high coral cover on some modern Caribbean reefs indicates that sponges are somehow prevented from overwhelming the corals. With hawksbill populations seriously depleted, predation by other spongivores — fish, especially parrotfish (Wulff, 1997; Dunlap and Pawlik, 1998; Hill, 1998), and invertebrates — has apparently played this role. Redundancy in ecosystems can mask the effect of species removal until all species performing a given function are lost (Jackson et al., 2001). As humans "fish down the food web" (Pauly et al., 1998), and spongivorous fish populations are depleted, the role of all sponge predators in maintaining the structure and function of coral reef ecosystems may become more apparent.

10.5 CONCLUSIONS

We present three general conclusions:

1. All species of sea turtles in the Caribbean were once extremely abundant. Despite enormous uncertainties, we can conclude that they occurred in the millions or tens of millions. These are conservative estimates.
2. Past sea turtle populations consumed large quantities of prey species, many of which are consumed only to a limited extent by other species. Sea turtles in the Caribbean were once the major consumers of seagrasses, sponges, and jellyfish.
3. Therefore, the virtual ecological extinction of sea turtles in the Caribbean must have resulted in major changes in the structure and function of the marine ecosystems they inhabited.

The roles of sea turtles in the evolution and maintenance of the structure and dynamics of marine ecosystems have gone largely unrecognized because their populations were seriously depleted long ago. Their ecological functions have been essentially unstudied, although sea turtles were an integral part of the interspecific interactions in marine ecosystems as prey, consumer, competitor, and host; served as significant conduits of nutrient and energy transfer within and among ecosystems; and substantially modified the physical structure of marine ecosystems. Research effort should be directed to these ecological questions as a high priority. Sea turtles should be integrated into models of trophic interactions and restoration plans for marine ecosystems.

ACKNOWLEDGMENTS

This work was conducted as part of the Long-Term Ecological Records of Marine Environments, Populations and Communities Working Group supported by the National Center for Ecological Analysis and Synthesis (funded by NSF grant DEB-0072909, the University of California, and the University of California, Santa Barbara). We thank Alan Bolten and Jeffrey Seminoff for their constructive comments on the manuscript.

REFERENCES

Alcolado, P. 1994. General trends in coral reef sponge communities of Cuba. Pages 251–255 in R.W.M. van Soest, T.M.G. van Kempen, and J.C. Braekman, editors. *Sponges in Time and Space*. A.A. Balkema, Rotterdam.

Antonius, A. and E. Ballesteros. 1998. Epizoism: a new threat to coral health in Caribbean reefs. *Rev. Biol. Trop.* 46(Suppl. 5):145–156.

Bjorndal, K.A. 1980. Nutrition and grazing behavior of the green turtle, *Chelonia mydas*. *Mar. Biol.* 56:147–154.

Bjorndal, K.A. 1982. The consequences of herbivory for the life history pattern of the Caribbean green turtle, *Chelonia mydas*. Pages 111–116 in K.A. Bjorndal, editor. *Biology and Conservation of Sea Turtles*. Smithsonian Institution Press, Washington, D.C.

Bjorndal, K.A. 1990. Digestibility of the sponge *Chondrilla nucula* in the green turtle, *Chelonia mydas*. *Bull. Mar. Sci.* 47:567–570.

Bjorndal, K.A. 1997. Foraging ecology and nutrition of sea turtles. Pages 199–231 in P.L. Lutz and J.A. Musick, editors. *The Biology of Sea Turtles*. CRC Press, Boca Raton, FL.

Bjorndal, K.A. In press. Roles of loggerhead sea turtles in marine ecosystems. Pages in A.B. Bolten and B.E. Witherington, editors. *Biology and Conservation of the Loggerhead Sea Turtle*. Smithsonian Institution Press, Washington, D.C.

Bjorndal, K.A., A.B. Bolten, and M.Y. Chaloupka. 2000. Green turtle somatic growth model: evidence for density dependence. *Ecol. Appl.* 10:269–282.

Bouchard, S.S. and K.A. Bjorndal. 2000. Sea turtles as biological transporters of nutrients and energy from marine to terrestrial ecosystems. *Ecology* 81:2305–2313.

Carlson, L.A. 1999. Aftermath of a feast: human colonization of the southern Bahamian Archipelago and its effects on the indigenous fauna. Ph.D. dissertation. University of Florida, Gainesville, FL. 279 pp.

Corredor, J.E. et al. 1988. Nitrate release by Caribbean reef sponges. *Limnol. Oceanogr.* 33:114–120.

Dampier, W. 1729. *A New Voyage Around the World*. James and John Knapton, London. Reprinted 1968, Dover Press, New York.

Dayton, P.K. et al. 1995. Environmental effects of marine fishing. *Aquat. Conserv. Mar. Freshwater Ecosyst.* 5:205–232.

Dayton, P.K. et al. 1998. Sliding baselines, ghosts, and reduced expectations in kelp forest communities. *Ecol. Appl.* 8:309–322.

Domning, D.P. 2001. Sirenians, seagrasses, and Cenozoic ecological change in the Caribbean. *Palaeogeogr. Palaeoclimatol. Palaeoecol.* 166:27–50.

Duerden, J.E. 1901. The marine resources of the British West Indies. *West Ind. Bull.* 2:121–163.

Dunlap, M. and J.R. Pawlik. 1998. Spongivory by parrotfish in Florida mangrove and reef habitats. *Mar. Ecol.* 19:325–337.

Eckert, K.L. 1995. Anthropogenic threats to sea turtles. Pages 611–612 in K.A. Bjorndal, editor. *Biology and Conservation of Sea Turtles*, revised edition. Smithsonian Institution Press, Washington, D.C.

Esquemeling, J. 1684. The buccaneers of America, translated from Dutch, edited by W.S. Stallybrass. George Routledge and Sons, London. Reprinted 1924, 480 pp.

Frick, M.G., K.L. Williams, and M. Robinson. 1998. Epibionts associated with nesting loggerhead sea turtles (*Caretta caretta*) in Georgia, USA. *Herpetol. Rev.* 29:211–214.

Goreau, T.F. and W.D. Hartman. 1963. Boring sponges as controlling factors in the formation and maintenance of coral reefs. Pages 25–54 in R.F. Sognnaes, editor. *Mechanisms of Hard Tissue Destruction*. American Association for the Advancement of Science, New York.

Groombridge, B. and R. Luxmoore. 1989. *The Green Turtle and Hawksbill (Reptilia: Cheloniidae): World Status, Exploitation and Trade*. CITES Secretariat, Lausanne, Switzerland. 601 pp.

Harvell, C.D. et al. 1999. Emerging marine diseases — climate links and anthropogenic factors. *Science* 285:1505–1510.

Hill, M. 1998. Spongivory on Caribbean reefs releases corals from competition with sponges. *Oecologia* 117:143–150.

Hirst, G.S.S. 1910. *Notes on the History of the Cayman Islands*. P. A. Benjamin Manuf. Co., Kingston, Jamaica; 412 pp. Reprinted in 1967 by Caribbean Colour, Grand Cayman, British West Indies.

Jackson, J.B.C. 1997. Reefs since Columbus. *Coral Reefs* 16:S23–S33.

Jackson, J.B.C. 2001. What was natural in the coastal oceans? *Proc. Natl. Acad. Sci. U.S.A.* 98:5411–5418.

Jackson, J.B.C. et al. 2001. Historical overfishing and the recent collapse of coastal ecosystems. *Science* 293:629–638.

King, F.W. 1982. Historical review of the decline of the green turtle and hawksbill. Pages 183–188 in K.A. Bjorndal, editor. *Biology and Conservation of Sea Turtles*. Smithsonian Institution Press, Washington, D.C.

LeBoeuf, B.J., K.W. Kenyon, and B. Villa-Ramirez. 1986. The Caribbean monk seal is extinct. *Mar. Mamm. Sci.* 2:70–72.

León, Y.M. and K.A. Bjorndal, in press. Selective feeding in the hawksbill turtle, an important predator in coral reef ecosystems. Marine Ecology Progress Series.

León, Y.M. and C.E. Diez. 1999. Population structure of hawksbill turtles on a foraging ground in the Dominican Republic. *Chelonian Conserv. Biol.* 3:230–236.

Lewis, C.B. 1940. Appendix: The Cayman Islands and marine turtle. The herpetology of the Cayman Islands. Bulletin of the Institute of Jamaica Science Series No. 2:56–65.

Lutcavage, M.E. et al. 1997. Human impacts on sea turtle survival. Pages 387–409 in P.L. Lutz and J.L. Musick, editors. *The Biology of Sea Turtles*. CRC Press, Boca Raton, FL.

Marine Turtle Specialist Group (SSC/IUCN). 1995. A global strategy for the conservation of marine turtles. IUCN Publications, Gland, Switzerland.

McNaughton, S.J. 1985. Ecology of a grazing ecosystem: the Serengeti. *Ecol. Monogr.* 55:259–294.

Meylan, A.B. 1999. Status of the hawksbill turtle (*Eretmochelys imbricata*) in the Caribbean region. *Chelonian Conserv. Biol.* 3:177–184.

Meylan, A.B. and M. Donnelly. 1999. Status justification for listing the hawksbill turtle (*Eretmochelys imbricata*) as Critically Endangered on *The 1996 IUCN Red List of Threatened Animals*. *Chelonian Conserv. Biol.* 3:200–224.

National Marine Fisheries Service and U.S. Fish and Wildlife Service. 1991a. Recovery plan for U.S. population of loggerhead turtle. National Marine Fisheries Service, Washington, D.C.

National Marine Fisheries Service and U.S. Fish and Wildlife Service. 1991b. Recovery plan for U.S. population of Atlantic green turtle. National Marine Fisheries Service, Washington, D.C.

O'Day, S.J. 2001. Change in marine resource exploitation patterns in prehistoric Jamaica: human impacts on a Caribbean island environment. Paper presented at the ICAZ Conference of the Fish Remains Working Group, New Zealand, 8–15 October 2001.

Parsons, J.J. 1962. *The Green Turtle and Man*. University of Florida Press, Gainesville, FL.

Parsons, J.J. 1972. The hawksbill turtle and the tortoise shell trade. Pages 45–60 in *Etudes de Géographie Tropicale Offertes á Pierre Gourou*. Mouton, Paris.

Pauly, D. 1995. Anecdotes and the shifting baseline syndrome of fisheries. *Trends Ecol. Evol.* 10:430.

Pauly, D. et al. 1998. Fishing down marine food webs. *Science* 279:860–863.

Pitcher, T.J. 2001. Fisheries managed to rebuild ecosystems? Reconstructing the past to salvage the future. *Ecol. Appl.* 11:601–617.

Preen, A.R. 1996. Infaunal mining: a novel foraging method of loggerhead turtles. *J. Herpetol.* 30:94–96.

Pritchard, P.C.H. and J.A. Mortimer. 1999. Taxonomy, external morphology, and species identification. Pages 21–38 in K.L. Eckert et al., editors. *Research and Management Techniques for the Conservation of Sea Turtles*. IUCN/SSC Marine Turtle Specialist Group Publication No. 4, Washington, D.C.

Richardson, J.I., R. Bell, and T.H. Richardson. 1999. Population ecology and demographic implications drawn from an 11-year study of nesting hawksbill turtles, *Eretmochelys imbricata*, at Jumby Bay, Long Island, Antigua, West Indies. *Chelonian Conserv. Biol.* 3:250–251.

Robblee, M.B. et al. 1991. Mass mortality of the tropical seagrass *Thalassia testudinum* in Florida Bay. *Mar. Ecol. Prog. Ser.* 71:297–299.

Ross, J.P. 1982. Historical decline of loggerhead, ridley, and leatherback sea turtles. Pages 189–195 in K.A. Bjorndal, editor. *Biology and Conservation of Sea Turtles*. Smithsonian Institution Press, Washington, D.C.

Rützler, K. 1978. Sponges in coral reefs. Pages 299–314 in D.R. Stoddart and R.F. Johannes, editors. *Coral Reefs: Research Methods*. Monographs on Oceanographic Methodologies (UNESCO), number 5, Paris.

Scudder, S.J. and I.R. Quitmyer. 1998. Evaluation of evidence for pre-Columbian human occupation at Great Cave, Cayman Brac, Cayman Islands. *Caribb. J. Sci.* 34:41–49.

Senties, G.A., J. Espinoza-Avalos, and J.C. Zurita. 1999. Epizoic algae of nesting sea turtles *Caretta caretta* (L.) and *Chelonia mydas* (L.) from the Mexican Caribbean. *Bull. Mar. Sci.* 64:185–189.

Sheppard, C. 1995. The shifting baseline syndrome. *Mar. Pollut. Bull.* 30:766–767.

Sinclair, A.R.E. 1995. Serengeti past and present. Pages 3–30 in A.R.E. Sinclair and P. Arcese, editors. *Serengeti II: Dynamics, Management, and Conservation of an Ecosystem.* University of Chicago Press, Chicago.

Sluka, R. et al. 1996. *Habitat and Life in the Exuma Cays, the Bahamas*. The Nature Conservancy, Arlington, VA. 83 pp.

Smith, R.C. 2000. *The Maritime Heritage of the Cayman Islands.* University Press of Florida, Gainesville, FL. 230 pp.

Stokes, A.V. and W.F. Keegan. 1996. A reconnaissance for prehistoric archaeological sites on Grand Cayman. *Caribb. J. Sci.* 32:425–430.

Suchanek, T.H. et al. 1983. Sponges as important space competitors in deep Caribbean coral reef communities. NOAA Symposium Series on Undersea Research 1:55–61.

Targett, N.M. and G.P. Schmahl. 1984. Chemical ecology and distribution of sponges in the Salt River Canyon, St. Croix, USVI. NOAA NMFS Technical Memorandum OAR NURP-1:1–60.

Thayer, G.W. et al. 1984. Role of larger herbivores in seagrass communities. *Estuaries* 7:351–376.

Thayer, G.W., D.W. Engel, and K.A. Bjorndal. 1982. Evidence of short-circuiting of the detritus cycle of seagrass beds by the green turtle, *Chelonia mydas* L. *J. Exp. Mar. Biol. Ecol.* 62:173–183.

Van Dam, R.P. and C.E. Diez. 1997. Predation by hawksbill turtles on sponges at Mona Island, Puerto Rico. *Proceedings of the 8th International Coral Reef Symposium* 2:1421–1426.

Vicente, V.P. 1990. Overgrowth activity by the encrusting sponge *Chondrilla nucula* on a coral reef in Puerto Rico. Pages 36–44 in K. Rützler, editor. *New Perspectives in Sponge Biology.* Smithsonian Institution Press, Washington, D.C.

Williams, N. 1970. *A History of the Cayman Islands*. The Government of the Cayman Islands, Grand Cayman. 94 pp.

Williams, S.L. 1988. *Thalassia testudinum* productivity and grazing by green turtles in a highly disturbed seagrass bed. *Mar. Biol.* 98:447–455.

Witherington, B.E. In press. The biological conservation of loggerheads: challenges and opportunities. In A.B. Bolten and B.E. Witherington, editors. *Biology and Conservation of the Loggerhead Sea Turtle*. Smithsonian Institution Press, Washington, D.C.

Witzell, W.N. 1983. Synopsis of biological data on the hawksbill turtle *Eretmochelys imbricata* (Linnaeus, 1766). FAO Fisheries Synopsis No. 137. Rome. 78 pp.

Wulff, J. 1997. Parrotfish predation on cryptic sponges on Caribbean coral reefs. *Mar. Biol.* 129:41–52.

Zieman, J.C. 1982. The ecology of the seagrasses of South Florida: a community profile. U.S. Fish and Wildlife Service, Office of Biological Services, Washington, D.C.

Zieman, J.C., J.W. Fourqurean, and T.A. Frankovich. 1999. Seagrass die-off in Florida Bay: long-term trends in abundance and growth of turtle grass, *Thalassia testudinum*. *Estuaries* 22:460–470.

Zieman, J.C., R.L. Iverson, and J.C. Ogden. 1984. Herbivory effects on *Thalassia testudinum* leaf growth and nitrogen content. *Mar. Ecol. Prog. Ser.* 15:151–158.

Zimmerman, L.C. and C.R. Tracy. 1989. Interactions between the environment and ectothermy and herbivory in reptiles. *Physiol. Zool.* 62:374–409.

11 Sea Turtle Population Ecology

Selina S. Heppell, Melissa L. Snover, and Larry B. Crowder

CONTENTS

11.1 INTRODUCTION: WHAT DRIVES SEA TURTLE POPULATION DYNAMICS?

Sea turtle population demographics reflect the effects of natural and anthropogenic stressors that include environmental variability, terrestrial habitat loss, terrestrial and aquatic habitat degradation, and direct and indirect fishing mortality (National Research Council, 1990; Lutcavage et al., 1997). New threats include increases in egg incubation temperatures (further skewing sex ratios, which are defined by incubation temperature) caused by global warming and loss of nesting habitat to rising sea levels on developed and armored beaches. In the marine system, accumulation of pollutants such as plastics, heavy metals, environmental estrogens, and oil products in pelagic nursery and demersal coastal habitats threaten juvenile and adult turtles of all species. In addition, sea turtles swim a gauntlet of fishing gear, including trawls, gill nets, pound nets, and longlines, as they migrate across

0-8493-1123-3/03/$0.00+$1.50

ocean basins and feed in nearshore areas. These stressors affect the survival and growth rates of each sea turtle life stage, which, in turn, influences population growth rates and dynamics.

The first requirement for analysis of sea turtle population dynamics is a long time series. Fortunately, several long-term studies on nesting beaches reveal both variability and directional change through time. Long-term monitoring has revealed dangerous declines in Pacific leatherbacks, encouraging trends in Kemp's ridleys, and low but steady nesting populations of hawksbills (Figure 11.1). Apparent cycles of nesting in sea turtles may occur over short or long periods, although the exact cause of such cycling in most species is unknown. Some species, such as green turtles, show marked periodicity that is a function of environmental variance and internesting (remigration) intervals (Hays, 2000; Chaloupka, 2001). In Australia, analysis of long time series of nesting green turtles shows that this species responds to environmental stochasticity and climate shifts such as the El Niño southern oscillation (ENSO) (Limpus and Nicholls, 1988; Chaloupka, 2001). Ridley turtles, at least, are capable of relatively rapid increase once critical mortality factors on large juveniles and adults have been removed and populations are augmented by egg protection programs (Peñaflores et al., 2000; Heppell et al., 2002a). Species that mature later, such as loggerheads and greens, may take much longer to recover from negative perturbations (Crowder et al., 1994; Chaloupka and Limpus, 2001). The time lags in population response for species that take decades to reach maturity are a daunting reality for conservation and management.

We can divide our discussion of sea turtle population dynamics into two components: factors that drive long-term population growth rates and factors that affect short-term variability in populations. Our understanding of these factors and their integration at the population level is limited temporally and spatially; most studies examine a single life stage in one location over a limited time frame. We rely on population models to put the pieces together and project how populations will respond to perturbations. Such extrapolations, although necessary, must be interpreted cautiously and updated continuously with new information.

Specific life histories of individual sea turtle species vary, but the common denominator in all of them is that sea turtles are long-lived, slow-growing species that use multiple habitats over their course of development (Meylan and Ehrenfeld, 2000). General characterizations also include temperature-dependent sex determination, low and variable survival in the egg and hatchling stage, and high and relatively constant annual survival in the subadult and adult life stages. Maximum intrinsic growth rates of sea turtles are limited by the extremely long duration of the juvenile stage in most species and fecundity that is limited by relatively large eggs and infrequent nesting (Heppell, 1998; Gibbs and Amato, 2000). Annual survival, stage duration (growth rates), and reproduction are *vital rates* that are influenced by environmental change and human impacts. These vital rates are the foundation of long-term population trends, and we can use models to assess how long-term trends may be affected by perturbations.

Numerous authors have recently highlighted the management and conservation issues that are critical to maintaining long-lived, slow-growing species (Congdon et al., 1993; Heppell, 1998; Crouse, 1999; Heppell et al., 1999; Musick, 1999). All of these articles emphasize the need for high survival rates in the large juvenile, subadult, and adult stages to achieve positive or stable long-term population growth. A general conclusion for sea turtles and species with similar life histories is that they are unlikely to be able to sustain even moderate levels of harvest, especially if the populations are already at reduced levels.

Variable remigration intervals (periods between nesting seasons), highly variable survival of nests and eggs, and fluctuating ocean conditions contribute to short-term variability in population abundance and population cycles (Limpus and Nicholls, 1988; Chaloupka and Limpus, 2001). These short-term changes are more difficult to predict, but are actually easier to measure in the field and can have a major impact on our assessment of population change.

Long-term population growth rates are also driven by density-dependent changes in vital rates. We know very little about how sea turtle populations respond to density, beyond a few examples of reduced egg survival on crowded nesting beaches (Bustard and Tognetti, 1969; Carr, 1986; Ballestero et al., 2000; Chaloupka, 2001) and evidence of slower growth rates of juvenile green turtles on crowded feeding grounds (Bjorndal et al., 2000a). Because most sea turtles are in low abundance relative to historical conditions (e.g., Jackson, 1997), some modelers have ignored the potential influences of density dependence on vital rates and population growth. However, as some populations recover, carrying capacity to growth and reproduction will become an important area of research. Likewise, the effects of species interactions on survival and growth are density dependent and may change dramatically through time.

In this chapter, we review the life history characteristics that are most relevant to sea turtle population dynamics, techniques used to assess critical vital rates and population trends, and application of population models to population dynamics.

11.2 LIFE HISTORY

11.2.1 Life Stages and Ontogenetic Shifts

Sea turtle species share a common life cycle composed of a series of stages. *Ontogenetic shifts*, or shifts in location and habitat that occur during the life cycle in response to changes in vital rates (Werner and Gilliam, 1984), have a major impact on where sea turtles of different sizes or stages occur and, subsequently, the human-caused hazards to which they are exposed. We know of at least one major ontogenetic shift that occurs in the hard-shelled sea turtles: the shift from pelagic to benthic feeding areas. There may be additional ontogenetic shifts in microscale habitat use or resource utilization of which we are not yet aware, but may have important ramifications for management.

The general life cycle and specifics for individual species are described by Musick and Limpus (1997). Briefly, adult females dig nest cavities on sandy, ocean-facing beaches and, depending on species, deposit anywhere from 50 to 130 eggs per nest (Van Buskirk and Crowder, 1994). After incubation, hatchlings emerge from the nests, crawl down the beach to the water, and swim out to the open ocean. Young juveniles remain pelagic for a length of time that varies by species and potentially by geographic location within species (Musick and Limpus, 1997). Following the

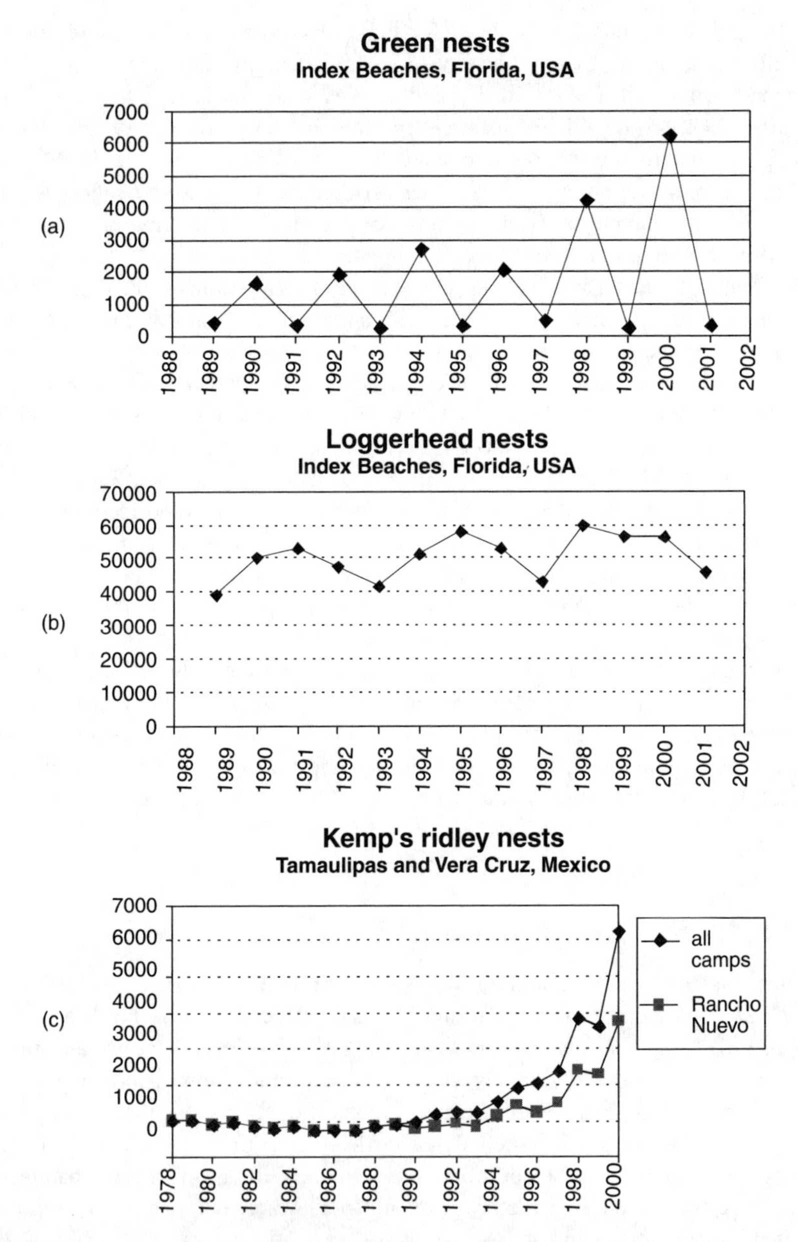

FIGURE 11.1 Examples of long time series of sea turtle abundance. (Data generously provided by B. Witherington, Florida Fish and Wildlife Conservation Commission, Florida Marine Research Institute, Index Nesting Beach Survey Program [greens and loggerheads]; the Gladys Porter Zoo and SEMARNAT/INE, Mexico [Kemp's ridleys].

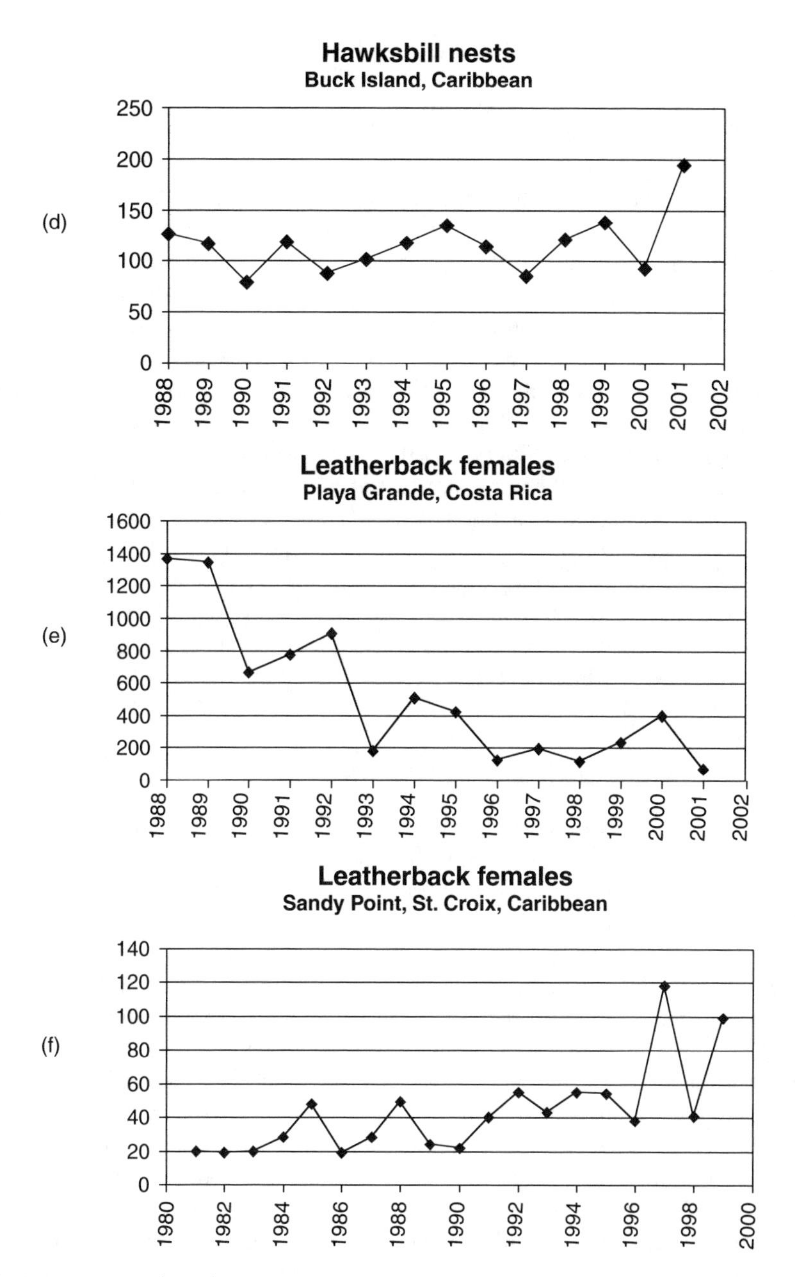

FIGURE 11.1 (continued) Examples of long time series of sea turtle abundance. (Data generously provided by Z. Hillis-Starr and B. Phillips, Buck Island Reef NM Sea Turtle Research Program [hawksbills]; D. Dutton and P. Dutton [leatherbacks, St. Croix], and R. Reina and J. Spotila [leatherbacks, Costa Rica].)

pelagic stage, juveniles of most species recruit to nearshore habitats and switch to feeding on benthic organisms. For juveniles found in temperate regions, there are usually migrations between summer and winter habitats, whereas migrations are not as extensive for more tropical species (Musick and Limpus, 1997).

A dramatic shift in habitat and diet occurs at least once in the life cycle of most juvenile sea turtles. This happens when juvenile sea turtles switch from an oceanic, pelagic habitat and epipelagic feeding to a neritic habitat and benthic feeding. For the green sea turtle, this habitat switch is accompanied by a shift from omnivory to herbivory (Bjorndal et al., 1995; 2000a). Most of the species of sea turtles appear to spend little time as pelagic juveniles, as they are first seen in coastal habitat at small sizes. The loggerhead is an exception to this; they do not recruit to nearshore habitats until they are 40–50 cm straight carapace length (SCL) for the southeastern U.S. population or >70 cm SCL in Australia (Limpus et al., 1994; Bjorndal et al., 2000b; 2001). Exceptions to the benthic habitat shift by juveniles occur in leatherback and, possibly, olive ridley sea turtles. Leatherbacks have been observed foraging in coastal waters; however, they are considered to remain pelagic throughout their lives (Eckert et al., 1989). Little information exists for olive ridleys, but they appear to remain epipelagic until they are adults, when they have been observed using both nearshore and pelagic habitats (Reichart, 1993; Plotkin et al., 1996). An additional exception to the general life cycle is the flatback sea turtle, which remains neritic throughout its life (Chaloupka and Musick, 1997).

11.2.2 Growth Rates and Stage Lengths

Age at sexual maturation (ASM) and growth rates are not definitively known for any species. Typically, age-based growth functions are applied to mark–recapture, length frequency, or skeletochronological data to infer ASM on the basis of size at maturity. We have summarized growth rate studies for sea turtles with the authors' estimates for ASM or the portion of the life cycle that they studied (Table 11.1).

Each habitat that sea turtles use over their ontogeny has different environmental parameters, such as food availability and temperature, that will influence growth rates. Little information is available on sea turtle growth rates in the pelagic environment (although see Zug et al., 1995; and Bjorndal et al., 2000b). In a skeletochronology study, Snover et al. (in review) found evidence of a shift in growth rates that corresponds to the ontogenetic shift from pelagic to benthic feeding. The shift is typified by a surge in growth followed by declining growth rates. Chaloupka and Limpus (1997) and Limpus and Chaloupka (1997) noted increasing growth rates after settlement in the hawksbill and green sea turtles. Growth rates increased until 50–60 cm curved carapace length (CCL) for the hawksbill and 60–63 cm CCL for the green. After these peaks, growth rates declined monotonically to adulthood in both species. Of course, it cannot be determined if the increasing growth rates postsettlement were continuing from the pelagic stages or if they were surges in growth after settlement. However, the results of these studies on sea turtle growth highlight the likelihood that sea turtle growth rates are compartmentalized and that shifts occur in conjunction with ontogenetic habitat shifts.

TABLE 11.1
Estimated Stage Durations for Sea Turtles

Species	Method	Location	Stage/Size	Duration (years)	Growth Function	Reference	N[a]
Loggerhead	Mark–recapture	SE U.S.	H–A/75 cm SCL	10–15	Linear	Mendonca, 1981	13[b]
	Mark–recapture	SE U.S.	H–A/74–92 cm SCL	12–30/30–47	von Bertalanffy/ logistic	Frazer and Ehrhart, 1985	28[b]
	Mark–recapture	Bahamas	BJ/25–75 cm SCL	3–4	Direct measurement	Bjorndal and Bolten, 1988	3
	Mark–recapture	SE U.S.	BJ/50–75 cm SCL	15+	von Bertalanffy/ logistic	Henwood, 1987	118[c]
	Mark–recapture	Australia	H–A/93 cm CCL	35.9	von Bertalanffy	Frazer et al., 1994	172[b]
	Mark–recapture	SE U.S.	BJ/45–75SCL		von Bertalanffy	Braun-McNeill et al., in review	32
	Mark–recapture	SE U.S.	BJ–A/49–90 cm SCL	32	von Bertalanffy	NMFS, 2001	111
	Length frequency	Azores	H–PJ/46–64 cm CCL	6.5–11.5	von Bertalanffy	Bjorndal et al., 2000b	1692
	Length frequency	SE U.S.	BJ/46–87 cm CCL	20	von Bertalanffy	Bjorndal et al., 2001	1234
	Skeletochronology	SE U.S.	H–A/80–90 cm CCL	13–15	Age estimates	Zug et al., 1986	19[b]
	Skeletochronology	SE U.S.	H–A/92.5 cm SCL	22/26	von Bertalanffy/ logistic	Klinger and Musick, 1995	83[b]
	Skeletochronology	Pacific	PJ/42 cm SCL	7.7–8.9	Estimated age	Zug et al., 1995	12
	Skeletochronology	SE U.S.	H–A/92.4 cm SCL	20–24	von Bertalanffy	Parham and Zug, 1997	98[b]

TABLE 11.1 (continued)
Estimated Stage Durations for Sea Turtles

Species	Method	Location	Stage/Size	Duration (years)	Growth Function	Reference	N[a]
Green	Mark–recapture	Australia	BJ/60–90 cm CCL	23	Estimated from growth rates	Limpus and Walter, 1980	40
	Mark–recapture	SE U.S.	H–A	25–30	Linear	Mendonca, 1981	12[b]
	Mark–recapture	Pacific	BJ–A/35–81 cm SCL	8.7–47.9	Estimated from growth rates	Balazs, 1995	35
	Mark–recapture	Pacific	BJ–A/35–92 cm SCL	10.8–59.4	Estimated from growth rates	Balazs, 1995	35
	Mark–recapture	SE U.S.	H–A/88–99 cm SCL	18–27/26–33	von Bertalanffy/ logistic	Frazer and Ehrhart, 1985	11[b]
	Mark–recapture	U.S. Virgin Is.	H–A/106–112 cm CCL	27–33	von Bertalanffy	Frazer and Ladner, 1986	8[b]
	Mark–recapture	Bahamas	BJ/30–75 cm SCL	17	von Bertalanffy	Bjorndal and Bolten, 1988	122
	Mark–recapture	Atlantic	H–A/104–130 kg	19–24	von Bertalanffy	Ehrhardt and Witham, 1992	52[b]
	Mark–recapture	Galapagos	BJ/40–67 cm SCL	92	Estimated from growth rates	Green, 1993	28
	Mark–recapture	Bahamas	BJ/30–70 cm SCL	11.96	von Bertalanffy	Bjorndal et al., 1995	524[c]
	Mark–recapture	Bahamas	BJ/30–70 cm SCL	13.05	Estimated from growth rates	Bjorndal et al., 1995	524[c]

(continued)

	Mark–recapture	Australia	H–A/95 cm CCL	≥40	Nonparametric regression	Limpus and Chaloupka, 1997	537
	Length frequency	Bahamas	BJ/30–70 cm SCL	13.04	Von Bertalanffy	Bjorndal et al., 1995	964[c]
	Skeletochronology	SE U.S.	H–BJ/70 cm SCL	14	Estimated age	Zug and Glor, 1998	59[b]
	Skeletochronology	Hawaii	H–BJ/80 cm SCL	25.3	Correction factor	Zug et al., 2002	104
	Skeletochronology	Hawaii	H–BJ/80 cm SCL	25.6	Spline integration	Zug et al., 2002	104
Kemp's ridley	Mark–recapture	Gulf of Mexico	H–A/60 cm SCL	10	Von Bertalanffy	Caillouet et al., 1995	117[b,d]
	Mark–recapture	SE U.S.	BJ/20–60 cm SCL	7–8	Von Bertalanffy	Turtle Expert Working Group, 2000	96
	Mark–recapture	SE U.S.	H–A/56–64 cm SCL	7–12	Von Bertalanffy	Turtle Expert Working Group, 2000	96[b]
	Skeletochronology	SE U.S.	H–A/60–65 cm SCL	13.2–15.7	Von Bertalanffy	Zug et al., 1997	69
Leatherback	Skeletochronology	Pacific	H–A/144.5 cm CCL	3.7–13.3	Logistic/Von Bertalanffy	Zug and Parham, 1996	16

Notes: H = hatchlings; PJ = pelagic juvenile; BJ = benthic juvenile; and A = adult. H–A represents age at sexual maturity. For H–A or H–PJ, the size distribution denotes the range in sizes for which the authors considered turtles to be adults or to be at the end of the pelagic stage. Otherwise, the size distribution is the range in size for which duration has been estimated. Captive growth rates are not included.

[a]Indicates actual number of turtles in study; some studies used multiple recaptures of the same turtle as additional data points.

[b]Results were extrapolated beyond the range of the data.

[c]Number of growth records used, actual number of turtles not reported.

[d]Animals from this study were kept in captivity for first year, then released as part of the head-start program for Kemp's ridleys.

Recent studies have highlighted the possibility of shifts in growth rates that do not occur in conjunction with previously defined ontogenetic habitat shifts. Chaloupka (1998) analyzed age data from a study by Zug et al. (1995) and found evidence of polyphasic growth within the pelagic stage of loggerheads in the Pacific Ocean. Similarly, with the Kemp's ridley, Chaloupka and Zug (1997) found evidence of polyphasic growth, with the first peak in growth rates occurring around 15 cm SCL, which is consistent with the size at which they begin to appear in coastal waters (Turtle Expert Working Group, 2000). However, this is difficult to interpret because the data set did not include any pelagic animals and the growth rate from hatchling to the first benthic animal in the sample was modeled as essentially linear, making the growth rates appear to slow after this period. The second peak in growth rates occurred at approximately 45 cm SCL (Kemp's ridleys mature at about 60 cm SCL). This observation is supported by Schmid (1998), who found that although not significant, average growth rates in the 40–50 cm SCL size class were higher than in 30- to 40-cm and 50- to 60-cm SCL size classes. These growth shifts do not relate to ontogenetic habitat shifts, but may be indicative of additional ontogenetic shifts, whether they be diet, habitat, or physiological in nature, that result in changes in growth rate.

The predominant method used to infer age-based growth rates in sea turtles has been the comparison of growth in carapace length between captures–recaptures of tagged individuals. This information is used in the interval forms of the von Bertalanffy and/or logistic growth equations to produce a size-at-age growth curve (Fabens, 1965; for example, see Frazer and Ehrhart, 1985). Because of the inaccessibility of all life stages, these curves are often prepared from data that span only a portion of the life stages. The stage most commonly not included is the pelagic (Mendonca, 1981; Frazer and Ehrhart, 1985; Frazer and Ladner, 1986; Frazer et al., 1994; Turtle Expert Working Group, 2000). As discussed previously, there is likely a shift in growth rates following the pelagic stage, and estimating pelagic growth rates from benthic juvenile growth rates is inappropriate. Growth rates from the adult life stage have also not been included when age-to-maturity has been estimated (Mendonca, 1981; Frazer and Ladner, 1986; Ehrhardt and Witham, 1992). Recently, authors have become more aware of this oversight and are estimating only the length of time it takes an animal to grow through the size classes for which they have data (Bjorndal and Bolten, 1988; Bjorndal et al., 1995; National Marine Fisheries Service [NMFS], 2001; Braun-McNeill et al., in review). Another potential problem with this application of the von Bertalanffy growth curve is that this growth function implies a monotonically declining relationship with growth and age (Chaloupka and Musick, 1997). As discussed previously, this is contradictory in some studies because their data indicate a nonmonotonic relationship (Chaloupka and Limpus, 1997; Limpus and Chaloupka, 1997; Chaloupka and Zug, 1997).

Skeletochronology uses growth marks found in bone tissue to estimate age. Numerous studies have applied this technique to sea turtles (Zug et al., 1986; 1995; 1997; Klinger and Musick, 1992; Zug and Parham, 1996; Parham and Zug, 1997; Bjorndal et al., 1998; Zug and Glor, 1998; Coles et al., 2001; Zug, 2002; Snover and Hohn, in review). Klinger and Musick (1992), Coles et al. (2001), and Snover and Hohn (in review) present evidence of the annual nature of the growth marks for loggerhead and Kemp's ridley sea turtles. In addition, Snover and Hohn (in review) demonstrate a constant proportionality between bone growth and somatic growth measured as SCL for the loggerhead and Kemp's ridley. This will allow the use of growth mark diameters for estimating growth rates in these species, making skeletochronology a very powerful tool in realizing actual sea turtle growth rates and in understanding how much individual variability influences our perceptions.

11.2.3 Age at Sexual Maturation

All species of sea turtles exhibit delayed maturity. The extent of this delay has only recently been appreciated by researchers. Initial estimates of ASM for loggerheads from the southeast U.S. were around 20 years (Frazer and Ehrhart, 1985; Mendonca, 1981) More recently, however, estimates exceed 30 years (Frazer et al., 1994, National Marine Fisheries Service, 2001). Exactly where the correct age or, more accurately, age range for reproductive maturity lies for this population or any other loggerhead population or sea turtle species remains to be definitively determined. Long-term population growth rates and responses to perturbations are strongly influenced by ASM (Heppell et al., 2000a). Heppell et al. (2002a) demonstrated the effect of increasing ASM on the effectiveness of turtle excluder devices (TEDs) for the southeast U.S. loggerhead population. The longer generation times caused by increased ASM result in slower long-term population growth.

There is a great deal of variability in ASM among those species that have been studied. For the hard-shell turtles, Kemp's ridleys mature earliest at 7–12 years (Table 11.1). Kemp's also have the smallest adult female size of 64.6 cm SCL (Van Buskirk and Crowder, 1994). For loggerheads and greens, the high-end estimates are similar, with values in the 30–35 year range (Table 11.1). Average adult female body sizes in these two species are similar (87.0 cm SCL for the loggerhead and 99.1 cm SCL for the green; Van Buskirk and Crowder, 1994). Given that postpelagic loggerheads are primarily carnivorous, whereas postpelagic greens are herbivorous, it is possible that greens have a longer benthic juvenile stage than do loggerheads simply because of the lower conversion potential of plant material.

The only other sea turtle for which ASM has been estimated from growth rate studies is the leatherback. Zug and Parham (1996) estimated an ASM of 13–14 years for Pacific leatherbacks using skeletochronology. Leatherbacks are the largest species of sea turtle (148.7 cm SCL for the adult female; Van Buskirk and Crowder, 1994), but this may not indicate that they should have the longest ASM. They have been shown to be distinct from the hard-shell sea turtles using different means of classification. They show a distinct phylogenetic separation from the rest of the sea turtle species (Bowen and Karl, 1997) and a distinct phenotypic separation based on reproductive traits (Van Buskirk and Crowder, 1994). In addition, leatherbacks are warm-blooded and might be expected to have a faster growth rate than poikilothermic species of similar mass (Musick, 1999). It is possible that growth rates in juvenile leatherbacks are extremely high in comparison to the hard-shell sea turtles, and that they do have a comparatively low ASM, but much more information is needed on this species.

11.2.4 Growth Rate Variability

The ranges in ASM values reported in studies are generally indicative of uncertainty in size-at-maturity (minimum or average size of nesters) or in the calculated growth curve (Table 11.1), rather than an attempt to capture the true variability in ages at which maturity occurs. There can be a great deal of year-to-year variability in growth rates of loggerheads within the same 10-cm juvenile size-class (Braun-McNeill et al., in review). The cumulative variability over nine 10-cm size classes can result in a wide range of predicted ASMs for individual turtles.

Variability in growth rates may be caused by a number of factors such as genetics, environmental conditions, and individual health. Limpus and Chaloupka (1997) found that immature, female green turtles in the southern Great Barrier Reef displayed significantly decreased growth rates at a time that coincided with a strong ENSO event in the early 1980s. Bjorndal et al. (2000a) measured growth rates of immature green sea turtles in the southern Bahamas. Over their 18-year study period, population densities increased by a factor of six, then decreased by a factor of three. They found a significant negative relationship between the estimated annual population density and the estimated mean annual growth rate, suggesting density-dependent effects on growth rates for this population.

There is also evidence of sex-specific growth rates (Chaloupka and Limpus, 1997; Limpus and Chaloupka, 1997). Female hawksbills from the southern Great Barrier Reef display faster growth rates at all benthic juvenile sizes than do males (Chaloupka and Limpus, 1997). In the same region, at sizes greater than 60 cm CCL, female green sea turtles grow faster than male green sea turtles in the southern Great Barrier Reef (Limpus and Chaloupka, 1997). This same study found a sexual dimorphism in adult body size for greens, indicating that breeding males are an average of 7 cm CCL smaller than breeding females. It has been demonstrated for reptiles that animals with fast growth mature to a larger size than animals with slow growth, and this may also be the case with sea turtles (Stamps et al., 1998). It is unclear how much of the variability observed in growth rates of similar-sized individuals may be attributable to sex-specific growth rates.

Most growth curves of sea turtles have been prepared with small sample sizes (Table 11.1). If the variability in growth rates for sea turtles is high, it will take large sample sizes to determine the "average" growth rate. With the possible exception of the data from Australia, from which large sample sizes are available, we can look at stage duration and ASM as only rough estimates.

11.2.5 Survival Rates

One risk of the delayed maturity representative of the sea turtle life history is the increased risk of dying before reproducing. ASM is extremely high in sea turtles (Table 11.1), and hatchling survival is extremely low; therefore, there must be high survival of juveniles and adults (Congdon et al., 1993). Crouse et al. (1987) determined that for loggerheads in the southeast U.S., the population intrinsic rate of increase (r) was most sensitive to proportional changes in the survival rate of large juveniles, which equates to the benthic juveniles. In other turtle species, large juveniles and/or adults have been identified as the stages with the highest proportional sensitivity (Heppell, 1998). This means, for example, that the long-term average growth rate of a sea turtle population is more sensitive to a 10% change in the survival rate of large juvenile turtles than it is to a 10% change in the hatchling or adult survival rate, in large part because most of the population consists of large juveniles when the population is at a stable stage distribution (Heppell et al., 2000a).

The egg and hatchling stage is the most easily accessible stage of sea turtles. Therefore, this stage has received the greatest amount of study and conservation effort. However, models demonstrate the low contribution of egg survival to mean population growth rate. Starting with a loggerhead population declining at about 5% per year, Crowder et al. (1994) showed that a 90% decrease in egg/hatchling mortality was not enough to prevent the population decline, whereas a 50% decrease in the benthic juvenile mortality alone resulted in positive population growth. This is not to say that the egg and hatchling stages are not important. Increased egg or hatchling survival cannot compensate for decreases in subadult and adult survival rates; however, inputs from these stages are critical in maintaining recruitment to the older stages. Egg harvest, coupled with fishing mortality, is thought to be the primary cause of population crashes in Kemp's ridley and Pacific leatherback turtles, and fox predation of loggerhead eggs may be the cause of recent recruitment failure in Australian loggerheads (Chaloupka and Limpus, 2001). Efforts to conserve nesting beaches and protect eggs from harvesting are important and need to be continued; however, conservation efforts that focus resources on this stage alone are not providing optimal benefits to population recovery.

Adult females exhibit a strong degree of nest site fidelity, allowing for the possibility of recapturing the same turtle when she returns to nest. Therefore, most estimates of sea turtle survival rates are for adult females (Table 11.2). An assumption of survival rates estimated from nesting females is that females who nest at the monitored beach will return to that beach to nest. It is known that nest site fidelity in sea turtles is not perfect (Miller, 1997) and that females may try out beaches before selecting a nesting beach. In analyses of nesting data, females that are tagged and never seen again are assumed dead, when they may have moved to another beach to nest. In most cases, then, survival rates estimated from nesting data would underestimate actual survival rates.

Studies to estimate juvenile survival rates are complicated by their multiple habitat use and highly migratory nature. Catch curves (Seber, 1982) have been used to calculate survival rates for the elusive juvenile stages. Frazer (1987) applied the technique to two cross-sectional data sets of loggerheads, one of dead strandings and one of trawl-caught turtles. Survival rates for U.S. loggerheads have recently been updated using new growth curves based on skeletochronology and mark–recapture (National Marine Fisheries Service, 2001). Catch-curve analysis assumes that populations are stable through time and that the data set used represents a cross section of the true population. Cohort analysis avoids the assumption of a stable population, and has been applied to Kemp's ridleys, where cohort strength is known (Turtle Expert Working Group, 2000). To prepare a catch curve or cohort analysis,

TABLE 11.2
Annual Survival Rates Estimated for Sea Turtles

Species	Method	Duration of Study (years)	Life Stage	Survival	Location	Reference
Loggerhead	Mark–recapture	12	Adult female	0.8091	SE U.S.	Frazer, 1983
	Catch curve	1	Benthic juvenile	0.70	SE U.S.	Frazer, 1987
	Catch curve	1	Benthic juvenile	0.68	SE U.S.	Frazer, 1987
	Mark–recapture	14	Adult female	0.9102	Australia	Heppell et al., 1996
	Mark–recapture	8	Adult female	0.782	Australia	Heppell et al., 1996
	CJS	14	Pubescent	0.8853	Australia	Heppell et al., 1996
	CJS	14	Subadults	0.8295	Australia	Heppell et al., 1996
	Catch curve	4	Benthic juvenile	0.893	SE U.S.	NMFS, 2001
	CJS	9	Benthic juvenile	0.918	Australia	Chaloupka and Limpus, 2002
	CJS	9	Adult	0.875	Australia	Chaloupka and Limpus, 2002
Green	Mark–recapture	20	Adult female	0.57–0.75	Costa Rica	Bjorndal, 1980
	CJS	9	Benthic juvenile	0.8804	Australia	Chaloupka, 2002; Chaloupka and Limpus, 1998
	CJS	9	Subadult	0.8474	Australia	Chaloupka, 2002; Chaloupka and Limpus, 1998
	CJS	9	Adult	0.9482	Australia	Chaloupka, 2002; Chaloupka and Limpus, 1998
Kemp's ridley	Catch curve	12	Pelagic juvenile	0.24–0.38	SE U.S.	Turtle Expert Working Group, 2000

	Catch curve	12	Benthic juvenile	0.83–0.92	SE U.S.	Turtle Expert Working Group, 2000
	Catch curve	12	Adult	0.83–0.92	SE U.S.	Turtle Expert Working Group, 2000
	Mark–recapture	15	Benthic juvenile	0.2–0.99	SE U.S.	Caillouet et al., 1995
Leatherback	Nest data	10	Nesting female	0.89	Malaysia	Chua, 1988
	Mark–recapture	11	Nesting female	0.66	Costa Rica	Spotila et al., 2000
Hawksbill	Mark–recapture	11	Nesting female	0.95	Antigua	Richardson et al., 1999

Note: CJS = Cormack-Jolly-Seber mark–recapture estimation.

it is necessary to estimate the ages of individuals, and growth curves such as von Bertalanffy curves are used to estimate age from size. This introduces a new source of error because the true relationship of size and age in sea turtles is not known and there is likely to be a great deal of variability in this parameter.

One of the most reliable means of estimating survival rates in juveniles is to apply Cormack-Jolly-Seber statistics to data obtained from the repeat observation of the same animals in a population, also known as mark–recapture or capture–mark–recapture (Cormack, 1964; Jolly, 1965; Seber, 1965). Unfortunately, this technique requires either a high recapture probability or a large number of captures over a long period of time. Few such data sets exist, and the studies are expensive to run. However, if we are to gain a better understanding of juvenile survival rates, resources must be applied to these studies.

11.2.6 Reproduction

Because sea turtles cannot provide parental care to offspring, a potential benefit conveyed by delaying reproduction is the ability to lay larger clutches with larger eggs. This is evidenced by a positive relationship between size of adult females and egg size in sea turtles (Hirth, 1980; Van Buskirk and Crowder, 1994). There is likely a great deal of variability in hatchling survival from year to year resulting from factors such as variable nest predation and environmental factors such as hurricanes. However, the overall survivorship of hatchlings to their first birthday is likely very low. To compensate for this, sea turtles produce large quantities of eggs in a nesting year and reproduce many times in a lifetime. There is evidence that females nesting for the first time may produce fewer nests with fewer eggs and have a longer period before nesting for a second time (Brooke and Garnett, 1983; Frazer and Richardson, 1985; Miller, 1997; Chaloupka, 2001). This may affect our estimates of population size and trends on nesting beaches where the number of females and hatchlings is extrapolated from nest counts.

Few sea turtles nest annually. The length of time between successive nesting migrations can vary among populations and individuals within a population. Evidence from tagging studies suggests that remigration intervals are not fixed and may be in response to environmental conditions such as ENSO events or ocean cycles of longer periodicity (Chaloupka, 2001). Good ocean conditions may influence survival and growth as well as nesting frequency, so periods of favorable ocean conditions may mask population declines caused by anthropogenic factors (Chaloupka, 2001). Green turtles, in particular, show marked periodicity in nest abundance, which may be due to productivity in seagrass beds (Bjorndal et al., 1999; Chaloupka and Limpus, 2001). Response to ocean conditions has also been suggested for black turtles (*Chelonia agassizi*, Fuentes et al., 2000). Broderick et al. (2001) speculated that green turtle nesting was more variable than loggerhead nesting at Ascension Island because fluctuating ocean conditions had a large impact on seagrass, the primary food source for herbivorous green turtles, but less impact on the invertebrate prey of carnivorous loggerheads.

The sex of sea turtle hatchlings is environmentally determined by a restricted range of nest incubation temperatures (Mrosovsky and Pieau, 1991). Pivotal and transitional ranges of temperatures determine whether a nest will produce males, females, or a combination of both. Many sea turtle populations appear to produce strongly female-biased offspring (Hanson et al., 1998; Godley et al., 2002; others). Freedberg and Wade (2001) theorize that female-biased sex ratios in reptiles with environmentally controlled sex determination are reinforced by natal nest site fidelity of females. Nest sites on beaches that produce predominantly female offspring will be perpetuated as the female offspring mature and return to the same sites. The authors do not provide an explanation for the adaptive significance of this in terms of population maintenance. Although long-term sex ratio skews in offspring are not predicted by life history theory (Fisher, 1930), they can be favored in spatially structured populations (Charnov, 1982). We are gradually learning about the mating systems of sea turtles through genetic studies (e.g., FitzSimmons, 1998), but it is difficult to determine the point at which a population could become male-limited. Potential effects of a shortage of adult males could include reduced eggs per nest, reduced clutches, or increased remigration interval. From a management perspective, temperature-dependent sex determination supports the need to protect and maintain populations in higher latitudes, such as the northern nesting subpopulation of loggerheads (NMFS, 2001).

In addition to reduced reproduction through sex ratio bias, small populations may suffer effects of *density depensation*, such as a reduction in nesting frequency caused by an inability of females to find potential mates. The problem could be severe for Pacific leatherbacks, which have wide geographic ranges and have suffered a decrease of as much as 99% over the last two decades (Chan and Liew, 1996; Spotila et al., 2000). Two centuries after the end of commercial exploitation, green turtles have failed to recover on the Cayman Islands; this population may have been driven to such a small size that recovery is no longer possible (Aiken et al., 2001).

11.3 POPULATION STRUCTURE

Understanding population structure is critical to the conservation and management of a species. Tagging studies initially demonstrated that adult female sea turtles exhibit a high degree of nest site fidelity in that they return to the same nesting region (e.g., Bjorndal et al., 1983). Studies of mitochondrial DNA (mtDNA) further indicate that female sea turtles are actually exhibiting natal nest site fidelity and returning to the area of beach at which they hatched (Bowen et al., 1992; 1993; Allard et al., 1994; Encalada et al., 1998). These studies demonstrate that there is female-mediated genetic differentiation between nesting areas that is maintained by natal nest site fidelity.

We have the best understanding of population structure for loggerheads in the Atlantic Ocean and Mediterranean Sea (Turtle Expert Working Group, 2000). Studies of mtDNA from rookeries in these regions have demonstrated at least eight genetically distinct nesting areas: (1) Greece/Cyprus; (2) Turkey; (3) Brazil; (4) Yucatán, Mexico; (5) Dry Tortugas, FL; (6) south Florida; (7) Florida panhandle; and (8) northeast Florida to North Carolina (Bowen et al., 1992; Encalada et al., 1998; Laurent et al., 1998; Francisco et al., in press).

The ramification of the genetic structuring is that if one of these nesting aggregations becomes extinct, it will not be recolonized on conservation-level time scales (Bowen et al., 1992). We do not yet fully understand the level of interaction between the nesting aggregations brought about by male-mediated gene flow. Studies have indicated that the genetic structuring based on nuclear DNA (nDNA) is not as strong as that of mtDNA, indicating that there may be male-mediated gene flow between proximate nesting regions (Karl et al., 1992; Francisco, 2001). If this is the case, the status of one nesting aggregation may not be independent of the status of an adjacent aggregation.

For example, consider temperature-dependent sex determination, which occurs in all species of sea turtles. Pivotal and transitional ranges of temperatures determine whether the nest will produce males, females, or both, with cooler temperatures producing males and warmer temperatures producing females (Mrosovsky and Pieau, 1991). Studies recording nest chamber temperatures near Cape Canaveral, FL, indicate that these nests produce nearly all females (Mrosovsky and Provancha, 1989; Hanson et al., 1998). The smaller nesting aggregation adjacent to the south Florida aggregation extends northward to North Carolina, and presumably produces a greater number of males. Although females from these two regions will return to nest in their region of birth, it is possible that males mate with females from both regions. We do not understand the mating structure of loggerhead sea turtles, but it may be that the males produced in the northern nesting areas are important to the continued health of the very large south Florida nesting aggregation.

Similar genetic structuring has been identified in the green (Bowen et al., 1992; Allard et al., 1994; Lahanas et al., 1994; Encalada et al., 1996) and hawksbill (Bass et al., 1996) sea turtles. Leatherbacks show somewhat less genetic structuring, whereby adjacent nesting regions are indistinguishable, suggesting either that the nesting areas were recently colonized or that there is less precise natal nest site fidelity for this species (Dutton et al., 1999).

Juvenile feeding grounds for sea turtles are of mixed stocks and consist of individuals from different nesting regions. An analysis of mtDNA from loggerheads that were caught in drifting longline fisheries within the Mediterranean demonstrated that 51–53% of the turtles originated from nesting aggregations within the Mediterranean, 45–47% originated from the south Florida nesting aggregation, and approximately 2% originated from the northeast Florida to North Carolina nesting aggregation (Laurent et al., 1998). Similarly, juvenile feeding grounds off Charleston, S.C., are composed of approximately 50% south Florida and 50% northeast Florida to North Carolina loggerheads (Sears et al., 1995). For hawksbills, Bowen et al. (1996) found that feeding grounds at Mona Island, Puerto Rico, were composed of turtles originating from throughout the Caribbean. Juvenile green sea turtles found foraging at Great Inagua, Bahamas, also originated from nesting colonies throughout the Caribbean (Lahanas et al., 1997). Hence, fisheries that incidentally take sea turtles do not just impact local nesting populations, but have a much broader influence. The spatial and genetic structure of sea turtles is complex and requires that we assess population dynamics on a global scale.

11.4 ASSESSING POPULATION CHANGE THROUGH TIME

Sea turtle populations exhibit both long- and short-term dynamics (Figure 11.1). Year-to-year changes in sea turtle abundance are caused by environmental stochasticity, periodicity in ocean conditions, sampling error, and the underlying population trend that is a cumulative result of density-dependent and density-independent forces that drive survival, growth, and reproduction. Nesting beach numbers are our primary response variable for assessing changes in sea turtle population size. Because female turtles exhibit strong site fidelity, a nesting beach survey can give a good assessment of changes in the adult female population, provided that the study is sufficiently long, and effort and methods are standardized (Meylan, 1995; Gerrodette and Brandon, 2000; Reina et al., 2002). However, nest and nesting female abundance is often highly variable from year to year because of environmental factors that affect female condition and the intrinsic variability caused by the variable remigration intervals of individual females (Meylan, 1995; Hays, 2000; Chaloupka, 2001). Simple regression analysis is generally insufficient for estimating trends because of high variability or cycling in nest abundance. Error in abundance estimation is likely when methods and effort are not standardized. Increased effort to locate nests, either by biologists or through improved public awareness, can result in apparent population growth or mask population decline. Surveys of index beaches over fixed time intervals can improve trend estimation; however, female nesting activity may exhibit shifts in time and space, making index surveys problematic (Godley et al., 2001). Trend analysis can be made more accurate by incorporating mark–recapture information in remigration probability (Kerr et al., 1999) and by statistical methods that incorporate uncertainty in extrapolation methods (Bjorndal et al., 1999; Reina et al., 2002).

A fundamental problem with nesting beach surveys is that they are unlikely to reflect changes in the entire population. This is because of the long time lag to maturity and the relatively small proportion of females on a nesting beach that are reproducing for the first time, at least in populations with high adult survival rates. Unknown adult and juvenile sex ratios also prevent extrapolation from nesting female abundance to population estimates (Meylan, 1995). A decrease in pelagic juvenile or benthic immature survival rates may be masked by the natural variability in nesting female numbers and the slow response of adult abundance to changes in recruitment to the adult population (Chaloupka and Limpus, 2001). Figure 11.2 illustrates how a hypothetical nesting population might respond to a decrease in pelagic juvenile survival. When random variability or a 5-year cycle of remigration rate is added to the simulation, it takes many more years to detect the change in recruitment (females nesting for the first time, a measure of cohort strength) on the nesting beach. Holmes (2001) found that extinction risk estimates based on time series of nesting females can be strongly biased, and advocated a weighted running sum method to reduce the variance in time series caused by stage-specific counts.

There are many ways to assess population size and trends in abundance beyond nesting beach surveys (Table 11.3). An alternative for assessment and monitoring is to combine beach surveys with in-water surveys and absolute abundance estimates with analyses of changes in survival rates (mark–recapture) and size or age distributions (e.g., Chaloupka and Limpus, 2001). All capture and census methodologies include biases, and many are highly variable (Table 11.3). However, using a variety of methods to assess changes in populations should improve our ability to detect problems that may lead to population declines. Likewise, we may be able to eliminate alternative hypotheses for our observations. For example, loggerhead turtles in North

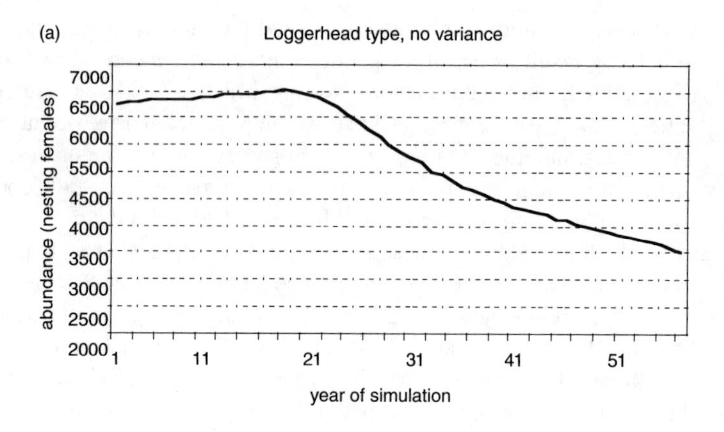

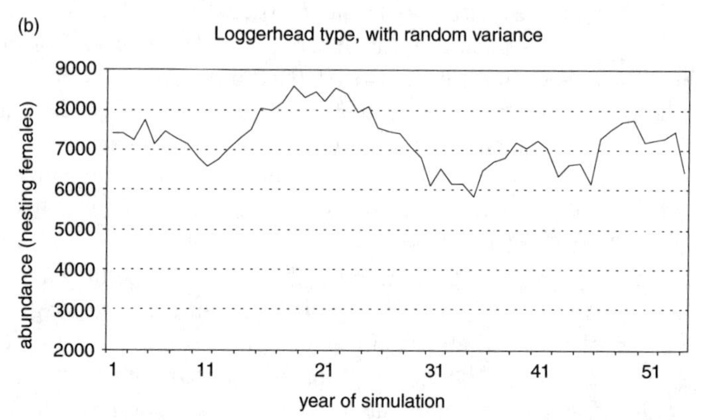

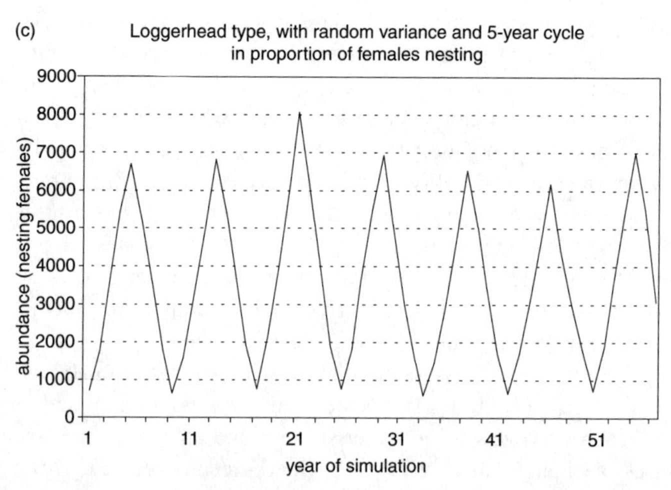

FIGURE 11.2 Changes in the abundance of nesting females following a 10% reduction in the survival rate of pelagic juveniles from a simulation of a loggerhead-type life history (ASM = 25; pelagic stage = age classes 1–9). (a) Deterministic projection. (b) Projection with random variance added to annual survival rates. (c) Projection with random variance and a 5-year cycle in the proportion of adult females nesting.

TABLE 11.3
Common Methods for Assessing Population Size and Trends in Sea Turtles

Census Type	Life Stage	Data	Examples
Nesting beach	Adult females, eggs, hatchlings	AT, eggs/nest, egg and nest survival	Greens: Bjorndal et al., 1999; Chaloupka, 2001 Loggerheads: Hays and Speakman, 1991 All species: Van Buskirk and Crowder, 1994
Nesting beach with tagging	Adult females	AT, nests/female, remigration interval, adult female survival	Hawksbills: Kerr et al., 1999 Loggerheads: Frazer and Richardson, 1985; Chaloupka and Limpus, 2002 Greens: Chaloupka and Limpus, 2001
In-water surveys with tagging	Pelagic juvenile	Growth rates, AT, size and age distribution	Loggerheads: Bolten, 2002
	Benthic juvenile	Growth rates, AT, size distribution, survival rates, maturation	Loggerheads and greens: Chaloupka and Limpus, 2001 Greens: Limpus and Chaloupka, 1997 Loggerheads: Braun-McNeill et al., in review; Chaloupka, 2001 Hawksbills: Chaloupka and Limpus, 1997 Kemp's ridleys: Schmid and Witzell, 1997
CPUE fishing recoveries	Benthic juvenile, adult	AT, size and age distribution	Loggerheads: Butler et al., 1987; Witzell, 1999 Various: Bjorndal and Bolten, 2000
Aerial or boat transects	Benthic juvenile, adult	AT	Various: Bjorndal and Bolten, 2000 Various: McDaniel et al., 2000 Loggerheads, Kemp's ridleys: Lutcavage and Musick, 1985

(continued)

TABLE 11.3 (continued)
Common Methods for Assessing Population Size and Trends in Sea Turtles

Census Type	Life Stage	Data	Examples
Power plant entrainment	Benthic juvenile, adult	Growth rates, size distribution	Turtle Expert Working Group, 2000
Strandings	Benthic juvenile, adult	Size and age distribution, trends in fishing mortality rate	Turtle Expert Working Group, 1998; NMFS, 2001

Notes: All techniques require standardized methods and are limited to specific segments of a population. AT = abundance and trends (specific to sampling location); NMFS = National Marine Fisheries Service.

Carolina, South Carolina, Georgia, and northern Florida (the northern nesting population) have failed to increase despite 15 years of TEDs in shrimp trawls. Model analyses predicted that if large juvenile, subadult, and adult turtles experienced higher survival rates because of TEDs, the population should show signs of recovery. Assuming that our assumptions about loggerhead life history are generally correct (i.e., the general model structure is correct), there are at least three reasons why loggerhead nesting populations are not increasing:

1. TEDs do not reduce mortality enough or do not benefit all loggerheads.
2. The time lag to maturity is preventing us from observing population response on the nesting beach.
3. Other mortality sources are reducing the realized benefits of TEDs.

None of these are mutually exclusive; in fact, all three could be operating simultaneously. Unfortunately, nesting beach counts alone cannot distinguish among these hypotheses. A mark–recapture study to estimate survival rates is needed to compare current survival rate estimates with those of Frazer (1987), which were calculated prior to TED implementation. A preliminary analysis using catch-curve estimation of stranded turtles suggested an increase in annual survival rates pre- and post-TED (NMFS, 2001). Population and size distribution estimates of juveniles could be used to determine whether this segment of the population is increasing and whether factors such as pelagic longlines could be affecting recruitment to the benthic feeding stage. Finally, an analysis of size distributions and locations of strandings could reveal whether TEDs are selectively excluding certain size classes of turtles, as postulated by Epperly and Teas (1999).

Few studies are long enough or over a broad enough spatial scale to estimate abundance or trends for entire populations; we are nearly always forced to extrapolate well beyond the data we have. Exceptions are the long-term mark–recapture and survey efforts for greens and loggerheads in Australia (Chaloupka and Limpus, 2001) and nesting beach data for Kemp's ridleys, in which the entire species was reduced to a few hundred females on a single nesting beach (Turtle Expert Working Group, 2000; Heppell et al., 2002a). Long-term studies of multiple life stages are essential to document population change through time.

Because there are so many extrinsic and intrinsic factors affecting population growth rates, it is difficult to assess the population-level impacts of our management efforts. Long time lags prevent us from attributing an observed increase or decline in abundance to a particular cause. For example, the recent increase in Kemp's ridleys is due to a combination of effects: nest protection, reduced trawling effort in Mexico, and TED regulations, and possibly other changes in vital rates (Turtle Expert Working Group, 1998; 2000). Egg protection efforts began in the 1960s and intensified in the 1980s, yet population recovery did not really begin until the late 1980s. Empirical evidence suggests that more hatchlings are currently surviving to maturity, likely because of an increase in survival rates afforded by TEDs (Heppell et al., 2002a).

11.5 POPULATION MODELS AS TOOLS FOR TESTING HYPOTHESES ABOUT POPULATION DYNAMICS

One of the key barriers to integration across scales in population biology has been our failure to put all the factors that we think might influence the status of a population into a common framework. Instead, we study what we know (e.g., behavior, physiology, toxicology, and fisheries interactions) and hope our hard-fought insights can be brought to bear on understanding the dynamics of these populations. This is particularly critical for research on threatened and endangered species such as sea turtles. If factor A reduces fecundity by 10%, is that important? Or, if factor B reduces juvenile survival by 5%, is that important? Is factor A more important than factor B? We must put these factors into a common currency to address these questions at the population level. Namely, we must estimate their age- or stage-specific effects on population vital rates (e.g., fecundity, growth, and survival). Integration of this sort is long overdue, and it is critical to forecasting the effects of environmental change and human activities on the future of sea turtles and other protected species.

Population models provide a useful framework to integrate what we know and to clearly identify what we do not know. Analyses of these models also allow us to compare management alternatives on the basis of their mode of action and impact to determine which are likely to contribute most to population recovery (Heppell et al., 2000b). Although the early models of sea turtle population dynamics were simplistic, we believe their qualitative results are robust for turtles (Heppell, 1998), and generally are robust for long-lived species (Heppell et al., 1999; 2000a). Model analyses not only identify poorly known parameters, but indicate those factors expected to have the greatest impact on population growth (Heppell et al., 2000b). In this way, research priorities can be more efficiently focused. Finally, these models are constrained by model assumptions and the available data; qualitative prediction based on life history constraints is generally reliable, but precise quantitative prediction awaits additional data collection (Heppell et al., 2002b). Still, in the context of ecological forecasting, qualitative predictions have provided useful guidance to managers responsible for the recovery of these endangered species.

Models are our primary method to evaluate alternative hypotheses for the causes of population decline and future trends caused by changes in vital rates. Extrapolation of measured trends can be useful, but a model that projects population growth as a function of ASM and vital rates may more accurately predict changes in that trend. For example, Spotila et al. (2000) extrapolated the decline of leatherback turtles nesting at Playa Grande, Costa Rica, but also estimated the response of the nesting population to increased adult recruitment through nest protection efforts. Models for Kemp's ridleys (Heppell et al., 2002a) project a decrease in the population growth rate starting around 2010 because of a decrease in the nest survival rate. These analyses were deterministic projections that can only generally predict changes in population size, but serve as a useful baseline to compare with actual population trends that we may witness in the future.

In Australia, more detailed data allowed development of more elaborate stochastic difference equation models for green and loggerhead sea turtles (Chaloupka, 2002; Chaloupka and Limpus, 2002). Chaloupka (2001) outlined many advantages of developing fully stochastic models, if the data are available. One basis for his analyses is that variation is probably greater in processes that influence fecundity and survival in the egg and early juvenile stages than it is for survival later in the life history. The qualitative effect of this variability on previous conclusions from deterministic models is that reproduction and early survival can have large effects on the *variability* in sea turtle abundance, although the *average* growth rate over long time periods may be similar in both models. This has been shown in comparisons of deterministic and stochastic models for fish populations (Quinlan and Crowder, 1999) and in viability analysis (Wisdom et al., 2000). Of course, one could use a variety of model structures from simple deterministic models to stochastic, individual-based, spatially explicit models (Letcher et al., 1998; Walters et al., 2002). Our approach has been to use the most appropriate modeling form as constrained by our questions and the available data. However, more complex models can be used as tools for the heuristic evaluation of population dynamics under a suite of assumptions about vital rates, variability, and density-dependent effects. The utility of such exercises to management will soon be apparent for new assessments of Pacific leatherback populations (M. Chaloupka, personal communication).

Population models are essential for conservation and management because sea turtles are such late-maturing species. A biologist who studies loggerhead, green, or hawksbill turtles might be lucky to see two generations in his or her lifetime. Long life and late ASM, coupled with variable abundance in space and time, decrease our ability to observe population changes that may lead to recovery or extinction. We are only beginning to understand how sea turtle population dynamics operate, through a combination of long-term surveys, mark–recapture studies, and population models. Much of our work is highly speculative, but has already contributed to the management and conservation of sea turtles worldwide.

REFERENCES

Aiken, J.J. et al. 2001. Two hundred years after a commercial marine turtle fishery: the current status of marine turtles nesting in the Cayman Islands. *Oryx* 35:145–151.

Allard, M.W. et al. 1994. Support for natal homing in green turtles from mitochondrial DNA sequences. *Copeia* 1994:34–41.

Balazs, G. 1995. Growth rates of immature green turtles in the Hawaiian Archipelago. Pages 117–126 in K.A. Bjorndal, editor. *Biology and Conservation of Sea Turtles*, 2nd edition. Smithsonian Institution Press, Washington, D.C.

Ballestero, J., R.M. Arauz, and R. Rojas. 2000. Management, conservation, and sustained use of olive ridley sea turtle eggs (*Lepidochelys olivacea*) in the Ostional Wildlife Refuge, Costa Rica: an 11 year review. Pages 4–5 in F.A. Abreu-Grobois, R. et al., compilers. *Proceedings of the Eighteenth International Sea Turtle Symposium*. U.S. Dept. of Commerce, NOAA Tech. Memo. NMFS-SEFSC-436, Miami, FL. 293 pp.

Bass, A.L. et al. 1996. Testing models of female reproductive migratory behavior and population structure in the Caribbean hawksbill turtle, *Eretmochelys imbricata*, with mtDNA sequences. *Mol. Ecol.* 5:321–328.

Bjorndal, K.A. 1980. Demography of the breeding population of the green turtle, *Chelonia mydas*, at Tortuguero, Costa Rica. *Copeia* 1980:525–530.

Bjorndal, K.A. and A.B. Bolten. 1988. Growth rates of juvenile loggerheads, *Caretta caretta*, in the southern Bahamas. *J. Herpetol.* 22:480–482.

Bjorndal, K.A. and A.B. Bolten, editors. 2000. Abundance and trends for in-water sea turtle populations. U.S. Dept. of Commerce, NOAA Tech. Memo. NMFS-SEFSC-445, Miami, FL. 83 pp.

Bjorndal, K.A., A.B. Bolten, and M.Y. Chaloupka. 2000a. Green turtle somatic growth model: evidence for density dependence. *Ecol. Appl.* 10:269–282.

Bjorndal, K.A., A.B. Bolten, and H.R. Martins. 2000b. Somatic growth model of juvenile loggerhead sea turtles *Caretta caretta*: duration of pelagic stage. *Mar. Ecol. Prog. Ser.* 202:265–272.

Bjorndal, K.A., A.B. Meylan, and B.J. Turner. 1983. Sea turtle nesting at Melbourne Beach, Florida. I. Size, growth, and reproductive biology. *Biol. Conserv.* 26:65–77.

Bjorndal, K.A. et al. 1995. Estimation of green turtle (*Chelonia mydas*) growth rates from length-frequency analysis. *Copeia* 1995:71–77.

Bjorndal, K.A. et al. 1998. Age and growth in sea turtles: limitations of skeletochronology for demographic studies. *Copeia* 1998:23–30.

Bjorndal, K.A. et al. 1999. Twenty-six years of green turtle nesting at Tortuguero, Costa Rica: an encouraging trend. *Conserv. Biol.* 13:126–134.

Bjorndal, K.A. et al. 2001. Somatic growth function for immature loggerhead sea turtles, *Caretta caretta*, in southeastern U.S. waters. *Fish. Bull.* 99:240–246.

Bolten, A.B. 2002. Active swimmers — passive drifters: the oceanic juvenile stage of loggerheads in the Atlantic system In A. Bolton and B. Witherington, editors. *Biology and Conservation of Loggerhead Sea Turtles*. Smithsonian Institution Press, Washington, D.C. In press.

Bowen, B.W. et al. 1992. Global population structure and natural history of the green turtle (*Chelonia mydas*) in terms of matriarchal phylogeny. *Evolution* 46:865–881.

Bowen, B. et al. 1993. Population structure of loggerhead turtles (*Caretta caretta*) in the northwestern Atlantic Ocean and Mediterranean Sea. *Conserv. Biol.* 7:834–844.

Bowen, B.W. et al. 1996. Origin for hawksbill turtles in a Caribbean feeding area as indicated by genetic markers. *Ecol. Appl.* 6:566–572.

Bowen, B.W. and S.A. Karl. 1997. Population genetics, phylogeography, and molecular evolution. Pages 29–50 in P.L. Lutz and J.A. Musick, editors. *The Biology of Sea Turtles*. CRC Press, Boca Raton, FL.

Braun-McNeill, J. et al. In review. An analysis of growth rates of immature loggerhead (*Caretta caretta*) sea turtles from North Carolina, U.S.A. *Copeia*.

Broderick, A.C., B.J. Godley, and G.C. Hays. 2001. Trophic status drives interannual variability in nesting numbers of marine turtles. *Proc. R. Soc. Lond. B Biol. Sci.* 268:1481–1487.

Brooke, M. de L. and M.C. Garnett. 1983. Survival and reproductive performance of hawksbill turtles *Eretmochelys imbricata* L. on Cousin Island, Seychelles. *Biol. Conserv.* 25:161–170.

Bustard, H.R. and K.P. Tognetti. 1969. Green sea turtles: a discrete simulation of density-dependent population regulation. *Science* 163:939–941.

Butler, R.W., W.A. Nelson, and T.A. Henwood. 1987. A trawl survey method for estimating loggerhead turtle, *Caretta caretta*, abundance in five eastern Florida channels and inlets. *Fish. Bull.* 85:447–453.

Caillouet, C.W., Jr. et al. 1995. Survival of head-started Kemp's ridley sea turtles (*Lepidochelys kempii*) released into the Gulf of Mexico or adjacent bays. *Chelonian Conserv. Biol.* 1:285–292.

Carr, A. 1986. *The Sea Turtle, So Excellent a Fishe. A Natural History of Sea Turtle*. University of Texas Press, Austin, TX. 280 pp.

Chaloupka, M.Y. 1998. Polyphasic growth in pelagic loggerhead sea turtles. *Copeia* 1998:516–518.

Chaloupka, M.Y. 2001. Historical trends, seasonality and spatial synchrony in green sea turtle egg production. *Biol. Conserv.* 101:263–279.

Chaloupka, M.Y. 2002. Stochastic simulation modeling of southern Great Barrier Reef green turtle population dynamics. *Ecol. Model.* 148:79–109.

Chaloupka, M.Y. and C.J. Limpus. 1997. Robust statistical modelling of hawksbill sea turtle growth rates (southern Great Barrier Reef). *Mar. Ecol. Prog. Ser.* 146:1–8.

Chaloupka, M. and C. Limpus. 1998. Modeling green turtle survivorship rates. Pages 24–26 in S.P. Epperly and J. Braun, editors. *Proceedings of the 17th Annual Symposium on Sea Turtle Biology and Conservation*. NOAA Tech. Memo. NMFS-SEFSC-415, Miami, FL.

Chaloupka, M.Y. and C.J. Limpus. 2001. Trends in the abundance of sea turtles resident in southern Great Barrier Reef waters. *Biol. Conserv.* 102:235–249.

Chaloupka, M.Y. and C.J. Limpus. 2002. Survival probability estimates for the endangered loggerhead sea turtle resident in southern Great Barrier Reef waters. *Mar. Biol.* 140:267–277.

Chaloupka, M.Y. and J.A. Musick. 1997. Age, growth and population dynamics. Pages 233–276 in P.L. Lutz and J.A. Musick, editors. *The Biology of Sea Turtles*. CRC Press, Boca Raton, FL.

Chaloupka, M.Y. and G.R. Zug. 1997. A polyphasic growth function for the endangered Kemp's ridley sea turtle, *Lepidochelys kempii*. *Fish. Bull.* 95:849–856.

Chan, E.H. and H.C. Liew. 1996. Decline of the leatherback population in Terengganu, Malaysia. 1956–95. *Chelonian Conserv. Biol.* 2:196–203.

Charnov, E. 1982. *The Theory of Sex Allocation*. Princeton University Press, Princeton, NJ.

Chua, T.H. 1988. Nesting population and frequency of visits in *Dermochelys coriacea* in Malaysia. *J. Herpetol.* 22:192–207.

Coles, W.C., J.A. Musick, and L. Williamson. 2001. Skeletochronology: validation from an adult loggerhead (*Caretta caretta*). *Copeia* 2001:240–242.

Congdon, J.D., A.E. Dunham, and R.C. van Loben Sels. 1993. Delayed sexual maturity and demographics of Blandings turtles (*Emydoidea blandingii*): implications for conservation and management of long-lived organisms. *Conserv. Biol.* 7:826–833.

Cormack, R.M. 1964. Estimates of survival from the sightings of marked animals. *Biometrika* 51:429–439.

Crouse, D.T. 1999. The consequences of delayed maturity in a human-dominated world. Pages 195–202 in J.A. Musick, editor. *Life in the Slow Lane: Ecology and Conservation of Long-Lived Marine Animals.* American Fisheries Society Symposium 23, Bethesda, MD.

Crouse, D.T., L.B. Crowder, and H. Caswell. 1987. A stage-based population model for loggerhead sea turtles and implications for conservation. *Ecology* 68:1412–1423.

Crowder, L.B. et al. 1994. Predicting the impact of turtle excluder devices on loggerhead sea turtle populations. *Ecol. Appl.* 4:437–445.

Dutton, P.H. et al. 1999. Global phylogeography of the leatherback turtle (*Dermochelys coriacea*). *J. Zool.* 248:397–409.

Eckert, S.A. et al. 1989. Diving and foraging behavior of leatherback sea turtles (*Dermochelys coriacea*). *Can. J. Zool.* 67:2834–2840.

Ehrdardt, N.M. and R. Witham. 1992. Analysis of growth of the green sea turtles (*Chelonia mydas*) in the western central Atlantic. *Bull. Mar. Sci.* 50:275–281.

Encalada, S.E. et al. 1996. Phylogeography and population structure of the Atlantic and Mediterranean green turtle *Chelonia mydas*: a mitochondrial DNA control region sequence assessment. *Mol. Ecol.* 5:473–483.

Encalada, S.E. et al. 1998. Population structure of loggerhead turtle (*Caretta caretta*) nesting colonies in the Atlantic and Mediterranean as inferred from mitochondrial DNA control region sequences. *Mar. Biol.* 130:567–575.

Epperly, S.P. and W.G. Teas. 1999. Evaluation of TED opening dimensions relative to size of turtles stranding in the western North Atlantic. U.S. Dept. Commerce, National Marine Fisheries Service SEFSC Contribution PRD-98/99–08; Sept. 1999, Miami, FL. 31 pp.

Fabens, A.J. 1965. Properties and fitting of the von Bertalanffy growth curve. *Growth* 29:265–289.

Fisher, R.A. 1930. *The Genetical Theory of Natural Selection.* Clarendon Press, Oxford.

FitzSimmons, N.N. 1998. Single paternity of clutches and sperm storage in the promiscuous green turtle (*Chelonia mydas*). *Mol. Ecol.* 7:575–584.

Francisco, A.M. 2001. Identification of loggerhead turtle (*Caretta caretta*) stock structure in the southeastern United States and adjacent regions using nuclear and mitochondrial DNA markers. M.S. thesis. University of Florida, Gainesville, FL.

Francisco, A.M. et al. In press. Stock structure and nesting site fidelity in Florida loggerhead turtles (*Caretta caretta*) resolved with mtDNA sequences. *Mar. Biol.*

Frazer, N.B. 1983. Survivorship of adult female loggerhead sea turtles, *Caretta caretta*, nesting on little Cumberland Island, Georgia, U.S.A. *Herpetologica* 39:436–447.

Frazer, N.B. 1987. Preliminary estimates of survivorship for wild loggerhead sea turtles (*Caretta caretta*). *J. Herpetol.* 21:232–235.

Frazer, N.B. and L.M. Ehrhart. 1985. Preliminary growth models for green, *Chelonia mydas*, and loggerhead, *Caretta caretta*, turtles in the wild. *Copeia* 1985:73–79.

Frazer, N.B. and R.C. Ladner. 1986. A growth curve for green sea turtles, *Chelonia mydas*, in the U.S. Virgin Islands, 1913–14. *Copeia* 1986:798–802.

Frazer, N.B., C.J. Limpus, and J.L. Greene. 1994. Growth and estimated age at maturity of Queensland loggerheads. Pages 42–46 in Bjorndal, K.A. et al., compilers. *Proceedings of the Fourteenth Annual Symposium on Sea Turtle Biology and Conservation.* NOAA Tech. Memo. NMFS-SEFSC-351, Miami, FL.

Frazer, N.B. and J.I. Richardson. 1985. Annual variation in clutch size and frequency for loggerhead turtles, *Caretta caretta*, nesting at Little Cumberland Island, Georgia, USA. *Herpetologica* 41:246–251.

Freedberg, S. and M.J. Wade. 2001. Cultural inheritance as a mechanism for population sex-ratio bias in reptiles. *Evolution* 55:1049–1055.

Fuentes, A.L. et al. 2000. Possible effects of El Niño-Southern Oscillation on the black turtle nesting population at Michoacan, Mexico. Pages 269–271 in H.J. Kalb and T. Wibbels, compilers. *Proceedings of the Nineteenth Annual Symposium on Sea Turtle Biology and Conservation.* U.S. Dept. of Commerce, NOAA Tech. Memo. NMFS-SEFSC-443, Miami, FL. 291 pp.

Gerrodette, T. and J. Brandon. 2000. Designing a monitoring program to detect trends. Pages 36–39 in Bjorndal, K.A. and Bolten, A.B., editors. *Proceedings of a Workshop on Assessing Abundance and Trends for In-Water Sea Turtle Populations.* U.S. Dept. of Commerce, NOAA Tech. Memo. NMFS-SEFSC-445.

Gibbs, J.P. and G.D. Amato. 2000. Genetics and demography in turtle conservation. Pages 207–217 in M.W. Klemens, editor. *Turtle Conservation.* Smithsonian Institution Press, Washington, DC.

Godley, B.J., A.C. Broderick, and G.C. Hays. 2001. Trophic status drives interannual variability in nesting numbers of marine turtles. *Proc. R. Soc. Lond. B Biol. Sci.* 268:1481–1487.

Godley, B.J. et al. 2002. Temperature dependent sex determination of Ascension Island green turtles. *Mar. Ecol. Prog. Ser.* 226:115–124.

Green, D. 1993. Growth rates of wild immature green turtles in the Galapagos Islands, Ecuador. *J. Herpetol.* 27:338–341.

Hanson, J., T. Wibbels, and R.E. Martin. 1998. Predicted female bias in sex ratios of hatchling loggerhead sea turtles from a Florida nesting beach. *Can. J. Zool.* 76:1850–1861.

Hays, G.C. 2000. The implications of variable remigration intervals for the assessment of population size in marine turtles. *J. Theor. Biol.* 206:221–227.

Hays, G.C. and J.R. Speakman. 1991. Reproductive investment and optimum clutch size of loggerhead sea turtles (*Caretta caretta*). *J. Anim. Ecol.* 60:455–462.

Henwood, T.A. 1987. Sea turtles of the southeastern United States, with emphasis on the life history and population dynamics of the loggerhead. Ph.D. dissertation. Auburn University, Auburn, AL.

Heppell, S.S. 1998. Application of life-history theory and population model analysis to turtle conservation. *Copeia* 1998:367–375.

Heppell, S.S., H. Caswell, and L.B. Crowder. 2000a. Life histories and elasticity patterns: Perturbation analysis for species with minimal demographic data. *Ecology* 81:654–665.

Heppell, S.S., D.T. Crouse, and L.B. Crowder. 2000b. Using matrix models to focus research and management efforts in conservation. Pages 148–168 in S. Ferson and M. Burgman, editors. *Quantitative Methods for Conservation Biology.* Springer-Verlag, Berlin.

Heppell, S.S., L.B. Crowder, and T.R. Menzel. 1999. Life table analysis of long-lived marine species with implications for conservation and management. Pages 137–148 in J.A. Musick, editor. *Life in the Slow Lane: Ecology and Conservation of Long-Lived Marine Animals.* American Fisheries Society Symposium 23, Bethesda, MD.

Heppell, S.S. et al. 1996. Population model analysis for the loggerhead sea turtle, *Caretta caretta*, in Queensland. *Wildl. Res.* 23:143–159.

Heppell, S. et al. 2002a. A population model to estimate recovery time, population size, and management impacts on Kemp's ridleys. *Chelonian Conserv. Biol.* In press.

Heppell, S.S. et al. 2002b. Population models of Atlantic loggerheads: past, present, and future. In A. Bolten and B. Witherington, editors. *Biology and Conservation of Loggerhead Sea Turtles*. Smithsonian Institution Press, in press.

Hirth, H.F. 1980. Some aspects of the nesting behavior and reproductive biology of sea turtles. *Am. Zool.* 20:507–523.

Holmes, E.E. 2001. Estimating risks in declining populations with poor data. *Proc. Natl. Acad. Sci.* 98:5072–5077.

Jackson. J.B. C. 1997. Reefs since Columbus. *Coral Reefs* 16(Suppl.):S23–S32.

Jolly, G.M. 1965. Explicit estimates from capture-recapture data with both death and immigration-stochastic model. *Biometrika* 52:225–247.

Karl, S.A., B.W. Bowen, and J.C. Avise. 1992. Global population genetic structure and male-mediated gene flow in the green turtle (*Chelonia mydas*): RFLP analyses of anonymous nuclear loci. *Genetics* 131:163–173.

Kerr, R.J., I. Richardson, and T.H. Richardson. 1999. Estimating the annual size of hawksbill (*Eretmochelys imbricata*) nesting populations from mark-recapture studies: the use of long-term data to provide statistics for optimizing survey effort. *Chelonian Conserv. Biol.* 3:251–256.

Klinger, R.C. and J.A. Musick. 1992. Annular growth layers in juvenile loggerhead turtles (*Caretta caretta*). *Bull. Mar. Sci.* 51:224–230.

Klinger, R.C. and J.A. Musick. 1995. Age and growth of loggerhead turtles (*Caretta caretta*) from Chesapeake Bay. *Copeia* 1995:204–209.

Lahanas, P.N. et al. 1994. Molecular evolution and population genetics of greater Caribbean green turtles (*Chelonia mydas*) as inferred from mitochondrial DNA control region sequences. *Genetica* 94:57–67.

Lahanas, P.N. et al. 1997. Genetic composition of a green turtle (*Chelonia mydas*) feeding ground population: evidence for multiple origins. *Mar. Biol.* 130:345–352.

Laurent, L. et al. 1998. Molecular resolution of marine turtle stock composition in fishery bycatch: a case study in the Mediterranean. *Mol. Ecol.* 7:1529–1542.

Letcher, B.H. et al. 1998. An individual-based, spatially-explicit simulation model of the population dynamics of the endangered red-cockaded woodpecker, *Picoides borealis*. *Biol. Conserv.* 86:1–14.

Limpus, C. and M. Chaloupka. 1997. Nonparametric regression modeling of green sea turtle growth rates (southern Great Barrier Reef). *Mar. Ecol. Prog. Ser.* 149:23–34.

Limpus, C.J., P.J. Couper, and M.A. Read. 1994. The loggerhead sea turtle, *Caretta caretta*, in Queensland: population structure in a warm temperate feeding area. *Mem. Queensl. Mus.* 37:195–204.

Limpus, C.J. and N. Nicholls. 1988. The Southern Oscillation regulates the annual numbers of green turtles (*Chelonia mydas*) breeding around Northern Australia. *Aust. J. Wildl. Res.* 15:157–161.

Limpus, C.J. and D.G. Walter. 1980. The growth of immature green turtles (*Chelonia mydas*) under natural conditions. *Herpetologica* 36:162–165.

Lutcavage, M.E. et al. 1997. Human impacts on sea turtle survival. Pages 387–409 in P.L. Lutz and J.A. Musick, editors. *The Biology of Sea Turtles*. CRC Press, Boca Raton, FL.

Lutcavage, M.E. and Musick, J.A. 1985. Aspects of the biology of sea turtles in Virginia. *Copeia* 1985:449–456.

McDaniel, C.J., L.B. Crowder, and J.A. Priddy. 2000. Spatial dynamics of sea turtle abundance and shrimping intensity in the U.S. Gulf of Mexico. *Conserv. Ecol.* 4:15. Document available online: http://www.consecol.org/vol4/iss1/art15.

Mendonca, M.T. 1981. Comparative growth rates of wild immature *Chelonia mydas* and *Caretta caretta* in Florida. *J. Herpetol.* 15:444–447.

Meylan, A. 1995. Growth rates of immature green turtles in the Hawaiian Archipelago. Pages 135–138 in K.A. Bjorndal, editor. *Biology and Conservation of Sea Turtles*, 2nd edition. Smithsonian Institution Press, Washington, D.C.

Meylan, A.B. and D. Ehrenfeld. 2000. Conservation of marine turtles. Pages 96–125 in M.W. Klemens, editor. *Turtle Conservation*. Smithsonian Institution Press, Washington, D.C.

Miller, J.A. 1997. Reproduction in sea turtles. Pages 51–82 in P.L. Lutz and J.A. Musick, editors. *The Biology of Sea Turtles*. CRC Press, Boca Raton, FL.

Mrosovsky, N. and C. Pieau. 1991. Transitional range of temperature, pivotal temperature and thermosensitive stages of sex determination in reptiles. *Amphibia-Reptilia* 12:169–187.

Mrosovosky, N. and J. Provancha. 1989. Sex ratio of loggerhead sea turtles hatching on a Florida beach. *Can. J. Zool.* 67:2533–2539.

Musick, J.A. 1999. Ecology and conservation of long-lived marine animals. Pages 1–10 in J.A. Musick, editor. *Life in the Slow Lane: Ecology and Conservation of Long-Lived Marine Animals*. American Fisheries Society Symposium 23, Bethesda, MD.

Musick, J.A. and C.J. Limpus. 1997. Habitat utilization and migration in juvenile sea turtles. Pages 137–163 in P.L. Lutz and J.A. Musick, editors. *The Biology of Sea Turtles*. CRC Press, Boca Raton, FL.

National Marine Fisheries Service. 2001. Stock assessments of loggerhead and leatherback sea turtles and an assessment of the impact of the pelagic longline fishery on the loggerhead and leatherback sea turtles of the western north Atlantic. U.S. Dept. of Commerce, NOAA Tech. Memo. NMFS-SEFSC-455, 343 pp. Document available online: http://www.sefsc.noaa.gov/SeaTurtles/TechMemo455/tm455.pdf.

National Research Council. 1990. *Decline of the Sea Turtles: Causes and Prevention*. National Academy Press, Washington, D.C. 259 pp.

Parham, J.F. and G.R. Zug. 1997. Age and growth of loggerhead sea turtles (*Caretta caretta*) of coastal Georgia: an assessment of skeletochronological age-estimates. *Bull. Mar. Sci.* 61:287–304.

Peñaflores, C.S. et al. 2000. Management, conservation, and sustained use of olive ridley sea turtle eggs (*Lepidochelys olivacea*) in the Ostional Wildlife Refuge, Costa Rica: an 11 year review. Pages 27–29 in F.A. Abreu-Grobois et al., compilers. *Proceedings of the Eighteenth International Sea Turtle Symposium*. U.S. Dept. of Commerce, NOAA Tech. Mem. NMFS-SEFSC-436, Miami, FL. 293 pp.

Plotkin, P., Byles, R.A., and Owens, D.M. 1996. Ups and downs in the life of the olive ridley turtle: breeding versus post-reproductive diving behavior. In Keinath, J.A. et al., editors. *Proceedings of the 15th Annual Sea Turtle Symposium*. NOAA Tech Mem. NMFS-SEFSC.

Quinlan, J.A. and L.B. Crowder. 1999. Searching for sensitivity in the life history of Atlantic menhaden: inferences from a matrix model. *Fish. Oceanogr.* 8(Suppl. 2):124–133.

Reichart, H.A. 1993. Synopsis of biological data on the olive ridley sea turtle *Lepidochelys olivacea* (Eschsholtz, 1829) in the western Atlantic. NOAA Tech. Mem. NMFS-SEFSC-336, 78, 4, Miami, FL.

Reina, R.D. et al. 2002. Nesting ecology of the leatherback turtle, *Dermochelys coriacea*, at Parque Nacional Marino Las Baulas, Costa Rica: 1988–89 to 1999–2000. *Copeia* 2002, in press.

Richardson, J.I., R. Bell, and T.H. Richardson. 1999. Population ecology and demographic implications drawn from an 11-year study of nesting hawksbill turtles, *Eretmochelys imbricata*, at Jumby Bay, Long Island, Antigua, West Indies. *Chelonian Conserv. Biol.* 3:244–250.

Schmid, J.R. 1998. Marine turtle populations on the west-central coast of Florida: results of tagging studies at Cedar Key, Florida, 1986–1995. *Fish. Bull.* 96:589–602.

Schmid, J.R. and W.N. Witzell. 1997. Age and growth of wild Kemp's ridley turtles (*Lepidochelys kempii*): cumulative results from tagging studies in Florida. *Chelonian Conserv. Biol.* 2:532–537.

Sears, C.J. et al. 1995. Demographic composition of the feeding population of juvenile loggerhead sea turtles (*Caretta caretta*) off Charleston, South Carolina: evidence from mitochondrial DNA markers. *Mar. Biol.* 123:869–874.

Seber, G.A.F. 1965. A note on the multiple recapture census. *Biometrika* 52:249–259.

Seber, G.A.F. 1982. *The Estimation of Animal Abundance and Related Parameters*, 2nd edition. MacMillan, New York. 600 pp.

Snover, M.L. et al. In review. The impact of an ontogenetic habitat shift on the growth of juvenile loggerhead sea turtles (*Caretta caretta*). *Ecology.*

Snover, M.L. and A.A. Hohn. In review. Validation of annual growth mark deposition and interpretation in loggerhead (*Caretta caretta*) and Kemp's ridley (*Lepidochelys kempi*) sea turtles. *Copeia.*

Spotila, J.R. et al. 2000. Pacific leatherback faces extinction. *Nature* 405:529–530.

Stamps, J.A., M. Mangel, and J.A. Phillips. 1998. A new look at relationships between size at maturity and asymptotic size. *Am. Nat.* 152:470–479.

Turtle Expert Working Group. 1998. An assessment of the Kemp's ridley (*Lepidochelys kempii*) and loggerhead (*Caretta caretta*) sea turtle populations in the western north Atlantic. NOAA Tech. Memo. NMFS-SEFSC-409, Miami, FL. 96 pp.

Turtle Expert Working Group. 2000. Assessment update for the Kemp's ridley and loggerhead sea turtle population in the western north Atlantic. U.S. Department of Commerce NOAA Tech. Memo. NMFS-SEFSC-444, 115 pp. Document available online: http://www.nmfs.noaa.gov/prot_res/readingrm/Turtles/tewg2000.pdf.

Van Buskirk, J. and L.B. Crowder. 1994. Life-history variation in marine turtles. *Copeia* 1994:66–81.

Walters, J.R., L.B. Crowder, and J.A. Priddy. 2002. Population viability analysis for red-cockaded woodpeckers using an individual-based model. *Ecol. Appl.* 12:249–260.

Werner, E.E. and J.F. Gilliam. 1984. The ontogenetic niche and species interactions in size-structured populations. *Annu. Rev. Ecol. System.* 15:292–425.

Wisdom, M.J., L.S. Mills, and D.F. Doak. 2000. Life stage simulation analysis: estimating vital-rate effects on population growth for conservation. *Ecology* 81:628–641.

Witzell, W.N. 1999. Distribution and relative abundance of sea turtles caught incidentally by the U.S. pelagic longline fleet in the western north Atlantic Ocean, 1992–1995. *Fish. Bull.* 97:200–211.

Zug, G.R., G.H. Balazs, and J.A. Wetherall. 1995. Growth in juvenile loggerhead sea turtles (*Caretta caretta*) in the north Pacific pelagic habitat. *Copeia* 1995:484–487.

Zug, G.R. et al. 2002. Age and growth in Hawaiian green sea turtles (*Chelonia mydas*): a metapopulation analysis based on skeletochronology. *Fish. Bull.* 100:117–127.

Zug, G.R. and R.E. Glor. 1998. Estimates of age and growth in a population of green sea turtles (*Chelonia mydas*) from the Indian River lagoon system, Florida: a skeletochronological analysis. *Can. J. Zool.* 76:1497–1506.

Zug, G.R., H.J. Kalb, and S.J. Luzar. 1997. Age and growth in wild Kemp's ridley sea turtles *Lepidochelys kempii* from skeletochronological data. *Biol. Conserv.* 80:261–268.

Zug, G.R. and J.F. Parham. 1996. Age and growth in leatherback sea turtles, *Dermochelys coriacea* (Testudines, Dermochelyidae): a skeletochronological analysis. *Chelonian Conserv. Biol.* 2:244–249.

Zug, G.R., A.H. Wynn, and C. Ruckdeschel. 1986. Age determination of loggerhead sea turtles, *Caretta caretta*, by incremental growth marks in the skeleton. *Smithson. Contrib. Zool.,* 1986, 427, 1–34.

12 Contemporary Culture, Use, and Conservation of Sea Turtles

Lisa M. Campbell

CONTENTS

0-8493-1123-3/03/$0.00+$1.50

12.1 INTRODUCTION

This chapter examines contemporary links between human culture(s) and sea turtle use and conservation. It is based on two central assumptions: (1) the value and role assigned to turtles as part of nature is culturally situated, and (2) the cultural context of human relations with sea turtles is critical to the success of conservation schemes. Key concepts and terms used are discussed in this introduction. Section 12.2 highlights various types and examples of sea turtle use and their cultural significance. In Section 12.3, the link between culture and conservation policy, and specifically, two contemporary conservation concepts — sustainable use and community-based conservation (CBC) — are discussed.

12.1.1 CULTURE

Williams (1981) describes two main senses of "culture." Culture is "a distinct 'whole way of life', within which, now, a distinctive 'signifying system' is seen not only as essential but as essentially involved in *all* forms of social activity." Culture in this sense mediates how we understand and make sense of the world around us. The more common sense of culture is "artistic and intellectual activity" and resulting products. The two senses converge, in that the former whole way of life incorporates the central interests and values of a people (Williams, 1981), and these are often manifested in products of material culture. Thus, sea turtles may be part of a whole way of life, and this may be reflected in art, crafts, or music.

Culture operates on a number of levels, and these levels interact (Seppälä and Vainio-Mattila, 1998). Although in Western society "culture" is a broad and encompassing term, subcultures (for example, corporate culture, culture tied to ethnic identity, and counterculture) can exist within and sometimes challenge dominant Western culture. Power is an issue in determining which cultures dominate, and domination by one culture implies subjugation of others. Culture is dynamic and in a constant state of change; change does not mean that people become cultureless, or that their cultures become meaningless.

There is a growing body of literature addressing cultural (and social) constructions of nature (Braun and Castree, 1998; Ellen and Fukui, 1996; Escobar, 1999). Via culture, society determines what constitutes nature and what role nature plays in cultural and social life. Via culture, priorities are set for conservation and development. As cultures differ across time and space, different cultures will place different priorities on the individual components of nature, and in some cultures, the concept of nature as something separate from humans does not exist. In wildlife conservation, culture is used to explain particular sets of human relations with various species (e.g., Nietschmann [1973; 1979] explained the cultural value of sea turtles to the Miskito Indians of Nicaragua; see Section 12.2.1.1).

In a conservation context, there are often two extreme positions on culture. For some, culture is sacred, something to be respected and revered, and deserving of conservation in its own right (e.g., Cultural Survival International, http://www.culturalsurvival.org/), especially if it is indigenous. For others, culture is a red herring raised to deter conservation efforts, the claims of which need to be thoroughly interrogated (Campbell, 2000). Culture is often most obvious when it is someone else's (Seppälä and Vainio-Mattila, 1998). Although a North American observer might see a Central American people's desire to consume turtle eggs as cultural, the same North Americans are less likely to explain their own desire to protect turtle eggs in the same way. Rather, protection is taken for granted as the desirable and correct outcome. Thus, rather than try to understand cultural meanings associated with sea turtles, force is sometimes used to make *them* do what *we* would, or education is used to get *them* to agree with *us*. These approaches underestimate the importance of cultural norms, and they can fall short of their long-term conservation goals as a result.

12.1.2 Valuing the Environment

As Williams' definition of culture implies, values are intricately related to and embedded in culture. Humans value the environment and wildlife in a variety of ways: for economic, recreational, scientific, aesthetic, historic, and philosophic or spiritual reasons (Rolston, 1994). Environmental values vary from place to place, and different environmental values can coexist within a particular place (e.g., Kempton et al., 1995). Sea turtles are valued in different ways by different people, and because of their international migrations that take them across geographic, political, and cultural boundaries, conflicts in values can frequently arise.

12.1.3 Culture, Values, and Conservation

If definitions of nature and environmental values are embedded in culture, then so too is conservation. For example, Western conservation has traditionally been pursued via the creation of parks and protected areas, and the national park model that emerged in the U.S. in the late 1800s reflects the culture of the time. national parks are physically delineated, the state is responsible for their creation and maintenance, and only certain nonextractive human activities are sanctioned within their borders. These features reflect the cultural beliefs that humans are separate from (and often above) nature, the state is responsible for and capable of protecting the public good, and nature can be contained in physically delineated areas. Key in the park movement were the U.S. "romantics," (primarily) men who reacted against the frontier mentality that characterized the settlement of the American West. The frontier mentality saw nature as dangerous and threatening, something to be tamed for productive purposes. The romantics saw nature as a purifier of the tarnished modern soul and needing protection (McCormick, 1989). Both opposing visions of nature were linked to the dominant culture of U.S. expansion and ideas of progress.

Over the last 20+ years, there has been a shift away from a traditional protected-areas approach toward an attempt to reconcile conservation with development needs, as reflected in the current definition of conservation adopted by the World Conservation Union (IUCN). Conservation is "...the management of human use of organisms or ecosystems to ensure such use is sustainable. Besides sustainable use, conservation includes protection, maintenance, rehabilitation, restoration, and enhancement of populations and ecosystems" (IUCN, 1980).

This shift arose as a number of shortcomings with protected areas became evident, particularly when applied outside of their cultural context in developing countries. The vision of humans as separate from nature, for example, can conflict with local visions of human–environment relations (Ghimire and Pimbert, 1997), and can undermine cultural norms and traditional or indigenous knowledge (Marks, 1984). The resulting cultural mismatch can sabotage conservation efforts; if local people do not support a conservation undertaking, encroachment and illegal harvesting activities may result. Two responses to the problems experienced with transferring protected areas to developing countries are sustainable use and CBC, and these are discussed in Section 12.3.

12.1.4 Other Key Terms

Other key terms in the discussion of culture and conservation are community, indigenous, traditional, and subsistence — terms that are often used to delimit sea turtle use. For example, a sea turtle egg-collecting project might be justified as a *traditional* activity of *indigenous* people undertaken for *subsistence* purposes. Such terms are rarely defined, and are thus problematic.

Community defines both actual social groups (i.e., the people of a district) and the quality of relations among people (i.e., holding something in common, or a sense of common interests or identity) (Williams, 1983). Community is also used to distinguish the more direct and total relations between people from the more formal, abstract, and instrumental relations of people with the state. "Community can be the warmly persuasive word to describe an existing set of relationships" (Williams, 1983) and is rarely used in a negative sense.

Communities are increasingly seen as critical to the success of conservation efforts, but there are difficulties associated with defining communities that may arise from the term's dual meaning. The people of a district may be assumed to equal the relations among people, and in conservation practice, communities are often treated as self-evident or generic, and homogeneous (Brosius et al., 1998; Leach et al., 1997; Wells and Brandon, 1992; Western and Wright, 1994). Communities are also assumed to share culture and related values of the environment. This is not always the case, and a clear sense of who and what the community is will be critical for conservation success at the local level.

Indigenous is defined as something "originating or occurring naturally in the place or country specified" (Avis, 1980). The cultural claims of indigenous peoples to use wildlife are often given greater weight than those of nonindigenous people. For example, Donnelly (1994) describes sustainable use of sea turtles as "...designed to promote controlled and renewable use of wildlife for the benefit of indigenous people and endangered species." Emphasis on indigenous assumes that use of sea turtles by indigenous peoples has different impacts from use by nonindigenous peoples, and that indigenous peoples have stronger cultural biases toward use, which may not always be the case.

Tradition refers to handing down knowledge, or passing on a doctrine, from one generation to another. Often, tradition is associated with a sense of ceremony, duty, and respect. The process by which certain elements of knowledge are passed down, whereas others are not, shows that traditions are selective (Williams, 1983). Like the term indigenous, tradition is used to explain or justify certain cultural practices. When evaluating traditional claims to resources use, some people claim that, to be traditional, an activity cannot have changed over time; for example, fishing for turtles with an outboard motor cannot be traditional when it was originally done using a dugout canoe (see Campbell, 2000). This interpretation of tradition focuses on the means for achieving, rather than the meaning of a tradition. Furthermore, it implies stasis that has never existed in human history. The addition of an outboard motor to a canoe, for example, is an incremental step in the evolution of technology, rather than a leap from traditional to nontraditional. Improved technologies do not always lead to increased resource exploitation (Lyver and Moller, 1999), and Berkes et al. (2000) warn against associating tradition with stasis.

Tradition has become more important in conservation because of the increasing popularity of traditional ecological knowledge (TEK). Studies of ecological change over time have sometimes challenged Western scientists' interpretations of environmental change and revealed the logic of indigenous or traditional management practices (Berkes et al., 2000; Leach and Mearns, 1996; Nader, 1996), and Miller (2000) discusses the links between traditional and nontraditional knowledge of sea turtles. The existence of TEK in communities does not equate automatically with desire or willingness to conserve, but there is nevertheless a need to recognize it.

Subsistence economies are those that extract the basis of existence from the natural environment, and that focus on satisfaction of existing food needs rather than on accumulation of surplus (Nietschmann, 1973). Such regimes existed prior to, or exist external to, the market economy, and are by and large devalued and destroyed by it (Escobar, 1992). With the widespread introduction of market economies throughout the world, few fully subsistence regimes exist, and the term has been "implicitly redefined as the individual producer's socio-biological survival under conditions of accumulation of capital" (Robert, 1992).

In a conservation context, depletion of sea turtles (and other species) is often blamed on the transition from a subsistence to market economy (e.g., Nietschmann, 1979; Spring, 1995; see also Section 12.2.1.1). Although such transitions have adverse environmental and cultural impacts, the reemergence of true subsistence economies is unlikely. Nonsubsistence use does not by definition imply large-scale, uncontrolled use, only that capital accumulation might result.

Discussions of culture and conservation are often restricted to "other" cultures that, in opposition to dominant Western culture, might be community based, subsistence, traditional, or indigenous. Although it is important to understand such terms, this narrow focus depicts culture as influencing conservation somewhere else. In the following examination of culture and sea turtle use and conservation, evidence from both other cultures and Western culture is included. Culture informs how all readers make sense of their worlds, including the world of sea turtle conservation.

12.2 CONTEMPORARY USES AND RELATIONS WITH SEA TURTLES

Consumptive use of sea turtles around the world has been documented recently (Thorbjarnarson et al., 2000a). Although some of the facts of use are repeated and expanded on here, the focus is on the links between culture and use, and on various kinds of use, including nonconsumptive. Thorbjarnarson et al. (2000a) consider some culture contexts of use. For example, they discuss the impacts of different religions on consumption. The listing of cultural influences on turtle use (Thorbjarnarson et al., 2000a), however, reflects the point made in Section 12.1.1: It is often easiest to see the influences of other cultures. Meanwhile, Western culture has impacted on contemporary use of sea turtles in two profound, and seemingly opposite, ways. First, the expansion of Western capitalism has shaped sea turtle consumption; economies that might previously have used turtles for subsistence purposes now have cash needs that may be met through selling sea turtles and their by-products. Second, the separation of humans from nature in Western culture, and the veneration of "charismatic megafauna," has created a demand in the West for the complete protection of sea turtles and their relegation to tourist spectacle. Some of these issues are discussed in more detail below.

12.2.1 Directed Take

12.2.1.1 Turtle Meat

The list of countries using sea turtles for meat (Thorbjarnarson et al., 2000a) includes the U.S. (Florida, Georgia, Louisiana, Mississippi, North Carolina, Texas, and Virginia), the Atlantic coast of Central America, Ecuador, Peru, Madagascar, Seychelles, India, Sri Lanka, Japan (fishing in other waters), Indonesia, Australia, Torres Strait, and Papua New Guinea. To this list can be added Bangladesh (Islam, 2001), Thailand (Aureggi et al., 1999), Liberia (Siakor et al., 2000), Egypt (Venizelos and Nada, 2000), Equatorial Guinea (Tomás et al., 1999), Guinea-Bissau (Fortes et al., 1998), Cuba (Carrillo et al., 1999), Nicaragua (Nietschmann, 1973; 1979; Lagueux, 1998), Costa Rica (Opay, 1998), Belize (Frazier, in press), Mexico (Nichols et al., 2000), and several Caribbean islands (Antigua and Barbuda, Bahamas, British Virgin Islands, Cayman Islands, Grenada, Haiti, Saint Kitts and Nevis, Trinidad and Tobago, and Turks and Caicos [Frazier, in press]). In some of these countries, use is illegal, but nevertheless continues (e.g., even in the U.S., illegal use occurs [Addison, 1995]).

Most accounts of sea turtle use are without reference to why turtles are used and what use means (beyond economic profit). There have been some studies of the cultural importance of sea turtles to communities, however, and some of these are described briefly here.

12.2.1.1.1 Nicaragua

The role that turtles play in the culture of the Miskito Indians of Nicaragua was made famous by Nietschmann (1973; 1979). At the time of Nietschmann's studies, a Miskito fisherman's ability to share green turtle meat among kin and friends was a critical component of social relations: "Meat shared in this way satisfied mutual obligations and responsibilities and smoothed out daily and seasonal differences in the acquisition of animal protein" (Nietschmann, 1979). These social relations took place in a wider cultural context, in which sea turtles were the Miskito's most important resource.

In later writings, Nietschmann (1979) described the impacts of the introduction of commercial turtling. As turtles gained cash value, people spent more time turtling, and the more time thus spent, the less time spent on other subsistence activities, and the greater the need for cash. With the introduction of nets (supplied by the manufacturers of turtle products), the traditions of fishing changed; nets made everyone fishermen, and the importance of skills as a striker diminished. As more people became turtlers, fewer turtles were easily captured and more time and effort had to be spent turtling.

The introduction of a cash value for turtles also created tension in the community. The need to sell turtles for cash in order to purchase goods meant that fishermen felt unable to fulfill their meat-sharing obligations. This was especially true in times of harvest scarcity. Nietschmann (1979) concludes that the introduction of a market economy contributed to both the reduction of the resource and the economic and cultural impoverishment of the community. Miskito Indians continue to take turtles; Lagueux (1998) estimates that 10,000 turtles are taken annually. However, the tradition of meat sharing has disappeared (C. Lagueux, personal communication, 2002).

12.2.1.1.2 Costa Rica

The importance of green turtles to the human community at Tortuguero, Costa Rica, has shifted over the years. The name Tortuguero (and sometimes Turtle Bogue) means turtle place, and Rudloe (1979) describes the link between *Cerro Tortuguero* (a hill at the mouth of the Tortuguero River) and the legend of the turtle mother, a rock believed to attract turtles to nest. Although humans settled at Tortuguero in the 1930s, turtles have long been fished there; the Miskito Indians fished offshore for hundreds of years prior to Tortuguero's settlement, and European explorers restocked food supplies with Tortuguero turtles as early as the seventeenth century (Lefever, 1992). Until the 1970s, residents of Tortuguero captured green turtles for consumption; cooking methods for green turtle meat and eggs have been described (Rudloe, 1979; Lefever, 1992). Residents and nonresidents also captured turtles onshore for sale to boats waiting offshore. Although commercial turtling was on the wane in the 1960s (Parsons, 1962), turtles remained an important local resource, especially because other enterprises in the region (a banana plantation and a sawmill) experienced boom and bust cycles. In an isolated rainforest with limited agricultural potential, turtles provided a dependable and free source of protein during the "bust." When conservation efforts began in the 1960s and 1970s, turtle exploitation was prohibited, with one exception: the community is theoretically allowed to slaughter one turtle a week (three according to Lefever [1992] and two according to Rudloe [1979]) for communal distribution. However, the criteria to be met for such harvest are stringent to the extent that this practice has stopped (S. Troeng, personal communication, 2000), although some longtime residents of Tortuguero would like to be able to eat turtle meat (Peskin, 2002). The role of turtles in the culture of Tortuguero continues to evolve with the growth of ecotourism (see Section 12.2.3.1).

In the Caribbean port city and provincial capital of Limon, Costa Rica, a sea turtle fishery that licensed the capture of 1800 green turtles a year operated until 1999, when environmentalists in Costa Rica and the U.S. challenged the constitutional legitimacy of the harvest and won (Taft, 1999). The "cultural basis for eating turtle" (Opay, 1998) in this area is recognized, and prior to the challenge, several Costa Rican biologists described the fishery as "justified" for cultural reasons (Campbell, 1997). The petition against the harvest illustrates how values within a country can clash. The conservation values held by the Costa Ricans petitioning against the harvest were given precedence over the cultural and economic values of turtles to Limon fishermen.

12.2.1.1.3 Mexico

The Seri (or Comcaac) of the Sonoran coast and islands of the Gulf of California are "one of the last indigenous cultures in North America able to withstand total integration into local European derived cultures" (Nabhan et al., 1999). Cultural links between Seri culture and sea turtles are evident; turtles are not just food, but "the symbolic foundation of their marine resource based culture" (Nabhan et al., 1999). The importance of sea turtles manifests in material culture, including songs and legends. With the introduction of laws preventing harvesting, the Seri have had to restructure their use of marine resources. They now focus on other fishing and supplement their livelihoods with products from "adjacent Mexican culture." The traditions of sea turtle harvesting are no longer passed on, and Nabhan et al. (1999) point out that when elders pass on, "we can assume much information will be lost."

12.2.1.1.4 Venezuela

For the Wayuu of Venezuela, sea turtles are related to fertility, and consuming meat and blood affects "masculine vitality." Dreaming of turtles is also related to sexual activity, and turtle craniums are often hung in fruit trees to encourage growth (Parra et al., 2000). Parra et al. (2000) identify the need to understand these beliefs, so that educational strategies to show people the "real valuation" of the resource can be designed.

12.2.1.1.5 Indonesia

Suarez and Starbird (1996) examined the cultural context of leatherback hunting by people living in the Kai Islands of Indonesia. They describe the traditions, rituals, and beliefs (known as *adat*) that guide the turtle hunt. Hunts are highly ritualized, and under adat, meat is used for subsistence purposes and cannot be sold. As population pressures grow and other subsistence resources are depleted, increased fishing for subsistence rather than ritual purposes has resulted. Suarez and Starbird (1996) believe increased fishing could be a sign of cultural erosion, and suggest provision of alternative sources of protein to reduce the need for subsistence leatherback fishing.

12.2.1.1.6 Papua New Guinea

Spring (1995) describes the cultural importance of turtles in Papua New Guinea, manifested in various products of material culture, including bride-price items made from shell, oral histories, and legends. Cultural rituals and traditions vary across the islands, and some contribute to conservation, while others do not. For example, in some clans, permission to hunt turtles must be sought from traditional authorities, and turtles are used only for feasts, both traditional and nontraditional. Traditional hunting techniques used by some clans limit the number of turtles caught. Certain clans who believe themselves descended from turtles do not eat turtles. In contrast, in one village with a strong cultural attachment to leatherbacks, every nesting female found is slaughtered. Spring (1995) expresses some concern that traditional authority is eroding, historically because of some colonial laws and practices, and more contemporarily among younger generations influenced by Western culture and in areas closer to urban centers. In more remote regions, traditions remain stronger.

12.2.1.1.7 The Caroline Islands

The cultural importance of turtle hunting by people in the Caroline Islands is described by McCoy (1995): "The turtles contribute much to their overall cultural stability, reinforcing their independence from the outside. The estimated maximum contribution to the protein … is not nearly as important as this cultural role." The tradition of travel by dugout canoe, a subsistence economy, and taboos and ceremonies that surround the hunt historically provided a buffer on the number of turtles taken. However, the introduction of a cash economy, government settlement programs that spread turtle fishing skills among islanders, and the erosion of traditional taboos have led to increased pressure on turtles. As part of their maritime culture, local people see turtles as part of the sea, the "provider for all things," and thus show little concern at evidence of decreasing populations. McCoy (1995) argues for conservation programs to be undertaken with the people of the area firmly in mind.

12.2.1.1.8 Australia

In February 2002, the Australian Broadcasting Corporation (ABC) posted a recipe for green turtle on its website, as part of an aboriginal television program that discussed traditional diets. The resulting debate that erupted on CTURTLE, an on-line discussion group, addressed many issues, including the ethics of the ABC in posting a recipe based on an endangered species, the rights of aboriginal people to hunt such species, and the legitimacy of claims to traditional culture (messages archived at www.lists.ufl.edu/archives/cturtle.html). The debate reflects some of the difficulties in dealing with issues of culture and use.

In Australia, aboriginal peoples are allowed to use sea turtles for noncommercial purposes. Kowarsky (1995) found that the cultural basis of sea turtle use varies between different groups and that, overall, the integration of aboriginals into modern Australia reduced the number of turtles hunted. This finding contrasts with other examples, where integration into Western economies led to increased exploitation. In Australia, it may be the rejection of modernization by aboriginal people and their return to traditional territories and lifestyles that ultimately increases turtle hunting (Kowarsky, 1995).

12.2.1.2 Eggs

The list of countries using sea turtle eggs (Thorbjarnarson et al., 2000a), both legally and illegally, includes countries on the Atlantic coast of Central America, Mexico, Iran, Saudi Arabia, India, Thailand, Malaysia, Indonesia, the Philippines, and Papua New Guinea. Countries that can be added to this list include Suriname (Mohadin, 2000), Costa Rica (Campbell, 1998), Guatemala (Juarez and Muccio, 1997), Panama (Evans and Vargas, 1998), Honduras (Lagueux, 1991), Nicaragua (Ruiz, 1994), Bangladesh (Islam, 2001), and Myanmar (Thorbjarnarson et al., 2000b).

There are few studies of egg use, and these have focused primarily on economic value and how use is regulated (e.g., Lagueux, 1991; Campbell, 1998). A common cultural reference is to beliefs about the aphrodisiac qualities of eggs, particularly in Central America. In Ostional, Costa Rica, however, Campbell (1997) found that, although people recognized the aphrodisiac reputation of eggs, they emphasized their nutritional and economic value to families. Women in particular dismissed the aphrodisiac claim. Lefever (1992) also found aphrodisiac doubters in Tortuguero, although Rudloe (1979) credits the claim based on his own experience. In Guatemala, Juarez and Muccio (1997) suggest that eggs are used for aphrodisiacal purposes, and as such are "not a basic need." However, the authors state that one nest of eggs earns an agricultural or farm laborer the equivalent of one fourth of a month's salary, suggesting significant importance. Eggs clearly fill an economic need, and culture and economy are not so easily separated.

12.2.1.2.1 Costa Rica

The legal, commercial egg collection project at Ostional, Costa Rica, is the best-known example of egg use (Cornelius et al., 1991; Campbell, 1998; Thorbjarnarson et al., 2000a). The economic value of this resource is well recognized in the community: 70% of households rely on the egg collection as their primary source of income (Campbell, 1998). However, Campbell's (1997) study of the egg project illustrates the subtle ways in which turtles play a part in culture. First, life in Ostional is organized around sea turtle nesting, and the work of the community cooperative extends beyond the egg project; for example, it implements village development activities. Second, the project has contributed to a sense of independence and pride in the community, and to a level of organization unseen in many comparable coastal villages (March, 1992). Third, the "sense of the world" meaning of culture is translating into material culture. Residents tell stories about the turtles, discuss in detail when they will arrive, and take pride in activities they do to protect hatchlings. There are several turtle carvers and one poet, even though the latter is not part of the egg-collecting cooperative. Fourth, many residents see themselves as lucky to have the turtles, whose presence on the beach is in the hands of God (Campbell, 1997). Turtles are thus intertwined in daily life and undoubtedly contribute to residents' understandings of their world.

12.2.1.3 Skin

The skin of olive ridley turtles has been used to fashion leather accessories. Turtle leather has been manufactured in Mexico and Ecuador, and leather products have been imported by Japan, France, Spain, Italy, and the U.S. (Thorbjarnarson et al., 2000a). The use of animal parts in fashion has a long tradition. In Victorian England, women decorated their hats with bird feathers, a trend that threatened some bird populations (McCormick, 1989). The use of animal fur in fashion has become highly politicized in the West, but fur continues to find a place in fashion (Anonymous, 2001). Any cultural significance of sea turtle leather accessories in fashion is undocumented.

12.2.1.4 Other Parts and Products

As reported in Thorbjarnarson et al. (2000a), oil from turtles has been used to cure wooden boats in the Persian Gulf (Ross and Barwani, 1982) and in India (Kar and Bhaskar, 1982). Turtle penis is used as an aphrodisiac on the Red Sea coast of Saudi Arabia (Miller, 1989), and turtle blood is used to treat ailments in India (Silas and Rajagopalan, 1984). In Togo, various parts of turtles are used for medicinal purposes (Hoinsoude et al., in press). The prevalence of wild animal parts in traditional medicine is a topic that concerns conservationists (Roberts et al., 1999). However, studies on cultural significance of, or attachment to, such remedies are lacking.

12.2.1.5 Taxidermy

Animal collection and taxidermy has had its place in Western culture. McCormick (1989) describes collectors in Victorian England, and how their amateur enthusiasm contributed to depletion of populations. The popularity of collecting reflected enthusiasm for science and particularly natural history. Turtle enthusiasts have also been collectors. In *The Windward Road* (Carr, 1967), Carr celebrates when a rare species of tortoise is captured and killed so that it can be added to his collection.

Taxidermy continues to serve the tourist trade: "'stuffed' turtle curios from Southeast Asia" have been available in Hawaii (Balazs, 1977). More recent reports of "whole, stuffed turtles and tortoiseshell products" in Vietnam (Thuoc et al., 2001), of subadult hawksbills in Bangladesh (Islam, 2001), and of turtle heads and carapaces in Uruguay (Lopez and Fallabrino, 2001) identify tourists as the target market. What drives current collection is unknown. Tourists purchasing such items risk fines if traveling to or from a Convention on International Trade in Endangered Species of Wild Fauna and Flora (CITES) signatory country, although some may be unaware of such risk.

12.2.1.6 Tortoiseshell

Tortoiseshell, traditionally obtained from the hawksbill turtle, has ranked among the world's luxury goods since earliest recorded times (Anonymous, 1977). Anthropologist Elizabeth Overton identified tourists purchasing hawksbill shell in the Maldives as contributing to "wiping out sea turtles" (Anonymous, 1977). Balazs (1977) included hawksbill shell on his list of items available in Hawaii. The World Society for the Protection of Animals (no date) reported that in a 12-month period,

FIGURE 12.1 Turtle shell items displayed in a souvenir shop, illustrating turtles as product. (From World Society for the Protection of Animals (WSPA) 1997 London. With permission.)

tortoiseshell was available to tourists in Barbados, Belize, Costa Rica, Cuba, the Dominican Republic, Fiji, Indonesia, Japan, Maldives, Mexico, Nicaragua, São Tomé, Sri Lanka, Thailand, and Vietnam (Figure 12.1). As with stuffed turtles, tourists risk fines in violation of CITES if caught transporting turtle shell between CITES signatory countries. Mortimer (1977) gives anecdotal evidence that tourists may be apprised of such risks while simultaneously encouraged to buy. Shell also has cultural importance for communities. Three cases of hawksbill shell use are described in further detail.

12.2.1.6.1 Japan

Japan has a long history of crafting hawksbill shell (bekko) into various decorative items, some of which have been found in ruins of a seventh-century city. Bekko is crafted using traditional techniques and tools thought to be the same as those used 300–1000 years ago. Families pass on skills from one generation to the next, and one of the oldest bekko families (Ezaki in Nagasaki) is traced to 1709 (Kaneko and Yamaoka, 1999). The popularity of bekko crafting in Japan depends on the status of trade relations, because hawksbill shell has to be imported, and the bekko industry is currently in decline because of trade restrictions. Although Japan formerly invoked an exception to CITES to import hawksbill shell, it withdrew this in 1994 under pressure from the U.S. The Japanese Bekko Association currently estimates that it would require 4 tons of hawksbill shell (compared to 20–30 tons imported in the 1980s) to allow the remaining family-owned bekko companies to continue to operate (Kaneko and Yamaoka, 1999). Japan's desire to import hawksbill has been at the center of two controversial Cuban proposals to CITES that would allow for limited trade in shell between the two countries (Campbell, 2002; Mrosovsky, 2000; Richardson, 2000).

12.2.1.6.2 Seychelles

"The people of Seychelles view turtles as an integral part of their culture and economy" (Mortimer and Collie, 1998). Hawksbill shell has been used for more than 200 years, exported to Europe and more recently Japan, and fashioned locally into items for sale to tourists. This ended in the late 1990s, when the government of Seychelles banned commercial trade in hawksbill products and slaughter within territorial waters. Artisans were compensated and sold their stockpiles to the government. Stockpiled shell was burned publicly in 1998.

At the Nineteenth Annual Symposium on Sea Turtle Biology and Conservation, a resolution was passed acknowledging "that by destroying its stockpile of raw hawksbill shell the Government of Seychelles made a statement to its citizens and to the world that it recognizes the beauty of the natural environment of Seychelles, of which healthy populations of free-living sea turtles are an inherent component…" (Plotkin, 1999). Not all sea turtle conservationists agreed with this position, as seen in an exchange on CTURTLE. Some people criticized the burning as a political gesture that did little to improve the fate of live hawksbills, or as wasteful; income earned via sale of the shell could have been used to compensate out-of-work artisans and to pay for conservation activities. The diversity of values and beliefs about conservation was highlighted in the exchange, and like many discussions of conservation taking place in internationalized cyberspace, the voices of local people were absent.

12.2.1.6.3 Palau

In Palau, the large costal hawksbill scutes are molded into polished bowls called *toluk* and used as an exchange valuable among women. This form of traditional money is circulated exclusively by Palauan women to give gratitude for services and courtesies offered among their families (Smith, 1983). The specific value of toluk is dependent on its aesthetic and historical characteristics, and is influenced by the changing economy of toluk circulation. Direct harvest of hawksbill turtles for toluk and the production of jewelry, mostly for sale to tourists, has led to continual pressure on local populations. In recognition of the marked declines in nesting and foraging turtles, the Palau women's association, Didil Belau, recently called for a 20-year moratorium on hawksbill harvesting (P.K. Mad and M.D. Guilbeaux, personal communication, 2002).

12.2.2 Incidental Take

12.2.2.1 Fisheries Interactions

Although turtle drownings in shrimp trawls have long been a concern to sea turtle biologists (Carr highlighted the impacts of shrimping on Kemp's ridleys in 1977 [Carr, 1977]), other types of fishing activities, for example longline and drift net, are of increasing concern. The link between culture and such an indirect use of sea turtles may seem tenuous. However, one of the best-known cases of fisheries interaction with sea turtles, i.e., via shrimping, and the debates about how to solve the bycatch problem via turtle excluder devices (TEDs), provides an example that can be examined through the lens of culture in three ways.

First, in an effort to reduce turtle drownings in shrimp trawls, use of TEDs was made mandatory under U.S. law in 1989, after 10 years of failed attempts to encourage voluntary use (Crouse, 1999; Margavio and Forsyth, 1996). Some shrimpers resisted the legislation, claiming that TEDs result in reduced catch and arguing that they should be compensated for property loss (Anonymous, 1995). However, resistance was more than economic. In their book, *Caught in the Net*, Margavio and Forsyth (1996) explore the cultural context of the TEDs conflict. For example, they describe traditional Cajun culture in Louisiana, its roots in fishing and hunting and its emphasis on family, and the role shrimping plays in supporting this. Resistance to TEDs was based on a desire to protect this culture. Resistance was also to regulation in general, the marginalization of shrimping in the face of other economic activities, and erosion of independence.

Second, TEDs can be described as the technocratic solution to the turtle drowning problem, and were designed as "an effective way to allow shrimping to proceed virtually unimpeded while protecting most sea turtles from drowning in trawls" (Crouse, 1999). Thus, TEDs are part of a solution that does not address the overcapitalization of the shrimp and other commercial fishing industries, and which is firmly imbedded in late-stage capitalism, a defining characteristic of Western culture.

Third, when U.S. shrimpers and environmentalists appealed to the U.S. government to expand the TEDs requirement to shrimp imports (which it did under Section 609 of Public Law 101–162), several Asian countries brought a dispute to the World Trade Organization (WTO). Even Thailand, a country that uses TEDs and would not have been embargoed, joined the dispute on principle. As Crouse (1999) points out, several issues were at stake in the dispute: sovereignty and rights to dictate fishing policy in territorial waters, rights of developed countries to dictate environmental policy in developing countries, rights of the U.S. to restrict access to its domestic market in the era of free trade, and obligations of a global trade system to protect endangered species. Values clearly play a role in this debate, and the prioritization of issues by individual countries reflects cultural biases and their coexistence in a global system (sea turtles have played a more general role in antiglobalization protests, as discussed in Section 12.2.3.4).

12.2.2.2 Habitat Use

In the mid-1990s, a conflict erupted in Volusia County, FL, and involved environmentalists, recreational drivers using a 28-mile stretch of sea turtle nesting beach, and county government officials. Because of impacts of beach driving and artificial lighting on sea turtles, environmentalists launched a lawsuit against the county for violation of the Endangered Species Act (ESA). The county responded by devising a management plan with the objective of obtaining a permit to "take" sea turtles and eggs (indirectly) under the ESA, and argued that beach driving constituted a cultural asset (Fletemeyer, 1996). Other issues that arose included the rights of people to hold local government accountable for actions impacting endangered species, and the tenth amendment of the U.S. Constitution that prevents the federal government from compelling states or localities to implement national programs. Two legal think tanks with little stake in the turtle conservation outcome became involved in the dispute to argue for upholding the tenth amendment (Kostyack, 1999), and this demonstrates how conservation conflicts are often about more than conservation. Although Fletemeyer (1996) might dismiss cultural claims to beach driving as euphemistic, several cultural issues are clearly at stake in this conflict.

More generally, development of sea turtle nesting beaches for private homes, industry, and tourism infrastructure has been a highly contentious issue in the southern U.S., particularly in Florida. The development itself, and subsequent efforts to protect the investment against damage, constitutes an indirect use of sea turtles through alteration and sometimes elimination of nesting habitat. Beliefs about private property, free market development, and the role of government regulation — all central components of contemporary Western culture — drive this competition.

12.2.3 Nonconsumptive Uses

12.2.3.1 Tourism and Ecotourism

Tourists interact with turtles in the U.S. (Balazs, 1995; Johnson et al., 1996), Honduras (Dempsey, 1996), Costa Rica (Campbell, 1999; Campbell, 2002b; Gutic, 1994), Brazil (Marcovaldi and Marcovaldi, 1999), Trinidad and Tobago (Fournillier, 1994), Greece (Dimopoulos, 2001), Turkey (Yerli and Canbolat, 1998), Taiwan (Cheng, 1995), and Australia (Wilson and Tisdell, 2001). There are also many relatively undocumented instances of tourist–turtle interactions.

The impacts of tourism on turtles include tourists' purchasing souvenirs made from turtle products (Section 12.2.1.6), loss of habitat through resort development, competition for use of beach with tourists and infrastructure (chairs, umbrellas), direct interference with turtles by tourists (taking pictures, sitting on turtles), turtle strikes by motorized water vehicles, hatchling disorientation by development lighting, and hatchling trampling by tourists on beaches at night (World Society for the Protection of Animals, 1997). Many of the locations listed above have experienced some if not all of these problems. Conflicts have erupted in some locales (e.g., Zakynthos, Greece) where the tourism industry and tourists themselves value coastal resources differently than do turtle conservationists.

Ecotourism is an alternative form of tourism, defined as "responsible travel to natural areas that conserves the environment and sustains the well-being of local people" (Ecotourism Society, no date). Ecotourism is often promoted as nonconsumptive use of wildlife and a means of reconciling conservation with economic development, particularly in developing countries. By providing income to local residents, ecotourists provide incentives to protect the resources they wish to view.

Turtle-based tourism activities have been introduced at a variety of nesting beaches around the world, often to serve different purposes (Figure 12.2). In the U.S., turtle walks are an educational tool designed to gain public support for protecting nesting beaches (Johnson et al., 1996). In Australia, ecotourism is promoted to provide an economic rationale for conserving the species (Wilson and Tisdell, 2001). In Tortuguero, Costa Rica, income earned by ecotourism can replace money earned formerly via a green turtle fishery (Jacobson and Robles, 1992; Peskin, 2002).

FIGURE 12.2 Viewing nesting leatherbacks in French Guiana, illustrating turtles as tourist activity. (From M. Godfrey and O. Drif. 2001. Developing sea turtle ecotourism in French Guiana: perils and pitfalls. *Mar. Turtle Newsl.* 91:1–4. With permission.)

At Playa Grande, Costa Rica, guided turtle walks generate income and awareness needed to justify maintaining a protected nesting beach in an area of dense tourism development (Campbell, 2002b). Turtle-based activities, however, do not equate with true ecotourism. In the case of Playa Grande, for example, tourism development may be partly responsible for declining numbers of nesting leatherbacks. On the other hand, tourism at Tortuguero appears to be partially meeting the objectives of ecotourism (Campbell, 2002b; Peskin, 2002).

Ecotourism in practice has often fallen short of its environmental, economic, and social objectives (Ross and Wall, 1999). Nevertheless, its perceived potential is high, and Godfrey and Drif (2001) suggest that "it is almost axiomatic to present the idea that developing ecotourism is a desirable goal" when undertaking sea turtle conservation projects. Enthusiasm for ecotourism is often linked to a lack of enthusiasm for consumptive turtle use (Campbell, 2000; Campbell, 2002b). For the purposes of this chapter, it is not the actual success or failure of ecotourism ventures that is of interest, but the cultural context of and the values associated with turtle-based ecotourism ventures.

In any ecotourism scenario, turtles are valued in a variety of ways. In the example of Tortuguero, Costa Rica, there are several stakeholder groups interested in turtles. Tourists travel to Tortuguero to view turtles (Jacobson and Robles, 1992), guides leading turtle tours earn income (Peskin, 2002), scientists value the species conservation and the research opportunities afforded, and research assistants and participants working for the Caribbean Conservation Corporation (CCC) value their turtle experience for a variety of reasons (Smith, 2002). Prior to the establishment of Tortuguero National Park, turtles were a food, economic, and cultural resource for local people (Lefever, 1992). Although multiple values of turtles are coexisting in Tortuguero, some residents (primarily older, original inhabitants) would like to use limited numbers of turtles as a source of food (Peskin, 2002).

Turtles are also used to advertise or promote tourism destinations, as discussed in Section 12.2.3.4.

12.2.3.2 Education

Turtles are charismatic megafauna, and as such are a flagship species. Public education regarding sea turtle issues may have wider spillover effects, because successful conservation of sea turtle habitat, for example, has benefits for other less charismatic species. This argument has been made with TEDs; although they were designed specifically to release turtles from shrimp trawls, they also reduce general bycatch. Education is also an objective of turtle-based tourism or volunteer activities, both by the promoters and the by participants. For example, in Tortuguero, Costa Rica, many individuals participating in the CCC's volunteer research programs are specifically motivated by the educational opportunity of working with turtles in the wild, and some are looking for materials they can use in their own teaching (Smith, 2002). Turtles are also kept in educational facilities; for example, Project Tartarugas Marinhas (TAMAR) in Brazil keeps some turtles in various stages of captivity to serve as direct educational tools (Marcovaldi and Marcovaldi, 1999).

12.2.3.3 Research

Research with wildlife often involves the use of the species itself. Normally, and particularly in the case of endangered species, this use is nonconsumptive, i.e., the species is not permanently removed from a population, and research protocols are designed to minimize any long-term effects. Many of the contributors to this volume use turtles in their research, and have careers based on such work.

12.2.3.4 Turtles as Symbols

Although little research has been done in this field, it appears that, for many people, sea turtles are symbols of the marine environment, of environmentalism, and of the historic struggles between humans and nonhuman beings. Turtles are the subjects of material culture; songwriters, poets, painters, carvers, photographers, and sculptors use turtles as their subjects. The annual Symposium on Sea Turtle Biology and Conservation relies on attendees' willingness to purchase such material products at an auction as a means of fundraising.

Sea turtles appear on postage stamps (Figure 12.3). Stamps from 163 countries can be found at http://www.2xtreme.net/nlinsley/. In this role, marine turtles are a symbol of national identity and environmental consciousness. Sea turtles are also used as marketing tools, and lend enviro-credibility to products (e.g., www.greenturtle.com). Some research has been done on the use of sea turtles to promote tourism (Cosijn, 1995; Schofield et al., 2001). Although critical of using sea turtles to market mass tourism, the studies have focused on the correctness of information provided in tourist brochures. Cosijn (1995) sees this use of turtles as a tool to "seduce people into buying their travel product." The attractiveness of turtles is speculated on, rather than studied, but the researchers and the tour companies recognize the symbolic role of sea turtles in the imaginations of potential tourists.

FIGURE 12.3 Tuvalu postage stamp, illustrating turtles as a national symbol.

FIGURE 12.4 Protesters at the WTO meetings in Seattle illustrating turtles as a symbol of the antiglobalization movement. (From Jen Rinick, Animal Welfare Institute. With permission.)

In 1999, sea turtles were a focal point of protests at WTO meetings in Seattle, WA. A subgroup of protesters who focused on the issue of TED use in shrimp trawls costumed themselves as sea turtles, and their images were picked up by the media (Figure 12.4). The turtle protesters became a symbol of an antiglobalization movement in general, one that can be considered a form of counterculture, challenging the promotion of global free trade and associated global values (Yuen et al., 2001). (Whether the original turtle protesters identify themselves with the wider movement and all its values is unknown.) In Seattle and at other related protests, the clash of values was paramount and encompassed nothing less than the very structure of economic, political, and social life.

12.3 CULTURE AND CONSERVATION: CULTURES OF CONSERVATION

Section 12.2 illustrates some of the uses of sea turtles (consumptive and nonconsumptive) and the ways that culture mediates such use. As discussed in Section 12.1.3, where the cultural context of the national park model was described, cultures also influence conservation. The conservation concepts of sustainable use and CBC, their application to sea turtles, how they reflect the interaction of culture and conservation, and how they fit into cultures of conservation are examined below.

12.3.1 Sustainable Use

12.3.1.1 Concept

According to the IUCN, sustainable use is central to contemporary conservation (Section 12.1.3). Sustainable use is generally defined as the managed use of resources in a way and at a rate that does not compromise their long-term existence. Use can be either consumptive or nonconsumptive, providing subsistence or commercial benefits (Freese, 1998). Sustainable use is often an objective of management rather than a certainty, because of the difficulties in determining definitively the outcomes of use schemes, and because of the reality that sustainable use programs are often implemented as alternatives to uncontrolled exploitation.

Sustainable use is based on the argument that wildlife and biodiversity must be valued by those expected to conserve it, and that value is often derived through use: "if wildlife has no value, then wildlife and its habitat will be destroyed to make way for other land uses" (Robinson and Redford, 1991). In the case of impoverished rural peoples, the most compelling value is assumed to be economic, and although this reduction of local values for wildlife to economic ones is oversimplified, it reflects the pervasiveness and reality of the market economy.

Sustainable use is the subject of much debate. The number of successful cases of sustainable use, and particularly of commercial, consumptive use, of wildlife is low (Freese, 1998). Difficulties are related to biological sustainability (e.g., the inability of resources to sustain even low levels of use, or predicting correctly the response of a population to use), and long-lived animals with slow reproductive rates and low levels of density dependence pose particular challenges for use regimes

(Robinson, 1993; Musick, 1999). Other difficulties relate to socioeconomic sustainability (e.g., establishing incentives that encourage long- rather than short-term views to use). Even when managed use schemes are believed to be biologically sound and return economic benefits to local people, they may fail to gain support for conservation if control over resources is not devolved to local users. This lack of support can translate into illegal use of managed resources and undermine overall sustainability. CBC, which in part arose in response to issues of control, is discussed in Section 12.3.2.

12.3.1.2 Sustainable Use and Sea Turtle Conservation

Sustainable use projects that involve consumptive use of sea turtles for conservation purposes are few. One example is egg collection at Ostional, Costa Rica (Campbell, 1998; Cornelius et al., 1991). Similar but less documented collections take place in Panama (Evans and Vargas, 1998) and Nicaragua (Ruiz, 1994). These collections are based on *arribada* nesting by olive ridleys, i.e., mass nesting that destroys many eggs, and the Ostional project is the only olive ridley collection where commercial sale of eggs is legal. In Suriname, leatherback and green turtle eggs that would otherwise be inundated and/or washed by an eroding shoreline are collected by both communities and the government agency responsible for national parks. Collected eggs are sold, and generate public support for conservation as well as income for conservation activities (Mohadin, 2000).

In other places, use of turtles or eggs is allowed, but at such low levels that biological sustainability is not a prime concern. These are instances of what might be termed minimal compensatory use. For example, a small harvest of leatherback eggs by the community at Gandoca was allowed in the Gandoca and Manzanillo Wildlife Refuge, Costa Rica, until the late 1990s. Such compensatory use programs seek to ensure community support for broader conservation goals:

> Information collected from Gandoca residents shows that community support for the project will increase, and poaching by locals will be reduced if local residents are permitted to consume moderate numbers of eggs in a controlled manner. In many cases it appears that what matters is not so much whether a given family obtains eggs as that they do not feel prohibited from doing so.
>
> **ANAI, 1995**

This type of compensatory collection has also been seen with some adult fishing operations. Examples from Tortuguero National Park and Limon, Costa Rica, have been discussed (Section 12.2.1.1). Similarly, in 1980 in one region of Mexico, the government switched from a closed turtle fishing season to a quota system that allowed 250 male green turtles to be used by a local cooperative. This had the impact of changing the attitudes of coop members, who stopped illegal use and began to accept the turtle recovery program (Clifton et al., 1995).

Sustainable use is controversial in sea turtle conservation. The Marine Turtle Specialist Group (MTSG) of the IUCN has promoted a no-use stance for most of its history (Campbell, 2002a). For example, the 1979 World Conference on Sea

Turtle Conservation cited "the use of sea turtles as food by people who live where sea turtles are found" and "differing attitudes toward conservation in different countries" as factors contributing to sea turtle decline (Bjorndal, 1981). The 1979 meeting did make a concession to use, when it was "a traditional way of obtaining food practiced by aboriginal people who are not yet part of a cash economy or technological society" (Bjorndal, 1981). As the discussion of key terms in Section 12.1.4 suggests, limiting use of marine turtles along these lines, where culture and economy are static, is problematic. Nevertheless, many marine turtle experts support such limits on use (Campbell, 2000).

These 1979 sentiments reflect the dominant culture of conservation of the time, one that relied on a traditional approach via the national park model. (A later manual on conservation techniques recommends that poachers be kept away from nesting beaches by patrolling "with assistance from military organizations, conservation officials and interested amateurs" [Pritchard, 1983].) Early proponents of local use existed within the MTSG, however, and Hughes (1979) argued that marine turtles are "extremely resilient" to exploitation: "Where total protection is feasible, let us have it; where not, let us not close the door to survival by ignoring a valid *conservation* technique — utilization." Similarly, Mrosovsky (1979), Reichart (1982), and Bustard (1980) made the argument for use where socioeconomic, political, and cultural conditions warranted it.

These early views in support of consumptive use of marine turtles reflect contemporary discussions of sustainable use. Over time, the MTSG's position on sustainable use has shifted slightly. For example, the 1995 *Global Strategy for the Conservation of Marine Turtles* (the *Strategy* [MTSG, 1995]) recognizes that turtles play a role in the cultural and social lives of coastal people and are an important source of protein. However, it stops short of accepting sustainable use:

> Too frequently…wide use by a growing human population, coupled with the migratory nature and slow rates of natural increase of these animals, has resulted in most utilization being non-sustainable....Although this Strategy recognizes that utilization of marine turtles occurs in many areas and does not oppose all use, it does not support non-sustainable use.
>
> **MTSG, 1995**

To date, there are no accepted guidelines for marine turtle use, although there was an opportunity for the MTSG to provide egg-collecting guidelines in a recently published techniques manual (Eckert et al., 1999), as suggested in the *Strategy* (MTSG, 1995).

Campbell (2000, 2002a) examined attitudes of sea turtle conservation experts toward sustainable use. Although experts agree that biological characteristics of marine turtles constrain the extent to which sea turtles can be used with certainty, their views on how to proceed from this starting point differ. It is on moving from this point that the cultures of conservation become most evident. Although almost all experts believe their views on sustainable use are informed by science, other values clearly play a role. For example, views on local rights to use resources and on local socioeconomic and cultural need impact on expert positions on use (Campbell, 2000). Very few experts recognize their own emotional response to sea turtles; they see their views as informed by science and value-free, whereas they characterize the views of others, particularly those who disagree with them, as value laden (Campbell, 2002a). Finally, the way that experts address issues of scientific uncertainty influences their views on use (Campbell, 2002a). Science as a foundation of Western culture in general, and its influence on conservation policy specifically, are themselves subjects of study (Leach and Mearns, 1996; Nader, 1996; Pepper, 1984).

12.3.2 Community-Based Conservation (CBC)

12.3.2.1 Concept

Like sustainable use, CBC assumes that support of local people is critical to conservation success. The terms differ in their foci; although sustainable use focuses on the use of the species itself, CBC is concerned with the local economic, social, and cultural context in which conservation takes place, and with the role of communities in conservation projects (regardless of whether they have a use component). There is no one definition of CBC, but it is commonly seen as having two objectives: to enhance conservation and to provide social and economic gains for local people. Ownership of conservation activities is a critical concern in CBC, and Little (1994) suggests that CBC implies "at least some of the following: local-level, voluntary, people-centered, participatory, decentralized, village based management." Nevertheless, there is a wide spectrum of views on CBC, and the mix of components and prioritization of objectives vary according to the definer.

Like sustainable use, CBC has experienced mixed success in practice. Some of the major obstacles are, first, that CBC implementers fail to operationalize community participation in project identification, design, and management. Participation is instead seen as a means to get people to support predetermined conservation programs (Hackel, 1999; Songorwa, 1999). Second, CBC projects have been undertaken without an adequate understanding of the local social, economic, and cultural context and by environmental nongovernmental organizations (NGOs) with limited experience in community development (Wells and Brandon, 1993). Third, CBC has not learned from the related field of participatory development, where organizations primarily interested in human development have struggled to implement successful participation (Little, 1994). Community itself is emerging as a problematic term, and communities are too often assumed to be homogeneous (Section 12.1.4). Although conservation can function in heterogeneous communities, understanding community structure is necessary to determine appropriate and realistic incentives for conservation (Campbell, 1998).

12.3.2.2 CBC and Sea Turtle Conservation

There are several accounts of community participation in sea turtle conservation, including stories of former poachers turned conservationists, of fishermen collecting data for conservation projects, and of communities participating in educational programs (e.g., Schulz, 1975; Nichols et al., 2000; Lima, 2001). The extent to which such projects are community-based is unknown, because research on the extent of community support for conservation is lacking, and descriptions of success are often provided by conservation organizations themselves. Because "the goal of true CBC facilitators is to work themselves out of a job" (Frazier, 1999), the ultimate test of CBC's success is whether conservation efforts continue in the absence of conservation organizations — a test most organizations are unlikely to face. Four projects that show indications of, or are often cited as, being community based are described briefly below.

12.3.2.2.1 Costa Rica

In the egg collection project at Ostional, Costa Rica, a community development association is responsible for almost all aspects of the conservation program. Community management is mandated by law, and although the University of Costa Rica and various government agencies are involved in the project, none is permanently present in the village. Campbell (1998) describes project regulation, economic value, and dependence on the project; use of profits in support of conservation and development projects; and willingness of community members to undertake additional sea turtle conservation efforts. An important conclusion from the work of Campbell (1998) is that CBC should not be romanticized. In the Ostional case, high levels of intracommunity conflict detract only marginally from achievement of overall objectives.

12.3.2.2.2 Mexico

Community-based research and conservation have been promoted in Baja California. Nichols et al. (2000) describe the need for a community approach in an isolated region with limited enforcement. They point out that researchers often focus on the ways in which local people detract from conservation, and that local values are oversimplified by outsiders: "local fishers have demonstrated an interest in conservation for ecological and aesthetic reasons, as well as to preserve a source of their traditional livelihood and an occasional source of food" (Nichols et al., 2000). In-water work at Bahía Magdalena by Nichols et al., for example, has relied on cooperation of local fishermen to show where turtles are regularly found and captured. Cultural motivations for exploitation are also recognized as critical to devising conservation schemes in the region. For example, because most turtles are eaten for special occasions and ceremonial feasts, attempts at substitution may be limited (Bird, in press). Although research and conservation efforts were initiated by outsiders in Bahía Magdalena, a grassroots organization has since formed to promote sea turtle recovery in the region. A recent decision by fishermen to exclude those who hunt turtles from participating in research activities reflects the extent to which conservation ideas are becoming community based (Nichols, personal communication, 2002).

12.3.2.2.3 Australia

In northern Australia, a collaboration between an indigenous community (the Yolngu of northeast Arnhem Land of the Northern Territory, represented by the Dhimurru Land Management Aboriginal Corporation) and university and government researchers has been undertaken to develop a strategy for sustainable subsistence use of turtles. The project is based on a recognition both of the rights and responsibilities of Australian aboriginals in managing the sea turtle resource and that their involvement in research and management activities will be critical to the survival of sea turtles in the region. The project combines traditional knowledge and law with contemporary scientific methods (Kennett et al., 1997). Activities have included recording traditional knowledge, a turtle stranding and rescue program, heavy metal analysis, habitat mapping, tagging and nesting studies, satellite tracking, genetic sampling, studies of temperature and sex determination, and quantifying indigenous harvests (R. Kennett, personal communication, 2002).

12.3.2.2.4 Brazil

Project TAMAR covers a large portion of the Brazilian coast and has been described in detail by Marcovaldi and Marcovaldi (1999). The geographic spread of the project means that many communities are impacted, and their experiences undoubtedly differ. Nevertheless, TAMAR's approach to CBC has generally been two-pronged: environmental education and provision of alternative economic activities. In one example, TAMAR set up a field station in the indigenous settlement of Almofala, where incidental catch of marine turtles in fishing operations was a concern. In addition to extensive environmental education activities, TAMAR assisted the local community to develop alternative economic activities, including artificial reefs for fishing, a community vegetable garden, and embroidery and lace making, as identified at a meeting between TAMAR and a community association (Lima, 2001).

Communities and their importance to conservation undertakings are increasingly included in the dominant culture of sea turtle conservation. The annual Symposium on Sea Turtle Biology and Conservation has expanded in scope over the last 8 years and now includes several sessions devoted to the interaction of conservation and communities. Nevertheless, messages about community involvement remain mixed in official policy. Although the MTSG's *Strategy* (1995) calls for local participation because local people are "a strong force in the depletion of marine turtle populations and the destruction of their habitats," the 1999 *Techniques Manual* includes a chapter on CBC that depicts local people as potential partners with vested interests in the continued existence of resources (Frazier, 1999). The recently negotiated Inter-American Convention for the Protection and Conservation of Sea Turtles has been criticized for its "top-down" approach and its failure to incorporate CBC (Campbell et al., 2002). In contrast, the Santo Domingo Declaration resulting from the regional meeting, Marine Turtle Conservation in the Wider Caribbean Region — A Dialogue for Effective Regional Management, calls for "greater community participation in the identification of management priorities and actions, as well as in the development, implementation and evaluation of activities directed at the conservation of sea turtles and their habitats" (Eckert and Abreu, 2001). It recognizes that "sea turtles comprise a unique part of the biological diversity of the region and an integral part of the cultural, economic, and social aspects of the societies found therein" (Eckert and Abreu, 2001). In a study of marine turtle experts and their views on conservation, Campbell (2000) found that although experts were highly supportive of local participation, their definitions of participation were often limited to people being employed by a conservation program, educated, and listened to. In such cases, objectives of conservation programs were assumed, and experts were generally opposed to relinquishing control over conservation programs to local people. Given the mixed treatment of community in the dominant culture of conservation, claims of projects being community based need to be carefully considered, and research on the extent of support for conservation among communities is needed.

12.4 CONCLUSIONS

On the basis of this review, three conclusions are highlighted:

1. The ways that culture influences the use and conservation of sea turtles are varied, and beyond the generalization that culture needs attention, there are currently few rules to share across regions and peoples. For example, the common observation that market infiltration undermines culture and makes use unsustainable is not true in all cases. Further site-specific research is needed, because the number of research-based assessments of the culture–conservation link are few; many of the authors cited in this chapter reflect on cultural issues rather than study them. Such research will serve to improve conservation in specific contexts, and accumulation of research might yield more generalizable results.
2. The notion that only certain kinds of cultures (indigenous, traditional, or subsistence) are relevant for sea turtle use is misleading. The importance of these issues in influencing use, and their relevance in a global economy, is not entirely clear. For example, insistence that if local people are in the market they lose any claims to tradition or culture, regardless of how marginalized they are within a global economic system, is an inaccurate characterization of culture as static, and may only serve to further impoverish rural peoples in developing countries. Instances of use need to be assessed on their cultural, social, economic, and environmental impacts, rather than on the extent to which they fulfill Western notions of what is traditional.
3. Culture is not only something that influences use and conservation of marine turtles in some other, more exotic locale. Cultural values are evident in all uses of marine turtles, and conservation, although often depicted as value-free, is itself a product of culture. Even the importance placed on science in conservation policy is a reflection of Western culture (Pepper, 1984). Recognizing the diversity of cultural relations with sea turtles and the cultural context of conservation policy may make such policy more flexible and dynamic, and more acceptable to and appropriate for the diversity of people living with sea turtles.

ACKNOWLEDGMENTS

The author's research is supported by the Canadian Social Sciences and Humanities Research Council. Jack Frazier contributed to the intellectual foundations of this chapter and provided detailed comments on various drafts. Matthew

Godfrey commented on two drafts, and tracked down and shared needed references. M. Guilbeaux, R. Kennett, C. Laguex, N. Linsley, P.K. Mad, J. Nichols, and S. Troëng provided additional information used in this chapter. Thanks to the editors for providing the opportunity for this contribution and for their patience with its production.

REFERENCES

Addison, D.S. 1995. Poaching in Everglades City, Florida, trivialized by sentences. *Mar. Turtle Newsl.* 69: 16–17.

ANAI. 1995. *Conservation of the Leatherback Turtle in Gandoca/Manzanillo National Wildlife Refuge: Volunteers Manual.* ANAI. San Jose, Costa Rica.

Anonymous. 1977. Indian Ocean tourists wiping out sea turtles. *Mar. Turtle Newsl.* 7: 4–5.

Anonymous. 1995. Guest editorial: endangered species vs. property rights. *Mar. Turtle Newsl.* 70: 15–16.

Anonymous. 2001. Fur's dirty dozen. *Anim. Agenda.* 21: 10–11.

Aureggi, M., G. Gerosa, and S. Chantrapornsyl. 1999. Marine turtle survey at Phra Thong Island, South Thailand. *Mar. Turtle Newsl.* 85: 4–5.

Avis, W.S., ed. 1980. Indigenous. In *Funk & Wagnalls Standard College Dictionary.* Fitzhenry & Whiteside Limited, Toronto. 1590 p. 1590.

Balazs, G.H. 1977. Sale of turtle products promoted in Hawaii. *Mar. Turtle Newsl.* 4: 4.

Balazs, G.H. 1995. Hawaiian Islands: promoting sea turtle watching in coastal waters of the Hawaiian Islands to enhance conservation and ecotourism. In *Tourism and Marine Turtles: Can We Live Together?* IUCN-MISG Committee on the Impact of Tourism on Marine Turtles. 1–2.

Berkes, F., J. Colding, and C. Folke. 2000. Rediscovery of traditional ecological knowledge as adaptive management. *Ecol. Appl.* 10: 1251–1262.

Bird, K. In press. Integrating local knowledge and outside knowledge in sea turtle conservation: a case from Baja California, Mexico. In *Proceedings of the 22nd Annual Symposium on Sea Turtle Biology and Conservation,* Miami.

Bjorndal, K.A. 1981. *Biology and Conservation of Sea Turtles.* Smithsonian Institution Press, Washington, D.C. 583 pp.

Braun, B. and N. Castree. 1998. *Remaking Reality.* Routledge, London. 312 pp.

Brosius, J.P., A. Lowenhaupt Tsing, and C. Zerner. 1998. Representing communities: histories and politics of community-based natural resource management. *Soc. Nat. Resour.* 11: 157–168.

Bustard, H.R. 1980. Should sea turtles be exploited? *Mar. Turtle Newsl.* 15: 3–5.

Campbell, L.M. 1997. International conservation and local development: the sustainable use of marine turtles in Costa Rica. Ph.D. thesis, University of Cambridge, Cambridge. xiii + 347 pp.

Campbell, L.M. 1998. Use them or lose them? The consumptive use of marine turtle eggs at Ostional, Costa Rica. *Environ. Conserv.* 24: 305–319.

Campbell, L.M. 1999. Ecotourism in rural developing communities. *Ann. Tourism Res.* 26: 534–553.

Campbell, L.M. 2000. Human need in rural developing areas: perceptions of wildlife conservation experts. *Can. Geogr.* 44: 167–181.

Campbell, L.M. 2002b. Conservation narratives and the "received wisdom" of ecotourism: case studies from Costa Rica. *Intl. J. Sustainable Dev.* S:3:300–325

Campbell, L.M. 2002a. Sustainable use of marine turtles: views of conservation experts. *Ecol. Appl.* 12:4:1229–1246.

Campbell, L.M., M.H. Godfrey, and O. Drif. 2002. Community based conservation via global legislation? Limitations of the Inter-American Convention for Protection and Conservation of Sea Turtles. *J. Int. Wildl. Law Policy* 5:121–143.

Carr, A. 1967. *The Windward Road: Adventures of a Naturalist on Remote Caribbean Shores,* First edition. Alfred A. Knopf, New York. 258pp.

Carr, A. 1977. Crisis for the Atlantic ridley. *Mar. Turtle Newsl.* 4: 2–3.

Carrillo, C.E., G.J. Webb, and S.C Manolis. 1999. Hawksbill turtles (*Eretmochelys imbricata*) in Cuba: an assessment of the historical harvest and its impacts. *Chelonian Conserv. Biol.* 3: 264–280.

Cheng, I.-J. 1995. Tourism and the green turtle in conflict on Wan-An Island, Taiwan. *Mar. Turtle Newsl.* 68: 4–6.

Clifton, K., D.O. Cornejo, and R.S. Felger. 1995. Sea turtles on the Pacific Coast of Mexico. In *Biology and Conservation of Sea Turtles*, revised edition. K. Bjorndal, ed. Smithsonian Institute Press. Washington, D.C. 199–209.

Cornelius, S.E. et al. 1991. Management of olive ridley sea turtles (*Lepidochelys olivacea*) nesting at Playas Nancite and Ostional, Costa Rica. In *Neotropical Wildlife Use and Conservation*. Robinson, J.G. and K.H. Redford, eds. University of Chicago Press, Chicago. 111–135.

Cosijn, J. 1995. Using sea turtles for tourism marketing. *Mar. Turtle Newsl.* 71: 12–14.

Crouse, D. 1999. Guest editorial: the WTO shrimp/turtle case. *Mar. Turtle Newsl.* 83:1.

Dempsey, M. 1996. Turtles and tourists get special attention. In *Profiles: The Magazine of Continental Airlines* (April:17).

Dimopoulos, D. 2001. The National Marine Park of Zakynthos: a refuge for the loggerhead turtle in the Mediterranean. *Mar. Turtle Newsl.* 93: 5–9.

Donnelly, M. 1994. *Sea Turtle Mariculture: A Review of Relevant Information for Conservation and Commerce.* Center for Marine Conservation, Washington, D.C. 113 pp.

Eckert, K.L. et al, eds. 1999. *Research and Management Techniques for the Conservation of Sea Turtles*. Publication No. 4, IUCN/SSC Marine Turtle Specialist Group, Washington, D.C. 235 pp.

Eckert, K.L. and F.A. Abreu. 2001. Marine turtle conservation in the wider Caribbean region: a dialogue for effective regional management. *Mar. Turtle Newsl.* 94: 12–13.

Ecotourism Society. Frequently asked questions; available online at http://www.ecotourism.org/tiessvsfr.html.

Ellen, R. and K. Fukui. 1996. *Redefining Nature: Ecology, Culture and Domestication*. Berg, Oxford. 664 pp.

Escobar, A. 1992. Planning. In *The Development Dictionary: A Guide to Knowledge as Power.* Sachs, W., ed. Zed Books, London. 6–25.

Escobar, A. 1999. After nature. *Curr. Anthropol.* 41: 1–16.

Evans, K.E. and A.R. Vargas. 1998. Sea turtle egg commercialization in Isla de Canas, Panama. In *Proceedings of the 16th Annual Symposium on Sea Turtle Biology and Conservation.* Byles, R. and Y. Fernandez, eds. p. 45.

Fletemeyer, J.R. 1996. Guest editorial: the shot heard around the world — Volusia sea turtle suit. *Mar. Turtle Newsl.* 72: 16–17.

Fortes, O., A.J. Pires, and C. Bellini. 1998. Green turtle, *Chelonia mydas*, in the Island of Poilão, Bolama-Bijagós Archipelago, Guinea-Bissau, West Africa. *Mar. Turtle Newsl.* 84: 4–6.

Fournillier, K. 1994. Integrating endangered species conservation and ecotourism: marine turtle management in North-East Trinidad. In *Tourism and Marine Turtles: Can We Live Together?* IUCN-MTSG Committee on the Impact of Tourism on Marine Turtles. 3–6.

Frazier, J. 1999. Community-based conservation. In *Research and Management Techniques for the Conservation of Sea Turtles*. Publication No. 4. Eckert, K.L. et al., eds. IUCN/SSC Marine Turtle Specialist Group, Washington, D.C. 15–18.

Frazier, J. In press. Science, conservation and sea turtles: what's the connection? In *Proceedings of the 21st Annual Symposium on Sea Turtle Biology and Conservation*. Philadelphia.

Freese, C.H. 1998. *Wild Species as Commodities: Managing Markets and Ecosystems for Sustainability*. Island Press, Washington, D.C. 319 pp.

Ghimire, K.B. and M.P. Pimbert. 1997. *Social Change and Conservation*. Earthscan, London. 342 pp.

Godfrey, M.H. and O.D. Drif. 2001. Guest editorial: developing sea turtle ecotourism in French Guiana: perils and practicalities. *Mar. Turtle Newsl.* 91: 1–4.

Gutic, J. 1994. Sea turtle eco-tourism brings economic benefit to community. *Mar. Turtle Newsl.* 64: 10–12.

Hackel, J.D. 1999. Community conservation and the future of Africa's wildlife. *Conserv. Biol.* 13: 726–734

Hoinsoude, G.S. et al. In press. Plan for sea turtle conservation in Togo. In *Proceedings of the 22nd Annual Symposium on Marine Turtle Biology and Conservation, Miami, FL.*

Hughes, G.R. 1979. Conservation, utilization, antelopes and turtles. *Mar. Turtle Newsl.* 13: 13–14.

Islam, M.Z. 2001. Notes on the trade in marine turtle products in Bangladesh. *Mar. Turtle Newsl.* 94: 10.

IUCN. 1980. *The World Conservation Strategy*. World Conservation Union (IUCN), Gland, Switzerland. 77 pp.

Jacobson, S.K. and R. Robles. 1992. Ecotourism, sustainable development, and conservation education: development of a tour guide training program in Tortuguero, Costa Rica. *Environ. Manage.* 16: 701–713.

Johnson, S.A., K.A. Bjorndal, and A. Bolten. 1996. A survey of organized turtle watch participants on sea turtle nesting beaches in Florida. *Chelonian Conserv. Biol.* 2: 60–65.

Juarez, R. and C. Muccio. 1997. Sea turtle conservation in Guatemala. *Mar. Turtle Newsl.* 77: 15–17.

Kaneko, Y. and H. Yamaoka. 1999. Traditional use and conservation of hawksbill turtles: from a Japanese industry's perspective; available online at http://www.iwmc.org/sustain/2ndsymposium/aquatic/aquatic-22–1/htm.

Kar, C.S. and S. Bhaskar. 1982. Status of sea turtles in the eastern Indian Ocean. In *Biology and Conservation of Sea Turtles*. Bjorndal, K.A., ed. Smithsonian Institution Press, Washington, D.C. 365 pp.

Kempton, W., J.S. Boster, and J.A. Hartley. 1995. *Environmental Values in American Culture.* MIT Press, Cambridge, MA. 320 pp.

Kennett, R.M. et al. 1997. Nhaltjan Nguli Miwatj Yolngu Djaka Miyapunuwu: Sea turtle conservation and the Yolngu people of north east Arnhem Land, Australia. In *Principles of Conservation Biology*. Meffe, G., ed. Sinauer Associates, Sunderland, MA. 426–432.

Kostyack, J. 1999. FL turtle lawsuit; available online at www.lists.ufl.edu/cgi-bin/wa?A2 = ind9904&L = cturtle&D = 0&P = 11746, CTURTLE archives.

Kowarsky, J. 1995. Subsistence hunting of sea turtles in Australia. In *Biology and Conservation of Sea Turtles*, revised edition. Bjorndal, K., ed. Smithsonian Institution Press, Washington, D.C. 305–314.

Lagueux, C. 1991. Economic analysis of sea turtle eggs in a coastal community on the Pacific coast of Honduras. In *Neotropical Wildlife Use and Conservation*. Robinson, J.G. and K.H. Redford, eds. University of Chicago Press, Chicago. 136–144

Lagueux, C. 1998. Marine turtle fishery of Caribbean Nicaragua: human use patterns and harvest trends. Ph.D. thesis, University of Florida, Gainesville, FL. 215 pp.

Leach, M. and R. Mearns. 1996. Introduction. In *The Lie of the Land: Challenging Received Wisdom on the African Environment*. Leach, M. and R. Mearns, eds. The International African Institute, Oxford, U.K. 1–33.

Leach, M., R. Mearns, and I. Scoones. 1997. Editorial: community-based sustainable development: consensus or conflict? *IDS Bull.* 4: 1–3.

Lefever, H.G. 1992. *Turtle Bogue: Afro-Caribbean Life and Culture in a Costa Rican Village*. Susquehanna University Press, Selinsgrove, PA. 249 pp.

Lima, E.H.M., 2001. Helping the people help the turtles: the work of projeto TAMAR-IBAMA in Almofala, Brazil. *Mar. Turtle Newsl.* 91: 7–9.

Little, P. 1994. The link between local participation and improved conservation: a review of issues and experiences. In *Natural Connections: Perspectives in Community-Based Conservation*. Western, D. and M.A. Wright, eds. Island Press, Washington, DC. 347–372.

Lopez, M. and A. Fallabrino. 2001. New kind of illegal trade of marine turtles in Uruguay. *Mar. Turtle Newsl.* 91: 10.

Lyver, P.O. and H. Moller. 1999. Modern technology and customary use of wildlife: the harvest of sooty shearwaters by Rakiura Maori as a case study. *Environ. Conserv.* 26: 280–288.

March, E. 1992. Diagnostico sobre situacion social de la poblacion de Ostional, Provincia de Guanacaste, Asociación de Desarollo Integral de Ostional, Ostional, Costa Rica. 40 pp.

Marcovaldi, M.A. and G. Marcovaldi. 1999. Marine turtles of Brazil: the history and structure of Projeto TAMAR-IBAMA. *Biol. Conserv.* 91: 35–41.

Margavio, A. and C. Forsyth. 1996. *Caught in the Net: The Conflict Between Shrimpers and Conservationists*. Texas A&M University Press, College Station, TX. 156pp.

Marks, S. 1984. *The Imperial Lion: Human Dimensions of Wildlife Management in Central Africa*. Bowker, Epping, U.K. 196 pp.

McCormick, J. 1989. *The Global Environmental Movement: Reclaiming Paradise*. Belhaven, London. 259 pp.

McCoy, M. 1995. Subsistence hunting of turtles in the Western Pacific: the Caroline Islands. In *Biology and Conservation of Sea Turtles*, revised edition. Bjorndal, K., ed. Smithsonian Institution Press, Washington, D.C. 275–280.

Miller, J.D. 1989. *Marine Turtles. Vol. 1, An Assessment of the Conservation Status of Marine Turtles in the Kingdom of Saudi Arabia*. Meteorology and Environmental Protection Administration. Coastal and Marine Management Series.

Miller, J. 2000. Editorial: listening to the elders. *Mar. Turtle Newsl.* 88: 1–2.

Mohadin, K. 2000. Sea turtle research and conservation in Suriname: history, constraints and achievements. In *Proceedings of the 3rd Meeting on the Sea Turtles of the Guianas*. Kelle, L. et al., eds. 5–8

Mortimer, J. 1977. Final approach to Bali airport. *Mar. Turtle Newsl.* 5:8.

Mortimer, J.A. and J. Collie. 1998. Status and conservation of sea turtles in the Republic of Seychelles. In *Proceedings of the 17th Annual Sea Turtle Symposium*. Epperly, S.P. and J. Braun, eds. Orlando, FL. p. 70.

Mrosovsky, N. 1979. Editorial. *Mar. Turtle Newsl.* 13: 1–4.

Mrosovsky, N. 2000. *Sustainable Use of Hawksbill Turtles: Contemporary Issues in Conservation.* Key Centre for Tropical Wildlife Management, Darwin, Australia. 107 pp.

MTSG. 1995. *A Global Strategy for the Conservation of Marine Turtles.* World Conservation Union (IUCN), Gland, Switzerland. 25 pp.

Musick, J. A. 1999. *Life in the Slow Lane; Ecology and Conservation of Long-Lived Marine Animals.* American Fisheries Society Symposium 23. Bethesda, MD.

Nabhan, G. et al. 1999. Sea turtle workshop for indigenous Seri tribe. *Mar. Turtle Newsl.* 86: 14.

Nader, L. 1996. *Naked Science: Anthropological Inquiry into Boundaries, Power, and Knowledge.* Routledge, New York. 318 pp.

Nichols, W.J., K.E. Bird, and S. Garcia. 2000. Community-based research and its application to sea turtle conservation in Bahía Magdalena, BCS, Mexico. *Mar. Turtle Newsl.* 89: 4–7.

Nietschmann, B. 1973. *Between Land and Water: The Subsistence Ecology of the Miskito Indians, Eastern Nicaragua.* Seminar Press, New York. 279 pp.

Nietschmann, B. 1979. *Caribbean Edge: The Coming of Modern Times to Isolated People and Wildlife.* Bobbs-Merrill, Indianapolis, IN. 280 pp.

Opay, P. 1998. Legal action taken to stop the hunting of green turtles in Costa Rica. *Mar. Turtle Newsl.* 79: 12–16.

Parra, L. et al. 2000. The sea turtle and its social representation in the Wayuu indigenous culture, Zulia State, Venezuela. In *Proceedings of the 19th Annual Symposium on Sea Turtle Biology and Conservation.* Kalb, H.J. and T. Wibbels, eds. p. 207.

Parsons, J. 1962. *The Green Turtle and Man.* University of Florida Press, Gainesville, FL. 126 pp.

Pepper, D. 1984. *The Roots of Modern Environmentalism.* Croom Helm, London. 246 pp.

Peskin, J. D. 2002. Local guides' attitudes toward ecotourism, sea turtle conservation, and guiding in Tortuguero, Costa Rica. M.S. thesis, University of Florida, Gainesville, FL. 86 pp.

Plotkin, P. 1999. Resolutions of the participants at the 19th Annual Symposium on Sea Turtle Biology and Conservation. *Mar. Turtle Newsl.* 85: 20–24.

Pritchard, P. et al. 1983. Protecting nesting beaches. In *Manual of Sea Turtle Research and Conservation Techniques*, 2nd edition. Bjorndal, K. and G.H. Balazs, eds. Center for Environmental Education, Washington, D.C. 85–88

Reichart, H.A. 1982. Farming and ranching as a strategy for sea turtle conservation. In *Biology and Conservation of Sea Turtles.* Bjorndal, K.A., ed. Smithsonian Institution Press, Washington, D.C. 465–471.

Richardson, P. 2000. Guest editorial: obstacles to objectivity: first impressions of a CITES CoP. *Mar. Turtle Newsl.* 89: 1–4.

Robert, J. 1992. Production. In *The Development Dictionary: A Guide to Knowledge as Power.* Sachs, W., ed. Zed Books, London. 177–191.

Roberts, A.M., G. Gabriel, and J. Robinson. 1999. Dying to heal: The use of animals in traditional medicine. *Anim. Agenda* 19: 30–31.

Robinson, J.G. 1993. The limits to caring: sustainable living and the loss of biodiversity. *Conserv. Biol.* 7: 20–28.

Robinson, J.G. and K.H. Redford. 1991. The use and conservation of wildlife. In *Neotropical Wildlife Use and Conservation.* Robinson, J.G. and K.H. Redford, eds. University of Chicago Press, Chicago. p. 3–5.

Rolston, H.I. 1994. *Conserving Natural Value.* Columbia University Press, New York. 259 pp.

Ross, J.P. and M.A. Barwani. 1982. Review of sea turtles in the Arabian area. In *Biology and Conservation of Sea Turtles.* Bjorndal, K.A., ed. Smithsonian Institution Press, Washington, DC. 373–384.

Ross, S. and G. Wall. 1999. Ecotourism: towards congruence between theory and practice. *Tourism Manage.* 20: 123–132.

Rudloe, J. 1979. *Time of the Turtle.* E.P. Dutton, New York. 272 pp.

Ruiz, G.A. 1994. Sea turtle nesting population at Playa La Flor, Nicaragua: an olive ridley "arribada" beach. In *Proceedings of the 14th Annual Symposium on Sea Turtle Biology and Conservation.* Bjorndal, K.A. et al., eds. NOAA Technical Memorandum NMFS-SEFSC-351. p. 129.

Schofield, G., K. Katselidis, and S. Hoff. 2001. Eastern Mediterranean "holiday hotspots" versus sea turtle "nesting hotspots." *Mar. Turtle Newsl.* 92: 12–13.

Schulz, J.P. 1975. Sea Turtles Nesting in Surinam. *Zoologische Verhandelingen,* 143:1–143

Seppälä, P. and A. Vainio-Mattila. 1998. *Navigating Culture: A Road Map to Culture and Development.* Ministry of Foreign Affairs, Department of International Development Cooperation. Helsinki. 60 pp.

Siakor, S.K. et al. 2000. Liberia sea turtle project. *Mar. Turtle Newsl.* 88: 9.

Silas, E.G. and M. Rajagopalan. 1984. Recovery programme for olive ridley, *Lepidochelys olivacea* (Eschscholtz, 1829), along Madras Coast. In *Sea Turtle Research and Conservation.* Bulletin 35. Central Marine Fisheries Research Institute, Cochin, India. 9–21.

Smith, C. 2002. Valuing and Volunteering for Wildlife Conservation in Tortuguero, Costa Rica. M.A. thesis, University of Western Ontario, London, ON. 166 pp.

Smith, D.V.R. 1983. *Palauan Social Structure.* Rutgers University Press, New Brunswick, NJ. 348 pp.

Songorwa, A.N. 1999. Community-based wildlife management (CWM) in Tanzania: are the communities interested? *World Dev.* 27: 2061–2079.

Spring, C.S. 1995. Subsistence hunting of marine turtles in Papua New Guinea. In *Biology and Conservation of Sea Turtles,* revised edition. Bjorndal, K., ed. Smithsonian Institution Press, Washington, D.C. 291–295

Suarez, M. and C.H. Starbird. 1996. Subsistence hunting of leatherback turtles in the Kai Islands, Indonesia. In *Coriacea, Chelonian Conservation Biology,* 2(2):190–195.

Taft, C. 1999. Lawsuit bans sea turtle killing in Costa Rica. *Velador* Spring: 2.

Thorbjarnarson, J. et al. 2000a. Human use of turtles: a worldwide perspective. In *Turtle Conservation.* Klemens, M.W., ed. Smithsonian Institution Press, Washington, DC. p. 33–84.

Thorbjarnarson, J.B., S.G. Platt, and S.T. Khaing. 2000b. Sea turtles in Myanmar: past and present. *Mar. Turtle Newsl.* 88: 10–11.

Thuoc, P. et al. 2001. Training workshop on marine turtle research and conservation in Viet Nam. *Mar. Turtle Newsl.* 94: 14–15.

Tomás, J., J. Castroviejo, and J.A. Raga. 1999. Sea turtles in the south of Bioko Island (Equatorial Guinea). *Mar. Turtle Newsl.* 84: 4–6.

Venizelos, L. and M.A. Nada. 2000. Exploitation of loggerhead and green turtles in Egypt: good news? *Mar. Turtle Newsl.* 87: 12–13.

Wells, M. and B. Brandon. 1992. *People and Parks: Linking Protected Area Management with Local Communities.* IBRD, Washington, D.C. 99 pp.

Wells, M.P. and K.E. Brandon. 1993. The principles and practice of buffer zones and local participation in biodiversity conservation. *Ambio* 22: 157–162.

Western, D. and M.A. Wright. 1994. The background to community-based conservation. In *Natural Connections: Perspectives in Community-Based Conservation.* Western, D. and M.A. Wright, eds. Island Press, Washington, D.C. 1–12

Williams, R. 1981. *Culture.* Fontana Press, Glasgow. 248 pp.

Williams, R. 1983. *Keywords: A Vocabulary of Culture and Society.* Fontana Press, Glasgow. 341 pp.

Wilson, C. and C. Tisdell. 2001. Sea turtles as a non-consumptive tourism resource especially in Australia. *Tourism Manage.* 22: 279–288.

World Society for the Protection of Animals (WSPA). 1997. *Turtle Alert! How the World's Biggest Industry Can Help Save One of the World's Oldest Species.* WSPA, London. 10 pp.

Yerli, S.V. and A.F Canbolat. 1998. Results of a 1996 survey of *Chelonia* in Turkey. *Mar. Turtle Newsl.* 79: 9–11.

Yuen, E., G. Katsiaficas, and D.B. Rose. 2001. *The Battle of Seattle: The New Challenge to Capitalist Globalization.* Soft Skull Press, New York. 393 pp.

13 Fisheries-Related Mortality and Turtle Excluder Devices (TEDs)

Sheryan P. Epperly

CONTENTS

13.1 INTRODUCTION

Sea turtles are subject to human-induced mortality during all life stages. On land, nesting females, incubating eggs, and emerging hatchlings may be impacted. The impact may be incidental, such as by disorientation by lights and disturbance on or of the beach, or it may be intentional by directed harvest of the adults and eggs. Once turtles are in the water, a vast variety of new sources of impact are brought to bear. These include pollution and marine debris, habitat degradation, directed harvest, and incidental capture or entrainment by a variety of sources, including fishing and dredging.

0-8493-1123-3/03/$0.00+$1.50

In 1988, the U.S. Congress mandated a study of the causes and significance of turtle mortality in the coastal waters of the country. A study team was convened by the National Research Council's Board on Environmental Studies and Toxicology and Board on Biology. After a comprehensive review, the study team concluded that the largest human-associated source of mortality was incidental capture in shrimp trawls, associating that activity with more turtle deaths than all other human activities combined (Magnuson et al., 1990). The team estimated that as many as 44,000 turtles were killed annually by the U.S. fleet. Although other fishing gears used worldwide, including longlines, gill nets, and pots and traps, also are significant sources of mortality (Allen, 2000; Castroviejo et al., 1994; Gerosa and Casale, 1999; Gribble et al., 1998; Julian and Beeson, 1998; Lagueux, 1998), this chapter will focus on the interaction of turtles with trawl fisheries, especially those for shrimp.

13.2 THE PROBLEM

Trawls are used worldwide to catch many species of aquatic invertebrates and vertebrates. In warm waters, where turtles are most likely to occur, shrimps, or prawns, are the main species sought with trawls. Prior to the twentieth century, shrimp harvesting probably did not significantly impact turtles because the main gear, haul seines, which allow turtles to surface and breathe, was pulled by hand in very shallow coastal waters (Klima et al., 1982). Trawling for shrimp is relatively recent, beginning with the introduction of the otter trawl in the early 1900s. Trawls allowed the fishery to expand beyond shallow coastal waters, and enabled fewer workers to efficiently harvest much more than a haul seine crew. The fishery, at least in the U.S., expanded in earnest after World War II (Klima et al., 1982). The relationship between trawling effort and sea turtle mortality has been well documented (Caillouet et al., 1991; 1996; Henwood and Stuntz, 1987; Poiner and Harris, 1996; Robins, 1995). Trawls forcefully submerse the air-breathing turtles and are responsible for the drowning deaths of many; as tow duration increases so does mortality (Henwood and Stuntz, 1987).

13.3 SOLUTIONS

Mortality of sea turtles may be decreased by closing an area to trawling or by reducing tow times. Area closures impact the fishery to the greatest extent. Decreased tow times also may impact the fishery, but reduce turtle mortalities only if there is compliance with regulations. With shortened tow times, the number of hauls within a given time increases, providing shorter respites for the crew and increasing wear on deck machinery. In the U.S., tow times have not been regulated under most conditions because they are difficult to enforce, and when compliance has been evaluated, it generally has been poor. (National Marine Fisheries Service [NMFS], 2001;* Epperly et al., 1995b).

* National Marine Fisheries Service. Unpublished data. Southeast Regional Office. 9721 Executive Center Dr., St. Petersburg, FL 33702.

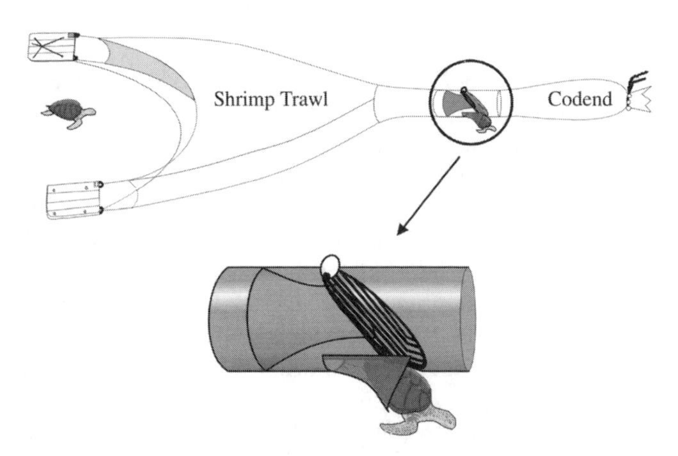

FIGURE 13.1 Schematic of a turtle excluder device showing the position of the TED in the trawl, just before the codend or tailbag. (Figure provided by NOAA Fisheries, Mississippi Laboratories, Harvesting Systems and Engineering Division.)

FIGURE 13.2 A loggerhead turtle, *Caretta caretta*, exiting a bottom-opening TED during TED trials in the eastern Gulf of Mexico. (Photo provided by NOAA Fisheries, Mississippi Laboratories, Harvesting Systems and Engineering Division.)

A technological solution to separate turtles from shrimps in trawls was available by the early 1980s. A turtle excluder device (TED) incorporates a trap door in the trawl, just before the tailbag or codend, to allow sea turtles to escape from the nets (Seidel and McVea, 1982) (Figure 13.1). Shrimp pass through the bars or webbing of the TED into the tailbag. Turtles and large bycatch are blocked by the TED and exit the opening (Figure 13.2). When installed properly, TEDs reduce turtle mortality and allow the fishery to continue unimpeded. Some shrimp loss may occur, but this usually is not significant (Renaud et al., 1993; Robins-Troeger et al., 1995), and the loss may be offset partially by increased efficiency realized by reduced bycatch. Decreased bycatch results in a higher quality product for the market and less work sorting catch (High et al., 1969; Brewer et al., 1998).

13.4 IMPLEMENTATION OF TEDS IN U.S. FISHERIES

13.4.1 The Process and the Shrimp Fishery*

TEDs were first developed and implemented in the U.S., but the process was protracted and contentious. Between 1970 and 1978 all six sea turtle species in U.S. waters were recognized as endangered or threatened and protected under the U.S. Endangered Species Act. Shrimp trawling was identified as a principle source of mortality in sea turtles as early as 1973 (Pritchard and Márquez, 1973); other documentation followed (Henwood and Stuntz, 1987; Hillestad et al., 1978). These studies, as well as the large numbers of loggerhead turtles, *Caretta caretta* (the most common species found in U.S. waters), washing up dead on ocean beaches caused NMFS to search for a solution. NMFS' first innovation was to install a large mesh panel of webbing over the mouth of the trawl. The panel allowed shrimp to pass and excluded most sea turtles, but the loss of shrimp was significant and sometimes turtles became entangled in the large meshes of the panel (Oravetz and Grant, 1986). The next approach was to allow everything to enter the trawl, but separate the target species from the bycatch near the codend. The design was based on a device already used by many shrimpers to exclude jellyfish when coelentrates were especially abundant. NMFS, in cooperation with commercial shrimpers, experimented with several configurations and found that turtle catch could be eliminated almost completely with little or no reduction in shrimp catches. By 1980 NMFS had a TED (Watson et al., 1986).

NMFS initially was reluctant to require TEDs for fear that additional regulations would exacerbate existing economic problems in the shrimp fishery (Oravetz and Grant, 1986). Also of concern was the agency's ability to enforce any TED regulations given the geographic scope and the number of vessels involved. NMFS anticipated that industry, with encouragement, would voluntarily accept and use the device, and over the next several years purchased and distributed a few hundred TEDs to industry. Throughout the 1980s, NMFS, Sea Grant, and conservation organizations worked with industry to encourage the use of TEDs and to improve upon the original NMFS design, which was heavy and unwieldy. As a result, several lighter designs were developed. Ultimately, however, fishermen could not relate their individual trawling activities to the turtle mortality caused collectively by the entire fleet because catch rates of turtles were relatively low: one turtle per 322.5 h fished in the Gulf of Mexico and one turtle per 20.6 h fished off the southeast U.S. (Henwood and Stuntz, 1987). Hence, they perceived TEDs as an unnecessary solution to an unfounded problem. When, in 1986, it became apparent that industry would not voluntarily adopt TEDs to save protected species — less than 1% were using them in 1985 (Oravetz and Grant, 1986) — the U.S. Fish and Wildlife Service and conservation organizations requested that the Gulf of Mexico Fishery Management Council (the organization with the authority to manage the shrimp fishery in the Gulf of Mexico) require TEDs to prevent the continued drowning of sea turtles in shrimp trawls (Center

* Much of the discussion on the implementation of TEDs in the U.S. was taken from Oravetz and Grant (1986) and Center for Marine Conservation (1995).

for Marine Conservation, 1995). They were particularly concerned about the plight of the Kemp's ridley, *Lepidochelys kempii*, which nests only in the western Gulf of Mexico. In the late 1940s, nests were estimated to be in the tens of thousands annually (Hildebrand, 1963), but by 1985 the annual number of nests had dropped below 800 (Márquez-M et al., 1989). The Council did not act to require TEDs.

Later in 1986 NMFS informally proposed TED regulations, but because the proposal exempted trawlers in particular areas of concern, the Center for Environmental Education (later the Center for Marine Conservation) notified the Department of Commerce of their intent to sue to protect sea turtles under the Endangered Species Act. Mediation meetings ensued, involving NMFS, nongovernmental organizations, and industry. When negotiations ended without agreement, NMFS drafted regulations and published a proposed rule to phase in TEDs in ocean areas seasonally over the next several years (Department of Commerce, 1987a). Public hearings on the proposed regulations began in early 1987 and by summer NMFS published final regulations (Department of Commerce, 1987b). TED opponents effectively delayed full implementation of TED regulations until 1990 by appealing to federal courts in Texas, Louisiana, and North Carolina, and to Congress. Invariably the courts eventually ruled in NMFS' favor. The U.S. Congress and the incumbent administration were more receptive to indusry complaints, at times ordering TED regulations or the enforcement thereof suspended. During this period, some individual states, such as South Carolina, Georgia, and Florida, promulgated regulations implementing TEDs in their territorial waters; sometimes these regulations also were challenged in state courts. By October 1989, however, the TED regulations were in force, but because the regulations were seasonal, they were not yet in effect in all areas.

Just as these regulations finally were implemented fully, the National Research Council's study team's report was published, conclusively supporting the need to reduce mortality by shrimp trawls (Magnuson et al., 1990). Furthermore, the study team recommended that TEDs be used in bottom trawls at most places and at most times of the year, from Cape Hatteras to the U.S.–Mexico border. Newly implemented federal regulations required TEDs seasonally and were limited to ocean trawlers. Trawlers in inshore waters were permitted to restrict tow times in lieu of using TEDs. By 1992, armed with the study team report, with evidence of significant use of inshore waters by turtles (Epperly et al., 1995a), and with the knowledge that trawlers working in inshore waters capture sea turtles (NMFS*), NMFS expanded the regulations, phasing them in over a 2-year period (Department of Commerce, 1992b). TEDs were now required in shrimp trawls throughout the year in all areas south of Cape Hatteras. Thus, by December 1994, with few exceptions, all shrimp trawlers in inshore and offshore waters were required to use TEDs in their nets at all times. Except for this major change, most other modifications of the regulations since 1987 have addressed technical design details and certification protocols.

* National Marine Fisheries Service. Unpublished data. Galveston Laboratory, 4700 Avenue U, Galveston, TX 77551.

13.4.2 Leatherback Conservation

Leatherback turtles, *Dermochelys coriacea*, are the largest of the sea turtles and lead a pelagic, mostly oceanic existence. Most of the animals in coastal waters are too large to fit through TED openings. Strandings of leatherbacks along the southeast U.S. coast predictably occur in the winter and spring, and shrimp trawling was linked to the episodic events (Department of Commerce, 1995). In response NMFS, in cooperation with the U.S. Fish and Wildlife Service and the states of South Carolina, Georgia, and Florida, developed a leatherback conservation plan for the southeast U.S. Atlantic coast (Department of Commerce, 1995), establishing a framework for short-term closures. This and subsequent rules made it illegal for shrimp trawlers to operate January–June in zones identified as having a high concentration of leatherbacks, unless the trawls were equipped with TEDs capable of excluding leatherback turtles. NMFS approved modifications to several TED designs that would allow leatherback turtles to escape the trawls. Since adopting the plan, NMFS also has required the large-opening leatherback TEDs during times of high strandings outside the conservation area (Spring, 2000 off Texas) or conservation time (December 1999 and December 2001 off northeast Florida). NMFS is considering requiring the leatherback TED modifications at more times and in more areas (Department of Commerce, 2000).

13.4.3 Exemptions

There have been a number of exemptions granted to TED requirements; all but one were temporary. When tropical storms have battered the southeastern U.S. coast and afterwards a significant amount of debris washed downstream into coastal waters, NMFS granted a reprieve from TEDs because debris can clog a TED making the escape of turtles difficult or impossible. The exemptions allowed fishermen to substitute limited tow times for TEDs.

A small area of live bottom off central North Carolina is a productive area for shrimp trawling by local fishermen. When macroalgae on the reef are dense, detached marine algae clog the TEDs. NMFS has granted an extended exemption from TEDs in this area, allowing tow times to be used in lieu of TEDs. The area is easily policed from the beach and enforcement is handled by the state; compliance has been high (North Carolina Division of Marine Fisheries*).

13.4.4 TEDs in Other Fisheries

Bottom trawls are used in other fisheries and where those fisheries overlap with turtles, turtles also are captured. Since 1992 TEDs have been required in the winter trawl fishery for summer flounder, *Paralichthys dentatus,* while operating between Cape Charles, Virginia and the North Carolina–South Carolina border (Department of Commerce, 1992a). Data collected November 1991–February 1992 showed that the fishery incidentally captured significant numbers of turtles (>1000) during the turtles' seasonal southward migration along the Atlantic coast of the U.S. (Epperly et al., 1995b). The National

* North Carolina Division of Marine Fisheries. Unpublished annual reports to NMFS for incidental take permit no. 1008. P.O. Box 769, Morehead City, N.C. 28557.

Research Council (NRC) study team previously had identified the fishery as a source of mortality. Although unpopular with local fishermen, TED regulations implemented in this finfish fishery were not contested in the courts.

Other bottom trawl fisheries in the western North Atlantic have been linked to the mortality of sea turtles. In South Carolina, whelk trawling is allowed only when water temperatures are less than 18°C (64°F) and since December 2000 whelk trawls used in Georgia waters are required to have TEDs (NMFS Southeast Fisheries Science Center, 2001).

13.4.5 Turtle Exclusion

TED designs must be certified by NMFS, based on a specific protocol (Department of Commerce, 1987b, 1990, 1992b). Foremost among the criteria for certification is the requirement that a prospective design releases 97% of the turtles tested or alternatively that it performs as well as a previously certified control TED. Certified designs, therefore, are assumed to reduce mortality of sea turtles 97%. A number of studies, however, indicate that the actual reduction realized is substantially less. Strandings on South Carolina and Georgia beaches were reduced 37–58% (Crowder et al., 1995; Royle and Crowder, 1998* as cited in Turtle Expert Working Group [TEWG], 2000; Royle, 2000**, as cited in TEWG, 2000) and estimates of a post-1990 multiplier of instantaneous mortality for benthic Kemp's ridley turtles range from 0.45 to 0.56, indicating a decrease in mortality coincident with the implementation of TEDs (Heppell et al., in press, a).

A recent analysis of strandings data revealed that the minimum opening of TEDs, measured as the height and width of a taut triangle, is too small to exclude the larger individuals of several species (Epperly and Teas, 2002). A significant proportion of loggerheads stranding along the east coast of the U.S. and in the Gulf of Mexico (33–47% annually), and a small proportion of green turtles, *Chelonia mydas* (1–7% annually), were too large to fit through the opening; their body depths exceeded the height of the openings. Loggerhead turtles were too large to fit through the opening at a size where most are still immature. Thus, not all sizes of loggerhead and green turtles are benefiting from TEDs since they cannot escape from the nets. Early population models evaluating the potential effect of TEDs on the loggerhead population trajectory assumed that TEDs benefited all benthic life stages (Crowder et al., 1994). TED size opening is not an issue for Kemp's ridleys because they do not attain a large size and all can fit through the existing minimum openings.

Genetic studies have indicated multiple demographic units within sea turtle species and these are regarded as management units (Bowen, 1995; Fitzsimons et al., 1995). Some of the loggerhead management units of the western North Atlantic do not appear to be increasing and, given current population trajectories, will not reach recovery goals formulated under the U.S. Endangered Species Act. Popula-

* Royle, J.A. and L.B. Crowder. 1998. Estimation of a TED effect from loggerhead strandings in South Carolina and Georgia strandings data from 1980-97. Unpublished report, U.S. Fish and Wildlife Service, Laurel, MD, 12 p.

** Royle, J.A. 2000. Estimation of TED effect in Georgia shrimp strandings data. Unpublished report, U.S. Fish and Wildlife Service, Laurel, MD, 11 p.

tion models have demonstrated a need to further decrease mortality in the benthic and oceanic pelagic stages of loggerheads in order to move these stable or declining subpopulations towards recovery (NMFS Southeast Fisheries Science Center, 2001; Heppell et al., in press, b). One identified management action that would result in fewer deaths of large benthic sea turtles would be to require larger TED openings. NMFS issued an advance notice of proposed rulemaking in April 2000 and is pursuing a final rule to require larger openings in all areas (Department of Commerce, 2000).

13.5 IMPLEMENTATION OF TEDS IN SHRIMP FISHERIES WORLDWIDE

During the development of TEDs in the U.S., scuba-diver observations indicated behavorial differences in fish and shrimp that could be used to separate shrimp from finfish (Watson et al., 1986). Design modifications of the original TED were effective in reducing finfish bycatch, and TED also became the acronym worldwide for trawl efficiency device. While the U.S. was grappling with implementation of TEDs, other countries were beginning to investigate their use, often for their potential to reduce finfish bycatch. One of the first countries to require TEDs was Indonesia (Oravetz and Grant, 1986). After learning about TEDS, rather than ending their joint venture with the Japanese, as contemplated, Indonesia decided to allow the Japanese fishermen to fish inside Indonesian waters with TED-equipped nets and Indonesian crews. At the time, other countries throughout the world were inquiring about TEDs and some, like Australia, were beginning to experiment with them.

13.5.1 Implementation of TEDs in Australia*

All marine turtles occurring in Australian waters are listed as either endangered or vulnerable on the Australian Endangered Species Act of 1992. Australian trawlers had used systems to reduce jellyfish and had tested various experimental designs, but still resisted the introduction of TEDs, citing handling and safety concerns and a concern about loss of prawns (Mounsey et al., 1995). In the early 1990s Australia introduced TEDs into the prawn fisheries of at least three regions. In the northern fisheries, sea turtle bycatch was driving the implementation of TEDs and groups outside the fishing community (e.g., conservationists and government) were pushing the issue. In 1989–1990, 5000–6000 turtles were estimated captured annually in the northern fishery (Poiner and Harris, 1996). In another fishery off the east coast of Queensland, annual captures were estimated to be 5295 ± 1231 (Robins, 1995). In New South Wales finfish bycatch was the driving factor and reduction in bycatch was being pushed by representatives of other fisheries, commercial and recreational, as well as by the government. In Gulf St. Vincent, a southern fishery, catch of small prawns and fish was the driving factor and bycatch reduction was championed by the fishermen themselves. The approach to introduce the TEDs differed in each area, and many designs were tested or developed (e.g., AusTED and AusTED II) (Brewer et al., 1998).

* Much of the discussion concerning Australia's implementation of TEDs is from Kennelly (1999).

Similar to the experience in the U.S., most fishermen would not voluntarily use TEDs to protect sea turtles; in some northern ports use was 0–20% and in others it was as high as 50–80%. Fishers used TEDs when they perceived a benefit to their use (turtle or jellyfish exclusion) or when they were used with no adverse impact. When finfish bycatch was the issue, once the best devices were identified through testing, 50–100% of the fishers voluntarily used them. When undersized prawns were the issue, virtually all fishers adopted TEDs. When prawn fishers had a vested interest, such as in the southern fisheries, the problem did not first have to be quantified; the researchers could skip directly to the testing phase and involve the fishermen immediately and directly. When fishers did not have a vested interest, such as in the northern fisheries, even after the problem was quantified, most still were not all willing to use TEDs. In the case of New South Wales, the response was intermediate. Although bycatch reduction was being driven from outside the fishery, because industry was involved at an earlier stage than for the northern fishermen in gear development and testing, a greater proportion used TEDs voluntarily. Kennelly (1999) concluded that the sooner and more fully industry is involved, the sooner and greater voluntary acceptance will be. In all cases, however, the government of Australia legislated, over a period of 3 years, the use of TEDs selectively, either for turtle exclusion or bycatch reduction. TED use in the Queensland east coast fishery began in selected areas in 1999 and their use in the northern prawn fishery was mandatory in 2000.

13.5.2 U.S. Public Law 101–162, Section 609

In 1988, at the urging of the shrimp industry, the U.S. Congress passed Public Law 101–162, Section 609. Section 609 prohibits the import into the U.S. of shrimp and shrimp products that were harvested in a manner that may adversely affect sea turtle species. Annually, the Department of State (DOS) certifies to Congress that the governments of certain harvesting nations have taken specific measures to reduce the incidental capture of sea turtles by their shrimp trawl fisheries, or that the fishing environment of those nations does not pose a threat to sea turtles. The latter situation can include fisheries that harvest shrimp manually or fisheries that occur only in cold water where they pose little or no risk to poikilothermic turtles. For 2002, 17 nations met the certification standards for sea turtle conservation (Belize, Colombia, Costa Rica, Ecuador, El Salvadore, Guatemala, Guyana, Honduras, Mexico, Nicaragua, Nigeria, Pakistan, Panama, Suriname, Thailand, Trinidad and Tobago, and Venezuela) and another 24 nations (Argentina, The Bahamas, Belguim, Canada, Chile, China, Denmark, Dominican Republic, Fiji, Finland, Germany, Iceland, Ireland, Jamaica, The Netherlands, New Zealand, Norway, Oman, Peru, Russia, Sri Lanka, Sweden, the United Kingdom, and Uruguay) and 1 economy (Hong Kong) were certified as having fishing environments that do not pose a danger to sea turtles (Department of State, 2002*). Such certifications are, in part, based on site visits by the DOS and NMFS. Currently the import of shrimp into the U.S. from other

* U.S. Department of State. April 30, 2002. Sea turtle conservation and shrimp imports. http://www.state.gov/r/pa/prs/ps/2002/9880pf.htm.

nations is prohibited unless the individual, clearly marked shipment meets special criteria, such as harvested by aquaculture, in cold waters, or by techniques that do not harm sea turtles (e.g., by nets using TEDs). Two nations, Brazil and Australia, had demonstrated that they had an enforcement and catch segregation system in place for making individual shipment certifications (Department of State, 2002).

One consideration in determining whether a sea turtle conservation program of a nation is comparable to that of the U.S. is their TED design. Currently there are two sets of regulations in the U.S. The required minimum size of the TED opening differs in the Gulf of Mexico and the Atlantic; the allowed opening is smaller in the Gulf (Department of Commerce, 1987b). It is the minimum regulation — the smaller opening for TEDs used in the Gulf of Mexico — that the DOS uses as their standard during comparisons, although there are situations where the leatherback modification is required. As the U.S. considers changes to their regulations to increase the size of the openings (Department of Commerce, 2000), it is with the knowledge that the changes could have impact worldwide.

Section 609, although apparently intended to protect the U.S. domestic shrimp industry rather than to protect sea turtles (U.S. Court of Appeals for the Federal Circuit, 2002)*, has the potential to have a very significant positive impact on sea turtle conservation worldwide. Section 609 was an innovative solution, using the markets to apply pressure worldwide. There still are many countries that do not export shrimp to the U.S., but do export them elsewhere, and some harvest shrimp without TEDs in areas where turtles are known to occur. One such market is the European Union, which currently does not have a law comparable to Section 609, and accepts shrimp harvested without TEDs.

13.5.2.1 U.S. Court of International Trade**

The DOS originally interpreted PL101-162, Section 609 to apply only within the Wider Caribbean and Western Atlantic region. Environmental and animal rights groups filed suit in the U.S. Court of International Trade, primarily to overturn the limited geographic scope of application of the law by the DOS. In December 1995 the Court ruled that Congress intended Section 609 to apply on a worldwide basis and ordered the department to comply. A request by DOS for a 1-year delay was denied by the Court. As a result, the importation of shrimp from many nations was prohibited on May 1, 1996. The plaintiffs reopened the litigation to reverse one aspect of the changes that DOS made — to allow an individual shipment from a noncertified country, if that shipment could be certified to contain only shrimp harvested under conditions that were not harmful to sea turtles. The plaintiffs argued that unless there was a program nationwide, all shrimp imports from the country should be prohibited according to Section 609. The Court originally ruled in favor of the plaintiffs, but the decision was vacated in 1998 by the U.S. Court of Appeals

* The U.S. Court of Appeals for the Federal Circuit and the World Trade Organization reached different conclusions about the purpose of the law. See discussion in Section 13.5.2.2.

** Information for this section is taken from U.S. Court of Appeals for the Federal Circuit. 2002. 00-1569, -1581, -1582, 36 pp. The full document can be viewed and downloaded at http://www.ll.georgetown.edu/Fed-Ct/Circuit/fed/opinions/00-1569.html.

for the Federal Circuit as they found that the plaintiffs had unilaterally and unconditionally withdrawn their original motion.

The plaintiffs filed suit and, in 1999, once again the Court found that importation of shrimp from noncertified countries violated the provisions of Section 609. The DOS issued the 1999 guidelines, still allowing importation of shrimp shipments from noncertified countries. In 2000, the Court held yet again that the shipment-by-shipment approach violated Section 609, but denied plaintiffs an injunction. The plaintiffs appealed and in March 2002 the U.S. Court of Appeals for the Federal Circuit concluded that the DOS's interpretation of Section 609 was correct and held that the plaintiffs were not entitled to injunctive relief.

13.5.2.2 World Trade Organization*

Claiming that the shrimp embargo was an "improper restriction on trade" and therefore violated the General Agreement on Tariffs and Trade (GATT), in September 1996, India, Malaysia, Pakistan, and Thailand, all recently affected by Section 609, brought a case against the U.S. in the WTO. The U.S. argued that specific sections of the WTO Agreement [Sections XX(b) and (g)] permitted members to take measures to protect life or conserve exhaustible natural resources, even if such measures were in conflict with other provisions of the Agreement. An arbitration panel ruled against the U.S. on most points in May 1998 (WTO, 1998a). The U.S. appealed the decision. On October 12, 1998, the WTO Appellate Body reversed the panel's findings on many issues, most notably finding that Section 609 qualified for provisional justification under Article XX(g), since it addressed the conservation of exhaustible natural resources (WTO, 1998b). However, the Appellate Body did find that some aspects concerning the way in which the DOS was implementing the Section were, in aggregate, in violation of U.S. obligations under the Agreement. The body determined that DOS should revise its implementation of Section 609. The DOS adopted those recommendations and (1) now considers any evidence that another nation presents that its sea turtle conservation program is comparable, not necessarily identical, to that of the U.S.; (2) instituted procedural changes so that the process is more transparent and predictable to nations and provides to governments not granted certification a full explanation for that decision; (3) facilitated a Memorandum of Understanding on the Conservation and Management of Marine Turtles and their Habitats of the Indian Ocean and South-East Asia; and (4) now offers technical training concerning TEDs to any requesting government.

Malaysia returned alone to the WTO with a new complaint that the U.S. had not fully complied with the original ruling. Three members of the original panel rejected that argument in June 2001, and, in October 2001, the WTO Appellate Body turned down a Malaysian appeal (WTO, 2001a, 2001b).

* The WTO Panel Reports, Appellate Body Reports, and Arbitrator's Reports can be viewed and downloaded at the WTO's website at http://www.wto.org/english/tratop_e/dispu_e/distab_e.htm#r58. They were the source material for the discussion of this section.

13.6 CONCLUSIONS

Trawl fisheries and sea turtles can coexist. Implementation of TEDs has been a protracted and contentious process. When properly installed in bottom trawls, TEDs can effectively exclude sea turtles and save significant numbers from drowning. When TEDs with large escape openings are used, virtually all turtles can escape with minimal impact on the fisheries. TEDs are in use throughout the world, in part, because of U.S. Public Law 101-162, Section 609, which uses market forces to provide an incentive for sea turtle conservation to all nations wishing to export shrimp to the U.S. However, a significant number of nations export their shrimp to other markets and do not use TEDs, despite operating fisheries in areas where turtles are likely to occur. Conservation communities worldwide are applying pressure to increase the number of nations employing TEDs in their warm water bottom trawl fisheries.

13.7 ACKNOWLEDGMENTS

Review comments by David Balton, Alex Chester, and Charles Oravetz were very much appreciated. Thanks, also, to David Bernhart, Harriet Corvino, and Joanne Braun-McNeill for providing several references. Both figures were kindly provided by NOAA Fisheries, Mississippi Laboratories, Harvesting Systems and Engineering Division.

REFERENCES

Allen, L.K. 2000. Protected species and New England fisheries: an overview of the problem and conservation strategies. *Northeast. Nat.* 7:411–418.

Bowen, B.W. 1995. Tracking marine turtles with genetic markers. *BioScience* 45:528-534.

Brewer, D., et al. 1998. An assessment of bycatch reduction devices in a tropical Australian prawn trawl fishery. *Fish. Res.* 36:195–215.

Caillouet, C.W., Jr. et al. 1991. Sea turtle strandings and shrimp fishing effort in the north-western Gulf of Mexico, 1986–1989. *Fish. Bull.* 89:712–718.

Caillouet, C.W., Jr. et al. 1996. Relationship between sea turtle stranding rates and fishing intensities in the northwestern Gulf of Mexico: 1986–1989 versus 1990–1993. *Fish. Bull.* 94:237–249.

Castroviejo, J. et al. 1994. Diversity and status of sea turtle species in the Gulf of Guinea Islands. *Biodiversity Conserv.* 3:828–836.

Center for Marine Conservation. 1995. *Delay and Denial. A Political History of Sea Turtles and Shrimp Fishing*. Center for Marine Conservation, Washington, D.C, 46 pp.

Crowder, L.B. et al. 1994. Predicting the impact of turtle excluder devices on loggerhead sea turtle populations. *Ecol. Appl.* 4:437–445.

Crowder, L.B., S.B. Hopkins-Murphy, and J.A. Royle. 1995. Effects of turtle excluder devices (TEDs) on loggerhead sea turtle populations. *Copeia* 1995:773–779.

Department of Commerce. 1987a. Sea turtle conservation; shrimp trawling requirements. *Federal Register* 52:6179–6199, Washington, D.C.

Department of Commerce. 1987b. Sea turtle conservation; shrimp trawling requirements. Federal Register 52:24244–24262, Washington, D.C.

Department of Commerce. 1990. Turtle excluder devices: adoption of alternative scientific testing protocol for evaluation. *Federal Register* 55:41092–41093, Washington, DC.

Department of Commerce. 1992a. Sea turtle conservation; restrictions applicable to fishery activities. *Federal Register* 57:53603–53606, Washington, D.C.

Department of Commerce. 1992b. Threatened fish and wildlife; threatened marine reptiles: revisions to enhance and facilitate compliance with sea turtle conservation requirements applicable to shrimp trawlers; restrictions applicable to shrimp trawlers and other fisheries. *Federal Register* 57:57348–57358, Washington, D.C.

Department of Commerce. 1995. Sea turtle conservation; restrictions applicable to shrimp trawl activities; leatherback conservation zone. *Federal Register* 60:25620–25623, Washington, D.C.

Department of Commerce. 2000. Endangered and threatened wildlife; sea turtle conservation requirements. *Federal Register* 65:17852–17854, Washington, D.C.

Department of State. 2002. Certifications pursuant to Section 609 of Public Law 101-162; relating to the protection of sea turtles in shrimp travel [sic] fishing operations. Federal Register 67:32078-32079, Washington, D.C.

Epperly, S.P. and W.G. Teas. 2002. Turtle excluder devices — Are the escape openings large enough? *Fish. Bull.* 100: 466–474.

Epperly, S.P., J. Braun, and A. Veishlow. 1995a. Sea turtles in North Carolina waters. *Conserv. Biol.* 9:384–394.

Epperly, S.P. et al. 1995b. Winter distribution of sea turtles in the vicinity of Cape Hatteras and their interactions with the summer flounder trawl fishery. *Bull. Mar. Sci.* 56:547–568.

FitzSimmons, N.N., C. Moritz, and S.S. Moore. 1995. Conservation and dynamics of microsatellite loci over 300 million years of marine turtle evolution. *Mol. Biol. Evol.* 12:432-440.

Gerosa, G. and P. Casale. 1999. *Interaction of marine turtles with fisheries in the Mediterranean.* Mediterranean Action Plan. United Nations Environment Programme, Regional Activity Centre for Specially Protected Areas, Tunisie, 59 pp.

Gribble, N.A., G. McPherson, and B. Lane. 1998. Effect of the Queensland shark control program on non-target species: whale, dugong, turtle and dolphin: a review. *Mar. Freshwater Res.* 49:645–651.

Henwood, T.A. and W.E. Stuntz. 1987. Analysis of sea turtle captures and mortalities during commercial shrimp trawling. *Fish. Bull.* 85:813–817.

Heppell, S. et al. In press, a. A population model to estimate recovery time, population size, and management impacts on Kemp's ridleys. *Chelonian Conserv. Biol.*

Heppell, S.S et al. In press, b. Population models for Atlantic loggerheads: past, present, and future. In *Synopsis of the Biology and Conservation of Loggerhead Sea Turtles.* Bolten, A. and B. Witherington, Eds. Smithsonian Institution Press, Washington, DC.

High, W.L., I.E. Ellis, and L.D. Lusz. 1969. A progress report on the development of a shrimp trawl to separate shrimp from fish and bottom dwelling animals. *Commer. Fish. Rev.* 31(3):20–33.

Hildebrand, H.H. 1963. Hallazgo del area de anidacion de la tortuga marina "lora" *Lepidochelys kempi* (Garman) en la costa occidental del Golfo de México. *Ciencia* 22:105–112.

Hillestad, H.O., J.I. Richardson, and G.K. Williamson. 1978. Incidental capture of sea turtles by shrimp trawlermen in Georgia. *Proceedings of the Annual Conference Southeastern Association of Fish and Wildlife Agencies* 32:167-178.

Julian, F. and M. Beeson. 1998. Estimates of marine mammal, turtle, and seabird mortality for two California gillnet fisheries: 1990–1995. *Fish. Bull.* 96:271–284.

Kennelly, S.J. 1999. The development and introduction of by-catch reducing technologies in three Australian prawn-trawl fisheries. *Mar. Technol. Soc. J.* 33:73–81.

Klima, E.F., K.N. Baxter, and F.J. Patella, Jr. 1982. A review of the offshore shrimp fishery and the 1981 Texas closure. *Mar. Fish. Rev.* 44(9–10):16–30.

Lagueux, C.J. 1998. Marine turtle fishery of Caribbean Nicaragua: human use patterns and harvest trends. Ph.D. dissertation, University of Florida, Gainesville, FL, 215 pp.

Magnuson, J.J. et al. 1990. *Decline of the Sea Turtles: Causes and Prevention*, National Academy Press, Washington, D.C., 259 pp.

Márquez-M., A. Villanueva-O., and P.M. Burchfield. 1989. Nesting population and production of hatchlings of Kemp's ridley sea turtle at Rancho Nuevo, Tamaulipas, México, pp. 16–19, in *Proceedings of the First International Symposium on Kemp's Ridley Sea Turtle Biology, Conservation and Management*. Calliouet, C.W., Jr. and A.M. Landry, Jr., Eds., Texas A&M University Sea Grant College Program, TAMU-SG-89–105, College Station.

Mounsey, R.P., G.A. Baulch, and R.C. Buckworth. 1995. Development of a trawl efficiency device (TED) for Australian prawn fisheries. I. The AusTED design. *Fish. Res.* 22:99–105.

National Marine Fisheries Service Southeast Fisheries Science Center. 2001. Stock assessments of loggerhead and leatherback sea turtles and an assessment of the impact of the pelagic longline fishery on the loggerhead and leatherback sea turtles of the western North Atlantic. U.S. Department of Commerce, NOAA Technical Memorandum NMFS-SEFSC-455, 342 pp.

Oravetz, C.A. and C.J. Grant. 1986. Trawl efficiency device shows promise. *Aust. Fish.* February, 1986:37–40.

Poiner, I.R. and A.N.M. Harris. 1996. Incidental capture, direct mortality and delayed mortality of sea turtles in Australia's northern prawn fishery. *Mar. Biol.* 125:813–825.

Pritchard, P.C.H. and M.R. Márquez. 1973. Kemp's ridley turtle or Atlantic ridley, *Lepidochelys kempi. IUCN Monograph 2: Marine Turtle Series*, 30 pp.

Renaud, M. et al. 1993. Loss of shrimp by turtle excluder devices (TEDs) in coastal waters of the United States from North Carolina through Texas: March 1988–August 1990. *Fish. Bull.* 91:129–137.

Robins, J.B. 1995. Estimated catch and mortality of sea turtles from the east coast otter trawl fishery of Queensland, Australia. *Biol. Conserv.* 74:157–167.

Robins-Troeger, J.B., R.C. Buckworth, and M.C.L. Dredge. 1995. Development of a trawl efficiency device (TED) for Australian prawn fisheries. II. Field evaluations of the AusTED. *Fish. Res.* 22:107–117.

Siedel, W.R. and C. McVea, Jr. 1982. Development of a sea turtle excluder shrimp trawl for the southeast U.S. penaeid shrimp fishery. In *Biology and Conservation of Sea Turtles*. Bjorndal, K.A., Ed., Smithsonian Institution Press, Washington, D.C., pp. 497–502.

Turtle Expert Working Group. 2000. Assessment update for the Kemp's ridley and loggerhead sea turtle populations in the western North Atlantic. U.S. Department of Commerce, NOAA Technical Memorandum NMFS-SEFSC-444, 115 pp.

United States Court of Appeals for the Federal Circuit. 2002. *Turtle Island Restoration Network, Todd Steiner, The American Society for the Prevention of Cruelty to Animals, The Humane Society of the United States, and the Sierra Club v Donald L. Evans, Colin L. Powell, Paul H. O'Neill, David B. Sandlaw, Penelope D. Dalton and National Fisheries Institute, Inc.,* 00-1569, -1581, -1582, 717 Madison Place N.W., Washington, D.C. 20439, 36 p.

Watson, J.W., J.F. Mitchell, and A.K. Shah. 1986. Trawling efficiency device: a new concept for selective shrimp trawling gear. *Mar. Fish. Rev.* 48(1):1–9.

World Trade Organization. 1998a. United States — import prohibition of certain shrimp and shrimp products. Report of the Panel. WT/DS58/R, 15 May 1998 (98-1710). rue de Lausanne 154, CH-1211 Geneva 21, Switzerland, 426 p.

World Trade Organization. 1998b. United States — import prohibition of certain shrimp and shrimp products. AB-1998-4. Report of the Appellate Body. WT-DS58/R 12, October 1998 (98-0000). rue de Lausanne 154, CH-1211 Geneva 21, Switzerland, 77 p.

World Trade Organization. 2001a. United States — import prohibition of certain shrimp and shrimp products. Recourse to article 21.5 by Malaysia. Report of the Panel. WT/DS58/RW, 15 June 2001 (01-2854). rue de Lausanne 154, CH-1211 Geneva 21, Switzerland, 108 p.

World Trade Organization. 2001b. United States — import prohibition of certain shrimp and shrimp products. Recourse to article 21.5 of the DSU by Malaysia. AB-2001-4. Report of the Appellate Body. WT/DS58/AB/RW, 22 October 2001 (01-5166). rue de Lausanne 154, CH-1211 Geneva 21, Switzerland, 51 p.

14 Social and Economic Aspects of Sea Turtle Conservation

Blair E. Witherington and Nat B. Frazer

CONTENTS

0-8493-1123-3/03/$0.00+$1.50

14.1 INTRODUCTION

For animals that spend most of their lives within the vast expanse of our planet's oceans, the lives of sea turtles and humans are remarkably interconnected. The history of these interconnections describes effects that are largely negative from a sea turtle perspective — effects that have resulted in declines in sea turtle abundance, extirpation of some populations, and the loss of some unique sea turtle phenomena. However, recently in the history of sea turtle and human interactions, conservationists have begun to observe some interconnections that are not all bad news for sea turtles.

Descriptions of the plight of sea turtles in the modern world focus principally on sea turtle–human associations that are either predator and prey or amensal (where one associate is harmed and the other receives no effect). Of these, the roles of predator and prey have been played the longest. For millennia, humans have preyed on sea turtles and have benefited from uses ranging from food and adornments to putative drugs and mystic talismans (Chapters 1 and 12, this volume). Although the period of human use of sea turtles is short by evolutionary standards, there has been sufficient time for this use to become incorporated into many human cultures. In the recent era of human industry, humans and sea turtles remain predator and prey, but within the broad reach of industrial human cultures, associations between sea turtles and humans have become amensal as well. In this amensal association, sea turtles are harmed by an expanding array of human activities for which sea turtle interactions are merely incidental (Lutcavage et al., 1997). These activities include fisheries that target other prey species, as well as many other activities that take place as part of commonplace industrial human habitation. This spoor of everyday living includes accidental spillage of petroleum, discard of plastics, errant artificial lighting, and placement of hardened shoreline-protection structures on sea turtle nesting beaches.

The arguable good news for sea turtle conservation is that sea turtles — not just harvested ones, but those living freely as well — have value to human beings. The extent to which this value is realized varies a great deal between cultures and individuals. Yet, it is a critical concept for sea turtle conservationists to consider, for it is the arguable driving force behind both the use of sea turtles and their preservation.

High value is both boon and bane to sea turtle conservation. Human-recognized value can benefit conservation because it serves as a reason to moderate our predatory and amensal relationships with sea turtles. That is, it can justify our regulatory decisions to use sea turtles without using them up. Furthermore, recognizing what we gain from sea turtles also might justify associations with sea turtles that are positive for them. Commensal associations, benefiting sea turtles and having no cost to humans, and mutualistic associations, having benefits for both parties, each seem possible within many conservation programs. The bane of value to sea turtles is that they are coveted for harvests that proceed despite a desire, even by harvesters, to conserve sea turtles. It is as if we cannot help ourselves. Through actions by well-meaning people, and with decisions that are perfectly rational from an economic perspective, high value can drive sea turtle species to extinction.

This chapter discusses the value of sea turtles and, in doing so, touches on how the social and economic aspects of sea turtle conservation problems can help direct effective conservation solutions. We outline a number of ways to describe what sea turtles are worth, and we present some case studies showing how worth has played a role in both conservation problems and their solutions. We also discuss how many sea turtles are needed to carry out biological phenomena and to allow frequent positive interactions with the humans who value sea turtles. Our goal is to provide at least a partial answer to the question, "Why should we save sea turtles?"

14.2 THREATS FROM USE: DIRECT AND INDIRECT

Thorough descriptions of the threats to sea turtles from direct consumptive use by humans can be found in Parsons (1962), Lutcavage et al. (1997), and Thorbjarnarson et al. (2000). An incomplete list of consumptive uses includes sea turtle meat and eggs used as food; oil used for medicines, lamp fuel, and boat caulking; skins used for clothing and accessories; shells and scutes used for various adornments; whole animals stuffed as curios; and various parts used for fishing bait, domestic animal feed, and fertilizer. The majority of these uses seem trivial to conservationists, and many uses probably have numerous adequate (even superior) substitutes that do not require the consumption of sea turtles. However, the assessments of value for these products are often made at a local level and with a cultural weighting toward authentic sea turtle products. Consumption of sea turtles is driven by more than simple utilitarian values.

Three consumptive uses of sea turtles deserve special mention because of their magnitude: uses of eggs, meat, and shell. Eggs of all species are eaten as food, and across multiple cultures, are believed to be an aphrodisiac. Consumption of eggs has been both subsistence and commercial, and is believed to have been the principal cause of severe declines in many sea turtle populations (Thorbjarnarson et al., 2000). A second use deserving special mention is the use of meat as food. Although all species are eaten, green turtles have been harvested to the greatest extent, for both subsistence and commercial use. Nearly everywhere green turtles occur, there is a history of green turtle harvest. This use is the principal suspect in the extirpation of green turtle nesting colonies, such as those once occurring at Alto Velo, Brazil; Bermuda; Grand Cayman; Hong Kong; Israel; Mauritius; and Reunion (Parsons, 1962; King, 1982; Groombridge and Luxmoore, 1989; Marcovaldi and Marcovaldi, 1989; National Research Council, 1990). A third important use is of scutes from the carapace and plastron of hawksbill turtles. This "tortoiseshell" is used by many cultures for small household items, jewelry, and ceremonial pieces, but the greatest use of tortoiseshell is driven by commercial production of bekko products in Japan (Meylan and Donnelly, 1999). Trade in hawksbill scutes is believed to have caused worldwide declines in hawksbill populations. Although much of the open trade has been halted by the signatories of the Convention on International Trade in Endangered Species of Wild Fauna and Flora (CITES), provisions are being debated to resume trade involving populations in Cuba, and subsequently other populations, at a level proposed to be sustainable (Mrosovsky, 2000).

Indirect consumption of sea turtles comes from incidental interactions between sea turtles and human activity (Lutcavage et al., 1997; Meylan and Ehrenfeld, 2000). The list of activities that indirectly consume sea turtles has become astonishing in its length. The list compiled by Meylan and Ehrenfeld (2000) includes activities both on land and in the water, and encompasses activities that result in lethal and sublethal effects, and in loss of sea turtle habitat. The greatest indirect sea turtle consumption for most species is by mortality from incidental capture by either neritic or oceanic fisheries (Chapter 13, this volume).

14.3 WHAT ARE SEA TURTLES WORTH?

Value is surely an elusive measure in openly traded goods and services, but especially for an environmental resource such as sea turtles. Of course, sea turtles are often openly traded, as are many other commodities, and this trade can help demonstrate a measurable economic value. However, superimposed on this trade value are other important values.

14.3.1 Use Values

14.3.1.1 Consumptive Use Value

Consumptive use value is the most directly measured value from an economic standpoint. Consumptive value can be measured by the market price of whole turtles or eggs, the per-unit price of sea turtle parts such as meat, calipee, carapace scutes, oil, and skins, or the products from sea turtles to which value has been added, such as green turtle soup, bekko jewelry, cosmetics, and leather products.

14.3.1.2 Nonconsumptive Use Value

Sea turtles can be used without consuming them, and compared to other organisms, this nonconsumptive value for sea turtles is high. With their foray onto beaches during reproduction and their large size and conspicuity in water, sea turtles lend themselves well to being watched. A principal nonconsumptive use of sea turtles is by ecotourism in the form of turtle watches. On turtle watches, individuals or groups, either guided or autonomous, locate sea turtles on beaches and watch the nesting process. Other nonconsumptive uses include scientific study and the collection of information, including genetic information that could be useful for technological applications.

14.3.1.3 Option Value

An option value comes from an anticipated or delayed use of sea turtles. The nature of this use may be unknown at the time that value is appraised, or a known use may be postponed given an anticipated increase in either consumptive or nonconsumptive value.

14.3.2 Nonuse Values

14.3.2.1 Existence Value

Many people would agree that there is value in just knowing that sea turtles exist. This concept of cognitive value is more than a starting point for philosophical debate; it can be an economic force that is measurable by methods revealing stated preference (see Section 14.6). People are willing to pay in order to keep sea turtles around. Existence value (Kramer and Mercer, 1997; Larson, 1993) is a term that is often used to capture a varied range of nonuse values to include bequest and intrinsic values (defined in Sections 14.3.2.2 and 14.3.2.3), and ethical, moral, and social values. Existence value also includes ecological value generated by effects of sea turtles on ecosystems shared with humans.

14.3.2.2 Bequest Value

Bequest value is the value of sea turtles as a resource for our kin or for future society. Bequest value might be considered to be part of existence value.

14.3.2.3 Intrinsic Value

Intrinsic value is not easily appraised by humans. Although we might debate its importance in assigning a total value to sea turtles, there is little sense in arguing over the value's magnitude. There are no units for intrinsic value common to other value assessments. Some of the argument that sea turtles have at least some intrinsic value comes from the assertion that all living things do. Commonly, animals that are large and charismatic, like sea turtles, are publicly perceived to have a high intrinsic value (Wilson and Peter, 1988). Sea turtles also may receive intrinsic-value points for longevity. The perception that sea turtles are ancient animals (with sea turtles extant for approximately 100 million years, this perception is accurate) often adds to discussions in the media about why sea turtles should be saved (authors' personal observations).

14.3.3 Total Value

Total value is the sum of all use and nonuse values. It will be difficult to include an assessment of total value in any evaluation of sea turtles. It is important to keep in mind that sea turtles almost always have unreported value to groups of people we fail to include when we focus our attention too narrowly on any particular use or nonuse value.

14.4 THE VALUE OF SEA TURTLE PARTS AND SUMS

Value can be placed on sea turtles at several levels: on sea turtle genes, individuals, populations, species, and phenomena, and on the functions of communities and ecosystems that have sea turtles as component species. Part of the total value of sea turtles comes from their genetic diversity, but human beings have only begun to experience the value of sea turtle genes. As yet, no genetic material from sea turtles has been marketed or has been used with recombinant technology to produce an economic product. In recent years, however, genetic material from sea turtles has become an integral part of scientific studies as a nonconsumptive use of sea turtles (Bowen and Karl, 1997). Much of this noneconomic value comes from genetic tools that allow a greater understanding of sea turtle populations and their relatedness (Bowen, 1999) as well as sea turtle mating (FitzSimmons, 1998; Francisco-Pearce, 2001) and migratory behavior (Meylan et al., 1990). Arguably, these genetic tools also assist in acquiring information valuable from a wider perspective, as information applied toward understanding the genetic consequences of a wide distribution for a long-lived animal and applied toward conserving highly migratory international species (Bowen, 1997). In all animals, this value at the genetic level is reduced by the loss of diversity and rare genotypes.

Sea turtles are most commonly thought of as having value as individuals, and most of the consumptive use of sea turtles drives value at this level. An individual green turtle captured from the water would typically be marketed by weight, or would be butchered and sold by the component weight of its meat. Some parts of a green turtle's anatomy are more valuable than others. For instance, the fatty and gelatinous tissue lining the interiors of the plastron and carapace (calipee and calipash) is boiled for green turtle soup stock and has historically brought a greater price by weight than green turtle meat (Carr, 1967; Parsons, 1962). Anatomical variation in value may be greatest in the hawksbill, a sea turtle having the majority of its consumptive-use value in the scutes covering its shell (Meylan and Donnelly, 1999).

Life stage plays a role in the valuing of individual sea turtles. For most species, the greatest rate of consumptive use of individual sea turtles comes from their harvest as eggs. It is argued that eggs have a high economic value relative to their reproductive value (reproductive value is the chance of an egg resulting in a turtle that produces more eggs; Crouse et al., 1987), and eggs are believed by some to have a higher potential for sustainable harvest than other sea turtle life stages (Mrosovsky, 1983). Consumptive use value of individual sea turtles that are eaten or are sold as food increases with the weight (and thus, age) of the turtle, but this increase seems unlikely to keep up with the high reproductive value of maturing turtles. For example, Frazer (1983) estimated that the reproductive value of an adult female loggerhead (average straight carapace length = 92 cm, approximately 110 kg) is approximately twice that of an 85-cm immature female (approximately 85 kg) and 8–10 times that of a 65-cm immature female (approximately 50 kg).

Much of the consumptive use of individual sea turtles is for subsistence, defining the ecological value of sea turtles as human prey. Unfortunately, many of the examples of predator–prey relationships between sea turtles and people are short-lived. There is a clear modern tendency for human beings to overexploit sea turtles (Parsons, 1962; Thorbjarnarson et al., 2000) — a tendency that underscores the recent nature of our predatory relationship with sea turtles.

Individual sea turtles also have a nonconsumptive use value that is generally recognized. Watching sea turtles and learning from them involves encounters with individual animals at sea or on nesting beaches. A relatively high existence value for individual sea turtles is reflected in human behavior, to the extent that people not only value sea turtles in general, but also may place value on an individual turtle's well being. Many sea turtles that are encountered by people are given names, an action that seems to clearly demonstrate a recognized existence value. Sea turtles briefly encountered on turtle watches, individuals kept in aquarium and rehabilitation facilities, and even turtles outfitted with telemetry and tracked by biologists all have acquired names as individuals. The empathy represented in and generated by a name may extend beyond the people who encounter the individual turtle directly and may multiply a turtle's existence value among many thousands of people vicariously tracking the welfare of a turtle on the Internet or through other media (Godfrey, 1998).

One example of how media attention and perceived value can focus intensely on an individual turtle involves a loggerhead that stranded in Florida with front-flipper injuries. Following an unusual amount of press coverage (many sea turtles with similar injuries strand each year in Florida), the turtle's amputated front flippers were replaced with rubber prosthetics fashioned by an international tire company and attached by a local orthopedic surgeon. The estimated cost of the procedure was $35,000 U.S. dollars (USD) and the reported cost of the prosthetics was $200,000 USD in 1984 (Miami Herald, 25/9/83, 4/10/83, 13/10/83, 17/1/84, 18/1/84). Unfortunately for the loggerhead, named Lucky, the artificial flippers fell off, attempts to reattach them failed, and the turtle remained in a captive rehabilitation facility (Miami Herald, 25/1/84, 29/1/84, 23/3/84, 9/4/84).

Although economic value is most easily placed on the individuals in a population, the population itself, as a potentially sustainable group of interbreeding individuals, should be more valuable than the sum-value of its members. The consumptive use value of populations is widely recognized by governmental agencies and nongovernmental watchdog groups who monitor many species of harvested organisms. Governmental regulation of sea turtle harvests (and prohibition of harvest) attests that sea turtles too are valued as populations. In early seventeenth-century Bermuda, where commercial harvest of green turtles first began (Parsons, 1962), colonists promulgated the first governmental mandate valuing live members of the green turtle population by issuing a fine for catching turtles smaller than 18 in. (Carr, 1952). Unfortunately, the nesting colony at Bermuda was extirpated by the English within the next two centuries (Carr, 1954).

The nonconsumptive use and existence values of sea turtle populations and species is recognized by governments that have set near prohibitions on taking sea turtles. One example of a near prohibition is the Endangered Species Act of 1973 (U.S.). This act contains provisions for protecting habitat for endangered species (including sea turtles), for removing species from protection when populations recover, and for limited take of individuals that benefits their populations (as from conservation science).

Additional existence value of sea turtle populations comes from the role of sea turtles in communities within ecosystems. The role of sea turtles in ecosystems is just beginning to be understood (see Chapter 10, this volume). For example, green turtles are reasoned to play an important role in the cycling of nutrients and in enhancing productivity within seagrass communities, even though this role must have diminished with orders-of-magnitude decreases in green turtle abundance (Jackson, 1997). With a sufficiently diminished ecological role and with the potential for "ecological extinction" (Jackson et al., 2001), the ecological value we gain from sea turtles also diminishes. In another important ecological role, loggerhead eggs have been shown to be the transport vehicle for high amounts of energy, nitrogen, and phosphorus assimilated at sea and brought to nutrient-starved beach sands (Bouchard and Bjorndal, 2000). The anthropocentric value assessment of these and lesser-known ecological roles is that with fewer turtles, or with no turtles, there would be ecological change in the appearance and function of beaches, of seagrass pastures, and of reefs, and changes in the abundance of other organisms valued by human beings.

The phenomena in which organisms take part have been recognized to have value. Brower and Malcolm (1991) discussed value for the spectacle of monarch butterfly (*Danaus plexippus*) migrations and overwintering aggregations and introduced the concept that these biological phenomena could be endangered by diminished butterfly abundance or by loss of behavioral traits. In a broad definition of endangered biological phenomena (Brower, 1997), one might include the mass nesting arribadas of *Lepidochelys*, the conspicuous passing-of-the-fleet migrations of green turtles (Carr, 1954), and terrestrial (Whittow and Balazs, 1982; Garnett et al., 1985) and shallow-water basking behavior (Felger et al., 1976) that may have once been more widespread among green turtle populations. Arguably, these phenomena have nonconsumptive use value in their conspicuity. Thousands of olive ridleys knocking together in their mass nesting, dense conspicuous passage of migrating green turtles, and the prominence of turtles lying about on land can each provide human observers with a grand experience, valuable either to those witnessing it in person or to those viewing the event through broadcast media. These sea turtle phenomena certainly have existence value. Many may place value in just knowing that these phenomena take place, but another consideration is of the ecological value from the effects of the phenomena themselves. For instance, the massive arribadas of the olive ridley (Cornelius, 1986) and the formerly massive arribadas of the Kemp's ridley (Carr, 1967) provide a food and nutrient source — eggs and hatchlings — that attracts and feeds multiple species of predators and brings nutrients for plant growth during this concentrated pulse of reproduction.

14.5 SEA TURTLE SUPPLY AND DEMAND

A fundamental concept in economics is that rarity and increased demand drive up the bid price of any given commodity. This concept has important consequences for sea turtle conservation that differ between consumptive and nonconsumptive uses. With consumptive use of sea turtles, use and marketing can create both a market for increased demand and rarity in supply that results in an unsustainable removal of turtles from the population. The resulting increase in consumptive use value may or may not result in a force that drives sea turtle conservation. In fact, with no system for regulated harvest among multiple harvesters, it is a rational economic decision to take part in maximum-level harvest in anticipation of a short-term gain (Clark, 1973). Clark described this decision for "optimal extinction" as being economically rational within the limited perspective of each individual harvester. In this perspective, profits from extirpated resources should be reinvested in other enterprises and the harvest of replacement resources (other turtle populations, other fisheries, etc.) should be expected to follow each anticipated extinction.

Of course, however economically attractive optimal extinction may appear in the short term, conservationists are not likely to consider it a viable option for sea turtles or any other renewable resource. The key limitation is in the eventual exhaustion of replacement resources. Examples of this limitation come from the harvest of Caribbean green turtles. As green turtle harvests from the Bahamas diminished to near extirpation, harvesting effort shifted to Florida, where green turtle numbers also plummeted (Carr, 1954). Harvesters then turned to Nicaraguan green turtles, a population that had also become the replacement resource for Cayman Island green turtles, a population extirpated following the arrival of turtle fishers formerly from Jamaica (Lewis, 1940; Parsons, 1962). Although the exploitation of green turtles aided a prolonged period of colonization throughout the Caribbean, this limited resource eventually gave out, with reduced numbers of green turtles too small for large-scale commercial turtling.

The conditions that bring about optimal extinction are important for sea turtle conservationists to understand. With conditions modified from Clark (1973), extinction is likely to occur when (1) turtles bring a high price relative to the cost of harvesting them (their harvest remains profitable because price rises with rarity even as rare turtles become difficult to catch); (2) there is a preference for short-term harvest in comparison to the generation time of turtle populations being harvested; and (3) there is no appropriate and complete allocation of harvesting rights (or those rights are contested among parties). Generally, more than one of these conditions is needed to drive overexploitation. For example, high profits may become an incentive for sustainable management if conditions 2 and 3 are not met. Unfortunately, conditions 2 and 3 seem to be readily met for most populations. In commercial operations driven by demand from consumers with little vested interest in the persistence of multiple sources of turtles they consume, high harvest rates would be expected to persist through indications of a failing resource (e.g., increased capture effort per turtle). Because most sea turtle species take decades to mature (Chaloupka and Musick, 1997), harvest rates, in units of turtles per generation, can be high even for small operations. Condition 3 may be the condition most easily met. The multinational distribution of most sea turtle populations (Bowen and Karl, 1997) involves many stakeholders who would compete for sea turtle resource rights. With multiple users, difficult harvest monitoring, and uncertainty over the identity of turtles being harvested (whose turtles are they?), it is currently difficult to achieve appropriate and complete allocation of sea turtle populations without contestation. Therefore, the use of sea turtles induces great temptation for optimal extinction, and this justifies skepticism for their sustained consumptive use. There has been considerable debate about how to determine sustainable harvest of sea turtles (Mrosovsky, 1997; 2000; Pritchard, 2000; Frazier, 1996).

Endangered species conservationists are on record with their dread of the additional demand brought about by expanding commercial markets for consumptive use (Barbier et al., 1990; Fischer, 2001), an increased demand that has been shown to elevate harvest. In his study of the Miskito turtle fishers of Nicaragua, Nietschmann (1973) monitored the harvest of green turtles from nearby Caribbean waters during the 1960s and 1970s. During this period, there was a pronounced switch from subsistence consumption to commercially driven harvest. The new demand for exports resulted in a several-fold increase in take of green turtles and, ironically, a decrease in turtle consumption within the local Miskito community.

Consumptive demand competes with nonconsumptive demand because turtles that are eaten are no longer available for people to experience in the wild. Thus, turtle harvesting enterprises can interfere with ecotourism enterprises. Decreases in abundance of sea turtles may also lower existence and bequest values considerably by making it more difficult to observe sea turtles and by reducing their function within ecosystems. Of course, the consumptive value relative to the nonconsumptive value varies a great deal among human cultures. International agreements made in the last 30 years, such as the CITES, the Bern Convention on the Conservation of European Wildlife and Natural Habitats, the Inter-American Convention for the Protection and Conservation of Sea Turtles, and the Bonn Convention on Migratory Species, appear to reveal a global tendency toward recognizing existence and option value in sea turtles. Many signatories of these agreements include countries where sea turtles were once harvested and traded in high numbers, but that have now found alternative sources of food and income from sea turtle products.

The cultural tendencies of the developed world to weigh nonconsumptive value in sea turtles over their consumptive value, and to weigh limited consumption over unlimited consumption, are strongly influenced by both living conditions and appreciation. Consuming sea turtles may be like a decision about what portion of one's library to burn for warmth. One could forgo this decision with the realization that there are alternative fuel sources. But without other combustibles, a freezing person is tempted to send even the most loved volumes into the fire, and for someone who does not appreciate reading, any book is as good as firewood.

14.6 MEASURING SEA TURTLE VALUE

Ideally, measures that are compared should have a common unit for their comparison. For value measurements, this unit is currency. Although for many, placing a dollar (or yen, euro, pound, etc.) value on sea turtles seems ludicrous, no comparison of different value appraisals (for instance, consumptive versus nonconsumptive use value) could be accurate without it. Currency value of sea turtles allows at least some attempt at cost–benefit analysis when appraising environmental policy and management projects, and it allows a justified assessment of environmental values (e.g., fines) in damage assessments and compensation. Oil spill damage may provide the most common example of compensation based on resource valuation including sea turtles (Hannah and Getter, 1981).

Some measures of consumptive use value for sea turtles come from commercial market prices. Sea turtles are priced by meat (Frazier, 1980; Nietschmann, 1982; Lagueux, 1998), eggs (Lagueux, 1991; Campbell, 1998; Chan and Liew, 1996), and tortoiseshell (Meylan and Donnelly, 1999). The value of whole commercial fisheries has been estimated as well (Parsons, 1962; Rebel, 1974; Woody, 1986). However, because these values do not include nonconsumptive use and nonuse value, they do not reveal the total value necessary to make informed economic decisions about sea turtles. Only rarely is one able to measure competing bids that stem from both use and nonuse value. In an example of just such a bid competition, Amiteye and Moller (2000) describe the purchase of an adult leatherback in Ghana for approximately $19 in order to keep it from being slaughtered for food.

The most widely used method to measure the total economic value of a resource is contingent valuation (CV; for a methodological review, see Bjornstad and Kahn, 1996). CV is described as a direct valuation method and as a stated-preference method because it involves asking people directly how much they would be willing to pay for a resource. Although widely used, CV is a controversial method. As a nonmarket assessment, CV does not measure actual human behavior in the marketplace, where economists turn for most economic indicators. The power of the method is that it can be used to estimate option, existence, and bequest values, as well as use values, and can also measure people's attitudes and motives behind their stated preferences. Much of the controversy about CV lies in the unsure art of survey design; that is, how to ask the questions (Spash et al., 2000). Only recently have CV studies included valuation of sea turtles (Milon et al., 1998; Tisdell and Wilson, 2001a; 2001b; in press; in review; Wilson and Tisdell, 2001; Shivlani et al., in press).

CV methods have been widely used to provide information on conservation activity. The method has provided value estimates for the diversity of ecosystems in which sea turtles play a role (Spash et al., 2000), and has been proposed for appraising the value of preserving endangered species (Burton, 1998) and of maintaining the abundance of charismatic species (Bulte and Van Kooten, 1999).

The measurement of value by CV methods varies greatly with the public's perception of supply and its expression of demand. Whitehead (1993) studied the public's willingness to pay for preserving nesting loggerheads in North Carolina in order to test how well a CV model would predict total economic value under turtle supply-and-demand uncertainty. As part of a public questionnaire that asked what dollar value each respondent placed on preserving loggerheads, respondents were also asked about their perceptions of demand (belief in the chances of them traveling to experience North Carolina loggerheads) and supply (belief in the probability that North Carolina loggerheads being extirpated). Whitehead concluded that when these measures of supply-and-demand uncertainty were factored into CV models, the models gave valid estimates of total economic value for nesting loggerheads. Accounting for this uncertainty, Whitehead found that North Carolina residents were willing to pay approximately $11/year/respondent to keep loggerheads extant for at least 25 years.

Many laws promulgated to protect sea turtles have provisions for fines with monetary values based on the number of sea turtles taken. Although the level of these fines may be arbitrarily set for many species of protected organisms, the fines provide an additional indication of how sea turtles are valued. Penalties for taking sea turtles can vary a great deal, even within one area of overlapping government jurisdictions. For example, in Florida, taking one sea turtle (to include harassment and injury as well as killing) costs between $100 and $600, and could include imprisonment with the fine [Florida Statutes Chapter 370.021(1),(2)(e)]. Taking the same turtle, as prosecuted by federal authorities of the U.S., would cost up to $25,000 (if prosecuted as a civil case) or up to $50,000 (if prosecuted as a criminal case with a higher burden of proof). Federal criminal prosecutions could include imprisonment up to a year in addition to the fine. In practice, these upper fine limits are seldom reached.

So do we now know, or will we ever know, what sea turtles are worth? Perhaps not. Any exercise in valuing sea turtles should consider the prospect that sea turtles have value that is immeasurable or unrealized. It is a common human social ideology that concern for intrinsic value can vastly outweigh concern based solely on monetary value (Tinker et al., 1982). Cash valuations underestimate total value in that they cannot adequately weigh intrinsic value of beauty, knowledge, and spirituality. Thus, the plea for conserving sea turtles may contain some elements that are measurable and some that are abstract — a mix that may never satisfy critics who rely solely on pecuniary judgment.

14.7 SEA TURTLE VALUE AND TRAGEDY OF THE OCEAN COMMONS

Sea turtle value has many facets and varies greatly by one's perspective and recognition of ownership. With the international distribution of most sea turtle populations, any one turtle's life path may pass through areas where it is valued for nonconsumptive use, valued for consumption, valued solely for its existence, or even considered a nuisance. Even within a single nation's waters these assessments can be mixed, but within a single political realm there is typically a sense of national "ownership" that can justify management, and if need be, laws to regulate human interactions with sea turtles. However, among the international regions of our planet there is either a greatly reduced or an intensely contested perception of ownership. In this vast international ocean commons, regulation of human activity has proven to be difficult.

A fundamental problem with the welfare of community and global resources was termed by Hardin (1968) as the "tragedy of the commons." The tragedy lies in the failure of individuals to act in the public's interest. It is a tragedy in the classic sense, with culpability lying not in mean-spirited villains but in well-meaning people who may very well share their communities' interests. For sea turtle conservation, a tragedy might begin with an individual person's decision to take a turtle from the ocean commons. This decision could be described as having one negative and one positive effect on the harvester. The positive effect is equal to the incremental benefit gained from one turtle eaten or sold; for example, +1. The negative effect is equal to the loss of one turtle in a population harvested by many people. Because the loss of value of that one turtle (whether calculated as option, existence, bequest, or even intrinsic value) is spread widely over all the people who value that resource, the negative effect of taking one turtle, experienced by the harvester, is only a small fraction of –1. Thus, every rational harvester in his cursory cost–benefit analysis decides to take a community turtle, and another, and so on, even as the extinction of the community's resource becomes evident. For the harvester, extinction of the turtles harvested is indeed a tragedy, but it is one that has very little effect on each individual decision to take just one more turtle.

A principal lesson offered by Hardin's (1968) description of the tragedy of the commons is that there is no technological escape from it. Rather, preventing the loss of our common resources will take what Hardin termed "mutual coercion mutually agreed upon"; that is, rule of law. The key to avoiding tragic loss of sea turtle populations and species will be to manage oceans not as a commons but as an area with turtles having specific identifiable ownership and legal responsibility (Eckert, 1991; Crowder, 2000).

14.8 CASE STUDIES OF SEA TURTLE VALUE AND CONSERVATION

14.8.1 Managing Light Pollution

Artificial lighting that is visible from the beach discourages females from emerging onto the beach to nest (Witherington, 1992) and disrupts the orientation of hatchlings in their attempt to find the sea from their nest (Witherington, 1997). The potential of light pollution to change spatial nesting patterns (Salmon et al., 2000) and cause mass mortality in misled hatchlings (Witherington, 1997) makes this an important conservation problem. The complexity of the problem is well represented in the case of lighted development on Florida nesting beaches.

Stakeholders in the problem of light pollution in Florida include resident beachfront property owners, beachfront businesses, the public at large with varied interests, governmental agencies at several levels and with varied interests, and nongovernmental organizations representing either business or sea turtle conservation. How the interests of these stakeholders compete can be described by the values sought by each of them. Resident property owners experience economic value from lighting that prevents theft and creates comfort (beautiful, comfortable homes are worth more), and businesses experience a similar value from lighting, but with additional value stemming from conspicuous lighting that attracts customers. Sea turtles on Florida beaches (principally loggerheads, green turtles, and leatherbacks) are a public resource. Because harvesting is illegal, they are presumed to be most valued for their nonconsumptive use, existence, and intrinsic value. Florida state and U.S. federal laws prohibiting sea turtle harassment are evidence that the value of Florida's sea turtles extends to people who live far from the state's beaches.

Although stakeholders are widely spread, the principal regulation of beach light pollution has come from local county and municipal laws. As of this writing, laws prohibiting light visible from the beach during the nesting season have been passed in 20 coastal counties (of 25 coastal counties where sea turtles nest) and 46 municipalities in Florida whose jurisdictions encompass approximately 95% of loggerhead and green turtle nesting in the state (Florida Fish and Wildlife Conservation Commission, unpublished data). When most of these ordinances were proposed to local governing commissions, arguments were made regarding costs and benefits of darkening the beach to protect sea turtles (B.E. Witherington, personal observation). Although the weight of the various arguments was not measured, many of the commissioners speaking during ordinance deliberations placed emphasis on light management that would maintain the value of lighting (i.e., allow it to be used so that light does not reach the beach) without diminishing the value of sea turtles (i.e., prevent harassment and mortality). Witherington and Martin (2000) argued that light management for sea turtle conservation, if properly conducted, has the potential to increase the utility and the aesthetics of beachside lighting.

A complete study of the economics of light management for sea turtle conservation is sorely needed for many beaches worldwide where there is electric light and turtle nesting. In Florida, there is limited information on the costs and benefits of light management, but there has been no economic survey to demonstrate how these costs and benefits are weighed by stakeholders. Light management can involve multiple options. An individual light that is turned off or aimed away from the beach may cost the light user almost nothing if the light illuminated an unimportant area. This action may actually have a benefit if the re-aiming puts more light onto off-beach property where it is desired or if either re-aiming or switching off a light reduces a stakeholder's total power consumption. Shielding a light source may cost approximately $1–50 USD plus labor (<1 h), with smaller lights requiring the lowest shielding cost and with lighting systems meant to light a broad area requiring the highest shielding cost. Where the demand for value gained by lighting is the greatest, either in aesthetics or in the extent of the area lighted, stakeholders may choose to replace existing light sources with new lighting that has better directional capabilities (Witherington and Martin, 2000). Although this option may cost $20–300 USD per light source, it is frequently chosen by stakeholders who wish to solve a lighting problem and are willing to pay higher costs in order to avoid any compromises on the appearance and utility of a lighting system. When light sources with a good directional design or with other attributes that allow good light management (Witherington and Martin, 2000) are chosen during the design and construction phase of a stakeholder's development, there is likely to be little or no added cost. Exceptions to this assessment come from debate over whether a stakeholder actually benefits from lighting the nesting beach itself, property that is commonly in public ownership. Value in lighting the beach may be greatest for beachside businesses that gain customers by being conspicuous. Even with this difficulty, there are some lights with long-wavelength spectral characteristics that can be used in conspicuous applications so that a business retains advertising value and effects on sea turtles are reduced (but not eliminated; Witherington and Martin, 2000).

In Florida, and elsewhere in the U.S., the cost of not managing light near nesting beaches includes potential fines from local, state, and federal laws. Many local light-management ordinances in Florida specify fines of hundreds of dollars per day, and in one example of the enforcement of the U.S. Endangered Species Act, a fine of $45,000 was levied against a condominium with lighting that caused hatchling mortality (Anonymous, 1995; the fine was later reduced to $15,000 but required corrective action taken by the condominium association).

14.8.2 Managing Mortality from Marine Fisheries

The impact of marine fisheries on sea turtle survival is a topic of great concern to conservationists (see Chapter 13, this volume). The use of turtle excluder devices (TEDs) can greatly reduce mortality in trawls, just as redesigning hooks may reduce future mortality rates of sea turtles from longlines. From the socioeconomic standpoint, the issue concerns who should pay for re-outfitting fishing gear to reduce sea turtle mortality. In the event that a suggested method is determined to be less effective than originally thought, who should be expected to absorb the costs of the second re-outfitting of gear? For example (Chapter 13), recent evidence indicates that current openings in some TEDs are not large enough to allow adults of some larger species to escape. Who will pay the cost if this knowledge results in a mandatory redesign and re-outfitting of trawls with new TEDs? Before attempting to answer this question, it is instructive to consider the context in which the issue must be addressed.

When sea turtle conservationists discuss incidental mortality in marine fisheries, we usually are concerned with the mortality of the turtles. It is important to remember that some 15,000,000 people work in marine fisheries (more than 90% on boats less than 24 m in length), in what may be the most dangerous occupation in the world. Workers' mortality rates among marine fishermen are 18–30 times higher than the national averages for countries such as the U.S., Australia, and Italy that keep such records (Food and Agriculture Organization of the United Nations [FAO], 2000). The FAO suggests that the mortality rates of marine fishermen may be even higher in countries that do not keep records of occupational mortality. Furthermore, marine fisheries are an important part of the global economy, with annual first-landing sales averaging between $75 and $80 billion in the late 1990s (FAO, 2000). Many of these first-landing sales are by independent fishermen who do not have a large profit margin from their catch and are unlikely to be able to increase the costs of their operation and remain economically viable. To exacerbate the problems faced by marine fishermen, several trends are already at work to drive their numbers down (FAO, 2000). Some countries are denying access to fisheries as a means of reducing fishing effort on overharvested stocks. Technological improvements are increasing productivity (fish caught per fisherman), thus reducing manpower needs. In addition, the total volume of fish landed has experienced a downward trend in the last decade of the twentieth century. All of these pressures will serve to force individuals to seek other occupations.

It is in this context of a dangerous and sometimes economically precarious livelihood that we may be asking or requiring marine fishermen to procure and adopt new, unfamiliar gear to help us protect sea turtles. There undoubtedly will be some social and economic costs associated with acquiring new TEDs or adapting old ones, or in outfitting longlines with alternative hook types. Who should pay the costs associated with changing gear? The fishermen, the consumer, or the public at large?

In the U.S., at least, the fishermen are unlikely to be required to pay the total cost of obtaining new gear simply on the basis of a scientific assessment of the environmental and ecological impact of their fishery activities and the predicted positive effect of changing gear. Recent case law and legislation have placed the Department of Commerce (including the National Oceanic and Atmospheric Administration [NOAA] and the National Marine Fisheries Service [NMFS]) under considerable pressure to give the social and economic impacts of new regulations equal weight in their ecological and scientific studies (Hendricks, 2000). In particular, legislation such as the 1996 Small Business Regulatory Enforcement Fairness Act and the 1996 amendments to the Magnuson-Stevens

Sustainable Fisheries Act require an analysis of sociocultural and economic impacts of proposed regulations that affect small businesses or communities that are dependent upon fisheries. It remains to be seen exactly what effect this trend might have on the ability of NMFS to impose future regulations to protect sea turtles in marine fisheries. However, some recent court rulings against NMFS on other issues have resulted from the agency's not having given adequate weight or rigor to the process for required reporting of sociocultural and economic impacts under the new requirements (Hendricks, 2000).

There is, of course, the possibility of passing costs on to the consumer, assuming that the increased costs are not sufficiently large to cause consumers to switch to another food item. For example, if the costs of implementing new TEDs increase the cost of shrimp relative to finfish, consumers may change preferences and buy more fish and less shrimp. In the event that these new preferred fish species are caught by longlines or gill nets, the switch in consumer preferences may exacerbate the impact on sea turtles caused by increased gear deployment in these fisheries. On the other hand, if the increased costs in both shrimp and finfish result from new TEDs and new hook types in the longline fishery, consumers might switch their preferences to beef. The increase in beef consumption could exacerbate the environmental problems caused by clearing tropical forests to increase the amount of grazing lands in tropical countries.

It may be the case, however, that consumers are willing to pay more for shrimp or for fish if they know that the increased cost results in turtles being protected. To our knowledge, no contingent valuation studies have been conducted to determine what increased costs consumers might be willing to incur for turtle-safe shrimp or fish. Nor do we know what costs members of the general public are willing to shoulder on behalf of government-subsidized retrofitting of fishing gear. However, there are indirect indications that some groups may be willing to assume higher costs in the cause of protecting sea turtles.

In a Sea Grant survey of Floridians' attitudes about the environment and coastal marine resources (Milon et al., 1998), almost 60% of the respondents indicated that funding for environmental protection should be increased. More than 75% felt that sea turtle populations, coral reefs, and coastal habitats in general were not in as good a condition as they had been in the past. There also was some direct evidence that attitudes about sea turtles and attitudes about increased funding for protection of the environment were linked. Those respondents who disagreed more strongly with the statement that "There are as many sea turtles living around Florida today than there ever were in the past" were more likely to support increased funding for environmental protection in general. In another study, respondents from several south Florida beaches were willing to pay a higher cost for parking fees at recreational beaches to support beach renourishment if the restoration would result in improved nesting habitat for sea turtles (Shivlani et al., in press). Tisdell and Wilson (2001a) found that tourists who had seen nesting or hatchling sea turtles on ecotour visits were more willing to pay to conserve them (see Section 14.8.3).

Although none of these studies is indicative of the general public's willingness to pay increased costs for shellfish or finfish to provide for sea turtle conservation, they do provide tantalizing hints that conservationists might deepen their understanding of the issue by conducting a thorough contingent valuation study. Contingent valuation studies have been criticized for creating the very values they attempt to measure, by forcing respondents to consider issues for which they previously might have had no opinions (Hannemann, 1944). Despite these criticisms, they could be a promising direction of future research if such studies are conducted properly.

In the final analysis, it may be simple economics and fisheries stock population dynamics rather than the activities of conservationists that have the most positive results on sea turtle populations. If present trends continue, at some point within the next 30 years, aquaculture operations will be providing more than half of all the fish consumed globally as capture fisheries continue to decline in importance (FAO, 2000). It remains to be seen whether the impact of aquaculture operations on coastal ecosystems (Masood, 1997) will offset any potential benefits that accrue to sea turtles from reduced incidental capture rates in marine fisheries.

14.8.3 Managing Ecotourism

Sea turtles lend themselves well to being watched, and in this admiration, there is an economic enterprise. The enterprise known as ecotourism can include sea turtles when visitors huddle around a female turtle on a beach to closely watch her nesting behavior, when tourists watch groups of hatchlings scrambling from nest to sea, or when boat or diving tours bring people within sight of turtles in the water.

A definition for ecotourism offered by Ceballos-Lascurain (1996) restricts this activity to "environmentally responsible, enlightening travel and visitation to relatively undisturbed natural areas in order to enjoy and appreciate nature ... that promotes conservation, has low visitor impact, and provides for beneficially active socio-economic involvement of local populations." The stipulations to this definition are structured to exclude entertainment activities that may harm the resource or are of little or no benefit to the resource's local stewards. Effects from tourism on sea turtle nesting are harmful when recreational equipment and visitation interferes with nesting (Arianoutsou, 1988), and when the lighting of tourist accommodations discourages nesting (Witherington, 1992) or misdirects hatchlings (Witherington, 1997). In many ecotourism enterprises, especially those in rural areas, there is a tendency for ecotourism revenue to leave with outside operators rather than remain in the local community (Campbell, 1999).

Although sea turtle ecotours are likely to vary in the degree to which they adhere to these stipulations, some of these enterprises are monitored. Guided nightly, nesting turtle watch programs at Tortuguero, Costa Rica (Jacobson and Robles, 1992), at Mon Repos, Queensland, Australia (Tisdell and Wilson, 2001b), and on Florida beaches (Johnson et al., 1996a) have been monitored by governmental or nongovernmental organizations to determine whether the conduct of the participants and the content of the educational program is in keeping with a conservation-based program. In the cases of green turtles watched at Tortuguero (Jacobson and Lopez, 1994) and of loggerheads watched in Florida (Johnson et al., 1996b), there has also been an assessment of the turtle's response to all this attention. Effects range from subtle changes in the time turtles spend nesting (Johnson et al., 1996b) to turtles being turned away in their attempts to emerge onto the beach (Jacobson and Lopez, 1994). A principal goal of this monitoring is to educate trained guides so that impacts on sea turtles are limited and so that the educational experience of visitors is maximized.

The total value of sea turtle ecotourism and its effects on sea turtle conservation have been most extensively measured for the turtle watch operation at Mon Repos, where between 14,000 and 24,000 tourists a year come to view nesting or hatchling sea turtles (Tisdell and Wilson, 2001a; 2001b; in press; in review; Wilson and Tisdell, 2001). The operation is regulated by the Queensland Parks and Wildlife Service and takes place on two adjacent protected beaches that serve the largest concentration of accessible (on continental beaches) nesting sea turtles (principally loggerheads) in eastern Australia. The operation has developed since 1985 to become a focus for both scientific research and tourism. In this development, Tisdell and Wilson (2001b) noted a change in visitation that fit Duffus and Dearden's (1990) hypothesis that the number of "novice generalists" (in this case, tourists who enjoy nature but who have a less specific interest in sea turtles) visiting an ecotourism attraction will increase with time more quickly than the number of "expert specialists" (in this case, sea turtle biologists). The importance behind this distinction between types of visitors is that, in comparison to specialists, novice generalists are likely to demand greater accommodations near the beach in the form of hotels, restaurants, and ancillary entertainment. However, contrary to Duffus and Dearden's hypothesis that novice generalists tend to increase with time along a logistic curve, Tisdell and Wilson found that visiting tourists varied annually and did not increase toward an asymptote. Thus, Mon Repos may be different from other ecotourism beaches where there may be a greater potential for an exploding visitation rate and increasing accommodation needs near the beach, and where the beach itself may not be completely within a protected reserve.

The nightly visitors to Mon Repos' nesting turtles pay a small fee that is only a small fraction of a visitor's total cost. Tisdell and Wilson (2001b) reported that tourists in the area to watch turtles spend money on transportation, lodging, food, and additional items, with the greatest part of this spending occurring within 25 km of the Mon Repos beach. The effect of turtle-watching visitors on the local economy was calculated to be approximately 800,000 AUD ($0.45 million USD) per year. The magnitude of this economy rivals whale-watching (another local ecotourism draw), sugarcane farming, and dairy and beef production — an economic effect that is especially high considering that the length of the season is only 4 months each year.

Tisdell and Wilson (2001a; 2001b) developed a contingent valuation of sea turtles at the Mon Repos turtle-watch operation by surveying participants' willingness to pay and their attitudes toward conservation. The resulting contingent valuation revealed that the average visitor who had just been on a turtle watch was willing to pay approximately 50% more for the entrance fee to watch turtles. Furthermore, approximately half of survey respondents reported that their experience at Mon Repos would affect their propensity to contribute money for sea turtle conservation. In an open-ended survey question asking visitors the maximum amount that they would be willing to pay to protect sea turtles in Australia over a 10-year period, respondents answered between approximately 2 AUD (including zero bids) and 2.50

AUD (excluding zero bids) per week. Beyond the economic impact of the Mon Repos turtle-watch operation, Tisdell and Wilson (2001a; 2001b) also measured effects on conservation behavior. Between 47 and 75% of survey respondents reported that their Mon Repos visit would influence their future behavior regarding the proper disposal of plastics and fishing gear, managing lighting near nesting beaches, and avoiding products made from sea turtles.

Another example of an operation that seems to fit the Ceballos-Lascurin (1996) ecotourism definition is Brazil's Projeto TAMAR (an abbreviation for marine turtle in Portuguese). TAMAR forms a network of 18 conservation stations along 1100 km of the Brazilian mainland coast and has objectives of sea turtle conservation, community development, and education (Marcovaldi and Marcovaldi, 1999). Tours to watch nesting sea turtles at night are only a small part of the TAMAR operation. On the tours, visitors pay a small fee to be guided by a trained biologist. Although nesting turtles are rarely seen, the interpretation given by educated guides and an occasional release of hatchlings keeps the attention of visiting tourists (Vieitas and Marcovaldi, 1997). Although the TAMAR program with its educational outreach is highly regarded for its effect on conservation (Mast, 1999), a detailed socioeconomic understanding of the operation has yet to be developed.

Many other examples of sea turtle ecotourism projects have been described, but their effects on conservation, economies, and sociology have not been measured. Examples include informally organized boat operators at Laganas Bay, Zakynthos, Greece, who take tourists from the local resorts to see loggerheads in the water off the adjacent nesting beaches (Margaritoulis, 1989), and local guides on nesting beaches in French Guiana who work within an area that would otherwise receive very few tourists and where few accommodations exist (Godfrey and Drif, 2001). These examples point out challenges both in directing ecotourism programs in densely visited areas and in managing the impacts of growing visitation in fragile habitats.

14.9 HOW MANY TURTLES?

As Harvard economists Metrick and Weitzman (1998) reminded us, "Decisions about endangered species reflect the values, perceptions, uncertainties, and contradictions of the society that makes them." Ever since Clark (1973) demonstrated 30 years ago that it is economically desirable under some circumstances to exploit natural populations to extinction in order to maximize profit, many conservationists have had an innate mistrust of economists and economic analysis. However, economists are not necessarily the enemies of conservation. A new generation of work is providing economic justification for conservation of wildlife and biodiversity (e.g., Tisdell, 1991; 2002). The simple fact is that economists and other social scientists are unlikely to be able to help us determine whether sea turtle conservation and management are feasible and affordable until we can provide them with a clear idea of our objectives (Metrick and Weitzman, 1998). This raises the questions of how many sea turtles we need and what sacrifices we are willing to make (or ask others to make) to obtain and keep that number of turtles. The answer to the first question will depend on what we think we need the turtles for; the answer to the second is beyond the scope of the present volume.

14.9.1 Preserving Genetic Diversity

When discussing the conservation of genetic diversity in a population, geneticists refer to the effective population size (Roughgarden, 1979). This number refers only to adults, and is typically smaller than the total size of the adult population for several reasons. First, if breeding is nonrandom, some individuals will contribute more to the genetic makeup of the next generation than will others. In fact, some adults may not breed at all, or very infrequently compared to others. This makes the effective population size smaller than the actual adult population numbers, because effectively fewer adults are contributing to the genetic makeup of the next generation. Some populations undergo fluctuations in numbers, rising and falling over time. In such cases, the effective population size over a long time period may be nearer that of the small, bottleneck populations (when genetic diversity has been reduced) than to the average or larger population sizes. As the species suffers a population crash, genetic diversity declines. When the numbers increase again, there almost certainly will be different allele and genotype frequencies in the larger population than there would have been if the crash had never occurred. Similarly, in populations with unequal sex ratios, the effective population size will be reduced by the rarer sex. For these and other reasons, the effective population size is almost always smaller than the actual population of adults. In addition, the actual total population size of adults and juveniles will then be much larger than the effective population size.

More than 20 years ago, Franklin (1980) and Soulé (1980) suggested that an effective population size of 500 adults was sufficient to maintain genetic variability in a species over the long term, and the so-called Franklin-Soulé number has been used in conservation and management plans ever since. More recently, Lande (1995) demonstrated that an effective population size of 5000 might be necessary to ensure the maintenance of adaptive genetic variance in quantitative characteristics. He further suggested that this might require average actual population sizes of wild species to be in the tens of thousands or even much higher to ensure their long-term viability.

In the case of sea turtles, it may be difficult to determine effective population size. We need information on the sex ratio of adults, on the proportion of females (and possibly males) that are actively breeding, on whether breeding is random, and if not, then on the specific breeding pattern. We also need information on what exactly constitutes a population, as opposed to a nesting aggregation. Information from the analysis of mitochondrial DNA suggests that a nesting aggregation may be genetically isolated from other rookeries because of female philopatry to their natal beaches (Bowen, 1997; Francisco et al., in press). If this is true, then we may be able to delineate populations geographically if we can assume that male turtles born to mothers from a particular nesting aggregation breed only with female turtles from the same natal beach. Some mitochondrial information indicates that male green turtles do display philopatry (FitzSimmons et al., 1997a). However, recent evidence from nuclear DNA studies on green turtles (FitzSimmons et al., 1997b) and loggerheads (Francisco-Pearce, 2001) indicates that males also may be important vectors, moving genes between and among nesting aggregations.

They apparently breed with females that pass by during their migrations to their nesting beaches, or on feeding grounds where turtles congregate from several different rookeries. This male-mediated gene flow will exacerbate the difficulty in determining effective population sizes with any accuracy, because we will have to quantify the effect and delineate the mating systems. In any case, it is likely that the maintenance of genetic variability in wild sea turtle populations will require effective population sizes of many thousands of adults and total population sizes of perhaps hundreds of thousands, if not millions, when we include young juveniles and neonates in our calculations.

14.9.2 Providing Ecological and Ecosystem Functionality

We have known for some time that sea turtles may play important roles in maintaining the structure of coral reefs and seagrass beds (see Chapter 10, this volume). Bouchard and Bjorndal (2000) demonstrated that nesting loggerheads are an important link in moving nutrients and energy between marine and terrestrial ecosystems. Using their figures for loggerhead nests, Frazer (2001) speculated on the magnitude of materials moved onto Caribbean beaches prior to the reduction of green turtle populations by European colonization of the New World. Using an estimate of 34,000,000 adult green turtles in the pre-Columbian Caribbean (see Jackson, 1997; Bjorndal et al., 2000) and assuming that half of them were females, Frazer (2001) estimated that they would make 23,800,000 nests per year (4.2 nests per female with a 3-year average remigration interval). That would have resulted in a contribution of some 1,600,000 kg of organic matter, 365,000 kg of lipids, 170,000 kg of nitrogen, 15,500 kg of phosphorus, and 44,500,000,000 kJ of energy to Caribbean beaches each year. Actual figures may have been higher, because green turtles tend to lay their eggs higher up on the beach than do loggerheads (Bouchard and Bjorndal, 2000). One can only wonder at the likely results and impact of greatly curtailing these previous levels of annual inputs into terrestrial systems.

Such speculation is an interesting pastime but does not really provide useful information in determining how many turtles we need. How many nesting turtles does it take to maintain a healthy level of nutrients on a beach? How many hawksbills are necessary for the proper maintenance of a productive coral reef (Meylan, 1988; see also Chapter 10, this volume)? How many green turtles are needed per hectare of seagrass bed to sustain optimal productivity of such systems? Although we admittedly do not know with any accuracy how many sea turtles of each species are necessary to restore or sustain their prehistorical ecological and ecosystem function, it is likely that it will require populations or meta-populations on the order of 10^4–10^7. However, if we desire these population levels, it may also be necessary to restore seagrass beds and coral reefs to their former levels of extent prior to human influences in order to support prehistorical numbers of turtles (Jackson 1997; Bjorndal et al., 2000). Such an undertaking is practically impossible; it would require the re-development or restoration of coastal habitats throughout the planet and displace millions of humans from their homes and livelihoods.

14.9.3 Supporting Ecotourism

Tisdell and Wilson (2001a; 2001b; in press; in review; Wilson and Tisdell, 2001) have provided the most thorough analysis available to date on the socioeconomics of sea turtle ecotourism. Their work at Mon Repos provides convincing evidence that properly managed sea turtle ecotourism is sustainable, that it can provide incentives for funding conservation, and that educating the public with encounters in the wild may be more effective than educational programs in zoos or aquaria in building support for conservation. In all of their analyses, they stressed the fact that the sustainability of ecotourism is dependent upon the successful management of the biological resource.

Recently, Tisdell and Wilson (in press) turned their attention to assessing how many turtles are needed to sustain an effective ecotourism program at Mon Repos. The sustainability of ecotourism is directly related to the probability of tourists observing either a nesting turtle or hatchlings emerging from a nest. They developed conceptual graphical models that depict (a) the relationship between the numbers of nesting turtles visiting the beach during the season and the number of visits for turtle watching by tourists, and (b) the abundance (population density) of the species being viewed and the cost–benefit to a tourist attempting to view the species. On the basis of their assessment, Tisdell and Wilson (in press) concluded that the approximately 200 nesting sea turtles on a 1-km beach at Mon Repos might be near the critical minimum threshold necessary to support current levels of tourism. In other words, if the numbers of nesting turtles were to drop much below present levels, the ecotourism program could decline dramatically.

Because this is the first study of its kind, it is difficult to generalize from these results. The sustainability of *in situ* ecotourism programs obviously is directly related to the numbers (or density) of turtles nesting and hatching on a particular beach. However, the exact number of turtles visiting and tourists accommodated will depend on other variables, such as the maximum numbers of tourists allowed in a turtle-watch group. There also is likely some flexibility if ecotours are willing to provide refunds, partial refunds, or rain checks to tourists who fail to encounter turtles (Tisdell and Wilson, 2001b). However, given that tour guides are not likely to be able to predict with certainty that tourists will see a turtle, and that many tourists currently fail to encounter a turtle during their visit to Mon Repos (Tisdell and Wilson, in press), the figure of 200 nesting turtles per km of beach per season may be a reasonable lower threshold until further studies can be conducted elsewhere. Given that numbers of nesting females on any particular beach can fluctuate widely from year to year, the total population of adult females necessary to support eco-tourism will of necessity be much greater than 200 turtles/km. Because no population can consist entirely of adult females, it is likely that a total sea turtle population of thousands, if not tens of thousands, including males and juveniles, will be necessary to support viable, sustainable ecotourism programs — even if nesting is restricted to only 1 km of beach!

14.9.4 Sustaining Harvests of Eggs or Turtles

Despite some excellent socioeconomic analyses of the harvest at Ostional, Costa Rica (Campbell, 1998; Hope, 2000), it is not currently possible to estimate the number of sea turtles necessary to support an economically sustainable egg harvest. To our knowledge, no economist has provided an estimate of the minimum number of eggs necessary to support a sustainable market. Even if such an estimate were available, it is clear that harvesting sea turtle eggs is probably not sustainable if the population is also being subjected to human-induced mortality at other life stages (Hope, 2000). For example, it is highly unlikely that a sea turtle population could support both harvesting of its eggs and incidental mortality in trawl fisheries or directed take of adults and juveniles in a turtle fishery. As indicated above, although it may be desirable in a strictly economic sense to exploit a biological resource to extinction in order to maximize a profit that would then be invested in other commodities (Clark, 1973), this is not acceptable if the goal is sustainability of the resource.

Similarly, to our knowledge no one has provided estimates of the minimum number of turtles (or kilograms of meat or tortoiseshell) that would be necessary for economic sustainability of a turtle fishery. Therefore, it is not possible to calculate the population size necessary to sustain such a fishery biologically. Most models of turtle population dynamics lead us to the conclusion that harvesting such long-lived, late-maturing species is not sustainable (Congdon et al., 1993; 1994; Crouse et al., 1987; Crowder et al., 1994; Heppell et al., in press). If that is the case, then there is no finite population size, however large, that would allow a sustainable take of adults or larger juveniles. However, from a purely monetary standpoint, we reiterate that it might be deemed desirable to exploit a species to extinction in order to maximize profits to reinvest in other markets (Clark, 1973). This option is likely to be unacceptable to anyone interested in the conservation of sea turtles or of biodiversity in general.

14.10 CONCLUSIONS AND RECOMMENDATIONS

Sea turtles have worth that is widely experienced yet seldom measured. Sea turtle value should guide our conservation efforts, and a comprehensive understanding of their total value should accompany any proposal for their use. Some sea turtle harvests may have important cultural roles and some incidental take may occur despite our most diligent application of mortality-reducing methods. However, even if this sea turtle consumption were sustainable, it could still diminish the value of competing enterprises such as ecotourism or reduce the ecological function of sea turtles and sea turtle phenomena. The effects of direct and indirect take on nonconsumption and nonuse values, coupled with the temptation for optimal extinction of sea turtle resources, mandate careful scrutiny of any proposal involving consumption of sea turtles.

Contingent valuation studies are needed to measure the total value of sea turtle resources so that all parties that value them can be represented in equitable resource-use decisions. Conserving sea turtles will always involve costs, which might be justifiable by measuring a willingness to pay among stakeholders. Many conservationists have realized the potential for generating willingness to pay by the media showcasing of sea turtle charisma. It remains a challenge for conservationists to gently focus the ensuing financial attention in biologically appropriate directions without ignoring the empathy of the public for individual turtles.

Sea turtles need stewardship. Wherever they occur there should be a consensual assessment of ownership that would prevent sea turtles from being considered part of a neglected ocean commons. This ownership could be multinational, but would require enforcement of agreements to manage sea turtles in international waters (as are currently being developed in some regions).

The questions of how many sea turtles to have and to use should be addressed in multidisciplinary studies of sea turtle resources. Addressing not only how many sea turtles we need, but also how many we want, will require coordinated input from population modelers, systems ecologists, environmental economists, and various stakeholders. Certainly, additional studies are needed to fully understand sea turtle survivorship, abundance, mating systems, and genetic exchange (do we have enough turtles?), and to understand the carrying capacity of nesting beaches and foraging grounds (can we have the turtles we want?). Yet, without fully understanding the biological limitations that sea turtles and their environment impose on us, we have applied both our demand to use sea turtles and our desire simply to see that they exist. Although we do not yet have a number of sea turtles that reflects a balance between our demands and desires, we might imagine that a goal of sea turtle conservation should be more than merely a future with some sustainable number of turtles. Instead, conservationists might seek a future in which all aspects of sea turtle value continue to contribute toward human happiness and well-being.

It is incumbent upon those consumptive users who wish to continue their predator–prey relationship with sea turtles to become "prudent predators" (Slobodkin, 1974) who do not deplete the source of their sustenance. Those seeking to use sea turtles nonconsumptively or to keep open our options for future uses also must act prudently to maintain or enhance the numbers of turtles. Even those who desire to focus solely on the existence, bequest, or intrinsic value of sea turtles must work with the "users" to minimize the negative effects of our amensal relationship with the turtles, reducing incidental mortality and other negative effects of modern industrial societies. It is clear that individuals and societies must make conscious decisions and take concerted action if we are to manage a planet that is habitable by both humans and sea turtles.

ACKNOWLEDGMENTS

We are grateful to Clem Tisdell, Manoj Shivlani, and Rob Hope, who provided us with prepublication copies of their manuscripts on contingent valuation and other aspects of sea turtle economics, and to the Sea Grant office at the University of Florida for pointing out and providing us with copies of several useful references. Madeline Broadstone helped gather additional references, and the manuscript benefited greatly from reviews given by Andrea Mosier, Michael Sorice, and Jim Quinn.

REFERENCES

Amiteye, B.T., and J. Moller. 2000. ¢70,000.00 (about US $19) saves a life of a leatherback turtle in Ghana. *Mar. Turtle Newsl.* 88: 13.

Anonymous. 1995. Legal briefs: condos fined for lighting violations. *Mar. Turtle Newsl.* 71: 26–31.

Arianoutsou, M. 1988. 1988. Assessing the impacts of human activities on nesting of loggerhead sea turtles, *Caretta caretta* L., on Zakynthos Island, Western Greece. *Environ. Conserv.* 15: 327–334.

Barbier, E.B. et al. 1990. *Elephants, Economics and Ivory.* Earthscan Publications., London.

Bjorndal, K.A., A.B. Bolten, and M.Y. Chaloupka. 2000. Green turtle somatic growth model: evidence for density dependence. *Ecol. Appl.* 10: 269–282.

Bjornstad, D.J. and J.R. Kahn. 1996. *The Contingent Valuation of Environmental Resources: Methodological Issues and Research Needs.* Edward Elgar, Brookfield, VT.

Bouchard, S.S. and K.A. Bjorndal. 2000. Sea turtles as biological transporters of nutrients and energy from marine to terrestrial ecosystems. *Ecology* 81: 2305–2313.

Bowen, B.W. 1997. Complex population structure and the conservation genetics of migratory marine mammals: lessons from sea turtles. Pp. 77–84 in: Dizon, A.E., Chivers, S.J., and Perrin, W.F. (Editors), *Molecular Genetics of Marine Mammals.* Society for Marine Mammalogy. Special Publication No. 3. Allen Press, Lawrence, KS.

Bowen, B.W. 1999. Preserving genes, species, or ecosystems? Healing the fractured foundations of conservation policy. *Mol. Ecol.* 8: S5–S10.

Bowen, B.W. and S.A. Karl. 1997. Population genetics, phylogeography, and molecular evolution. Pp. 29–50 in: Lutz, P.L. and J. Musick (Editors), *The Biology of Sea Turtles.* CRC Marine Science Series, CRC Press, Boca Raton, FL.

Brower, L.P. 1997. A new paradigm in conservation of biodiversity: Endangered biological phenomena. Pp. 115–118 in: Meffe, G.K. and Carroll, C.R. (Editors), *Principles of Conservation Biology,* 2nd edition. Sinauer Associates, Sunderland, MA.

Brower, L.P. and S.B. Malcolm. 1991. Animal migrations: endangered phenomena. *Am. Zool.* 31: 265–276.

Bulte, E.H. and G.C. Van Kooten. 1999. Marginal valuation of charismatic species: implications for conservation. *Environ. Resour. Econ.* 14: 119–130.

Burton, M. 1998. Contingent valuation and endangered species: methodological issues and applications. *Manchester School* 66: 609–610.

Campbell, L. 1998. Use them or lose them? Conservation and the consumptive use of marine turtle eggs at Ostional, Costa Rica. *Environ. Conserv.* 25: 305–319.

Campbell, L.M. 1999. Ecotourism in rural developing communities. *Ann. Tourism Res.* 26: 534–553.

Carr, A.F. 1952. *Handbook of Turtles: The Turtles of the United States, Canada and Baja California.* Comstock, Division of Cornell University Press, Ithaca, NY.

Carr, A.F., Jr. 1954. The passing of the fleet. *AIBS Bull.* 4: 17–19.

Carr, A. 1967. *So Excellent a Fishe.* Natural History Press, Garden City, New York.

Ceballos-Lascurain, H. 1996. *Tourism, Ecotourism and Protected Areas.* IUCN, World Conservation Union, Gland, Switzerland.

Chaloupka, M.Y. and J.A. Musick. 1997. Age, growth, and population dynamics. Pp. 233–276 in: *The Biology of Sea Turtles.* CRC Marine Science Series, CRC Press, Boca Raton, FL.

Chan, E.H. and H.C. Liew. 1996. Decline of the leatherback population in Terengganu, Malaysia, 1956–1995. *Chelonian Conserv. Biol.* 2: 196–203.

Clark, C.W. 1973. Profit maximization and the extinction of animal species. *J. Polit. Econ.* 81: 950–961.

Congdon, J.D., A.E. Dunham, and R.C. Van Loben Sels. 1993. Delayed sexual maturity and demographics of Blanding's turtles (*Emydoidea blandingii*): implications for conservation and management of long-lived organisms. *Conserv. Biol.* 7: 826–833.

Congdon, J.D., A.E. Dunham, and R.C. Van Loben Sels. 1994. Demographics of common snapping turtles (*Chelydra serpentina*): implications for conservation and management of long-lived organisms. *Am. Zool.* 34: 397–408.

Cornelius, S.E. 1986. *The Sea Turtles of Santa Rosa National Park.* Fundacion de Parques Nacionales, Costa Rica.

Crouse, D.T., L.B. Crowder, and H. Caswell. 1987. A stage-based population model for loggerhead sea turtles and implications for conservation. *Ecology* 68: 1412–1423.

Crowder, L. 2000. Leatherback's survival will depend on an international effort. *Nature* 405: 881.

Crowder, L.B. et al. 1994. Predicting the impact of excluder devices on loggerhead sea turtle populations. *Ecol. Appl.* 4: 437–445.

Duffus, D.A. and P. Dearden. 1990. Non-consumptive wildlife-oriented recreation: a conceptual framework. *Biol. Conserv.* 53: 213–231.

Eckert, K.L. 1991. Leatherback sea turtles: a declining species of the global commons. Pp. 73–90 in: Borgese, E.M., Ginsburg, N., and Morgan, J.R. (Editors), *Ocean Yearbook 9*. University of Chicago Press, Chicago.

FAO. 2000. *The State of World Fisheries and Aquaculture 2000.* Food and Agricultural Organization of the United Nations. Rome, Italy.

Felger, R.S., K. Cliffton, and P.J. Regal. 1976. Winter dormancy in sea turtles: independent discovery and exploitation in the Gulf of California by two local cultures. *Science* 191: 283–285.

Fischer, C. 2001. The complex interactions of markets for endangered species products. Resources for the future discussion paper 02–21. Resources for the Future, Washington, D.C.

FitzSimmons, N.N. 1998. Single paternity of clutches and sperm storage in the promiscuous green turtle (*Chelonia mydas*). *Mol. Ecol.* 7: 575–584.

FitzSimmons, N.N., C.J. Limpus, and C. Moritz. 1997a. Philopatry of male marine turtles inferred from mitochondrial DNA markers. *Proc. Natl. Acad. Sci. U.S.A.* 94: 8912–8917.

FitzSimmons, N.N. et al. 1997b. Geographic structure of mitochondrial and nuclear gene polymorphisms in Australian green turtle populations and male-biased gene flow. *Genetics* 147: 1843–1854.

Francisco, A.M. et al. In press. Stock structure and nesting site fidelity in Florida loggerhead turtles (*Caretta caretta*) resolved with mtDNA sequences. *Mar. Biol.*

Francisco-Pearce, A.M. 2001. Contrasting population structure of *Caretta caretta* using mitochondrial and nuclear DNA primers. Master's thesis, University of Florida, Gainesville, FL.

Franklin, I.R. 1980. Evolutionary changes in small populations. Pp. 135–149 in: M.E. Soulé and B.A. Wilcox (Editors), *Conservation Biology: An Evolutionary-Ecological Perspective*. Sinauer Associates, Sunderland, MA.

Frazer, N.B. 1983. Demography and life history evolution of the Atlantic loggerhead sea turtle, *Caretta caretta*. Ph.D. dissertation, University of Georgia, Athens, GA.

Frazer, N.B. 2001. Management and conservation goals for marine turtles. Pp. 69–74 in: Eckert, K.L. and F.A. Abreu-Grobois (Editors), *Proceedings of the Regional Meeting: Marine Turtle Conservation in the Wider Caribbean Region: A Dialogue for Effective Regional Management*. Santo Domingo, 16–18 November 1999. WIDECAST. IUCN-MTSG, WWF and UNEP-CEP.

Frazier, J. 1980. Exploitation of marine turtles in the Indian Ocean. *Hum. Ecol.* 8: 329–370.

Frazier, J. 1996. Marine cheloniphiles and sustainable development. Pp. 92–96 in: Keinath, J.A. et al. (Compilers), *Proceedings of the Fifteenth Annual Symposium on Sea Turtle Biology and Conservation*. NOAA Technical Memorandum NMFS-SEFSC-387.

Garnett, S.T., G.M. Crowley, and N. Goudberg. 1985. Observations of non-nesting emergence by green turtles in the Gulf of Carpentaria. *Copeia* 1985: 262–264.

Godfrey, D.B. 1998. Anatomy of a successful sea turtle conservation education program. Pp. 56–58 in: Epperly, S.P. and Braun, J. (Compilers), *Proceedings of the Seventeenth Annual Sea Turtle Symposium*. NOAA Technical Memorandum NMFS-SEFSC-415.

Godfrey, M.H. and O. Drif. 2001. Guest editorial: developing sea turtle ecotourism in French Guiana: perils and practicalities. *Mar. Turtle Newsl.* 91: 1–4.

Groombridge, B. and R. Luxmoore. 1989. The green turtle and hawksbill (*Reptilia*: *Cheloniidae*): world status, exploitation and trade. CITES (Convention on International Trade in Endangered Species) Secretariat, Lausanne, Switzerland.

Hannah, R. and Getter, C.D. 1981. Resources at risk. Pp. 85–103 in: Hooper, C.H. (Editor), The Ixtoc Oil Spill: The Federal Response, NOAA Special Report NOAA-SR-HMRP-1981.

Hannemann, W.M. 1994. Valuing the environment through contingent valuation. *J. Econ. Perspect.* 8: 19–43.

Hardin, G. 1968. The tragedy of the commons. *Science* 162: 1243–1248.

Hendricks, B. 2000. Stewardship and analysis: preserving nature and communities — an assessment of economics and the social sciences within the National Oceanic and Atmospheric Administration. A report to the Secretary of Commerce's Policy Office. 6 January. Washington, D.C.

Heppell, S.S. et al. In press. Population models for Atlantic loggerheads: past, present, and future. In: A. Bolten and B. Witherington (Editors), *The Biology and Conservation of Loggerhead Sea Turtles*. Smithsonian Institution Press, Washington, D.C.

Hope, R.A. 2000. Egg harvesting of the olive ridley marine turtle (*Lepidochelys olivacea*) along the Pacific coast of Nicaragua and Costa Rica: an arribada sustainability analysis. M.A. thesis, Institute for Development Policy and Management, University of Manchester, U.K.

Jackson, J.B.C. 1997. Reefs since Columbus. *Coral Reefs* 16(Suppl.) :S23–S33.

Jackson, J.B.C. et al. 2001. Historical overfishing and the recent collapse of coastal ecosystems. *Science* 293: 629–638.

Jacobson, S.K. and A.F. Lopez. 1994. Biological impacts of ecotourism: tourists and nesting turtles in Tortuguero National Park, Costa Rica. *Wildl. Soc. Bull.* 22: 414–419.

Jacobson, S.K. and R. Robles. 1992. Ecotourism, sustainable development, and conservation education: development of a tour guide training program in Tortuguero, Costa Rica. *Environ. Manage.* 16: 701–713.

Johnson, S.A., K.A. Bjorndal, and A.B. Bolten. 1996a. A survey of organized turtle watch participants on sea turtle nesting beaches in Florida. *Chelonian Conserv. Biol.* 2: 60–65.

Johnson, S.A., K.A. Bjorndal, and A.B. Bolten. 1996b. Effects of organized turtle watches on loggerhead (*Caretta caretta*) nesting behavior and hatchling production in Florida. *Conserv. Biol.* 10: 570–577.

King, F.W. 1982. Historical review of the decline of the green turtle and the hawksbill. Pp. 183–188 in: Bjorndal, K.A. (Editor), *Biology and Conservation of Sea Turtles*. Smithsonian Institution Press, Washington, D.C.

Kramer, R.A. and D.E. Mercer. 1997. Valuing a global environmental good: US residents' willingness to pay to protect tropical rain forests. *Land Econ.* May.

Lagueux, C.J. 1991. Economic analysis of sea turtle eggs in a coastal community on the Pacific coast of Honduras. Pp. 136–144 in: Robinson, J.G. and Redford, K.H. (Editors), *Neotropical Wildlife Use and Conservation*. University of Chicago Press, Chicago.

Lagueux, C.J. 1998. Marine turtle fishery of Caribbean Nicaragua: human use patterns and harvest trends. Ph.D. dissertation, University of Florida, Gainesville, FL.

Lande, R. 1995. Mutation and conservation. *Conserv. Biol.* 9: 782–791.

Larson, D.M. 1993. On measuring existence value. *Land Econ.* 69: 377–388.

Lewis, C.B. 1940. The Cayman Islands and marine turtle. Pp. 56–65 in: Grant, C. (Editor), *The Herpetology of the Cayman Islands*. Bulletin of the Institute of Jamaica Science Series no. 2.

Lutcavage, M.E. et al. 1997. Human impacts on sea turtle survival. Pp 387–409 in: Lutz, P.L. and Musick, J.A. (Editors), *The Biology of Sea Turtles*. CRC Marine Science Series, CRC Press, Boca Raton, FL.

Marcovaldi, M.A. and G.G. Marcovaldi. 1999. Marine turtles of Brazil: the history and structure of Projeto TAMAR-IBAMA. *Biol. Conserv.* 91: 35–41.

Margaritoulis, D. 1989. Successes and failures: conservation and tourism on the nesting beaches of Laganas Bay, Zakynthos, Greece, 1989. *Mar. Turtle Newsl.* 49: 13–14.

Masood, E. 1997. Aquaculture: a solution or source of new problems? *Nature* 386: 109.

Mast, R. 1999. Guest editorial: common sense conservation. *Mar. Turtle Newsl.* 83: 3–7.

Metrick, A. and M.L. Weitzman. 1998. Conflicts and choices in biodiversity preservation. *J. Econ. Perspect.* 12: 21–34.

Meylan, A.B. 1988. Spongivory in hawksbill turtles: a diet of glass. *Science* 239: 393–395.

Meylan, A.B., B.W. Bowen, and J.C. Avise. 1990. A genetic test of the natal homing versus social facilitation models for green turtle migration. *Science* 248: 724–727.

Meylan, A.B. and Donnelly, M. 1999. Status justification for listing the hawksbill turtle (*Eretmochelys imbricata*) as Critically Endangered on the 1996 IUCN Red List of Threatened Animals. *Chelonian Conserv. Biol.* 3: 200–224.

Meylan, A.B. and D. Ehrenfeld. 2000. Conservation of marine turtles. Pp. 96–125 in: Klemens, M.W. (Editor), *Turtle Conservation*. Smithsonian Institution Press, Washington, DC.

Milon, J.W., C.M. Adams, and D.W. Carter. 1998. Floridians' attitudes about the environment and coastal marine resources. Technical paper 95. Florida Sea Grant Program, Gainesville, FL.

Mrosovsky, N. 1983. *Conserving Sea Turtles*. British Herpetological Society, London.

Mrosovsky, N. 1997. A general strategy for conservation through use of sea turtles. *J. Sustain. Use* 1: 42–46.

Mrosovsky, N. 2000. Sustainable use of hawksbill turtles: contemporary issues in conservation. Key Centre for Tropical Wildlife Management, Northern Territory University, Darwin, NT, Australia.

National Research Council. 1990. *Decline of the Sea Turtles: Causes and Prevention*. National Academy Press, Washington, DC.

Nietschmann, B. 1973. *Between Land and Water, the Subsistence Ecology of the Miskito Indians*. Seminar Press, New York.

Nietschmann, B. 1982. The cultural context of sea turtle subsistence hunting in the Caribbean and problems caused by commercial exploitation. Pp. 439–445 in: Bjorndal, K.A. (Editor), *Biology and Conservation of Sea Turtles*. Smithsonian Institution Press, Washington, D.C.

Parsons, J.J. 1962. *The Green Turtle and Man*. University of Florida Press, Gainesville, FL.

Pritchard, P.C.H. 2000. A response to Nicholas Mrosovsky's sustainable use of hawksbill turtles: contemporary issues in conservation. *Chelonian Conserv. Biol.* 3: 761–767.

Rebel, T.P. 1974. *Sea Turtles and the Turtle Industry of the West Indies, Florida, and the Gulf of Mexico*. University of Miami Press, Coral Gables, FL.

Roughgarden, J. 1979. *Theory of Population Genetics and Evolutionary Ecology: An Introduction*. MacMillan, New York.

Salmon, M., B.E. Witherington, and C.D. Elvidge. 2000. Artificial lighting and the recovery of sea turtles. Pp. 25–34 in: Pilcher, N. and Ismail, G. (Editors), *Sea Turtles of the Indo-Pacific*. ASEAN Academic Press, London.

Shivlani, M.P., D. Letson, and M. Theis. In press. Visitor preferences for public beach amenities and beach restoration in south Florida. *Coastal Manage.*

Slobodkin, L.B. 1974. Prudent predation does not require group selection. *Am. Nat.* 108: 665–678.

Soulé, M.E. 1980. Thresholds for survival: maintaining fitness and evolutionary potential. Pp. 151–170 in: M.E. Soulé and B.A. Wilcox (Editors), *Conservation Biology: An Evolutionary-Ecological Perspective*. Sinauer Associates, Sunderland, MA.

Spash, C. et al. 2000. Lexicographic preferences and the contingent valuation of coral reef biodiversity in Curaçao and Jamaica. Pp. 97–117 in: K. Gustavson, R.M. Huber, and J. Ruitenbeek (Editors), *Integrated Coastal Zone Management of Coral Reefs: Decision Support Modeling*. The World Bank, Washington, D.C.

Thorbjarnarson, J.B. et al. 2000. Human use of turtles: a worldwide perspective. Pp. 33–84 in: Klemens, M.W. (Editor), *Turtle Conservation*. Smithsonian Institution Press, Washington, D.C.

Tinker, A.M., B.D. Merino, and M.D. Niemark. 1982. The normative origins of positive theories: ideology and accounting thought. *Account. Organ. Soc.* 7: 167–200.

Tisdell, C.A. 1991. *Economics of Environmental Conservation: Economics for Environmental and Ecological Management*. Elsevier, Amsterdam.

Tisdell, C.A. 2002. *The Economics of Conserving Wildlife and Natural Areas*. Edward Elgar Publishers, Cheltenham, U.K.

Tisdell, C. and C. Wilson. 2001a. Wildlife-based tourism and increased support for nature conservation financially and otherwise: evidence from sea turtle ecotourism at Mon Repos. *Tourism Econ.* 7: 233–249.

Tisdell, C. and C. Wilson. 2001b. Tourism and the conservation of sea turtles: an Australian example. Pp. 356–368 in: C. Tisdell (Editor), *Tourism Economics, the Environment and Development: Analysis and Policy.* Edward Elgar, Cheltenham, U.K.

Tisdell, C. and C. Wilson. In review. Perceived impact of ecotourism on environmental learning and conservation: turtle watching at an Australian site. *Environ. Conserv.*

Tisdell, C. and C. Wilson. In press. Ecotourism for the survival of sea turtles and other wildlife. *Biodivers. Conserv.*

Vieitas, C.F. and M.A. Marcovaldi. 1997. An ecotourism initiative to increase awareness and protection of marine turtles in Brazil: the Turtle by Night Program. *Chelonian Conserv. Biol.* 2: 607–610.

Whitehead, J.C. 1993. Total economic values for coastal and marine wildlife: specification, validity, and valuation issues. *Mar. Res. Econ.* 8:119–132.

Whittow, G.C. and G.H. Balazs. 1982. Basking behavior of the Hawaiian green turtle (*Chelonia mydas*). *Pac. Sci.* 36: 129–139.

Wilson, C. and C. Tisdell. 2001. Sea turtles as a non-consumptive tourism resource especially in Australia. *Tourism Manage.* 22: 279–288.

Wilson, E.O. and F.M. Peter, Editors. 1988. *Biodiversity.* National Academy Press, Washington, D.C.

Witherington, B.E. 1992. Behavioral responses of nesting sea turtles to artificial lighting. *Herpetologica* 48: 31–39.

Witherington, B.E. 1997. The problem of photopollution for sea turtles and other nocturnal animals. Pp. 303–328 in: J.R. Clemmons and R. Buchholz (Editors), *Behavioral Approaches to Conservation in the Wild.* Cambridge University Press, Cambridge, U.K.

Witherington, B.E. and R.E. Martin. 2000. Understanding, assessing, and resolving light pollution problems on sea turtle nesting beaches. FMRI Technical Reports TR-2. Florida Marine Research Institute, St. Petersburg, FL.

Woody, J.B. 1986. On the dollar value of the Oaxacan ridley fishery. *Mar. Turtle Newsl.* 36: 6–7.

15 Practical Approaches for Studying Sea Turtle Health and Disease

Lawrence H. Herbst and Elliott R. Jacobson

CONTENTS

0-8493-1123-3/03/$0.00+$1.50

15.1 INTRODUCTION AND BACKGROUND

Interest in health and disease of sea turtles has increased along with a general interest in wildlife and environmental health. Dramatic epizootic events such as marine turtle fibropapillomatosis (FP), regional coral die-offs, toxic algal blooms, and amphibian population declines as well as concern for the effects of pesticides, industrial contaminants, and climate change on human and wildlife populations have spurred an interest in incorporating health assessment and disease surveillance into population monitoring programs.

As these programs are developed and implemented, it will be important to gain an appreciation of the potential role that pathogens and infectious diseases may have as primary mortality factors in the population ecology of these species. For some wildlife ecologists, the concept of infectious disease is traditionally understood as an epiphenomenon or secondary process that follows a primary environmental stressor, such as resource depletion. The presumption is that through host–parasite (pathogen) coevolution, a normal unstressed host will tend to be resistant to disease from infectious agents.

Although this conceptual view may hold true for diseases caused by opportunistic pathogens, a broader understanding of host–pathogen interactions recognizes that there are theoretical conditions under which natural selection would not drive host and parasite coadaptations toward a less antagonistic relationship (Ewald, 1993; May and Anderson, 1983). Furthermore, even in situations where selection does drive the relationship toward low virulence, the relationship is probably not an evolutionarily stable strategy in that the system remains susceptible to invasion by highly virulent strains that gain a tremendous short-term fitness advantage (Maynard-Smith, 1976). Given that new and highly virulent strains can evolve and spread rapidly at a higher rate than a vertebrate host's ability to respond, there will always be the possibility that an infectious agent is a primary morbidity–mortality factor, stressing and killing otherwise healthy sea turtles. Furthermore, the human impact on our environment is greater today than ever before, and in both subtle and not such subtle ways, humans may be affecting the spread of pathogens throughout the world. Thus, it should be assumed that new diseases may appear and a condition that is sporadic one year may become catastrophic the next. Consequently, there is value in investigating the pathophysiology of disease (disease research), in monitoring for disease and health problems, and in preparing at some level to cope with disease outbreaks.

Health assessment of sea turtles is based upon methods and procedures used in evaluating other animals, including other chelonians. However, much work needs to be done to establish better methods for assessing health of individuals and populations of sea turtles. Parameters need to be defined to build a database that can be used in assessment. Although some good information is available on infectious and noninfectious diseases in sea turtles in captivity, relatively little is known about diseases in wild populations (George, 1997; Herbst and Jacobson, 1995; Lauckner, 1985). Overall, the pathophysiology and pathogenesis of sea turtle diseases have been poorly studied. Therefore, there remains a tremendous need for basic research involving health assessment and disease of sea turtles.

The purpose of this chapter is to provide a conceptual framework and some practical advice on how to approach health and disease problems in a logical and systematic manner. Any successful program depends upon carefully recorded systematic observations, data and sample gathering, preservation, and analysis and interpretation. The ability to assess health of sea turtles and determine causes of illness and death is highly tied to resources at hand. Our attempt here will be to identify those tools that are currently in use, and it is hoped that these can be adapted or modified by readers who may not have similar resources at their disposal. Limitations of current methodologies will be pointed out, and those that are in need of improvement will be mentioned. The tools and methods used in health assessment of any species will improve as we better understand the biology of the animal and as new technologies allow us to build upon our diagnostic repertoire.

This chapter is organized into three sections. The first section discusses various situations in which medicine or health assessment will be relevant. The second outlines and discusses general systematic approaches to health assessment and disease investigation. The third section discusses the cost–benefit considerations and other practical issues that must be taken into consideration before and during an investigation.

15.2 SITUATIONS INVOLVING SEA TURTLE MEDICINE

15.2.1 Health Assessment vs. Disease Investigation

Health is defined as the "overall condition of an organism at a given time" and as "freedom from disease or abnormality" (*Stedman's Medical Dictionary*, 2001). The state of being healthy is defined as "possessing good health." These definitions presume that there is some standard measure of overall condition, the means to determine "freedom from disease or abnormality," and a subjective judgment of what is "good." Health assessment, therefore, can mean different things to different people. Nevertheless, as mentioned above, there is value in trying to evaluate the health status of individuals and populations (herd health), and to make comparisons over time within and among populations. The purpose of a health assessment program is to evaluate the overall condition and to detect abnormalities and disease in individuals, and to detect changes in prevalence of disease or abnormalities in populations. This process can identify situations that merit further investigation, but its primary purpose is description and monitoring.

Implicit in the health assessment process is the establishment or availability of normative data, i.e., determining the range of conditions to be found in apparently healthy animals within a population, so that deviations can be recognized. This can include normal ranges for quantitative physical, physiologic, and biochemical parameters as well as background frequencies (prevalence) for infections or exposures — i.e., to what agents the population is exposed. Making an assessment requires familiarity both with disease and with what is normal. Some parameters such as blood biochemical values can be quantitated and can be statistically treated to define "reference ranges." Health assessment also has subjective aspects that are dependent on the experience of the person performing the assessment. Health assessment also is confined to a specific time point at which an animal is evaluated. Drawing inferences from these data about the future health of animals or populations also requires some knowledge about the risks associated with specific conditions.

There is no single currency for assessing health status, and therefore, assessment of health is circumscribed by how thoroughly the patient is examined, what parameters are evaluated, and which tests are conducted for specific conditions or diseases. Consequently, health assessments should be characterized in the most specific objective terms possible. Characterizations such as "healthy," "sick," or "stressed" are too vague and impossible to interpret or compare without knowing the parameters that were measured to define them. Furthermore, although the parameters that are selected will provide some useful information about health status, one must remember that much information relevant to this assessment will remain unknown.

In contrast to health assessment, disease investigations have very specific goals to further characterize disease processes and identify the cause(s), source, and contributory factors that are responsible for certain abnormal findings and diseases that are recognized in individuals and populations. Whereas health assessment may identify problems, disease investigation seeks to understand the basis for these problems.

15.2.2 Individual vs. Population Health

There is a distinction between health assessments of individuals versus health assessments of populations. When discussing health assessment, one usually is referring to individual health. Population health ultimately is dependent upon the health of individuals, but evaluating all individuals in a population is impossible. A population of turtles at any given time will include individuals that have never been exposed to a particular pathogen, toxin, or other disease-causing agent; individuals that have been exposed but were resistant to infection or toxicity; individuals that were infected or intoxicated but have fully cleared the infection or toxin and are no longer exposed; and individuals that are currently colonized, infected, or exposed to the toxin. In the last group of exposed individuals, some may not develop any pathology, others may develop a disease process or have tissue damage that remains subclinical, whereas others develop overt clinical disease, and some of these animals die. Understanding health at the population level requires being able to detect individuals in each of these categories, to describe their distribution over various age/stage classes at any given time, and to detect changes in their frequency distribution over time.

A critical component of population health is the overall abundance and age–stage structure of the population. This is information that population ecologists and conservation biologists need to determine whether there is adequate recruitment to the population and whether the population is stable, increasing, or declining. The population sampling methods and life history models that are needed for population assessment are beyond the focus of this chapter. Suffice it to say, however, that individual health and health risk assessments must be integrated into these studies to evaluate the true impact of disease on populations. The marine environment and life history of sea turtles make population assessment especially complex and difficult to monitor. Loss of individuals from the population may not be appreciated until there is sufficient decline to affect sample estimates. Increased mortality may be seen as increased numbers of stranded turtles, but one can only speculate on the true impact on the population unless monitoring can be performed in relatively confined areas.

15.2.3 Captive vs. Free-Ranging Turtles

The range of health problems that will be encountered in captive animals can differ greatly from those encountered in free-ranging animals. The clinical manifestations, magnitude, and severity of any particular health problem may also vary markedly between captive and wild animals. Both situations, however, have a role in turtle health and disease studies.

Compared to the free-ranging condition, captivity presents relatively confined living space and artificially high animal densities that, even with the best husbandry programs, will enhance the transmission of contagious infectious agents, in a density-dependent process. The confined living quarters can accumulate high levels of environmentally persistent parasites and pathogens as well. Confinement and crowding also contribute to stress, which can alter a turtle's resistance to disease. Captivity may also bring together animals from different parts of the world or species that may never come together in the wild. Where the animal husbandry program is suboptimal, poor nutrition, poor water quality, and poor sanitation and infection control procedures multiply the risks of transmission and disease.

Disease in all animals can exist in a subclinical state. That is, although an animal might appear to be healthy, a significant problem may be ongoing internally. Sea turtles with chronic illness that would probably die in the wild may live for extended periods in captivity. Thus, captivity provides a favorable environment for subclinical diseases (undetected in apparently healthy animals) to manifest themselves clinically (sick animals), for latent infections to recrudesce, and for otherwise innocuous opportunistic agents to cause disease. It is not surprising that many of the known sea turtle diseases and infectious agents were first observed and in some cases only observed in outbreaks among captive animals (Herbst and Jacobson, 1995). Examples include gray-patch disease (Rebell et al., 1975), lung–eye–trachea (LET) disease (Jacobson et al., 1986), and chlamydiosis (Homer et al., 1994).

Although the unnatural conditions of captivity can result in disease syndromes that are unlikely to be seen in the wild (e.g., growth anomalies resulting from imbalanced nutrition [George, 1997]) and therefore of limited interest to students of ecosystem and wild population health, it is equally likely that most of the infectious agents that will cause disease in captivity have their source in the wild and were introduced into captive collections through inapparently affected animals. Thus, what is learned from captive animals may become extremely valuable in the face of an epizootic in the wild population. For example, FP was first described in captive green turtles at the New York Aquarium in 1938, but was not recognized as a significant threat (Smith and Coates, 1938). In the mid 1980s, however, when FP emerged as a worldwide problem in green turtles, these early descriptions became extremely valuable for clinicians trying to understand the disease (Herbst, 1994). Similarly, LET disease was first described at Cayman Turtle Farm (Jacobson et al., 1986). The herpesvirus that was found to be associated with this disease in captivity has not yet been isolated in wild turtles with similar clinical signs. However, there is now a body of serologic evidence that wild green and loggerhead turtles are exposed to this virus (Coberley et al., 2001a; 2001b). Furthermore, marine turtles may be kept in zoos, aquaria, and rehabilitation centers as educational and tourist exhibits, and also in large numbers as part of captive breeding, farming, and "head-start" programs. In situations in which captive animals may be released to the wild, their health problems may directly impact wild populations (Jacobson, 1996).

Captivity provides a number of advantages in the study of marine turtle health and diseases. First, because diseases are likely to occur, and occur with high incidence, captivity provides an excellent opportunity for discovery and description of new diseases and infectious agents if the animal care program involves adequately trained and observant professional staff, including a consulting veterinary clinician and pathologist. Captive collections allow for ready access to animals, intensive monitoring with longitudinal observations and repetitive sampling of individual turtles, and thorough diagnostic workups that include access to sophisticated diagnostic tools. Thus, the opportunity for detailed investigation is very good. Second, turtles in captivity may provide access to life stages such as pelagic posthatchlings and juveniles that are very difficult to observe and sample in the wild. Infectious agents that may only cause clinical disease and mortality in a specific susceptible life stage may not be observed among free-ranging animals because of the improbability of recovering ill and dead animals in the field. Third, captive collections provide a resource for development and improvements in diagnostic tests and procedures, and improvements in treatments, either through planned clinical research or empirically through practice.

The study of disease processes occurring in wild marine turtle populations, on the other hand, is extremely important because conservation efforts are aimed at protecting and managing viable free-ranging stocks. Certain diseases and infections, especially parasitic infections, are more likely to be seen in wild populations because quarantine procedures and prophylactic treatments given to captive turtles may remove ecto- and endoparasites and disrupt complex parasitic life cycles. The natural environment also provides the full range of factors and variables that may be important in diseases that have complex etiologies. It is important for one to appreciate the extent and severity of diseases in sea turtles in their natural environment: to know what is "out there" as a reality check. One must always be aware, however, that biased observation and sampling of wild populations may reinforce the perception that primary disease is rare in wild populations.

Unfortunately, disease problems in wild sea turtles have been poorly studied. Those that have been best investigated are diseases that have a dramatic presentation or have resulted in epizootics (e.g., FP). Those animals that die in small numbers are probably never seen. Even with stranded turtles that offer a high potential for examination of ongoing background disease and detection of new problems that are emerging in a population, little money and resources have been expended on this valuable source of information.

15.2.4 Mass Morbidity–Mortality Events vs. Sporadic–Incidental Problems

In a mass morbidity–mortality event, it is easy to appreciate the potential for impact on a population or species, and investigation of these events takes on high priority. Investigations, aimed at characterizing the event and identifying causative and contributory factors, may be performed in a more systematic way, involving expert working groups and coordinated centralized data management, sample routing, and archiving. Such events, however, may quickly overwhelm the available resources, and opportunities may be lost because of lack of preparation or timely response. The magnitude of the event may also stimulate disjointed efforts by several independent groups which can result in poor information-sharing, duplication of efforts, incomplete workups, and use of different methodologies that make later data comparisons impossible. A mass event provides a series of animals and a range of clinical presentations and varying severities, which allow a more thorough characterization of the event and more opportunities to discover all the factors involved. Multiple opportunities exist to obtain specific samples and to perform diagnostics, although not always on the same animal.

Sporadic–incidental problems, on the other hand, may seem less important. However, these cases may provide the first opportunity to document a disease condition that may later cause a mass morbidity–mortality event. Furthermore, among free-ranging turtles, what may appear on the surface to be a sporadic, incidental, or mild condition may in fact be the "tip of the iceberg" — a condition that is having far more serious impact than appreciated because turtles with severe disease are lost to predation and only the less affected animals are observed. Limited accessibility to turtles in certain habitats and especially to early life history stages exacerbates this problem. Sporadic cases are a challenge because the primary observer may lack the training to recognize them, the understanding and experience to recognize their potential significance, or the interest to record observations and collect materials. Many of these cases therefore may be worked up in a very haphazard way, if at all, depending on the interest level and experience of the observer as well as the availability of funds and resources to conduct these investigations. These individual cases, however, sometimes provide the best material for thorough workup, especially if the animal can be brought to a clinic with appropriate facilities and expertise. The value of careful observation and documentation, and a systematic approach, is as great for these infrequent cases as for mass events.

15.3 SYSTEMATIC APPROACHES

15.3.1 Health Assessment

An individual and population health assessment program can provide very useful information, if it is conducted in a systematic manner. As stated above, the purpose of health assessment is to describe the condition of an organism or group of organisms at a specific time. Obviously, by definition any health assessment program should identify individuals that are exhibiting clinical illness or injury. However, although turtles with overt disease may be easy to recognize, those with low-grade and subclinical disease processes are often a challenge to identify. What other observations, measurements, and tests can be included in health assessment, and how are the data and results interpreted? Condition indices have been attempted and promoted for use in assessing health of chelonians, but these can be used as only one method in an array of diagnostics routinely employed in health assessment (Jacobson et al., 1993).

15.3.1.1 Goals and Limitations

There is always a desire to make a health assessment program as comprehensive as possible, but this is rarely feasible; it is important to develop a rationale for including certain types of evaluation and excluding others. It is important to recognize up front that it will not be possible to evaluate all body systems, both functionally (physiology) and structurally (anatomy). It is generally more valuable to do few things well than to try to do too many things, all poorly. At the outset, the purpose and goals of the health assessment program should be defined. Knowing why things are being done helps to guide selection of methods and tests.

The following major goals should be considered when designing a health assessment program.

1. Establish normative reference ranges for the species or population for any of the anatomic and physiologic parameters and analytes of interest. These values will show both interspecific and intraspecific variation. Intraspecific variation may occur with age, sex, season, and diet, and reference ranges may need to be established for each subpopulation.
2. Establish a pathologic database (including serology and toxicology) for the species or population being studied. This will allow an estimation of the background prevalence of specific disease conditions, toxin levels, and infections in the population at a given time. This provides a reference for recognizing the most significant lesions in dead or stranded turtles and for recognizing changes over time.
3. Establish a surveillance program to monitor the population through time, including trends and spikes in prevalence (epizootics) or the introduction of new pathologic agents to a population.
4. Evaluate the relationships between various environmental and demographic factors and specific health parameters and pathologic conditions. Testing hypotheses about the association of specific abnormalities, diseases, and

pathologic conditions, either with environmental factors such as habitat type, diet, water temperature, and season or with specific known events such as oil spills and algal blooms, will indicate areas for further research to investigate possible pathophysiologic mechanisms.

15.3.1.2 Test Selection

Decisions regarding what tests and procedures to include in a health assessment program are critical because, as stated, these parameters define the depth of the assessment. Health assessment will be as good as the diagnostic tools that are used, the reference ranges that are available for the species being studied, and the skills of the investigator at recognizing turtles with abnormal signs and interpreting test results. The range of diagnostic tools that can be used will be narrower in the field situation than in a laboratory of a veterinary clinic.

Minimally, any health assessment program should include baseline morphometric data and a physical examination (discussed in Section 15.4.4). Screening tests should be included if possible. When the purpose of the study is to establish reference ranges for specific parameters, these basic observations and data are needed in evaluating individuals for inclusion in or exclusion from the reference population, and the definition of the reference population will include the criteria used to select them as "normal" (Walton, 2001b). It is difficult to give specific recommendations beyond this because test selection will be based on the specific health questions and hypotheses of interest.

There are, however, general considerations in selecting tests and parameters, study design, and interpretation. One should have a basic physiological understanding of the value and limitations of a specific test — i.e., what the results can indicate about the animal and, equally important, what they cannot. No single test will give a complete answer regarding the health status of an animal. Although each test may provide specific objective information, at best, results will indicate a range of possible explanations. One should be aware of other tests that may be needed to confirm a test result or to support a particular interpretation, and consider incorporating these in a tiered approach. In a disease investigation, the significance of individual test results will be integrated with the results of other supporting data and interpreted in light of the animal's clinical condition. Interpreting health parameters in a population of apparently healthy individuals is more problematic.

15.3.1.3 Interpretation of Out-of-Range Data and Positive Test Results

For tests that yield quantitative data, such as cell counts, enzyme activities, and analyte concentrations, results are interpreted relative to a reference range for that population. A critical factor in interpretation is that reference ranges should be representative of the population being assessed (Walton, 2001b). There is a high probability of misinterpreting a result as abnormal if the reference range is inappropriate. For example, available reference ranges for blood biochemistry parameters for all turtle species are quite limited, so interpretation of blood values from an individual turtle is often based on extrapolation from other species and limited data sets. In addition to species differences, distinct normal populations may be discriminated by differences in age, sex, season, reproductive condition, and genetic background. For example, Bolten and Bjorndal (1992) found that among juvenile green turtles, several plasma analytes varied significantly with body size, whereas others such as uric acid and cholesterol differed between the sexes. Similarly, the normal values for plasma calcium of adult female sea turtles vary depending on their reproductive condition. As the number of samples tested increases, the ability to find statistical significance in small differences between means and variances also increases (Zar, 1974). These differences may or may not be biologically relevant.

How samples were collected, transported, stored, and processed; the analysis method and specific laboratory procedures, equipment, and reagents used; and how well the assay was optimized and validated for the species being tested all affect the interpretability and comparability of test results (Meyer et al., 1992; Walton, 2001a; 2001b). Values for several plasma biochemistry parameters, for example, varied significantly when duplicate samples from loggerhead turtles were analyzed on two different automated machines (Bolten et al., 1992). Thus, it is important for a study that all samples be collected, handled, processed, and analyzed in the same way, preferably in batches in the same laboratory using the same equipment and reagents, and sometimes even analyzed by the same technician. Each laboratory should develop its own reference ranges for each species. The issues and methodologies involved in establishment of reference ranges and validating assays are discussed in depth by Walton (2001a; 2001b).

Reference ranges are statistical constructs, defined as the maximum and minimum values between which a specified proportion of the population frequency distribution will be found. Inevitably, this means that some individuals in a normal population will fall outside the reference range by chance alone. For example, for data that have a gaussian (normal) distribution and a reference range defined as two standard deviations above and below the mean, only about 95% of the population will fall within the reference interval. Thus, in a sample of 100 turtles, 5 animals can be expected to have values more extreme (either greater or less) than these limits, and yet be completely normal, healthy individuals with respect to that parameter.

For tests that yield categorical positive or negative readouts such as serology, microbiological culture, and polymerase chain reaction (PCR), the performance characteristics of the test on the basis of its ability to discriminate true positive from true negative samples (specificity and sensitivity) must be considered (Weisbroth et al., 1998). The sensitivity of a test is the ability of the test to detect the true positives in a population. It is that proportion of the population that is truly positive that yields positive test results. The proportion that tested negative is false negative. The more sensitive the test, the fewer false negatives will result. The specificity of the test measures the ability of the test to recognize the true negatives in a population, and is the proportion of the population that is truly negative that is detected as negative by the test. The more specific the test, the fewer false positives will result. When either of these values is less than 100%, the predictive value of the test (i.e., how much confidence can be placed in the result being true) will vary, depending on the true prevalence of the condition in the population. Predictive value of a positive result is the proportion of all animals that test positive that really are positive. In general, the less common the condition, the less predictive value a given test has and the less confidence can be placed in the result. For example, if the true prevalence of a given condition is 50%, a test with 95% specificity and 100% sensitivity will yield 2.5 false positives among 100 animals tested, and the predictive value of the test will be 95%. If, however, the true prevalence in the population is only 5%, then 4.75 false positives are expected and the predictive value declines to only 51%. That is, only 51% of the positive test results can be interpreted as being correct.

These statistical artifacts are amplified when a battery of independent tests are performed. Because each test has its own independent probability of being found out of range or false positive, the overall probability of finding at least one normal individual that will have abnormal test results increases with the number of tests performed. Similarly, when comparing different sample populations to one another or to a reference distribution, the chances of finding a statistically significant difference increases with the number of independent pair-wise comparisons that are made. Thus, interpreting the sporadic positive test, out-of-range result, or statistically significant difference between sample populations becomes somewhat of an intuitive skill, and is especially difficult when one is surveying an apparently healthy population for conditions that are rare. A strong argument can be made for using the best tests (high specificity and sensitivity), testing the most closely matched reference population possible, and employing confirmatory tests when available to help distinguish false positives from true positives (Weisbroth et al., 1998). When a diagnostic test is used to monitor a population for the introduction of a known disease or infection, or to maintain some level of confidence that the population is free of a specific disease, it is especially critical to employ confirmatory tests if the surveillance data will be used to support management decisions involving the culling of positive animals or quarantine of populations.

15.3.1.4 Interpretation of Within-Range and Negative Results

When quantitative test values are compared to an inappropriate reference population, values that are actually abnormal may be misinterpreted as being within range. Interpretation of within-range and negative test results also must consider the sensitivity of the test — its ability to identify all the abnormal individuals (true positives) in the population. Many tests that are used as screening tools are set up to maximize sensitivity, thereby minimizing false-negative results. Nevertheless, test results that fall within the normal reference range do not necessarily mean that there is not a problem. Some tests, such as certain blood biochemistry assays, are relatively insensitive to the underlying disease processes. In many cases, a threshold level of ongoing tissue damage or loss of function must be reached before abnormalities are detected on a particular test parameter (Meyer et al., 1992). Because many organ systems have redundant physiologic capacity, significant pathology and loss of organ function may go undetected when certain tests are used. For tests that yield categorical results, there are limits of detection inherent to the method that affect sensitivity. For example, PCR in theory may be able to detect a single virus genome in a sample, but in practice, it may require ten or more viral particles to be present (Persing,

1993). Negative-staining electron microscopy, on the other hand, is unlikely to detect viruses when there are fewer than 10^4 particles per microliter of sample.

When diagnostic tests such as serology are used to monitor populations to ensure that they are free of a particular agent, interpretation of negative test results must take into account the probability of detection (Weisbroth et al., 1998). Even when a test is able to detect every positive animal (100% sensitive), sample sizes must be adequate to ensure that a population is negative. The overall chances (P) of detecting a single positive animal will be a function of sample size (n) and prevalence (p) described by the equation, $P = [1 - (1 - p)^n]$. Thus, one can calculate the sample size needed for a particular level of probability of detection when the agent has a specific prevalence. For example, to have a 99% chance of detecting even a single turtle that is positive for antibodies to the FP-associated herpesvirus in a population that has a true prevalence of 40% requires that at least ten turtles be tested. If the true prevalence is only 10%, at least 40 turtles must be tested for the same degree of confidence. Presented another way, if only ten turtles are tested in a population that has a true prevalence of 10%, the herpesvirus would have a 35% chance of going completely undetected. Thus, the more rare the disease condition in the population, the more animals must be sampled to have a reasonable chance of detecting it. If one accounts for lower test sensitivities, the required sample sizes increase.

There are also several biologically important reasons why a test may fail to detect an abnormality or disease agent. The time that the diagnostic procedure was performed and the sample collected relative to the disease course is important. For example, it takes a certain period of time for turtles to mount an immune response against a pathogen. Thus, early in the course of infection, pathogen-specific antibodies may not be detected serologically. Some infectious agents replicate only during specific stages of the disease and sometimes can be found in different tissues at different stages. Therefore, tissue samples collected too early or too late in the course may yield negative results. Furthermore, in severe disease under certain circumstances, values for a particular assay that is typically a sensitive indicator of a disease condition may be found to be within normal limits. For example, the white blood cell count, a sensitive indicator of an active inflammatory response to infection, may yield counts within the normal range if a turtle is losing cells from the circulation faster than it is able to replace them. Similarly, the elevation of certain liver enzymes in blood indicates liver cell damage, but the levels could be within normal limits in chronic active liver disease if sufficient liver parenchyma has already been lost.

Many factors related to sample quality, preparation, storage, handling, and contamination could affect test results in either direction. For example, exposure of a plasma sample to light degrades bilirubin, falsely lowering its measured concentration. Contamination of plasma with hemolyzed blood causes marked elevation in several enzymes and interferes with colorimetric measurements of some analytes (Meyer et al., 1992). Plasma samples that have been repeatedly thawed and refrozen have decreased enzyme activities and lower specific antibody titers.

There is a significant additional problem in interpreting the biological and clinical relevance of some tests (especially certain blood biochemistry values) for sea turtles. Many of the analytes tested in blood biochemistry panels were selected for their clinical relevance to humans and some domestic species. Even among different species of mammals, the utility of specific plasma enzymes as biomarkers of function or injury in particular organs or tissues varies (Loeb and Quimby, 1989; Meyer et al., 1992). This is partly related to the tissue origin of the predominant isozymes found in the blood and the degree to which these blood levels change in response to tissue injury. In dogs and cats, for example, aspartate aminotransferase (AST) and alanine aminotransferase (ALT) are useful markers for liver status, because the isozymes expressed in liver contribute 90% of the circulating enzyme activity. Conversely, in horses and ruminants, the predominant source of plasma AST and ALT is skeletal muscle (Meyer et al., 1992). Basic research into the clinical relevance of available tests for each species of sea turtle is needed.

15.3.2 A Basic Health Assessment Program

Given the complexities and caveats discussed above, there is still a strong rationale for developing health assessment programs and including health assessments routinely in other field studies that involve the capture and handling of turtles, even if the primary purpose of the study is not health assessment. Because sea turtles are encountered and handled frequently, the turtle biologist is an essential front-line person in a general surveillance program for emerging health problems. Some fairly straightforward and field-friendly techniques are required that will not be burdensome to the field researcher, but will provide useful information that can be compared broadly across studies. Outlined below is what we consider to be both important and feasible for most field studies. More sophisticated programs can build upon this basic foundation.

15.3.2.1 Capture Data

Certain field data that are collected routinely in any turtle study provide important background information in health assessment. These data include locality, date, and time of effort; observation/capture methods used; weather; water conditions (temperature, tide); time and location of observation or capture of individual turtles; species; age–size class (based on size measurements); and sex (if adult). Important summary data for each sampling session include duration of effort, total number of turtles of each species that were captured or observed, and number that were considered to have a health problem (below).

15.3.2.2 Behavioral Evaluation

It is important to record the turtle's behavior prior to capture, if possible. For example, was the turtle swimming, basking, or crawling normally, or was it found floating or entangled? Did the turtle make a vigorous effort to elude capture or escape, or was it "listless"? After being captured and landed, was it alert and responsive to stimuli or weak and unresponsive? Did the turtle have symmetrical use of its head and limbs? A basic neurologic examination can be performed to assess both peripheral and central nervous system (Chrisman et al., 1997).

15.3.2.3 Body Mass

We strongly recommend that body mass be measured along with routine morphometric measurements, such as carapace length (CL) (Bolten, 1999). For health assessment, these objective data can be used to produce body condition indices that can provide a broad measure of how the animal is faring, and can be compared across studies and field sites. For example, either the ratio of body mass to CL or the ratio of mass to estimated volume such as (CL^3) could readily be compared among individuals and across studies.

15.3.2.4 Physical Examination

While the turtle is handled for measurements and tagging, a thorough external physical exam should be performed. The limbs, skin, carapace, and, plastron should be examined to determine whether they are intact or have defects (e.g., cuts or scars). For example, is the shell smooth or does it appear to have delaminating or missing scutes, which could be a sign of either shell infection or serious systemic disease? The skin and shell should be examined for lumps and abnormal growths. The abundance and types of epibiota (commensals and ectoparasites) should be noted. The eyes should be examined to determine whether they are intact and clear. Any obvious indications of entanglement or other fisheries interactions should be described. The cloaca and oral cavity should be examined to identify hooks, line, or lesions. Color and amount of mucus present as well as any odor should be noted. Abnormalities should be described using the most objective and precise terms possible, and illustrated with drawings and measurements. For example, a large, raised, firm, and smooth swelling on the skin of a turtle should not be identified as FP; even though the word "tumor" may be appropriate, it commonly evokes an interpretation of neoplasia. Such a raised mass could be neoplasia, an abscess, granuloma, cyst, scar, or other anomaly. If the turtle is tagged and released, at least there is documentation and objective description of the abnormality.

Because the accurate description and documentation of suspected abnormalities will be the most important component of any health assessment program, it is worth discussing data records. A field data sheet has been developed for health assessment of the desert tortoise, *Gopherus agassizzi* (Berry and Christopher, 2001), and a basic data sheet for stranded sea turtles is available (Shaver and Teas, 1999) that can be modified. Data sheets that are designed as questionnaires with clear "yes," "no," or categorical (multiple choice) answers for physical and behavioral examinations facilitate coding and data entry (Berry and Christopher, 2001). Categorical choices help keep descriptive data as objective as possible, and coding these data allows data management and development of descriptive statistics and a reference database. They also serve as mnemonic devices, prompting the investigator to look for specific details.

Data sheets should be designed and used in a way that clearly indicates whether a part was examined and whether an abnormality was observed. Missing data should not be misinterpreted as negative findings. For example, were the eyes and the oral cavity examined? The data sheet should also record whether specific samples (e.g., blood, biopsy, or ectoparasites) were collected. Data sheets containing line drawings that depict dorsal, ventral, and side views of a turtle can be used for noting location and relative size of specific external abnormalities and lesions. Pritchard and Mortimer (1999) provide excellent line drawings of each sea turtle species. Photographs can be valuable, but should not replace line drawings, because it is sometimes easier to interpret line drawings. Film photography may be problematic because of the additional notes needed to link photos to field notes, and because the quality of the photograph may not be readily apparent. Digital photography has made photography relatively simple and inexpensive. Images can be circulated electronically to individuals who may be able to render an opinion when an abnormality is recognized or when a question arises about an animal's appearance. In addition to the basic descriptive and morphometric data collected, the following additional procedures can be performed in the field and should be considered for incorporation in routine studies.

15.3.2.5 Blood Samples

Blood collection has not always been part of routine fieldwork, but because blood samples are easily obtained and can provide much valuable information, we strongly recommend that samples be collected. With the ability to determine the sex of immature turtles using plasma steroid hormone assays (Owens, 1999; Wibbels, 1999) and to perform genetic analysis on DNA derived from blood cells (FitzSimmons et al., 1999), it has become more commonplace for blood to be collected and archived. Additional health information can be obtained from this blood with a little extra effort. For example, blood smears can be prepared by spreading a drop of whole blood on a microscope slide. Blood smears provide a way to evaluate blood cell morphology and relative cell abundance, and smears can also be examined for blood-borne parasites. If adequately dried and fixed, the smears can be stored indefinitely at room temperature and examined at a later date. Blood cell counts can be performed on whole blood samples if they are transported on ice to the clinical pathology laboratory within 12 h. Preservative solutions need to be developed that maintain cell morphology and integrity for longer periods of time.

If blood is collected for any reason, an aliquot should be centrifuged to separate blood cells from plasma. The pelleted blood cells are a source of DNA, and the plasma can provide a resource for biochemistry and serology screening assays. Plasma should be removed from whole blood immediately to prevent artifacts, such as elevated potassium from cell leakage or decreased glucose because of cell metabolism, and either transported on ice for immediate analysis or archived at ultralow temperatures.

When blood is separated, measurement of the packed cell volume (PCV) is a simple-to-perform procedure that can provide additional health data (Herbst, 1999). PCV is the proportion of cells by volume in blood. A clinical benchtop centrifuge or microhematocrit centrifuge provides rapid separation of blood cells and plasma, and PCV can be measured in straight-walled tubes using a ruler or calipers. PCV is a robust indicator of health status, although the causes of low PCV may not be apparent. Because PCV will decrease in chronic debilitating diseases such as neoplasia, severe parasitism (leeches), and prolonged anorexia, the finding of a series of animals with low PCV could be reason to initiate a disease investigation.

More elaborate and specialized tests can be performed on fresh or frozen blood plasma and cells, but whether to incorporate these into an assessment program would depend on the specific goals of the study, costs, and feasibility.

15.3.2.6 Biopsy

In many cases, histological evaluation of a biopsy is the only way to distinguish various lesions. A biopsy may be especially important for sporadic or rare cases that are unlikely to be seen again. For example, the first suspected cases of FP in a population or species might provide compelling reason to collect a biopsy. Under routine field conditions, some cutaneous lesions can be safely biopsied for histological evaluation. One should obtain some rudimentary training and be prepared to attempt this procedure if it becomes necessary. A few small containers of 10% buffered formalin, disinfectant (povidone iodine), and sterile biopsy packs containing scalpel blades, forceps, and scissors should be kept on hand. Biopsies involving tissues that have higher risk of permanent damage (such as eyes, cloaca, or glottis) or deeper tissues should be attempted only by those with more specialized training and experience. A guide for performing biopsies and necropsies has been published (Jacobson, 1999) and is available on line at www.vetmed.ufl.edu/sacs/wildlife/seaturtletechniques.

15.3.2.7 Imaging

Techniques such as radiology, ultrasound, and laparoscopic imaging have been adapted to field use and have been most commonly used to evaluate reproductive status of turtles (Owens, 1999; Wibbels, 1999). With training and experience in recognizing normal anatomic structures, these techniques can certainly be adapted to evaluate other organ systems. For example, investigators who use laparoscopy to visualize the ovaries and ovary ducts of turtles could begin to examine the kidneys and intestinal surface for cysts, masses, adhesions, and perforations. It is unlikely, however, that these techniques would be routinely incorporated into field studies solely to evaluate health, because of the expense and expertise needed.

15.3.3 Systematic Approach to Disease Investigations

Disease investigation in sea turtles, whether sporadic (individual) or mass-event (population), and whether among captive or free-ranging animals, uses a basic approach that is used in all medical investigation, and constitutes a major component of the practice of veterinary medicine. Consequently, medical professionals, specifically veterinarians with training and experience in wildlife, reptile, and marine turtle medicine, should be involved in this process because they are trained in the art and science of diagnosis. An overview of the approach is presented so that nonmedical professionals can gain a perspective about the process. As illustrated in Figure 15.1, the approach is an iterative process that involves description and prioritization of problems, diagnostic planning (selection of tests), assessment, and integration of results, so that at each level, the pathologic processes are better characterized and possible alternative explanations are eliminated until a diagnosis is reached.

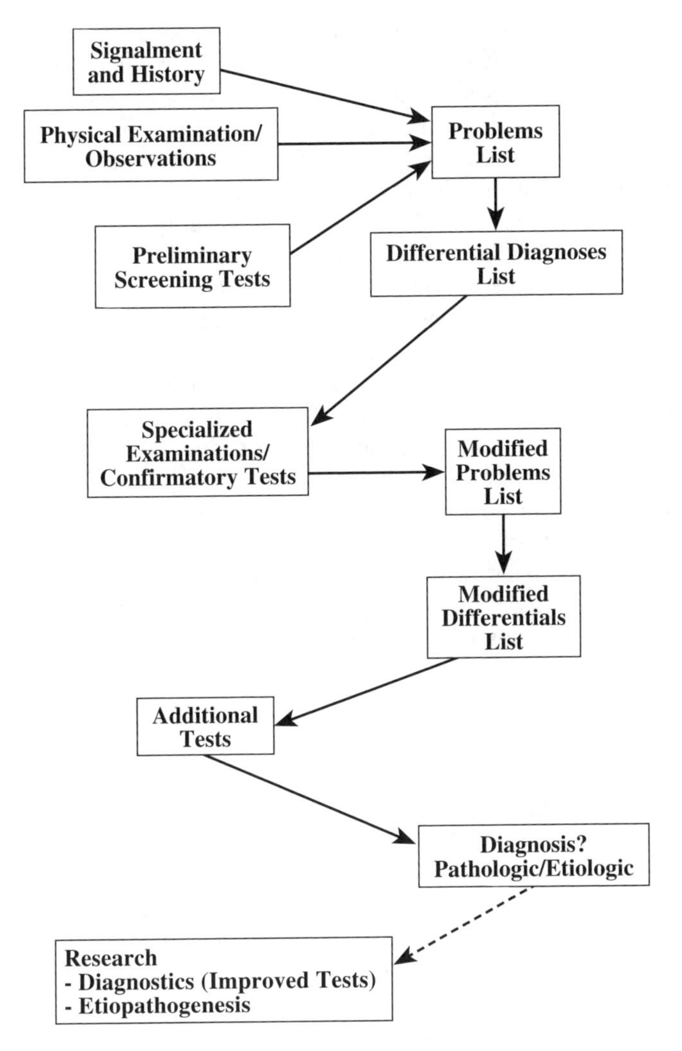

FIGURE 15.1 Outline of the disease investigative process.

The initial stages are the same as for health assessment (above), involving observation, description, and basic data collection, except that in this case there is a presenting problem, such as stranding or an abnormal finding on routine health assessment, that triggers the diagnostic investigation to determine what is causing animals to be sick or to die.

15.3.3.1 Signalment, Presenting Problem, and History

Signalment, presenting problem, and history are the preliminary data upon which all later data interpretation will rest and may suggest whether, on the basis of the clinician's experience, certain findings represent primary or secondary problems. Signalment is the specific information about the individual patient, including species, size, age, and sex (if known). The presenting problem is the abnormality that was observed. For example, the turtle may have a wound or a lump, have fishing line trailing from its cloaca, appear lethargic, or have a buoyancy problem. History refers to all the background information about the individual and how it was encountered. How many other turtles were affected out of those that were examined? Is this an ongoing problem within the population, and if so, for how long? If the turtle has been previously observed or has been recaptured, when was it last observed to be apparently normal? Has this individual had any other problems in the past? For population events, are other species affected concurrently in a similar way? Is there a known or recognized environmental event associated with the problem, such as a cold front or the opening of shrimp trawling season?

Knowledge of the timing between the environmental event and the stranding event or discovery of the presenting problem can help guide interpretation. For example, during a cold-stunning event, mass stranding and death of apparently healthy animals will occur and examinations and tests may show few abnormalities. Several weeks following such an event, however, stranding turtles may be ill and have a high prevalence of fungal pneumonias (George, 1997), and diagnostic tests may show numerous abnormalities (Carminati et al., 1994; Walsh, 1999). A similar mass stranding mortality event in summer might coincide with a toxic algal bloom or increased aerial spraying of pesticides.

15.3.3.2 Physical Examination (External)

As discussed under Section 15.3.1, careful observation, precise description, and documentation are essential. Even if such an assessment was performed in the field, this should be repeated by the health professional involved in the investigation upon presentation and periodically during the course of the study if the animal remains alive, to monitor for changes and identify new problems. If the turtle is alive, its attitude and behavior in and out of the water should be evaluated. Body condition (body weight versus size) provides clues about how long the animal has been ill. A turtle in good body condition probably is acutely ill or died acutely. Emaciated turtles have been chronically sick or are starving. For dead turtles, the physical exam is extended to include a complete gross necropsy (Jacobson, 1999). Any person involved in performing necropsies of sea turtles should gain basic training in turtle anatomy. *The Anatomy of Sea Turtles* (Wyneken, 2001) provides an excellent reference resource for this purpose.

15.3.3.3 Preliminary Screening Tests

These tests may include those that could routinely be performed in the field as part of a health assessment program. However, any screening tests that were performed in the field should be repeated in the clinic. Screening tests that should always be performed as part of the preliminary diagnostic workup include plasma biochemistry, hematology and differential white cell counts, survey radiographs, and fecal analysis. Preliminary assessments of these screening tests provide some indication of the organ systems that may be involved, and may identify problems such as ingested fish hooks and internal masses.

15.3.3.4 Problems List

From the history, physical examination, and preliminary screening test results, all the recognized problems are listed. These are the abnormalities that the veterinary clinician will seek to understand and treat. The problems list is usually prioritized so that those problems that are most threatening to the survival of the animal are addressed first, in terms of previously treatment and diagnostic workup. This requires some clinical judgment and experience.

15.3.3.5 Differential Diagnoses List

The next step involves the development of a list of alternative possible causes for each of the problems identified. In many cases, the cause of the problem may not be obvious, and there may be numerous possibilities. For example, ingestion of foreign bodies or debris, bowel perforation, bowel impaction, neoplasia, infections, and toxicity all can result in nonspecific problems such as weakness, emaciation, and floating. Even when there is an obvious factor such as trauma or massive FP to suggest the cause, there may be predisposing factors or ultimate causes that should be considered. For example, boat trauma may have occurred because of buoyancy problems associated with an infection, which may be secondary to cold stunning weeks earlier. Among the general categories of disease processes (developmental, autoimmune/allergic, metabolic, nutritional, infectious, trauma, toxicity [DAMNIT]), the clinician lists known conditions, which can be targeted for testing. Information from signalment, history, and the nature of the presenting problem can guide the clinician in prioritizing this list and deciding which processes are more likely to be involved, and which to try to rule out by further testing. In making these judgments, the veterinary clinician tries to integrate all the information so that if possible, most of the problems are explained by a single disease process.

15.3.3.6 Specialized Examinations, Procedures, and Secondary Tests

The goal of additional evaluations is to gather additional data to further characterize specific problems and to support one or another differential diagnosis (hypothesis) over others. An experienced medical professional will be able to determine what supporting data, additional diagnostics, and confirmatory tests are needed to rule in or rule out alternate hypotheses or to confirm a diagnosis. As discussed for health assessment, the interpretation of individual test results is always problematic; however, here all test results are integrated with the turtle's clinical presentation and abnormal findings can be further investigated. Specialized examinations and diagnostic procedures are selected to further assess specific organ systems. Procedures may include additional radiological imaging with or without contrast media, other forms of imaging such as magnetic resonance imaging and ultrasound, endoscopy, laparoscopy, and exploratory surgery.

15.3.3.7 Assessment of Results, Amended Problems and Differentials Lists, and Decisions

Throughout this process, decisions must be made about what additional tests to perform, and whether to begin supportive therapy, attempt to treat and rehabilitate, or euthanize and necropsy the animal. The goals of the investigation must be considered; i.e., is one trying to cure the individual turtle or learn more about its pathologic condition to help the population? These goals can be in extreme conflict. In some cases, euthanasia of a mildly affected turtle in the early stages of a disease can yield more information about the pathogenesis and etiology of the disease. In some cases, treatment may eliminate the etiologic agent or introduce artifact. On the other hand, response to the specific therapy may aid in diagnosis. Before a decision is made to treat, the investigative team should consider whether they have performed the tests and obtained the samples needed for further evaluation. For example, in cases of bacterial septicemia, blood microbiological culture performed on an aseptically collected blood sample is an important diagnostic test. Administration of systemic antibiotics prior to sample collection, however, could lead to false-negative culture results and a missed opportunity to isolate the bacterial pathogen.

Ideally, as results are evaluated and diagnosis proceeds, the list of differential diagnoses is amended and shortened until a definitive diagnosis is reached. In reality, however, the process may not get this far. In single or sporadic cases and turtles that are already dead, the stage of disease at the time of presentation may be dominated by secondary processes and particular samples may not have been collected at the optimal time or in the appropriate manner to achieve a diagnosis. In some cases, resources and availability limit the extent to which a case can be worked up. In many cases, several alternative explanations or hypotheses will remain because various tests fail to differentiate them, such tests do not exist for turtles, or the etiology is complex or the specific etiology is unknown.

Diseases generally fall into two broad categories. First are those that are relatively straightforward and easily elucidated. However, the elucidation may require evaluation of a series of affected individuals, even necropsy of several animals. This is most successful in situations where a large case series is available for examination or where the disease process is fairly well described in the literature, and recognized by its clinical presentation and diagnostics. In the second category are those diseases that are complex and have several causes that may work in concert to produce the clinical presentation seen. For many clinical problems and pathologic processes that can be described, the causes are yet to be identified. For example, algal toxins are suspected to cause die-offs in marine turtles, but this has not yet been substantiated in the literature for any marine turtle. Similarly, the full range of marine toxins that could be involved has yet to be identified.

At the very least, however, the systematic approach outlined yields a collection of objective data and observations (including the problems list), and a list of alternate possibilities. The descriptive or pathologic diagnosis will at least characterize the case so that future cases can be compared to it and one can plan how to proceed with similar case presentations in the future, so that answers that are more thorough can be obtained. The results of this process can also identify major questions for future research, including what types of diagnostic tests need to be developed to improve diagnostic capabilities.

15.4 COSTS–BENEFITS

In an ideal world, one would want to do the most thorough evaluation of every available animal and completely work up every necropsy or illness case. The reality is that resources (money, equipment, personnel, and, most important, time) are limited. Therefore, decisions must be made on the basis of time, money, and materials — how to get the most information using the resources that are available. The level of investigation often mirrors the extent of the problem. Historically, however, causes of morbidity and mortality in sea turtles have not been perceived as being as important as other aspects of their population biology. Basic research into sea turtle pathophysiology and improving disease diagnosis has often received low priority. When available, more resources are invested in major epidemics versus individuals that are sporadically found as stranded animals. However, often these resources are mobilized too late, and lack of sustained investment in pathologic evaluations of sporadic cases and strandings may represent lost opportunities to gather information and perspective about background disease problems.

Questions to consider that are relevant to cost–benefit decisions in developing health assessment programs and conducting diagnostic investigations include the following:

1. Is the health problem relevant to population or species conservation? Because resources are limited, priority should be given to studies for which the answer to this question is clearly "yes." For example, diseases that occur in high prevalence and are known to cause mortality probably warrant intensive investigation. Discrete events that could have significant health impacts, such as a documented chemical or oil spill, a cold snap, or an algal bloom, provide important opportunities to characterize these impacts. This does not, however, diminish the potential value of other studies even if the benefits are harder to appreciate.
2. Is the project feasible under the current logistic–funding constraints? In-depth diagnostic studies will require that a captured or stranded free-ranging turtle be taken into a specialized veterinary facility. This involves holding and transporting the animal, as well as maintaining it in captivity for a period of time. This will disrupt other important field research activities and may require extra personnel and vehicles to deal with the turtle. Even routine basic health screening and diagnostic tests can be very expensive. Routine plasma biochemistry and hematology panels cost about $20–$40 per sample. Histological processing of a single biopsy or necropsy specimen that yields a paraffin block and a single hematoxylin and eosin stained slide presently costs between $10 and $20, and additional slides with special stains or unstained for immunohistochemistry may cost $2–$5 each. Screening histopathology of representative tissues

resulting from a single necropsy could easily exceed $200. If a pathologist examines these slides and produces a histopathology report, the costs will increase. Toxin residue analyses can cost hundreds to thousands of dollars depending on how many different classes of compound and their congeners are assayed. A full workup of a dead animal, including necropsy, histopathologic examination, toxicology screen (organic residues, metals), microbial cultures, and serology could easily exceed several thousand dollars per individual. These costs combined with funding constraints and poor study design may lead to reductions in sample sizes that become inadequate for statistical analysis and interpretation. Unless these small sample sizes can be added to and integrated with other studies, so that there is a cumulative sample database, these studies may be a waste of time and money.

3. Are support facilities and diagnostic services available and accessible? Many specialized diagnostic assays can be performed in only one or a few laboratories that have the appropriate reagents (e.g., cell lines, antibodies, or molecular probes) and validated assays. Diagnostic laboratories and medical facilities should be contacted during the design stages of the project and at least prior to beginning the study to determine feasibility. Diagnostic laboratories and medical facilities may have limited capacity to handle numbers of turtles or to process and analyze large numbers of samples. These facilities may require time and money to set up or scale up operations, especially if assays have to be validated and optimized for various sea turtle species. Many samples may be sensitive to transport time and storage conditions and must be transported to a receiving laboratory promptly. A diagnostic laboratory may have specific days and times that it can receive samples and may have preferred methods for sample preservation and transport. As discussed previously, each laboratory that can analyze samples needs to establish its own set of reference values for each sea turtle species. For large-scale or regional studies, selection of one or a few laboratories in advance is important. Data comparisons among laboratories and between methods may be a serious problem. If more than one laboratory must be used, it should participate in a performance quality assurance program that involves routine assay of a common set of standard reference samples and cross-checking of the results for consistency between laboratories (Walton, 2001a).
4. Are specialized reagents and diagnostic tests available to perform a valid study in sea turtles? Although a question may be of great interest and importance to sea turtle health, the appropriate tests may not be available. Many diagnostic tests and reagents are highly species-specific and do not perform reliably in a different species. Biochemical assays designed for humans or mammals may not function properly when applied to reptiles. The analyte being detected may have completely different structural and functional properties that affect its performance in an assay. A classic example is quantification of plasma albumen using the dye-binding method (Walton, 2001a). Each test must be optimized and validated for

each species. Serologic tests that detect antibody responses to particular antigens require species-specific reagents. Furthermore, interpretation of many available biomedical assays relies upon mammalian pathophysiology. We cannot be as certain in reptiles or in each species of sea turtle that these tests have the same biological and clinical relevance. There is a tremendous need for basic biomedical research to improve turtle-specific testing. Sustained investment is required to encourage the development and improvement of assays for sea turtles and maintain their availability for comparative studies.

5. Are the investigational materials (biologic samples, carcasses, etc.) of adequate quality to yield useful results? One must evaluate the cost of analysis versus the information to be gained when dealing with poor-quality or inappropriately handled specimens. Many diagnostic assays and tests are sensitive to the conditions under which the sample was preserved, handled, and processed, and may yield spurious results. For example, plasma that obviously contains hemolyzed red cells will not be very useful for many biochemical and hematological analyses because the out-of-range values will reflect hemolysis rather than any disease process (Meyer et al. 1992). Carcasses that are autolyzed (rotting) may be necropsied and tissues examined grossly, but histological evaluation may not be informative enough to justify the cost. Similarly, submission of samples for microbial culture would likely provide spurious results. Plasma and tissue samples that are collected for certain biochemical assays such as enzyme activity (e.g., cholinesterase) must be frozen or analyzed quickly, or activity levels will change. Similarly, samples for RNA analysis must be immediately frozen at ultracold temperatures or otherwise protected against degradation with specialized preservatives. Turtles that were dead when found are poor sources of RNA. Tissues that have been frozen are difficult to evaluate histologically, and whole blood that has been frozen prior to separation will have no intact blood cells and will yield a hemolyzed plasma sample. Tissue specimens that have been frozen at –20°C will be less likely to produce successful virus isolation than samples stored at 5 or –70°C.

Other miscellaneous practical issues must be taken into consideration as well. These include permits, preparedness, and long-term maintenance of sample archives, records, and data management. In the U.S., state and federal permits are required to capture, handle, or sample any sea turtle species (which are protected), and to possess sea turtle tissues or parts. Any activity that results in a "take," the death or removal from the free-ranging population, also requires special permits. This includes euthanasia of moribund and catastrophically injured animals, which could provide valuable tissues. In many instances, the best material for analysis is obtained from a freshly euthanized sick animal. Therefore, even though an interesting disease case may be found, it would be illegal to collect blood or a skin lesion biopsy unless specifically permitted to do so. Thus, permit issues should be settled before undertaking health studies. In addition, supplies and materials needed to support health studies and sample collection must be kept in stock and in date for use when required.

Archiving samples properly for future analysis is important but costly. For certain materials, archival samples allow retesting, confirmatory testing, and retrospective studies based on new information and hypotheses and using analyses that were not available at the time samples were collected. This is especially important for sea turtles, for which there are likely to be many more unknown diseases and pathologic agents yet to be discovered. Samples archived for biochemical and molecular assays and virus isolation must be held at –70°C or below. This requires an ultracold freezer with a temperature-monitoring and alarm system and provisions for backup power or alternative freezer space. It is important to consider the effects of repetitive thawing and refreezing, and archived specimens should be subdivided and stored in aliquots to avoid this problem. Specimen redundancy in backup freezers also helps reduce the risk of loss due to inevitable freezer failures and other disasters. Tissue specimens for histopathology can be preserved indefinitely at room temperature in fluid preservatives such as 10% formalin or 70% ethanol, but leakage and evaporation can lead to specimen loss. Histology specimens can be embedded in paraffin blocks and stored efficiently, but costs of processing and embedding must be considered. Management of the archive, specimens, and data is an essential feature and long-term commitment. Adequate records of archive contents and specimen locations are needed, as well as a relational database that cross-references field data with clinical evaluations, pathology reports, and laboratory and diagnostic test results. Even the most meticulously organized and maintained archive will be useless if information cannot be searched and samples cannot be retrieved efficiently.

15.5 CONCLUSION

Although there is tremendous benefit to be gained by incorporating the art and science of health assessment and systematic disease investigation into sea turtle biology, this should not be undertaken lightly. We hope that this chapter has helped provide some perspective on the process and its limitations.

REFERENCES

Berry, K.H. and Christopher, M.M., Guidelines for the field evaluation of desert tortoise health and disease, *J. Wildl. Dis.*, 37, 427, 2001.

Bolten, A.B., Techniques for measuring sea turtles, in *Research and Management Techniques for the Conservation of Sea Turtles*, Eckert, K.L. et al., Eds. IUCN/SSC Marine Turtle Specialist Group Publication No. 4, 110–114, 1999.

Bolten, A.B. and Bjorndal, K., Blood profiles for wild population of green turtles (*Chelonia mydas*) in the southern Bahamas: size-specific and sex-specific relationships, *J. Wildl. Dis.*, 28, 407, 1992.

Bolten, A.B., Jacobson, E.R., and Bjorndal, K.A., Effects of anticoagulant and autoanalyzer on blood biochemical values of loggerhead sea turtles (*Caretta caretta*), *Am. J. Vet. Res.*, 12, 2224–2227, 1992.

Carminati, C.E. et al., Blood chemistry comparison of healthy vs. hypothermic juvenile Kemp's ridley sea turtles (*Lepidochelys kempii*), in *Proceedings of the 14th Annual Symposium on Sea Turtle Biology and Conservation*, Bjorndal, K.A. et al., Compilers. N.M.F.S. Tech. Memo. NOAA-TM-NMFS-SEFSC-351, Miami, FL, 203, 1994.

Chrisman, C.L. et al., Neurologic examination of sea turtles, *J. Am. Vet. Med. Assoc.*, 211, 1043–1047, 1997.

Coberley, S.S. et al., Detection of antibodies to a disease-associated herpesvirus of the green turtle, *Chelonia mydas*, *J. Clin. Microbiol.*, 39, 3572, 2001a.

Coberley, S.S. et al., Survey of Florida green turtles for exposure to a disease-associated herpesvirus, *Dis. Aquat. Org.*, 47, 159, 2001b.

Ewald, P.W., *Evolution of Infectious Disease*, Oxford University Press, Oxford, U.K., 1993.

FitzSimmons, N., Moritz, C., and Bowen, B.W., Population identification, in *Research and Management Techniques for the Conservation of Sea Turtles*, Eckert, K.L. et al., Eds. IUCN/SSC Marine Turtle Specialist Group Publication No. 4, 72–179, 1999.

George, R.H., Health problems and diseases of sea turtles, in *The Biology of Sea Turtles*, Lutz, P.L. and Musick, J.A., Eds. CRC Press, Boca Raton, FL, 1997, 363–385.

Herbst, L.H., Fibropapillomatosis of marine turtles, *Ann. Rev. Fish Dis.*, 4, 389, 1994.

Herbst, L.H., Infectious diseases of marine turtles, in *Research and Management Techniques for the Conservation of Sea Turtles*, Eckert, K.L. et al., Eds. IUCN/SSC Marine Turtle Specialist Group Publication No. 4, 208–213, 1999.

Herbst, L.H. and Jacobson, E.R., Diseases of marine turtles, in *Biology and Conservation of Sea Turtles*, revised edition, Bjorndal, K.A., Ed. Smithsonian Institution Press, Washington, DC, 1995, 593–596.

Homer, B.L. et al., Chlamydiosis in mariculture-reared green sea turtles (*Chelonia mydas*), *Vet. Pathol.*, 31, 1, 1994.

Jacobson, E.R., Marine turtle farming and health issues, *Mar. Turtle Newsl.*, 72, 13, 1996.

Jacobson, E.R., Tissue sampling and necropsy techniques, in *Research and Management Techniques for the Conservation of Sea Turtles*, Eckert, K.L. et al., Eds. IUCN/SSC Marine Turtle Specialist Group Publication No. 4, 214–217, 1999.

Jacobson, E.R. et al., Conjunctivitis, tracheitis, and pneumonia associated with herpesvirus infection in green sea turtles, *J. Am. Vet. Med. Assoc.*, 189, 1020, 1986.

Jacobson, E.R. et al., Problems with using weight versus carapace length relationships to assess tortoise health, *Vet. Rec.*, 132, 222, 1993.

Lauckner, G., Diseases of reptilia, in *Diseases of Marine Animals*, volume IV, part 2, Kinne, O., Ed. Biologische Anstalt Helgoland, Hamburg, Germany, 1985, 553–626.

Loeb, W.F. and Quimby, F.W., Eds. *The Clinical Chemistry of Laboratory Animals*, Pergamon Press, New York, 1989.

May, R.M. and Anderson, R.M., Parasite–host coevolution, in *Coevolution*, Futuyma, D.J. and Slatkin, M., Eds. Sinauer Associates Inc., Sunderland, MA, 1983, 186–206.

Maynard-Smith, J., Evolution and the theory of games. *Am. Sci.*, 64, 41–45, 1976.

Meyer, D.J., Coles, E.H., and Rich, L.J., *Veterinary Laboratory Medicine, Interpretation and Diagnosis*, W.B. Saunders, Philadelphia, 1992.

Owens, D.W., Reproductive cycles and endocrinology, in *Research and Management Techniques for the Conservation of Sea Turtles*, Eckert, K.L. et al., Eds. IUCN/SSC Marine Turtle Specialist Group Publication No. 4, 119–123, 1999.

Persing, D.H., In vitro nucleic acid amplification techniques, in *Diagnostic Molecular Microbiology: Principles and Applications*, Persing, D.H. et al., Eds. American Society for Microbiology, Washington, DC, 1993, 51–87.

Prichard, P.C.H. and Mortimer, J.A., Taxonomy, external morphology, and species identification, in *Research and Management Techniques for the Conservation of Sea Turtles*, Eckert, K.L. et al., Eds. IUCN/SSC Marine Turtle Specialist Group Publication No. 4, 21–38, 1999.

Rebell, G., Rywlin, A., and Haines, H., A herpesvirus-type agent associated with skin lesions of green sea turtles in aquaculture. *Am. J. Vet. Res.*, 36, 1221, 1975.

Shaver, D.J. and Teas, W.G., Stranding and Salvage Networks, in *Research and Management Techniques for the Conservation of Sea Turtles*, Eckert, K.L. et al., Eds. IUCN/SSC Marine Turtle Specialist Group Publication No. 4, 152–155, 1999.

Smith, G.M. and Coates, C.W., Fibro-epithelial growths of the skin in large marine turtles, *Chelonia mydas*, *Zoologica*, 23, 93, 1938.

Stedman's Medical Dictionary, Houghton Mifflin, Boston, 2001.

Walsh, M., Rehabilitation of sea turtles, in *Research and Management Techniques for the Conservation of Sea Turtles*, Eckert, K.L. et al., Eds. IUCN/SSC Marine Turtle Specialist Group Publication No. 4, 202–207, 1999.

Walton, R.M., Validation of laboratory tests and methods, *Semin. Exotic Pet Med.*, 10, 59–65, 2001a.

Walton, R.M., Establishing reference intervals: health as a relative concept, *Semin. Exotic Pet Med.*, 10, 66–71, 2001b.

Weisbroth, S.H. et al., Microbiological assessment of laboratory rats and mice, *ILAR J.*, 39, 272–290, 1998.

Wibbels, T., Diagnosing the sex of sea turtles in foraging habitats, in *Research and Management Techniques for the Conservation of Sea Turtles*, Eckert, K.L. et al., Eds. IUCN/SSC Marine Turtle Specialist Group Publication No. 4, 139–143, 1999.

Wyneken, J., *The Anatomy of Sea Turtles*, U.S. Department of Commerce, NOAA Technical Memorandum NMFS-SEFSC-470, 2001.

Zar, J.H., *Biostatistical Analysis*, Prentice-Hall, Englewood Cliffs, NJ, 1974.

16 Sea Turtle Husbandry

Benjamin M. Higgins

CONTENTS

0-8493-1123-3/03/$0.00+$1.50

16.1 INTRODUCTION

Sea turtle husbandry is the care and maintenance of sea turtles through scientific and judicious use of resources. Caring for turtles in captivity presents some problems, whether rearing them for research and conservation, public display (zoos and aquaria), or other commercial purposes. Sea turtles, in general, are sensitive to temperature variation; can be aggressive when crowded; are long-lived; and can reach great sizes, requiring large accommodations. Even if the sea turtles' natural physical environment can be artificially duplicated in captivity, general biological information is still lacking. For instance, little is known about the wild pelagic (early) life stages of all species, including basic information such as diet and feeding, growth, activity levels, and natural survival, all of which are fundamental parameters if one is to maintain turtles in captivity from hatchlings. Despite the lack of basic biological knowledge on sea turtles, many facilities have reared sea turtles in captivity with varying degrees of success. Much of what is known was learned through trial and error over decades of work. Available information on specific rearing practices is limited. Even with missing biological information, there is no reason to believe that sea turtles cannot be successfully reared and maintained in captivity by simply following sound animal husbandry practices.

The largest biological obstacles to sea turtle rearing are diet and disease. Green (*Chelonia mydas*) (Wood, 1991; Huff, 1989; Lebrun, 1975), loggerhead (*Caretta caretta*) (Caillouet, 2000; Buitrago, 1987), Kemp's ridley (*Lepidochelys kempii*) (Caillouet, 2000; Fontaine et al., 1985; 1988), and hawksbill (*Eretmochelys imbricata*) (Glazebrook and Campbell, 1990; Brown, 1982; Gutierrez, 1989) turtles have all successfully been reared in captivity. Attempts have been made to rear leatherback hatchlings (*Dermochelys coriacea*) in captivity with limited success (Jones et al., 2000; Voss et al., 1988), and little information is available on captive rearing of the olive ridley (*Lepidochelys olivacea*) (Rajagopalan et al., 1984) and flatback (*Natator depressus*).

Many facilities, throughout the world, have experimented with rearing sea turtles in captivity, some with more success than others (Wood, 1991; Stickney, 2000). In the 1960s and 1970s, rearing sea turtles in captivity was synonymous with farming or ranching, primarily as a response to new laws protecting the wild take of sea turtles (Stickney, 2000). In the 1980s, there was a shift in focus from farming to research (Huff, 1989; Caillouet, 2000; Caillouet et al., 1997). Most facilities are now rearing sea turtles for public display or conservation (Ross, 1999), with efforts directed toward wild stock enhancement.

In 1977, the National Marine Fisheries Service, Sea Turtle Facility (NMFS STF), was established in Galveston, TX. The NMFS STF is a U.S. federal government facility dedicated to rearing sea turtles for research, specifically aimed at reducing sea turtle bycatch in the U.S. commercial fisheries (Mitchell et al., 1989). Large sample sizes are required for certifying and evaluating potential sea-turtle-saving measures, thus necessitating the rearing of hundreds of sea turtles each year. The NMFS STF also rears loggerheads and Kemp's ridleys for research on physiology, tagging, and genetic and population dynamics. This chapter uses the NMFS STF as a model facility to describe successful sea turtle husbandry techniques.

16.2 REARING FACILITIES

Creating a suitable environment in which to raise sea turtles requires the ability to house the turtles in a controlled environment. In the U.S., all sea turtle species are protected animals, and there are specific state (Florida Fish and Wildlife Conservation Commission [FWC], 2002) and federal (U.S. Fish & Wildlife Service, 1973) government guidelines regulating sea turtle holding and rearing operations, including tank dimensions, feed, and environmental requirements. Tank layout and water delivery systems used to hold sea turtles are varied and include ocean pens constructed along shorelines; large concrete tanks with flow-through water delivery; or many small tanks connected to complex recirculating biofilter systems. The layout of the NMFS STF has been described previously in several publications (Caillouet, 1988; 2000; Fontaine et al., 1985; 1988; 1990) and consists of a static water system containing twenty 5940-l fiberglass raceways. The raceways are contained in a temperature-controlled warehouse-style building.

16.2.1 TANK SELECTION

The physical dimensions and material in which the turtles are contained are determined by the size and activity of the species cultured. Smooth-surfaced, unfurnished containers that are large enough to allow for unimpeded movement and complete submersion of the turtle are the minimum requirements. Sea turtles will eat artificial corals, fish, standpipes, plumbing, and other tank furnishings. Great care should be exercised when placing a turtle in a tank to ensure that it cannot be injured through impact with or ingestion of tank furnishings. Plexiglas is easily scratched by sharp claws, and this should be taken into consideration when a turtle in placed into a tank for public display in a zoo or aquarium. Provision for separation of turtles and their waste products should be addressed either by using a physical barrier or by constant mechanical removal of waste. As turtles grow, they require more space, necessitating progressively larger accommodations (Table 16.1). Turtles must be reared in individual containers to prevent injuries from contact with other turtles; this may include separate tanks or common compartmentalized tanks. Aggressiveness varies among species. All species can and will bite each other when housed together in the same tank (Glazebrook and Campbell, 1990; Leong et al., 1989; Klima and McVey, 1982). The NMFS STF maintains sea turtles in a variety of independent rearing containers housed in fiberglass-reinforced, polyester resin, gel-coated fiberglass tanks and raceways (Caillouet, 2000) (Figure 16.1).

16.2.2 Tank and Raceway Preparation

Prior to stocking, raceways are drained and thoroughly hand-scrubbed using Scotch-brite®-type (3M Home Care Division, St. Paul, MN) nylon abrasive pads. The raceways are then filled completely with seawater. Two gallons (7.58 l) of bleach (sodium hypochlorite) is added to the approximately 6814 l of water. The bleach is allowed to disinfect the tank for 24 h. Raceways are then drained and rinsed with freshwater. If the surface of the tank is porous or scratched, high-pressure washing (freshwater at 1500–1800 psi) may also be done to remove algae and other detritus imbedded in the gel coat. The raceways are refilled with seawater and allowed to soak for 24 h. Raceways are again drained, rinsed with freshwater, and are then ready for stocking.

16.2.3 Container Preparation

16.2.3.1 Hatchling Rearing Containers

Plastic flowerpots are used to house hatchlings for the first 60 days at the NMFS STF. Flowerpots are cleaned and disinfected prior to the arrival of new hatchlings. The pots are allowed to soak in a bleach solution (2:1 of sodium hypochlorite and 115:1 of freshwater) for 15–30 min. Each pot is hand scrubbed inside and out with a Scotchbrite-type nylon abrasive pad to remove all traces of dirt and algae. The pots are dipped into a bleach solution (1:15), rinsed in freshwater, and allowed to air dry. Clean pots are stored in an insect- and dust-free container until they are ready for use. Just prior to stocking, the pots are soaked in seawater for 24 h followed by a freshwater rinse.

16.2.3.2 Post-Hatchling and Juvenile Rearing Containers

Modified milk crates are used to house turtles from 60–90 days until 10–11 months, and custom-built hanging cages are used from 11–22 months at the NMFS STF (Caillouet, 2000). Crates and cages are removed from the facility, and every surface is cleaned with a high-pressure washer (freshwater at 1500–1800 psi) to remove all traces of dirt and algae. The containers are placed back into the raceways and are

TABLE 16.1
A Comparison of Actual and Recommended Rearing Space Sizes and Stocking Densities

Facility/Agency	Turtle Size	Turtle Age (months)	Tank/Pen/Container Size (m^2)	Tank Volume (l)	Number Turtles/Tank	Surface Area/Turtle (m^2)	Water Vol./Turtle (l)	Stocking Density (g/l)
NMFS STF	10–60 g	0–3	Cont., 0.02	2195	200	0.02	11	1.0–5.5
	60–500 g	3–11	Cont., 0.11	3292	80	0.11	41	1.5–12.2
	0.5–7.0 kg	12–24	Cont., 0.46	5486	14	0.46	392	1.3–17.9
	7.0–25.0 kg	25–48	Cont., 1.3	5940	7	1.30	849	8.2–29.4
	5.0–25.0 kg	20–48	Pen,[a] 142.7	178,734	10–30	4.7–14.2	5958–17,873	0.8–1.4
CTF	4.1 kg[b]	14[b]	Tank[b]	130,000[b]	100[b]		1300	3.2
	4.1 kg[b]	14[b]	Tank[b], 7.1	3000[b]	50[b]	0.14	60	68.3
	21.6 kg[b]	44[b]	Tank[b], 7.1	3000[b]	10[b]	0.71	300	72
	90–200 kg[c]		Pond/pen[c]			4.0–6.5[c]		
	24.7 kg[d]	36–84[j]	Pond, 207[d]	289,800	37[d]	6[d]		3.2
Los Roques	Hatchling	0–6[e]	Tank[e], 1.3[e]	260	30[e]	0.04	32.8	8–10[e]
	Post-hatchling	>6[e]	Pen[e]		1[e]	3.0[e]		0.5[e]
FWC	<5 cm[f,g]	0–1[g]	0.09[g]	29[g]	1[g,i]	0.09[g]	29	0.7[h]
	>5 cm and <10 cm[f,g]	1–6[g]	0.09–2.22[g]	29–698[g]	1[g,j]	0.09–2.22[g]	29–698	0.7–16.5[h]

(continued)

	<60 cm[f,g]	6–40[g]	2.32–5.01[g]	1816–4709[g]	1[g,j]	2.32–5.01[g]	1816–4709	6.5–1.8[h]
	>90 cm[f,g]	>40[g]	4.74–11.42[g]	4709–14,168[g]	1	4.74–11.42[g]	4709–14,168	
Costa Rica	4–24 cm[k,l]							
	15g–24.2 kg			21,000[l]	60[l]		333	0.05–9.9[l]

Note: Unreferenced figures were calculated based on referenced data. NMFS STF (*Caretta caretta* [Cc], *Lepidochelys kempii* [Lk]).

[a]NMFS Panama City, Florida facilities (Cc).

[b]Data from CTF (*Chelonia mydas* [Cm]). (From Ross, J.P. 1999. Ranching and captive breeding sea turtles: Evaluation as a conservation strategy, pp. 197–201, in: *Research and Management Techniques for the Conservation of Sea Turtles*. K.L. Eckert et al. (eds.). IUCN/SSC Marine Turtle Specialist Group Publication No. 4. With permission.)

[c]Data from CTF (Cm). (From Wood, F. 1991. Turtle culture, in: *Production of Aquatic Animals*. C.E. Nash (ed.). World Animal Science, Elsevier, Amsterdam. With permission.)

[d]Data from CTF (Lk). (From Wood, J.R. and F.E. Wood. 1988. Captive reproduction of Kemp's ridley *Lepidochelys kempii*. *J. Herpetol.* 1:247–249. With permission.)

[e]Data from Los Roques (*Eretmochelys imbricata* [Ei], Cm). (From Buitrago, J. 1987. Rearing, with aim of repopulating, of three marine turtle species at Los Roques, Venezuela. *Mem. Soc. Cienc. Nat. La Salle.* 127–128:169–201. With permission.)

[f]Straight carapace length (SCL). (From Bolten, A.B. 1999. Techniques for measuring sea turtles, pp. 110–114, in: *Research and Management Techniques for the Conservation of Sea Turtles*. K.L. Eckert et al. (eds.). IUCN/SSC Marine Turtle Specialist Group Publication No. 4. With permission.)

[g]Data were calculated using NMFS STF 1996 year-class Texas loggerhead data. (From Florida Fish and Wildlife Conservation Commission. 2002. Unpublished. Marine turtle conservation guidelines: Section 4 — holding turtles in captivity. Tallahassee, Florida. With permission.)

[h,i]Add 25% surface area for each additional turtle.

[j]Add 50% surface area for each additional turtle.

[k]Method of carapace measurement is unknown.

[l]Ei data from Costa Rica. (From Gutierrez, W. 1989. Experiences in the captive management of hawksbill turtles (*Eretmochelys imbricata*) at Isla Uvita, Puerto Limon, Costa Rica, pp. 324–326, in: *Proceedings of the Second Western Atlantic Turtle Symposium*, Oct. 12–16. L. Ogren (ed.). NOAA Tech. Mem. NMFS-SEFC-226. With permission.)

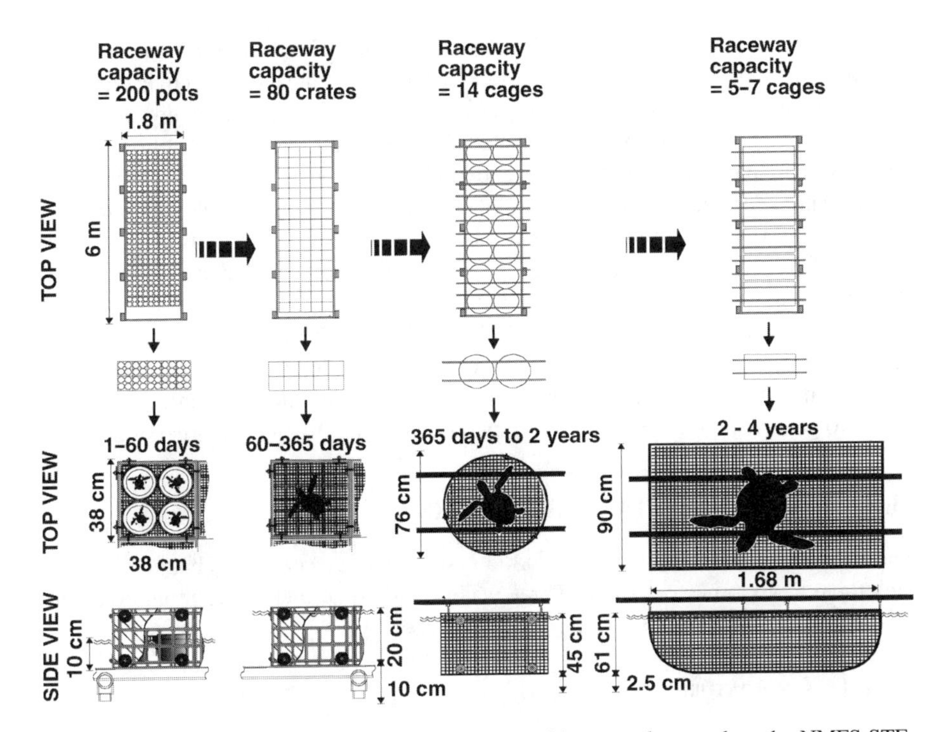

FIGURE 16.1 Progression of rearing container size with sea turtle growth at the NMFS STF. A common raceway tank is used to house 5–200 turtles. (Modified from C.W. Caillouet, Jr. 2000. Sea turtle culture: Kemp's ridley and loggerhead turtles, pp. 788–798, in: *Encyclopedia of Aquaculture*. R.R. Stickney (ed.). John Wiley & Sons, New York, 2000, 786. With permission.)

bleached at the same time the raceways are disinfected. The crates are soaked in raceways filled with the bleach solution for 24 h. The crates are rinsed with freshwater followed by an additional 24 h soak in seawater. The containers are again rinsed with freshwater and stored dry until they are ready for use. Just prior to use, the containers are soaked in seawater for 24 h followed by a freshwater rinse. A new batch of hog rings (Caillouet, 2000) is applied to the bottom of the cages annually to replace those that rusted off or became loose during the previous year. New nylon cable ties (Caillouet, 2000) are used to suspend the cages. Crates and cages are stored dry until they are ready for use.

16.3 SEAWATER SYSTEM

The NMFS STF relies on a natural seawater system consisting of a beach pump, sump, and water storage tanks (Caillouet, 2000; Fontaine et al., 1985). Water is drawn directly from the Gulf of Mexico.

16.3.1 Water Treatment and Storage

The NMFS STF uses four 26,000-l and two 38,000-l insulated fiberglass tanks to store seawater (Caillouet, 2000). Each of the four small tanks contains a quartz immersion heater (14,000 W). Each heater is connected to a temperature-control unit, allowing the tanks to be adjusted independently. There is no active filtration or treatment of any kind in the seawater system. The well points below the sand remove large particles from the water. Settling in the sump and large holding tanks removes most particulate matter, suspended algae, larvae, and some bacteria from the seawater. Further settling in the eight smaller water storage tanks removes the remainder of suspended sediment. The NMFS STF uses approximately 37,854–68,137-l of new seawater daily. Wastewater is discarded into the city of Galveston sanitary sewer system.

From late September through April, the NMFS STF heats seawater. Seawater is heated to approximately 38–43°C in three of the four 26,000-l storage tanks. Hot water is mixed with ambient water 10–26°C by manipulating hot- and cold-water valves to achieve an incoming water temperature of 26–30°C with a target of 28.5°C.

16.4 ENVIRONMENTAL PARAMETERS

The STF uses natural seawater in a static system where water is exchanged in each tank three to six times per week. Three water quality parameters are monitored and recorded daily: temperature, salinity, and pH.

16.4.1 TEMPERATURE

Maintaining a constant and acceptable temperature is critical for growth and for preventing disease in sea turtles (Haines and Kleese, 1977; Leong et al., 1989; Caillouet et al., 1997). Water temperature at the NMFS SFT is maintained within the range of 26–30°C (Figure 16.2). Water temperature is maintained by mixing warm (heated) and cold (ambient) seawater to the desired temperature. Air temperature over the tanks is also controlled using forced-air heaters in cool months and ventilation fans in hot months. The air temperature in the facility is 29–32°C at night (maintained by heaters in cool months) and is reduced to 24–26°C during the day to provide a more comfortable environment for captive rearing staff. In months where heating the air is not required, the facility remains at a constant 28°C day and night with the assistance of exhaust fans and cross-flow ventilation. Temperature is measured with a thermometer, accurate to 0.5°C.

When the temperature falls below 22°C, turtles that are normally maintained at 26–30°C will slow or cease feeding. At temperatures above 32°C, water quality becomes an issue because algae and bacteria populations can rapidly multiply in the raceways. Sea turtles that are normally maintained at temperatures 24–25°C may tolerate temperatures as low as 20°C before exhibiting signs of reduced metabolic activity. Sea turtles should be maintained at 20–30°C (FWC, 2002), preferably in the range of 25–30°C. Even short periods of water temperatures below 22°C combined with shorter photoperiods in winter months can trigger carapace lesions in loggerheads (Higgins, unpublished data). The carapace lesions can be characterized by a white fluffy exudate that appears to grow from the eroding neural and postmarginal scute spines. Outbreaks are directly related to water temperature and water quality. The lesions, if left untreated, result in keratin

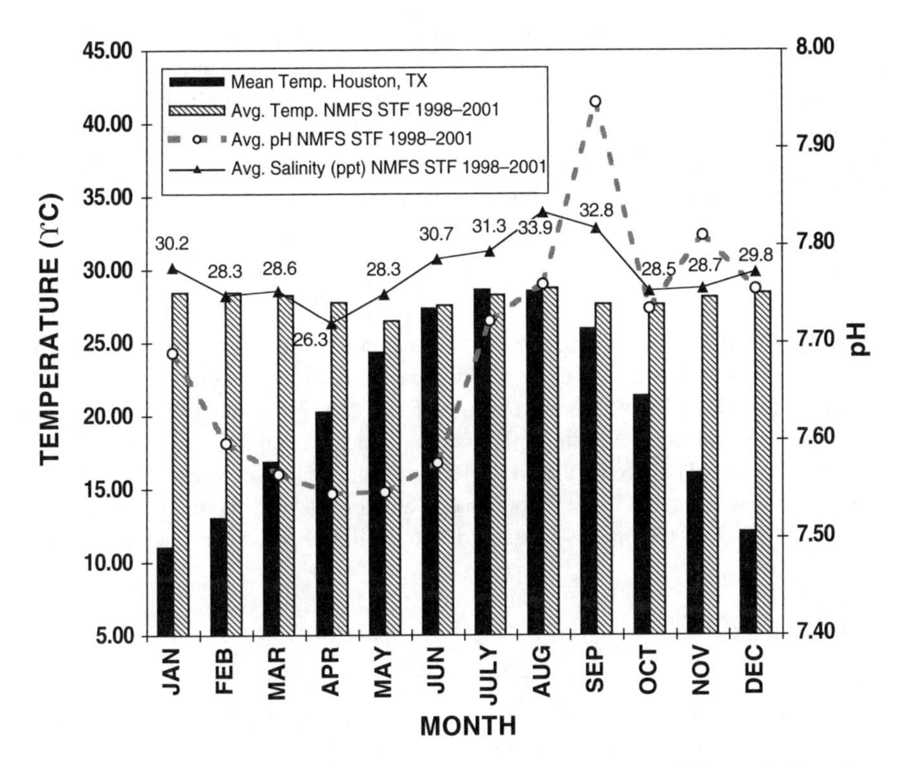

FIGURE 16.2 Graphical comparison of average water temperature, salinity, and pH readings over a 4-year period (1998–2001) in the NMFS STF. Water temperature remains at a constant 27–29°C despite wide-ranging ambient outside air temperature (Data from NOAA, National Climatic Data Center Archives). pH fluctuates with biomass in rearing tanks, whereas salinity remains in the range of 26–34 ppt. Typically, the biomass peaks in the NMFS STF in early May when more than half of the loggerheads are shipped to Florida for research. Note the steady decline in pH from January through May, followed by a rapid increase in pH mid-May, which correlates to the departure of the large turtles. pH peaks in September just prior to the arrival of loggerhead hatchlings.

loss and eventual bone erosion and degeneration. Histology results on carapace samples showed no infectious agents. Similar lesions have been reported on captive-reared loggerheads with bacterial and fungal organisms present (Neiffer et al., 1998; Leong et al., 1989).

16.4.2 Salinity

Salinity at the NMFS STF is maintained between 14 and 32 ppt. Normal natural salinity of NMFS STF incoming water is 26–30 ppt. When salinity exceeds 34 ppt, freshwater is added to the water storage tanks to dilute hyper-saline water. Optimal salinity for maintaining healthy captive sea turtles is 20–35 ppt (FWC, 2002). Sea turtles may be temporarily maintained at salinities outside the normal range for therapeutic purposes as prescribed by a veterinarian. Low salinity may be helpful for removing parasites and fouling epibiota such as barnacles. Sea turtles can withstand short periods of freshwater (0 ppt) (Walsh, 1999). Freshwater treatments of up to 2 weeks have been used to treat floating–bloating problems with limited success, with no observable detriment to the turtles' health. A nonprescribed salinity of less than 14 ppt for a period of more than 2 weeks would require the addition of salt to the storage tanks to increase salinity. Salinity should be measured with a refractometer accurate to 1 ppt.

16.4.3 pH

pH is an indicator of water quality. As water quality degenerates from the accumulation of turtle waste products, pH decreases. Normal incoming NMFS STF seawater has a pH of 7.9–8.2. A pH reading of less than 7.4 indicates that a raceway is in need of cleaning. The normal pH of a clean raceway containing turtles ranges from 7.5 to 8.1. pH is controlled by cleaning and changing water. Average pH decreases as biomass and bioload increase in a raceway. Optimal pH for sea turtles is 7.5–8.5 (FWC, 2002). pH is measured with a digital pH meter accurate to two decimal places.

16.4.4 Light

The majority of light in the NMFS STF comes from the translucent fiberglass panels that make up a portion of the roof (15–1.2 m × 2.4 m panels). Fluorescent lighting (2 × 40 W × 15 fixtures) is used approximately 48 h/week to supplement sunlight. The daily amount of light the turtles receive is dependent on the natural available sunlight in Galveston, TX. The amount of actual ultraviolet (UV) light that reaches the sea turtles in the NMFS STF through the fiberglass panels is unknown. No health problems have been identified and associated with a lack of suitable light. Kemp's ridleys held in captivity for more than 1 year and loggerheads held in captivity for more than 2 years tend to be lighter in complexion than their wild counterparts. Sunlight is important in reptiles for the synthesis of vitamin D_3. A lack of suitable light may require dietary supplementation.

Experimentation with different quality and quantities of artificial light (both full-spectrum fluorescent lights, limited-spectrum lights [grow lights], and metal halide lights [5000 K]) as treatments for carapace lesions or infections and floating–bloating has been tried without success. Short periods of direct sunlight may help treat topical fungal lesions of the skin (Fontaine et al., 1988). Natural diurnal light patterns should be replicated for turtles housed in captivity. Excess light and nontherapeutic direct sunlight should be avoided to control algae growth, and to prevent elevated water temperature and sunburn.

16.5 HATCHLING SELECTION

Every attempt should be made to acquire captive stock bearing good genetic lineage, ideally from many different nests. Avoiding physical deformities from the onset will pay dividends in the end. Turtles with visible deformities may exhibit stunted and slow growth, and feeding and behavioral problems. In the U.S., both federal and state laws may prevent the public display, release, or euthanization of congenitally deformed turtles. When hatchlings are selected, three criteria need to be addressed: physical deformities, weight and size, and activity.

16.5.1 Physical Deformities

Each hatchling should be carefully inspected for eye deformity (blindness, lesions), cross-beak, curvature of the spine, and carapace deformities or abnormalities (extra or missing scutes). Hatchlings with visual physical deformities should be avoided. Spinal curvature may be very subtle in hatchlings but can develop into crippling deformities in older turtles.

16.5.2 Weight and Size

Hatchlings that are light in weight or excessively heavy may have genetic abnormalities and should be avoided. Small, underweight hatchlings are an indication of poor development or dehydration.

16.5.3 Activity

Hatchlings that are selected should be active and exhibit vigorous climbing and crawling activity. Lethargic hatchlings should be avoided.

16.5.4 Quarantine

Hatchlings are quarantined for 60 days when they arrive at the NMFS STF. The new hatchlings are housed in a raceway separate from the other turtles and care is taken to prevent any cross contamination between raceways. When space permits, each clutch of hatchlings is maintained together to monitor variations in survival and growth. Staff members wear latex surgical gloves when handling hatchlings for the first 30 days. Captive-rearing staff members are required to wash their hands before handling turtle feed and any measuring equipment is cold sterilized before and after use on new hatchlings. After 60 days, hatchlings can be combined in raceways. The NMFS STF also maintains a full quarantine facility for chronically and terminally ill sea turtles. A minimum quarantine period of 90 days is required before a rehabilitated turtle is placed in the captive-rearing facility. The introduction of a rehabilitating turtle into a raceway containing healthy captive reared stock is not recommended.

16.6 DIET, FEEDING, AND GROWTH

16.6.1 Diet

Sea turtles are opportunistic omnivores, consuming whatever is available. Wild turtles have a highly varied diet that changes with life stage (Bjorndal, 1997). Hatchlings and pelagic turtles typically consume what is available at the surface, whereas older, larger benthic turtles consume food thoughout the water column with emphasis on benthos (Bjorndal, 1997). Gut content analysis studies have been done on all sea turtle species cataloging food items consumed (Bjorndal, 1997), but little is known about actual wild feeding rates. In captivity, overfed turtles and turtles fed *ad libitum* are prone to obesity, fatty degeneration of the liver (Solomon and Lippett, 1991), and bloating.

Captive sea turtle diets vary considerably, ranging from natural foods (whole fish) to commercially prepared dry pelleted diets. Blends of natural foods (i.e., fish, shrimp, squid, crab, scallops), often supplemented with vitamins, are popular diets in zoos and aquaria. Blends may be prepared fresh each day or prepared in bulk and kept frozen until needed. Mixtures of natural foods may contain gelatinous binders to keep food from dispersing in the water prior to consumption by the sea turtles (Jones et al., 2000; Cong and Wang, 1997). To produce the same growth as dry pelleted foods, wet food is offered at a rate of up to five times that of dry food to compensate for the difference in moisture content. Wet feed rates vary from maintenance diets of 1% body weight per week (Higgins, unpublished data) to production–growth diets of 12–15% per day (Sumano Lopez et al., 1980). Feeding to satiation has also been reported with leatherback hatchlings (Jones et al., 2000).

The NMFS STF feeds a natural maintenance diet (whole mackerel) to loggerheads at 1% body weight per week when trying to maintain turtles at a specific size for research. Fasting days are common in aquaria. One or two days without food helps promote an appetite and may help maintain water quality. The NMFS STF feeds six days per week. Turtles maintained in aquaria with other animals may become very aggressive and compete for food. Turtles may attempt to consume all food introduced into the tank, resulting in overfeeding and leading to obesity. Feeding demonstrations are popular attractions at aquaria, and aggressive feeding by turtles may put divers at risk of injury. It may be necessary to distract or isolate turtles while the rest of the tank is fed.

Most large captive-rearing facilities feed some form of commercially prepared pelletized feed consisting of 25–45% crude protein, and 3.9–12% fat, 3.22–8.58% fiber (Caillouet et al., 1989; Wood and Wood, 1980; Wood, 1980). Higher protein levels result in greater growth (Wood, 1980; Caillouet et al., 1989). The protein source and content of the feed typically drives the feed cost; higher protein content commands a higher price, and protein from fish meal is more expensive than plant protein. Although weight gains of turtles fed fish versus plant protein sources are similar (Higgins, unpublished data), those turtles fed the fish meal source appeared to be more robust (Higgins, unpublished data). Diets containing soy products, especially those utilizing soy as the primary protein source, may increase estrogen levels in the blood (Shaw et al., 1989). Soybeans contain high levels of phytoestrogen. Increased phytoestrogen levels in humans have been linked to liver disease and reproductive problems (Shaw et al., 1989).

Several commercially available diet formulations have been successfully used over the years (Wood, 1991; Wood and Wood, 1980; Caillouet et al., 1989). Currently, the NMFS STF turtles are fed Aquamax® 500 Grower (PMI Nutrition International, Inc., Brentwood, MO). This is a dry, 4.7-mm (3/16 in.) diameter extruded pellet, which is small enough to be consumed by all sized turtles. The pellets float and remain intact in seawater. Pellet food should be bought in small quantities and kept frozen until use to preserve freshness, and prevent mold and rodent and insect infestation. Wet diets such as squid and fish create water quality problems and should be avoided for small turtles. Diets composed primarily of squid are high in phosphorus and may not contain sufficient calcium to meet the needs of the animal (Goldman et al., no date); some form of calcium supplementation may be required. Larger pellet feed is available for larger turtles, but often, feeding whole fish or squid is more convenient. Squid should be avoided as a long-term diet for sea turtles. However, squid is readily accepted by most species of turtles in captivity and is particularly good for coaxing wild rehabilitating turtles to eat. Several wet–semiwet gel diets have been developed specifically for hatchling leatherbacks (Jones et al., 2000; Cong and Wang, 1997), with limited success.

16.6.2 Feeding

The NMFS STF uses floating pellets for the first 2 years, and switches to a natural fish-based diet for larger turtles. Starting on day 10, each hatchling is offered one pellet, which represents approximately 2% of its body weight. Pellet feeding rates vary from 1.19 to 1.99% body weight/day (Figure 16.3). Each week the number of pellets is increased by one until each turtle receives 12 pellets twice per day. Starting in week 2, turtles are fed twice per day. Two smaller feedings are superior to one large single feeding, from a growth and water quality standpoint (Caillouet et al., 1989). Starting in week 25, calibrated feeders are made based on a percentage of

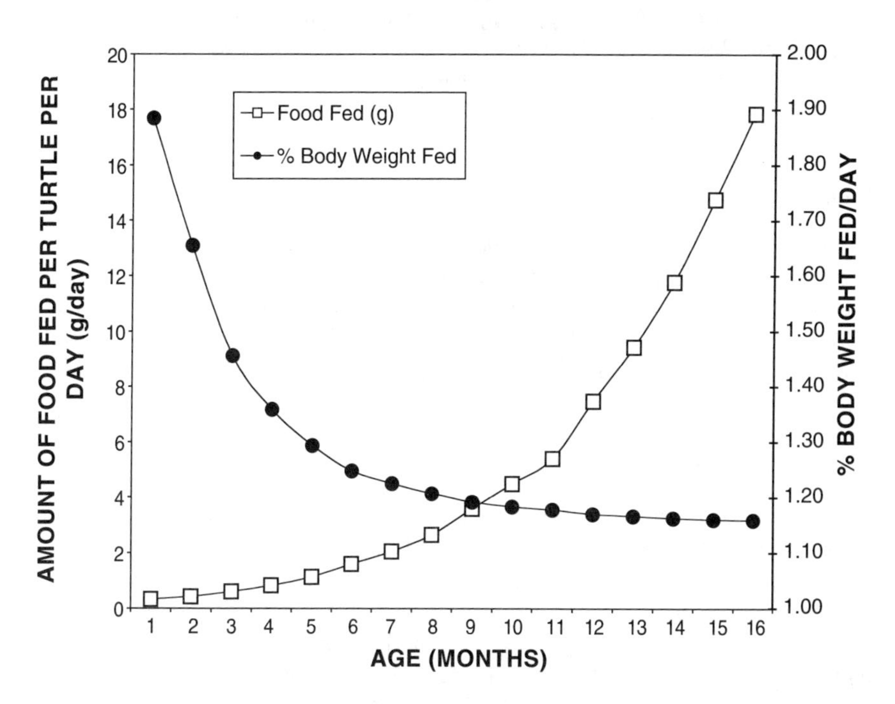

FIGURE 16.3 Graphical comparison of amount of food given and percentage of body weight fed per day. Percentage of body weight fed decreases with growth and amount of food given. Feeding rate was calculated using the NMFS STF feeding formula on the basis of the average weight of the 1995–2000 year class of Kemp's ridley sea turtles.

body weight as determined by a formula developed by the NMFS STF for Kemp's ridleys. The formula is also used for loggerheads. The food is scooped from a bucket by hand, with the feeder being the unit of measurement. Each turtle is given one scoop, twice per day. The NMFS STF feeding formula is

$$Y = 0.12515 + [(11.502x)/1000]$$

where x = turtle weight in grams and Y = amount of feed per turtle per day in grams.

Each turtle is individually fed, ensuring that all turtles get approximately the same amount of food.

16.6.3 Hatchlings

NMFS STF hatchlings are not fed until they are 10 days of age. Often, hatchlings are physically excavated from the nest, thus, they have not expended any energy emerging, crawling, or swimming. At rearing facilities, hatchlings are housed in relatively confined containers where swimming motion and energy expenditure are minimal; a longer time period is required for them to completely absorb the internal yolk sac. A delay in feeding gives the hatchlings an opportunity to partially absorb the internal yolk before they take in external nutrition. Feeding hatchlings prior to yolk sac absorption can result in constipation, lethargy, dehydration, and sudden death (Leong et al., 1989; Fontaine and Williams, 1997). Necropsy results on hatchlings fed prior to yolk sac absorption show compaction of food in the gut caused by the yolk sac displacement of internal organs (Leong et al., 1989; Fontaine and Williams, 1997).

16.6.4 Post-Hatchlings

Turtles are fed twice per day, at the beginning and end of the normal business day, at 7:30 a.m. and 4:00 p.m. The turtles are allowed 15–30 min to consume the morning feeding before the tanks are drained and cleaned. Hatchlings typically take longer to consume their food than the larger turtles, and are given a minimum of 30 min to consume food presented. On average, a healthy sea turtle will consume 100% of food offered within 15 min.

16.6.5 Growth and Survival

Sea turtle growth is directly related to food consumed and is exponential (Caillouet et al., 1989; 1997; Fontaine, unpublished data) (Figure 16.4). Survival rates are variable among different species and facilities and are a function of genetics and husbandry practices (i.e., water quality, feed, disease prevention and treatment). Most mortality occurs at the hatchling stage, and mortality levels off by month 6. Facilities with steep and rapid growth curves also have lower survival; this may be a function of water quality related to feeding. Much of the published data are from large-scale captive rearing operations, and the same growth–survival trend may not be present in smaller facilities such as zoos and aquaria. Sea turtle growth can be accelerated or maintained by varying the amount of food offered.

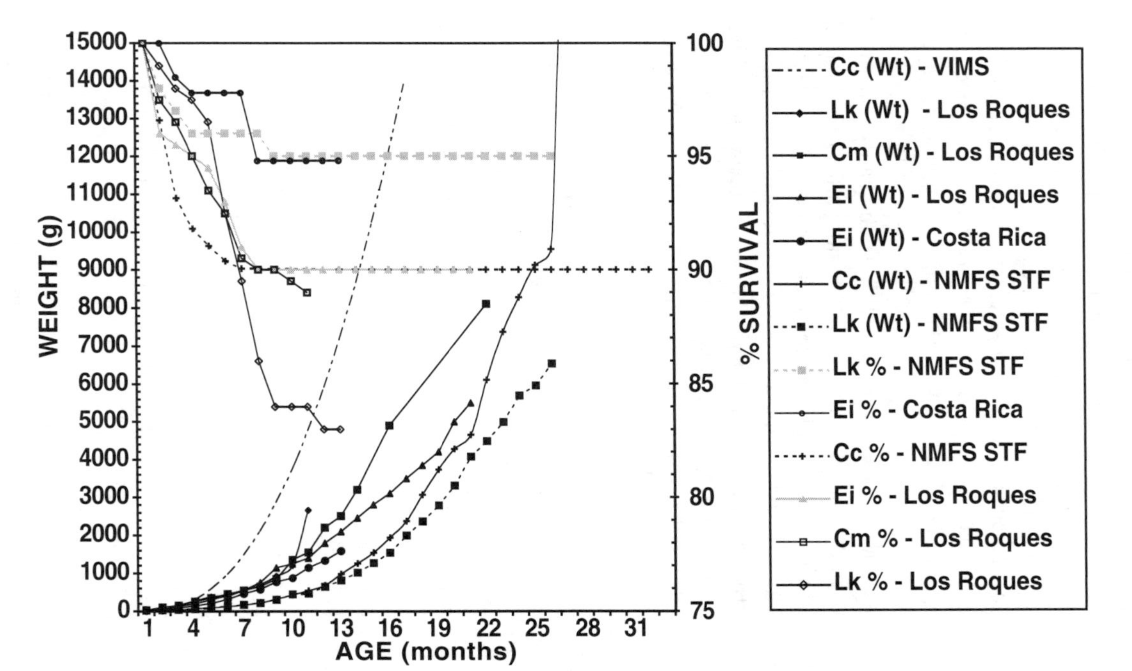

FIGURE 16.4 Comparison of growth and survival of captive sea turtles at different rearing facilities. The Virginia Institute of Marine Science (VIMS) curve was extrapolated from two data points (from Swingle, W.M. et al. 1993. Exceptional growth rates of captive loggerhead sea turtles, *Caretta caretta*. *Zoo Biol.* 12:491–497, with permission). The Los Roques data were extrapolated (from graphs in Buitrago, J. 1987. Rearing, with aim of repopulating, of three marine turtle species at Los Roques, Venezuela. *Mem. Soc. Cienc. Nat. La Salle.* 127–128:169–201, with permission). The Costa Rican data were calculated from a table (from Gutierrez, W. 1989. Experiences in the captive management of hawksbill turtles (*Eretmochelys imbricata*) at Isla Uvita, Puerto Limon, Costa Rica, pp. 324–326, in: *Proceedings of the Second Western Atlantic Turtle Symposium*, Oct. 12–16. L. Ogren (ed.). NOAA Tech. Mem. NMFS-SEFC-226, with permission). The NMFS STF data are an average of data taken on 6 year classes from 1995 to 2000. Cc = *Caretta caretta*; Lk = *Lepidochelys kempii*; Cm = *Chelonia mydas*; Ei = *Eretmochelys imbricata*.

16.6.6 Feeding Problems

Overfeeding of loggerheads and Kemp's ridleys with commercially prepared diets can cause buoyancy problems (bloating–floating). Bloating can be chronic, resulting in carapace–plastron deformities, rendering the turtle unfit for release and often leading to death. Even using the NMFS STF feeding formula, very careful monitoring of the turtles is required to ensure that an overfeeding situation does not arise. The amount of feed is increased or decreased from the formula amount on the basis of the appearance and activity of the turtles. Treatment of bloated turtles is still in the experimental phase and several parameters are being evaluated, including change of diet, change of water depth, increased rearing space, and devices to stimulate diving and benthic behavior (Higgins and Bustinza, unpublished data).

Food consumption is used as an index of turtle health. High-quality pellet feed floats and does not dissolve in the water, so any food remaining after 15–30 min is an indicator that feeding activity is abnormal. It is common for sea turtles to exhibit a period of slow eating activity in winter months or on overcast days. Sea turtles reared in a facility that is primarily lit by natural sunlight are photoperiod-sensitive. Slow eating behavior or a cessation of eating is a major concern and often is a symptom of a serious health problem. Turtles are observed during and after every feeding. Turtles that exhibit slow feeding are monitored, as are turtles that stop feeding. A turtle that shows no interest in food for 3 consecutive days is removed from the common tank and moved to an area of isolation, known as the STF first care area (FCA). The FCA is housed within the same building as the general rearing facility and is not a 100% quarantine area. The FCA contains a series of small individual tanks where turtles can be given specific medicinal treatments and water quality is therapeutically manipulated.

16.7 STOCKING DENSITIES

Stocking densities play a large role in sea turtle husbandry. Stocking densities combined with feeding ultimately determines water quality. Better water quality will most often be accompanied by a low stocking density. High stocking densities are usually indicative of tank culture, whereas low densities are found in pens. Four facilities were examined and stocking densities were determined directly from literature or calculated from available information. Stocking densities varied from 0.05–72 g/l (Table 16.1). Florida's FWC publishes guidelines for holding sea turtles in captivity (FWC, 2002). The FWC guidelines are primarily for facilities that maintain only a few turtles for research, public display, and rehabilitation. The FWC-recommended stocking densities are very conservative, and it would be difficult for large production facilities to economically meet these standards, even though they were created for the benefit of sea turtles and they should be viewed as an ultimate goal.

16.8 ROUTINE CLEANING

16.8.1 TURTLES

Every other day, each tank is completely drained, rinsed with freshwater and refilled with fresh seawater. Every inside surface of the tank and rearing containers is hosed off. The turtles are sprayed with freshwater directly on all exposed surfaces, with particular emphasis on the carapace to remove any loose algae or molting carapace material. For the first 90 days, one-half city water pressure (approximately 30 psi) is used when spraying the turtles. After 90 days, the turtles are sprayed with full city water pressure (approximately 60 psi). Turtles are out of water for up to 30 min every other day during the cleaning process, but are not allowed to dry. Prolonged periods out of the water will cause carapace desiccation leading to scute peeling. Damage to carapace from desiccation may provide an entry path for bacterial and fungal infection. Small loggerheads are more sensitive to desiccation and scute peeling than Kemp's ridleys, greens, or hawksbills.

Although wild sea turtles naturally display varying degrees of epibiota, captive-reared turtles are susceptible to heavy and rapid biofouling, requiring periodic removal. Typically, captive and wild pelagic turtles lack access to a substrate on which to scrape away accumulating epibiota. Epibiotic buildup leads to hydrodynamic drag and increased weight, and can potentially impact swimming performance. Epibiotic buildup may also create a medium for bacterial and fungal infection. Smaller turtles may be more sensitive to buildup than larger turtles, especially hatchlings. In the wild, algae and settling larvae are possibly kept in check and removed from pelagic turtles by epibiotic crabs (Frick et al., 2000), a relationship that is most likely absent in the captive environment. Algae must be mechanically removed from captive turtles with scrub brushes. Nylon bristle brushes are disinfected prior to use and between each tank or group of turtles in a solution of bleach (sodium hypochlorite 2%) (1:4) or chlorhexidine (2% chlorhexidine gluconate) (1:4) followed by a freshwater rinse. Every effort is made to maintain the integrity of the independent environment in each tank by reducing the chance of cross contamination.

16.8.2 REARING CONTAINERS

Uneaten food and waste that has settled on the bottom of the tank is directed by hose nozzle manipulation and city water pressure (60 psi) to the drain. To facilitate cleaning the raceway bottom, crates are moved along a polyvinyl chloride (PVC) support rack, which keeps the crates off the tank bottom (Caillouet, 2000). Cages are suspended off the bottom to facilitate cleaning and can be moved to gain access to the tank bottom. Once a month the inside tank surfaces and racks are hand scrubbed with a Scotchbrite-type nylon abrasive pad. Hand scrubbing is preferred to high-pressure washing. High-pressure washing tends to atomize waste material, creating a potential airborne tank-to-tank contamination problem. High-pressure washing (1500–1800 psi) may be used after tanks are disinfected.

16.9 DATA COLLECTION

Every hatchling is weighed and measured upon arrival at the NMFS STF. Every 28 days thereafter, a subsample of turtles from each year class is weighed and measured. At random, a minimum of 25 turtles are selected from each year class and those turtles are used to determine the feed ration. Four measurements are taken: weight, straight carapace length (notch to tip), straight carapace width (maximum), and body depth (maximum) (Bolten, 1999). An average weight is determined for the subsample and used to calculate the feeding rate for that year class. Weighing and measuring allows the captive-rearing staff a chance to carefully examine the sea turtles for any signs of problems. Each sea turtle is handled as little as possible to prevent injury to turtle and handler.

16.10 TURTLE TRANSPORT

16.10.1 Hatchlings, Post-Hatchlings, and Juveniles

Hatchlings are transported in commercially available plastic containers with lids. Shipping container size varies with the size of turtle being transported (Table 16.2). Ventilation holes are made three-quarters of the way up the side of the container. Size and number of ventilation holes vary with container size. The container is lined with a piece of solid, open-cell foam rubber and moistened with seawater. Carpet underpadding is the preferred foam type for smaller containers and it is available in large rolls, allowing custom-sized pieces to be cut. The amount of saltwater added to the foam varies with turtle and container size. If too little water is added, the turtle will desiccate. If too much water is added, it can cause vehicle motion-related injuries. Hatchlings are particularly sensitive to drowning from too much water, should turtles crawl and rest on top of one another. Shipping container size for turtles >1.0 kg should be large enough to comfortably hold the turtle, but small enough to prevent excessive motion and turning.

All turtles should be shipped in a climate-controlled vehicle (23–30°C) with temperature and moisture checked regularly and adjusted as necessary. Additional moisture can be added using a fine mist from a spray bottle. The number of turtles per shipping container should be reduced for long trips. Containers should be stacked not more than three high to prevent them from tipping over during normal vehicle movement. If possible, the use of straps and/or cargo nets is recommended when transporting turtles by aircraft.

16.10.2 Subadults and Adults (>15 kg)

Because of their size and strength, subadult and adult turtles can be difficult to transport.

16.10.2.1 Short Distances or Short Time Periods

Not many readily available containers can be purchased off the shelf to transport large turtles. Those that are available tend to be available in small quantities and

TABLE 16.2
Recommended Turtle Shipping Container Size and Water Volume

Turtle Size[a]	Turtle Age (months)[a]	Number Turtles/ Container	Shipping Container Volume (l)	Shipping Container Dimensions (cm)	Foam Dimensions (cm)	Water Volume (l)	Vent Hole Size (cm)/Number
<30 g	<1	80–100	38	50 × 36 × 20	33 × 50 × 1.27	1.0	3.8/8
30–60 g	0–3	1–30	38	50 × 36 × 20	33 × 50 × 1.27	1.0	3.8/8
60–500 g	3–11	1–6	38	50 × 36 × 20	33 × 50 × 1.27	1.5–2.0	3.8/8
0.5–1.0 kg	11–14	1–4	38	50 × 36 × 20	33 × 50 × 1.27	3.0	3.8/8
1.0–5.0 kg	14–22	1	38	50 × 36 × 20	33 × 50 × 1.27	4.0	3.8/8
5.0–25.0 kg	22–48	1	125	74 × 38 × 41	38 × 69 × 1.27	6.0	3.8/10
<40 kg	>48	1	240[b]	61 × 91 × 22[c]	60 × 90 × 5.01	15.0–20.0	5.01/16

[a]On the basis of NMFS STF average Cc and Lk data (1995–2000).
[b]Volume of two containers combined.
[c]Height of two containers combined.

are expensive. For turtles up to 35 kg and 65 cm straight carapace length, a relatively inexpensive shipping container can be made by fastening two plastic concrete mixing trays together with nylon cable ties. The tray that forms the bottom of the container is lined with a piece of saltwater-soaked foam cut to fit. A wet terry cloth towel can also be draped over the carapace and flippers to assist in keeping the turtle from drying out. Excessive evaporative cooling must be avoided, and wet towels should not be used when shipping during cool weather. Coating the carapace and skin with a petroleum-based (Walsh, 1999) (Vaseline®, Chese-borough-Pond's USA Co., Greenwich, CT) or water-based lubricant (KY®, McNeil-PPC, Inc., Skillman, NJ) may help prevent desiccation. The top of the tray is fastened to the bottom with nylon cable ties inserted through 6-mm holes drilled through the top and bottom lips of the two trays. This arrangement does not allow the shipper to easily inspect or replenish water lost during transport, and therefore this shipping method should be reserved for very short trips of 2 h or less, or for unaccompanied air travel.

16.10.2.2 Long Distances

Large turtles can be shipped long distances using readily available rental trucks. The cargo compartment of a rental truck can be subdivided into individual compartments with 19 mm plywood. The cargo bay is lined with a waterproof tarp or plastic to contain moisture and prevent damage to the vehicle. Ten- to 15-cm-thick foam is cut to fit the bottom of the compartment and is soaked with seawater to keep the turtle moist. Additional moisture can be added using a fine mist from a spray bottle directly onto the turtle, and new clean water may be required to replace water lost to evaporation. Transport in this fashion should be limited to trips less than 12 h and when the cargo compartment can be kept between 23 and 30°C. If possible, the turtles should be shipped in darkness to keep activity to a minimum. Some rental trucks have a translucent fiberglass roof, and light transmission through the roof will increase turtle activity.

16.11 GROW-OUT FACILITIES

The NMFS Panama City, FL (NMFSPC), sea turtle holding facilities consist of nine pens and a series of individual holding containers. Stocking densities of loggerhead and Kemp's ridley sea turtles need to be carefully monitored when captive turtles are allowed to interact with one another in a confined environment. Aggressive behavior has also been observed in green (Wood, 1991) and hawksbill sea turtles (Glazebrook, 1990). Although water quality is less of a concern in a pen-type environment, lesions caused by biting and fighting can seriously compromise turtle health. Sea turtles held in pens rarely venture from the sides, thus utilizing only a fraction of the available pen surface area. Construction of pens that maximize perimeter is preferred to pens that maximize open water (Higgins, unpublished data).

16.11.1 Semiwild Conditioning

Prior to use in research projects and before release, NMFS STF loggerhead sea turtles are given a period of semiwild conditioning. Conditioning introduces the turtles to a wild-type environment while keeping them in a controlled and observable space. Kemp's ridleys that are regularly exercised exhibit greater strength and stamina than turtles that are not exercised (Stabenau et al., 1992). Both loggerhead and Kemp's ridleys that are semiwild conditioned for a period of 2–4 weeks appear to have increased strength and stamina compared to unconditioned turtles (J. Mitchell, personal communication).

16.11.1.1 Conditioning–Rearing Pens

Large pens have been constructed at the NMFSPC facility to accommodate up to 200 sea turtles that are 2 years old. The pens range in size from 92 to 223 m^2 and 0.76 to 2.1 m deep. The pens are framed with treated wooden pilings and metal pipe covered with either vinyl-coated metal wire mesh or plastic Durethene® (ADPI, Philadelphia, PA) mesh. The pens are lined on the inside bottom perimeter with a skirt of vinyl-coated metal wire to keep the turtles from burrowing under the pen walls (Figure 16.5). Various walkways, bulkheads, and docks provide access to the pens. The pens contain a natural bay bottom, including sand, seagrass, shell rubble, and rock. The bulkhead walls and mesh sides contain a variety of biota, including algae, barnacles, oysters, sponges, soft corals, tunicates, and various species of cnidarians, echinoderms, fish, crabs, shrimp, and mollusks.

After transportation, the turtles are given a 24-h acclimation period prior to introduction into the pens. A series of independent basins equipped with a flow-through seawater system is used for acclimation, for holding aggressive turtles, and for treatment/observation of problem turtles. The basin system consists of plastic laundry tubs in groups of 10–21 connected to a common drain with water level maintained by an adjustable standpipe. Seawater is delivered by submersible (0.5 hp) pumps feeding an incoming line of valves. Each incoming water valve supplies water to two basins at a variable rate of 150–227 l/h. The basins are housed in a structure shading them from direct sunlight (Figure 16.6).

16.11.1.2 Temporary Holding Facilities

Unfiltered flow-through seawater systems should be avoided. Larvae in the water will settle in and on the pipes, rearing containers or tanks, and sea turtles. At NMFSPC, barnacle larvae settle on the carapace, plastron, and skin of the loggerheads in unfiltered flow-through seawater systems. This problem is confined to the holding tanks and is rarely observed on turtles in the pens. Although epibiota are common to wild loggerheads (Frick et al., 2000), turtles maintained in confined spaces are susceptible to rapid fouling. Rapid turtle growth is common during conditioning. As turtles grow, new carapace tissue is laid down between the scutes (suture lines). This new tissue is easily distinguished by color, and is often accompanied by an attractive striated pattern. Barnacles settling on this new tissue are able to penetrate deeper than on older tissue, and with heavy infestation will lift the edges

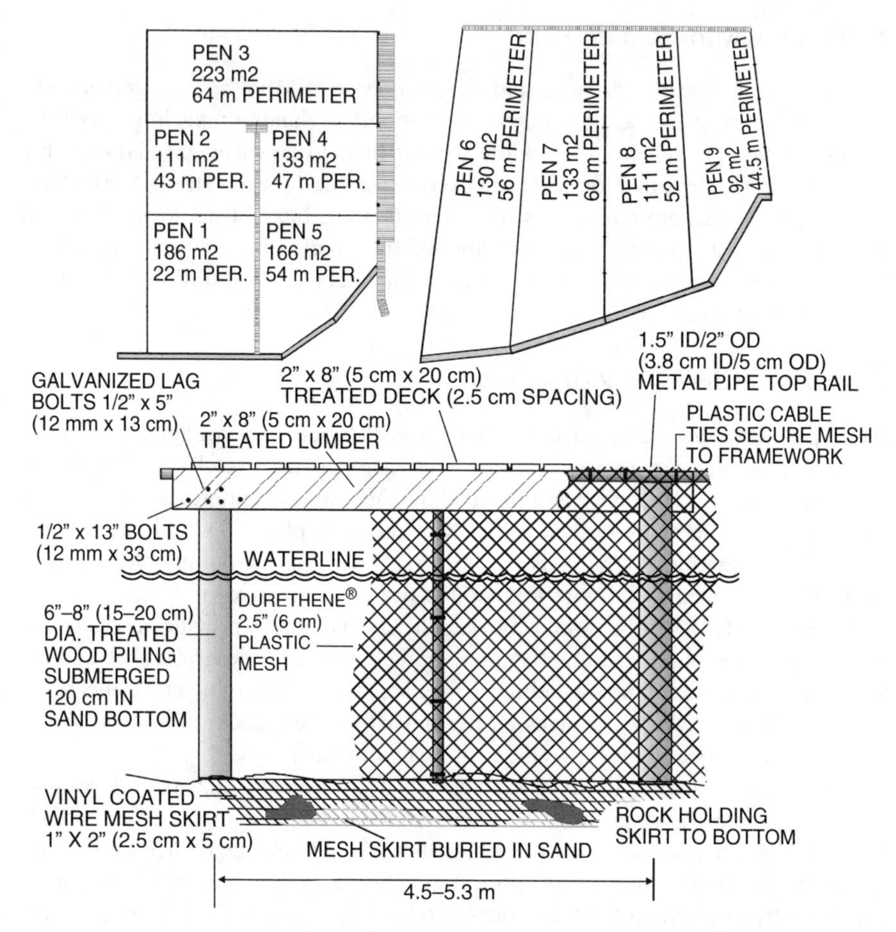

FIGURE 16.5 Panama City, FL, pen-holding facilities. Many other sea turtle rearing facilities use pens or fenced off areas for at least part of the rearing period.

of the scute, causing lesions and infection. Barnacle larvae can settle and lift scutes in as little as 10 days. Barnacles are removed from sea turtles with a stiff brush or plastic scraper as soon as they become visible to the naked eye. Lesions are treated topically with chlorhexidine (chlorhexidine gluconate 2%). Oyster larvae will settle on captive sea turtles, but without the same damaging effects as barnacles. Turtles that are maintained in shallow water and confined spaces in full sunlight will rapidly grow a thick coat of algae on all dorsal surfaces. Turtles should be given deep enough water and ample space to move about to minimize algae buildup. Algae are scrubbed from the turtles every 30 days during conditioning and just prior to release.

16.11.1.3 Feeding

Turtles are allowed to forage naturally in the pens, and they are offered thawed, frozen squid at a rate of 5–10% average body weight/day. Fish that are naturally

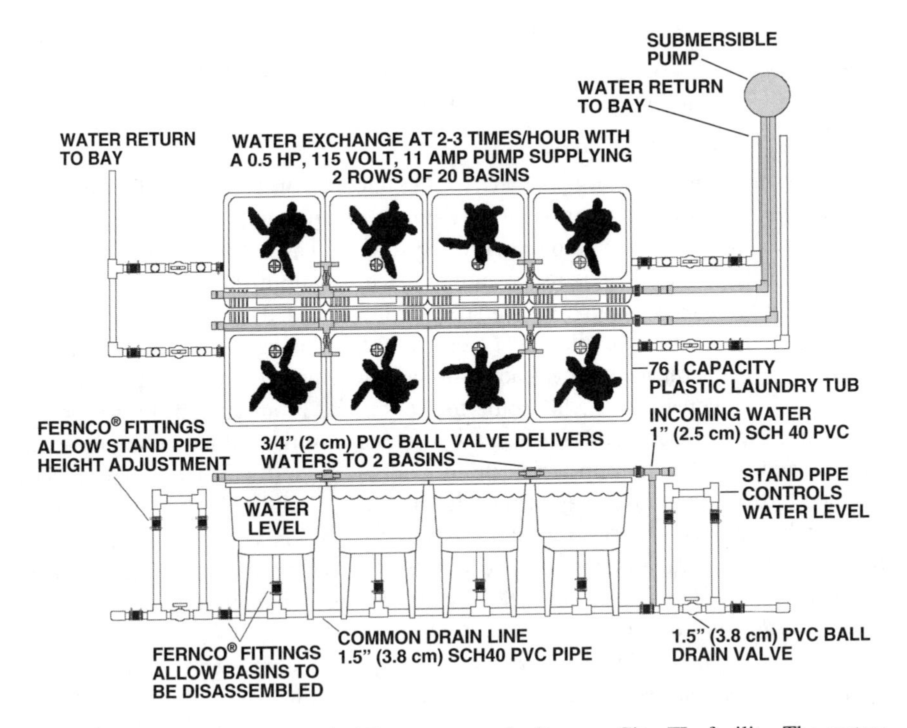

FIGURE 16.6 Temporary basin-holding system at the Panama City, FL, facility. The system can be broken down for storage and expanded to hold additional turtles (Fernco®, Fernco, Inc., Sparks, NV).

present in pens consume a significant portion of the food offered, depending on sea turtle activity. More than the target 5% is offered to compensate for competition from fish. Large squid are cut to facilitate feeding of smaller turtles. Turtles in pens have been observed catching and eating fish, barnacles, and crabs. Often, turtles kept in pens will be returned to a tank or basin system, if significant amounts of sand, rock, and shell debris are found in the feces. Whether the turtles consume the rock and sand along with food or prey items or whether they eat these items for another purpose is unknown.

16.11.1.4 Behavioral Problems

Occasionally, loggerheads and Kemp's ridleys reared in isolation, when placed together, will fight. What triggers the fighting is unknown. Stocking density, water clarity, water temperature, tidal level, wave action, and the presence of other animals in the pens have been investigated in an attempt to isolate the cause of fighting. Thermal stratification and temperature inversions in the pens can force the turtles into a single plane of water, crowding the turtles at the surface or on the bottom. In any large group of loggerheads, there will be a few turtles that cannot coexist with other turtles. If these aggressive turtles can be identified and removed from the general population early in the stocking phase, fighting can be reduced or even eliminated among the remaining turtles. Often, aggressive loggerheads removed for fighting are below average in size or weight, exhibit physical scute anomalies, and may be lighter in pigmentation (blonde loggerhead).

16.11.1.4.1 Aggressive Behavior

Fighting among juvenile loggerheads (20–48 months of age) is manifested as a progressive escalation of aggression. An aggressive turtle will follow another turtle, often at a distance at the bottom or below the surface. The distance soon closes, and the pursued turtle accelerates its swimming pace to gain distance. The aggressive turtle will accelerate to match the pursued turtle's speed. The lead turtle will start to circle in an attempt to elude the aggressive turtle. The aggressive turtle will bite at the posterior carapace of the lead turtle, and will progress to biting at the rear flippers. The pursued turtle will come to the surface in an attempt to escape the aggressive turtle. Once at the surface, the aggressive turtle will try to bite the front flippers and neck of the pursued turtle by coming up and over the carapace. The pursued turtle at this point may turn to defend itself, resulting in the turtles' facing one another, plastron to plastron, jaws locked with heads and fore flippers out of the water. Considerable splashing and flipper slapping on the surface accompany the final stages of fighting. If the turtles are not separated and removed, severe injury to both turtles will occur.

During the pursuit, incidental contact usually occurs between the two engaged turtles and other turtles in the pens. If the aggressive turtle is not removed at the circling phase, and fighting escalates to the surface, all the turtles in the pen, and occasionally adjacent pens, will become agitated. Agitation advances to contact, resulting in fighting among all turtles in the pen. Removal of all turtles from a pen is sometimes the only remedy to an escalating fighting situation. The majority of the time, if the aggressive turtle can be identified and removed at the circling stage, injuries and collateral fighting can be eliminated.

16.11.1.4.2 Passive–Aggressive Behavior

Aggression in a pen can be detected indirectly. A turtle that remains at the surface, with fore flippers folded on the carapace and head down, nervously scanning from side to side below the surface, is a sure sign that there is an aggressive turtle in the pen. Often, turtles in this defensive stance will not feed. Other turtles in the pen will avoid the turtle exhibiting passive–defensive behavior and may interpret this behavior as aggressive in itself. Early identification, removal, and isolation of aggressive and passive–aggressive turtles is the key to pen harmony.

16.12 HEALTH PROBLEMS OF CAPTIVE-REARED TURTLES

Although histopathological and bacteriological analytical services for sea turtle samples are readily available (Texas Veterinary Medical Diagnostic Laboratory, College Station, TX), it often is necessary to initiate treatment prior to receiving test results. Sea turtles may deteriorate rapidly once a health problem is manifested. On-the-spot diagnosis and initial treatment is often necessary to prevent an epizootic outbreak. Any facility rearing sea turtles should have access to a qualified veterinarian with experience in diagnosing and treating reptiles, specifically sea turtles.

16.12.1 Bacterial and Viral Infections

The most common bacteria species isolated from sea turtle infections include *Vibrio*, *Aeromonas*, *Pseudomonas*, and *Cryptophaga-Flavobacterium* (Glazebrook, 1990; Leong et al., 1989; George, 1997). *Streptococcus*, *Salmonella*, and coliform bacteria have also been identified as pathogens in green, hawksbill, loggerhead, and Kemp's ridleys (Glazebrook, 1990; Leong et al., 1989; George, 1997). Many opportunistic bacteria (*Vibrio, Flavobacterium*) are naturally present in seawater and become pathogenic only when the animals are stressed, injured, or the environmental conditions are compromised (Glazebrook, 1990). *Aeromonas* and *Pseudomonas* may be natural opportunistic flora of the sea turtle, becoming pathogenic when the turtle's health is compromised. Through careful attention to water quality, independent isolated rearing, and a suitable diet, bacterial infections of sea turtles can be reduced to an occasional occurrence or even eliminated. Slow feeding, cessation of feeding, and lethargy are sure signs of a primary or secondary bacterial or viral infection. Identifying, isolating, and treating sick turtles with injectable antibiotics in the early stages of infection greatly increase the odds of recovery.

16.12.1.1 Dermal Lesions

Skin lesions (traumatic ulcerative dermatitis) caused by biting and physical contact with the rearing tank are universal in the culture of all sea turtle species (Glazebrook, 1990; Leong et al., 1989). The most prominent areas of lesions include tips and trailing edges of flippers, neck, and tail. Dermal lesions quickly become infected with bacteria, with morbidity and mortality rates of 30–100% (Glazebrook, 1990). Hatchlings are particularly sensitive to secondary bacterial infections, which can approach epizootic levels in rearing systems where turtles are allowed contact with one another (Glazebrook, 1990). Rearing turtles in independent isolation, in appropriately surfaced, sized, and shaped containers, can virtually eliminate ulcerative dermatitis problems for all species except the leatherback. Captive-reared leatherbacks are susceptible to dermal lesions on the head and fore flippers through contact with rearing container walls (Jones et al., 2000). Attempts to keep leatherback hatchlings from contacting rearing containers to prevent self-inflicted dermal lesions have had limited success (Jones et al., 2000).

16.12.1.2 Eye Lesions

Keratoconjunctivitis–ulcerative blepharitis have been reported in green sea turtles, with symptoms ranging from yellow deposits on the eyelids and cornea to complete erosion of tissues, probably as a result of secondary bacterial infection associated with impact trauma or biting (Glazebrook, 1990). Injuries to the eyes are common when turtles are in contact with one another and when housed in tanks with sharp objects and abrasive surfaces.

16.12.1.3 Respiratory Infections

Bacterial and fungal respiratory infections are not as common as dermal and gastrointestinal infections (Glazebrook et al., 1993; Leong et al., 1989) in sea turtles, but should be suspected if a turtle is lethargic and floating on its side. Tilted-swimming or side-floating turtles should be quarantined immediately because respiratory infections (bacterial and fungal) are most often fatal and may be contagious. Leong et al. (1989) described mycobacterial pneumonia (MP) infections as a cause of tilted-swimming, and found that antibiotic treatment of MP with streptomycin was ineffective. The NMFS STF has been identifying and treating tilted-swimming turtles with 5 mg/kg injectable enrofloxacin (Baytril®, Bayer Corp., Shawnee, KS) with promising results. Although an MP or fungal infection is suspected, in most cases, limited attempts are made to isolate the infectious agent because of the intrusive nature of obtaining samples from the lungs. The MP or fungal infection may be secondary to a bacterial pneumonia infection.

16.12.1.4 Viral Infections and Gray-Patch Disease (GPD)

The herpesvirus is believed to cause cutaneous lesions on the flippers and neck of green sea turtle hatchlings. At the Cayman Turtle Farm, GPD has been reported to infect 65–95% of green turtle hatchlings. Although it appears to be cutaneous in nature, infecting only the epidermal layers, it can be fatal (Haines, 1978; Haines and Kleese, 1977). GPD occurs in two forms: pustular-like (blister) lesions that resolve spontaneously with time, and extensive gray-patch lesions that may spread to cover large areas of skin (Haines, 1978). The latter form is often lethal (Haines and Kleese, 1977).

Recovery from pustular GPD results from spontaneous healing by the time the turtles reach 12 months of age (Haines, 1978). Poor water quality, most notably elevated water temperature (>30°C), and overcrowding (stress) may trigger the manifestation of both forms of GPD. Infection rates and severity vary with age, with hatchlings being most susceptible (Glazebrook et al., 1990; Leong et al. 1989; Haines, 1978).

16.12.2 Bloating–Floating

Loggerheads fed a commercially prepared pellet diet can develop floating–bloating problems even at levels not considered to be overfeeding. Overfeeding Kemp's ridley hatchlings or post-hatchlings can cause bloating (Fontaine et al., 1985). Pelagic loggerhead hatchlings and yearlings are the most susceptible to the condition, and the condition is exacerbated by low water temperature and a short photoperiod, with the problem being more prevalent in the winter months. The exact cause is unknown, and no bacterial or viral pathogens have ever been isolated. Permanent carapace and plastron deformation occurs with growth. The problem can become chronic without treatment. Increasing the size of the rearing container volume (a minimum of three times), increasing water depth (a minimum of two times), temporarily switching the diet from pellet food to squid or fish, offering the turtle a tube (15 cm diameter PVC × 31 cm long) in which to hide and/or hold itself down on the bottom, increasing water temperature, and increased artificial illumination has proved to be an effective rehabilitation regime, with a success rate exceeding 90% (Higgins and Bustinza, unpublished data). Attempts have been made to treat the condition by manipulating just one or two treatment parameters without success (Higgins and Bustinza, unpublished data).

16.12.3 Carapace Lesions

Leong et al. (1989) describes two forms of carapace lesions: dull-white suture (DWS) syndrome, and shiny-white suture (SWS) syndrome, both common on young captive loggerheads. Both DWS and SWS are described as ribbons of white material along the suture lines. Microscopic examination of the white ribbon material identified debris, bacteria, and *Fusarium*-like fungal spores. No effective treatment was described by Leong et al. for SWS, aside from maintaining good water quality.

Periodically, NMFS STF loggerheads develop something similar to what Leong et al. describe as SWS, and for future reference, this condition will be referred to as fluffy-white suture syndrome (FWSS). FWSS is first manifested by fluffy-white material appearing on the posterior edge of the carapace scute spines. In severe cases, the suture lines become covered with the fluffy-white material similar to the description of Leong's SWS. Left untreated, the FWSS lesions expand to cover large areas at the center of the scute, resulting in a rough circle of damaged tissue. Keratin damage with exposure of the underlying epidermis and bone follows.

Histological examination of scrapings taken from infected turtles revealed no bacteria or fungal presence; the white material is described simply as an exudate-with debris. Oral and injectable treatments with enrofloxacin and topical applications of Vagisil® (Combe, Inc., White Plains, NY), Betadine® (the Purdue Frederick Company, Norwalk, CT), povidone iodine (10% topical solution, 1% available iodine), and Neosporin® (Warner-Lambert Consumer Healthcare, Morris Plains, NJ) proved to be ineffective, with the FWSS exudate returning in as little as 10–14 days (Higgins, unpublished data). Regular debriding of infected areas twice per week with a scrub brush, followed by the application of strong tincture of iodine (7%) with a paintbrush, followed by 15–30 min of air drying reduced the visible signs of the infection, but keratin and bone regeneration was slow. Debriding followed by applications of a 50% solution of chlorhexidine gluconate (2% solution) with a spray bottle or paintbrush has been very effective, followed by temporary isolation of the turtle and treatment of the rearing water with 10.5 ml/l chlorhexidine gluconate. Early detection of lesions and prompt treatment are critical when dealing with large numbers of turtles sharing a common water source. In serious cases of keratin loss and bone degeneration, topical treatment with antibiotic creams followed by covering the lesions with a protective epoxy coating has proved to be effective (Neiffer et al., 1998).

REFERENCES

Bjorndal, K.A. 1997. Foraging ecology and nutrition of sea turtles, pp. 199–231, in: *The Biology of Sea Turtles*. P.L. Lutz and J.A. Musick (eds.). CRC Press, Boca Raton, FL.

Bolten, A.B. 1999. Techniques for measuring sea turtles, pp. 110–114, in: *Research and Management Techniques for the Conservation of Sea Turtles*. K.L. Eckert et al. (eds.). IUCN/SSC Marine Turtle Specialist Group Publication No. 4. Consolidated Graphic Communications, Blanchard, PA.

Brown, R.A., G.C. Harvey, and L.A. Wilkins. 1982. Growth of Jamaican hawksbill turtles (*Eretmochelys imbricata*) reared in captivity. *Br. J. Herpetol.* 6:233–236.

Buitrago, J. 1987. Rearing, with aim of repopulating, of three marine turtle species at Los Roques, Venezuela. *Mem. Soc. Cienc. Nat. La Salle.* 127–128:169–201.

Caillouet, C.W., Jr. 2000. Sea turtle culture: Kemp's ridley and loggerhead turtles, pp. 789–798, in: *Encyclopedia of Aquaculture*. R.R. Stickney (ed.). John Wiley & Sons, New York.

Caillouet, C.W., Jr. et al. 1988. Can we save Kemp's ridley sea turtle? Believe it or not!, pp. 20–43, in: *10th and 11th International Herpetological Symposia on Captive Propagation & Husbandry.* K.H. Peterson (ed.). Zoological Consortium Inc., Thurmont, MD.

Caillouet, C.W., Jr. et al. 1989. Feeding, growth rate and survival of the 1984 year-class of Kemp's ridley sea turtles (*Lepidochelys kempii*) reared in captivity, pp. 165–177, in: *Proceedings of the First International Symposium on Kemp's Ridley Sea Turtle Biology, Conservation and Management*. C.W. Caillouet Jr., and A.M. Landry Jr. (eds.). TAMU-SG-89–105. Texas A&M University, Sea Grant College Program, College Station, TX.

Caillouet, C.W., Jr. et al. 1997. Early growth in weight of Kemp's ridley sea turtles (*Lepidochelys kempii*) in captivity. *Gulf Res. Rep.* 9:239–246.

Cong, S., and Z. Wang. 1997. Study on marine turtle and its raising along the coasts of Shandong. *Trans. Oceanol. Limnol./Haiyang Huzhao Tongboa.* 3:76–80.

Florida Fish and Wildlife Conservation Commission. 2002. Unpublished. Marine turtle conservation guidelines: Section 4 — holding turtles in captivity. Tallahassee, Florida.

Fontaine, C.T. 2002. Unpublished manuscript. Feeding and growth of Kemp's ridley (*Lepidochelys kempii*) and loggerhead (*Caretta caretta*) sea turtles reared in captivity.

Fontaine, C.T. et al. 1985. The husbandry of hatchling to yearling Kemp's ridley sea turtles (*Lepidochelys kempii*). NOAA Tech. Mem. NMFS-SEFC-158, National Technical Information Service, Springfield, VA.

Fontaine, C.T. et al. 1990. Kemp's ridley headstart experiment and other sea turtle research at the Galveston laboratory: annual report — fiscal year 1989. NOAA Tech. Mem. NMFS-SEFC-266, National Technical Information Service, Springfield, VA.

Fontaine, C.T., T.D. Williams, and D.B. Revera. 1988. Care and maintenance standards for Kemp's ridley sea turtles (*Lepidochelys kempii*) held in captivity. NOAA Tech. Mem. NMFS-SEFC-202, National Technical Information Service, Springfield, VA.

Fontaine, C.T. and T.D. Williams. 1997. Delayed feeding in neonatal Kemp's ridley, *Lepidochelys kempii*: a captive sea turtle management technique. *Chelonian Conserv. Biol.* 2:573–576.

Frick, M.G., K.L. Williams, and D. Veljacic. 2000. Additional evidence supporting a cleaning association between epibiotic crabs and sea turtles: how will the harvest of sargassum seaweed impact this relationship? *Mar. Turtle Newsl.* 90:11–13.

George, R.H. 1997. Health problems and diseases of sea turtles, pp. 363–385, in: *The Biology of Sea Turtles*. P. Lutz and J.A. Musick (eds.). CRC Press, Boca Raton, FL.

Glazebrook, J.S. and R.S.F. Campbell. 1990. A survey of the diseases of marine turtles in northern Australia. I. Farmed turtles. *Dis. Aquat. Org.* 9:83–95.

Glazebrook, J.S., R.S.F. Campbell, and A.T. Thomas. 1993. Studies on an ulcerative stomatitis–obstructive rhinitis-pneumonia disease complex in hatchling and juvenile sea turtles *Chelonia mydas* and *Caretta caretta*. *Dis. Aquat. Org.* 16:2–8.

Goldman, K.E., R. George, and M.W. Swingle. No date. No more squid for captive sea turtles. Virginia Marine Science Museum, Virginia Beach, VA.

Gutierrez, W. 1989. Experiences in the captive management of hawksbill turtles (*Eretmochelys imbricata*) at Isla Uvita, Puerto Limon, Costa Rica, pp. 324–326, in: *Proceedings of the Second Western Atlantic Sea Turtle Symposium*, Oct. 12–16. L. Ogren (ed.). NOAA-Tech. Memo. NMFS-SEFC-226, National Technical Information Service, Springfield, VA.

Haines, H. 1978. A herpesvirus disease of green sea turtles in aquaculture. *Mar. Fish. Rev.* 40:33–37.

Haines, H. and W.C. Kleese. 1977. Effect of water temperature on a herpesvirus infection of sea turtles, infection and immunity. *Am. Soc. Microbiol.* 15:756–759.

Higgins, B.M. Unpublished data (1997). Plans for the modification and construction of additional sea turtle holding pens for the fall 1997 TED certification trials, Panama City, Florida — recommended stocking densities.

Higgins, B.M. Unpublished data (1999). Comparison of growth and survival of Kemp's ridley and loggerhead sea turtles fed diets varying in fat and protein sources.

Higgins, B.M. Unpublished data (1999). Diagnosis and treatment of carapace shell lesions on captive reared loggerhead (*Caretta caretta*) sea turtles.

Higgins, B.M. and A.C. Bustinza. Unpublished data (2001). Diagnosis and treatment of floating/bloated loggerhead (*Caretta caretta*) sea turtles in captivity.

Huff, J.A. 1989. Florida (USA) terminates "headstart" program. *Mar. Turtle Newsl.* 72:13–15.

Jones, T.T. et al. 2000. Rearing leatherback hatchlings: protocols, growth and survival. *Mar. Turtle Newsl.* 90:3–6.

Klima, E.F. and J.P. McVey. 1982. Headstarting the Kemp's ridley turtle, *Lepidochelys kempii*, pp. 481–487, in: *Biology and Conservation of Sea Turtles*. K.A. Bjorndal (ed.). Smithsonian Institution Press, Washington, D.C.

Lebrun, G. 1975. Elevage a la Reunion de juveniles de la tortue verte *Chelonia mydas* (Linnaeus) 1758. *Sci. Peche.* 248:1–25.

Leong, J.K. et al. 1989. Health care and diseases of captive-reared loggerhead and Kemp's ridley sea turtles, in: *Proceedings of the First International Symposium on Kemp's Ridley Sea Turtle Biology, Conservation and Management*. C.W. Caillouet Jr., and A.M. Landry Jr. (eds.). TAMU-SG-89–105. Texas A&M University, Sea Grant College Program, College Station, TX.

Mitchell, J.F. et al. 1989. An alternative protocol for the qualification of new turtle excluder devices (TEDs), pp. 115–117, in: *Proceedings of the Ninth Annual Workshop on Sea Turtle Conservation and Biology*. S.A. Eckert, K.L. Eckert, and T.H. Richardson (compilers). NOAA Tech. Mem. NMFS-SEFC-232, National Technical Information Service, Springfield, VA.

Neiffer, D.L. et al. 1998. Shell lesion management in two loggerhead sea turtles, *Caretta caretta*, with employment of PC-7 epoxy paste. *ARAV* 8:12–17.

Rajagopalan, M., M. Vijayakumaran, and A.B. Fernando. 1984. Some health problems observed in the hatchlings and juveniles of sea turtles in captivity, in: *Sea Turtle Research and Conservation*. E.G. Silas, M. Vijayakumaran, and P.T.M. Sundaram (eds.). *CMFRI Bull.* 35:55–58.

Ross, J.P. 1999. Ranching and captive breeding sea turtles: evaluation as a conservation strategy, pp. 197–201, in: *Research and Management Techniques for the Conservation of Sea Turtles*. K.L. Eckert et al. (eds.). IUCN/SSC Marine Turtle Specialist Group Publication No. 4.

Shaw, S. et al. 1989. Possible effects of artificial foods on sea turtle health, in: *Proceedings of the Ninth Annual Workshop on Sea Turtle Conservation and Biology*. S.A. Eckert, K.L. Eckert, and T.H. Richardson (eds.). NOAA Tech. Mem. NMFS-SEFC-232.

Solomon, S.E. and R. Lippett. 1991. *Anim. Technol.* 42:77–81.

Stabenau, E.K., A.M. Landry Jr., and C.W. Caillouet Jr. 1992. Swimming performance of captive-reared Kemp's ridley sea turtles *Lepidochelys kempii* (Garman). *J. Exp. Mar. Biol. Ecol.* 161:213–222.

Stickney, R.R. 2000. Sea turtle culture: General considerations, pp. 784–786, in: *Encyclopedia of Aquaculture*. R.R. Stickney (ed.). John Wiley & Sons, New York.

Sumano Lopez, R. et al. 1980. Cultivo de tortugas marinas en Mexico, pp. 2113–2133, in: *Memorias del 2. Symposio Latinoamericano de Acuacultura*, vol. 1, Dep. Pesca, Mexico City, Mexico.

Swingle, W.M. et al. 1993. Exceptional growth rates of captive loggerhead sea turtles, *Caretta caretta*. *Zoo Biol.* 12:491–497.

United States Fish & Wildlife Service. 1973. Endangered Species Act. Washington, DC.

Voss, J. et al. 1988. Elevage d'une tortue luth (*Dermochelys coriacea* (Vandelli 1761)) a l'aquarium universitaire de Liege. *Cahiers d'Ethologie Appliquee* 8:457–466.

Walsh, M. 1999. Rehabilitation of sea turtles, pp. 202–207, in: *Research and Management Techniques for the Conservation of Sea Turtles*. K.L. Eckert et al. (eds.). IUCN/SSC Marine Turtle Specialist Group Publication No. 4.

Wood, F. 1991. Turtle culture, in: *Production of Aquatic Animals*. C.E. Nash (ed.). World Animal Science, Elsevier, Amsterdam.

Wood, J.R. 1980. Amino-acids essential for the growth of young green sea turtles (*Chelonia mydas*). *J. World Mariculture Soc.,* 5:233.

Wood, J.R. and F.E. Wood. 1980. Growth and digestibility for the green turtle (*Chelonia mydas*) fed diets containing varying protein levels. *Aquaculture* 25:269–274.

Wood, J.R. and F.E. Wood. 1988. Captive reproduction of Kemp's ridley *Lepidochelys kempii*. *J. Herpetol.* 1:247–249.

Index

A

B

C

D

F

I

J

K

L

M

N

O

P

Q

R

S

T

U

V

W

X

Y

Z